GW00394162

RED PANDA

BIOLOGY AND CONSERVATION OF THE FIRST PANDA

RED PANDA

Biology and Conservation of the First Panda

Edited By

ANGELA R. GLATSTON

Amsterdam • Boston • Heidelberg • London • New York • Oxford
• Paris • San Diego • San Francisco • Singapore • Sydney • Tokyo
Academic Press is an imprint of Elsevier

Academic Press is an imprint of Elsevier
32 Jamestown Road, London NW1 7BY, UK
30 Corporate Drive, Suite 400, Burlington, MA 01803, USA
525 B Street, Suite 1800, San Diego, CA 92101-4495, USA

First edition 2011

British Library Cataloguing-in-Publication Data
A catalogue record for this book is available from the British Library

Library of Congress Cataloging-in-Publication Data
A catalog record for this book is available from the Library of Congress

ISBN: 978-1-4377-7813-7

For information on all Academic Press publications
visit our website at www.elsevierdirect.com/rights

Typeset by MPS Limited. A Macmillan Company, Chennai, India
www.macmillansolutions.com

Printed and bound by CPI Group (UK) Ltd, Croydon, CR0 4YY

Transferred to Digital Printing, 2013

Contents

Foreword

There is a strong tendency at the moment for conservation to focus on ecosystems, and the services that the ecosystems provide to humans. There is, of course, nothing wrong with this tendency, providing that is not carried out to the exclusion of a simultaneous focus on species. Unfortunately, with modern trends, species are often ignored by those who are supposed to be conserving them. As a result, important species, particularly those less familiar to us, can slip under the radar.

The red panda is one such species, and therefore this book is important, first, because it makes this animal more familiar to us, and secondly because it draws attention to its current plight. The status of the red panda in the wild is relatively unknown, but it is often assumed to be more common than is in fact the case. There are reasonable numbers in zoos, but captive breeding programmes are failing to deliver.

So why the great interest in the red panda? Why is it special? First, it is undeniably a uniquely attractive species. It is one of the most strikingly patterned and vividly coloured of all mammals.

Scientists have long focused on the red panda's controversial taxonomy. Is it a bear, is it a procyonid (raccoon relative), is it a panda, or is it a unique family on its own? All these have been hypothesized. Does this matter? It is a unique species regardless of its true taxonomic status. It is either the only procyonid in the Old World, or a very aberrant bear, or the closest living relative to the giant panda, or just unique – a living fossil, the last of its kind.

The red panda is comparatively little known. Until comparatively recently, most of our knowledge of the species in the wild was derived from a 19th century naturalist, Brian Hodgson. Even today most of our knowledge of red panda biology has been obtained in captivity. There is an urgent need for more field studies to enable conservation in the wild. This volume provides an overview of current knowledge and indicates where further research is needed.

The red panda has a spectacular home – the Himalayas, one of the global hotspots for conservation. The Himalayas have a unique flora and fauna, and are also of great cultural significance to the local people. Himalayan rivers provide drinking water to a large number of people, but this region is under threat from global warming. Indeed, climate change is proceeding faster at higher elevations, and this could destroy this region if we do not act quickly. The red panda is a classic flagship species. It is well placed to function as a focus for the region and its problems.

There are a good number of red pandas in zoos around the world. These have the potential to become a valuable educational and public relations resource to teach people about the conservation needs of the important Himalayan region. Today, the future of the red panda hangs in the balance. Its survival in the wild is questionable and its survival in captivity is not secure. This unique species must not be allowed to become extinct; and its habitat must not be destroyed. I commend this book to the reader as a timely reminder of the problems confronting the red panda and the Himalayan regions, and of what we all stand to lose should it be allowed to die out. I thank the editor and authors for their dedication and thoroughness in producing this remarkable book.

Simon N. Stuart,
Chair of the IUCN Species Survival Commission

Prologue

I often tell people that my personal interest in the red panda dates back to my starting work in Rotterdam Zoo as a member of the research department. I had just finished my PhD work in London and was looking for a new area of interest. My attention was drawn to a group of animals held in an exhibit just by the entrance to the zoo. It was early in the year, the breeding season, and these animals were very active following each other around, marking trees and rocks and calling to each other with a strange bleating sound. I have always felt that if you are going to study the behaviour of an animal it must attract you on some level, and these beautiful creatures certainly caught my attention. I started to observe their behaviour and my interest was further piqued when I discovered how little had been published about the species. My then director, Dick van Dam, suggested that it might be a good idea if I were to produce an international studbook for the zoo population of red pandas. From the information I collected it became clear to me that red pandas were not doing very well in captivity. I thought something needed to be done about this and then the rest is history as they say. However, sometime later when I was looking through some childhood photographs, I came across some pictures taken during a primary school outing to Whipsnade Zoo and among them I discovered a grainy black and white image of a red panda peering at me from out of its nest box. It must be karma.

Red pandas in Whipsnade Zoo in the early 1960s.

List of Contributors

Mauricio Antón Departamento de Paleobiología, Museo Nacional de Ciencias Naturales-CSIC, Madrid, Spain

Kurt Benirschke University of California, San Diego, USA

Nancy Czekala Papoose Conservation Wildlife Foundation, Del Mar, California, USA

Ellen S. Dierenfeld Novus International, Inc., St. Charles, MO, USA

J.W. Duckworth INCN/SSC Small Carnivore Red List Authority Focal Point, Vientiane, Lao PDR

Pijush Kumar Dutta Landscape Coordinator, Western Arunachal Landscape, WWF-India, Tezpur, Assam, India

Rebecca E. Fisher Department of Basic Medical Sciences, University of Arizona, College of Medicine-Phoenix, Phoenix, Arizona; School of Life Sciences, Arizona State University; Tempe, Arizona, USA

Axel Gebauer Naturschutz–Tierpark Görlitz, Görlitz, Germany

Dipankar Ghose Head, Eastern Himalaya and Terai Program, WWF-India Secretariat, New Delhi, India

Angela R. Glatston Rotterdam Zoo, Rotterdam, The Netherlands

Colin Groves School of Archaeology & Anthropology, Australian National University, Canberra, Australia

Alankar K. Jha IFS, Padmaja Naidu Himalayan Zoological Park, Darjeeling, India

Marvin L. Jones Formerly of San Diego Zoological Gardens, San Diego, CA, USA

Kamal Kandel Red Panda Network–Nepal, Kathmandu, Nepal

Martin Kundrát Centre of Excellence for Integrative Research of the Earth's Geosphere, Geological Institute, Slovak Academy of Sciences, Banská Bystrica, Slovak Republic

Kristin Leus Copenhagen Zoo and IUCN/SSC Conservation Breeding Specialist Group (European regional office), Merksem, Belgium

Zhijin Liu Key Laboratory of Animal Ecology and Conservation Biology, Institute of Zoology, The Chinese Academy of Sciences, Chaoyang, Beijing, People's Republic of China

Kati Loeffler Veterinary and Scientific Consultant for International Animal Welfare and Conservation Programmes

Naveen K. Mahato Red Panda Network–Nepal, Kathmandu, Nepal

Jorge Morales Departamento de Paleobiología, Museo Nacional de Ciencias Naturales-CSIC, Madrid, Spain

Joeke Nijboer Rotterdam Zoo, Rotterdam, The Netherlands

Lesley E. Northrop Reproductive Medicine Associates of New Jersey, Morristown, New Jersey, USA

Stéphane Peigné Muséum national d'Histoire naturelle, Département Histoire de la Terre, USM 0203 – UMR 7207 CNRS/MNHW/UPMC Centre de Recherche sur la Paléobiodiversité et les Paléoenvironnements, Paris, France

Joost Philippa Wildlife Conservation Society, Field Veterinary Program, Viet Nam and

Indonesia, Bogor, Indonesia; Department of Small Animal Clinical Sciences, College of Veterinary Medicine, University of Tennessee, Knoxville, Tennessee; Knoxville Zoological Gardens, Knoxville, Tennessee USA

Brian Preece Formerly of Veterinary Survillance Department, Veterinary Laboratories, UK

Ed Ramsay Wildlife Conservation Society, Field Veterinary Program, Viet Nam and Indonesia, Bogor, Indonesia; Department of Small Animal Clinical Sciences, College of Veterinary Medicine, University of Tennessee, Knoxville, Tennessee; Knoxville Zoological Gardens, Knoxville, Tennessee, USA

Manuel J. Salesa Departamento de Paleobiología, Museo Nacional de Ciencias Naturales-CSIC, Madrid, Spain

Steven C. Wallace Department of Geosciences and Don Sundquist Center of Excellence in Paleontology, East Tennessee State University, Johnson City, Tennessee, USA

Fuwen Wei Key Laboratory of Animal Ecology and Conservation Biology, Institute of Zoology, the Chinese Academy of Sciences, Chaoyang, Beijing, People's Republic of China

Brian Williams Red Panda Network, Mountain View, California, USA

Zejun Zhang Institute of Rare Animals and Plants, China West Normal University, Nanchong, Sichuan, People's Republic of China

Acknowledgements

Many people have contributed to the realization of *Red Panda*, and most of them have been acknowledged individually by the authors of the various chapters. However there are two key people who have not been mentioned, and without whom this book would not exist. They are Jeri Wacher and Thane Johnson who were both instrumental in initiating this book. I would like to thank Jeri and Thane both for inviting me to produce a comprehensive book on my favourite species and also for having the faith that I would manage to do so; it has been an interesting and challenging experience.

I would also like to take this opportunity to thank all the contributors for taking the time and effort to make *Red Panda* a reality, without you there would have been no book. Thanks are also due to Rotterdam Zoo, not only for supporting my interest in red panda conservation but also for allowing me some time to work on this book.

CHAPTER

1

Introduction

Angela R. Glatston

Rotterdam Zoo, Rotterdam, The Netherlands

When the name "panda" is mentioned, the image that comes to most people's mind is that of a clumsy-looking, black and white, bear-like animal; that is the well-known and much-loved symbol of the Worldwide Fund for Nature (WWF), the giant panda. However, this creature is not the only, nor even the first, species to bear the charismatic name "panda". For nearly half a century a small, chestnut-coloured, cat-like animal was the only panda we knew which is why this book refers to the red panda as "the first panda". It was not until 1869 that Père Armand David eventually encountered the species that we know today as the giant panda, a creature that owes its very name to its perceived similarity to the original panda. The discovery of this giant panda meant that the real panda's name had to be changed to distinguish the two species. This is why the "first panda" is now better known as the red or lesser panda. Nevertheless, this panda remains unique in its own right. It is classified as a carnivore but it feeds almost entirely on bamboo, indeed, for a long time nobody was quite certain exactly to which family of mammals it belonged. It is a panda, which is not piebald but rather is flamboyantly clad in chestnut, chocolate and cream and is surely, as Frederic Cuvier is quoted as saying, "quite the most handsome mammal in existence" [1] (Figure 1.1; the original illustration of the red panda can be found elsewhere in this volume, see Figure 7.11).

It is not just the pandas themselves that are remarkable, their very name, "panda", is of itself interesting, as its origin is uncertain. There are a number of theories about how it was derived, however, the most likely theory is that it comes from one of the local names for the red pandas, *nigalya ponya*; *nigalya* is thought to come from *nigalo* meaning cane or bamboo. The source of *ponya* is less certain, although it may come from *ponja* meaning the ball of the foot or the claws, if this is correct then *nigalya ponya* may mean bamboo footed, an appropriate name for either panda [2].

It is abundantly apparent that the red panda is an extremely attractive species and charismatic in its own right. Cuvier, who published the first description of the red panda, was only one among many to have remarked on its beauty and its charm [3,4] (Figure 1.2). Nor is it without friends in high places; Jawaharlal Nehru, the former prime minister of India,

Red Panda. DOI: 10.1016/B978-1-4377-7813-7.00001-X

FIGURE 1.1 The most handsome mammal in existence. *(Photo: Axel Gebauer)*

FIGURE 1.2 Advertisement of the imminent arrival of a red panda to a zoo. *(Reprinted courtesy of Houston Zoo)*

kept red pandas in his home where they were childhood pets of his daughter, another Indian prime minister, Mrs Indira Ghandi [5]. Yet, despite its beauty and connections, the red panda has remained relatively unknown to both scientist and layman alike. On the other hand, it would be untrue to say that nothing is known about the species; there have been a number of publications dealing with the red panda and its biology which have appeared in various journals and books, over the last hundred or more years. However, most of this information seems to have faded from our collective radar. This book aims to redress this situation and to focus attention on this rare and poorly known animal by providing an up-to-date synthesis of red panda data gathered from captivity, field studies and the laboratory as well as from culture and tradition, and to publish this in a single volume which will provide the first comprehensive account of the red panda.

The history of the red panda in western science begins in controversy; the man who is generally acknowledged as the first person to describe the species was not accorded the usual honour of naming it. The discovery of the red panda is attributed to the Englishman, Major-General Thomas Hardwicke, who served in the Indian Service. Hardwicke was a keen naturalist who explored the areas where he was stationed. On 6 November 1821 (four years before the Cuvier publication), Hardwicke presented a paper to the Linnaean Society of London called "Description of a New Genus of the Class Mammalia, from the Himalayan Chain of Hills between Nepaul and the Snowy Mountains". Unfortunately, Hardwicke's paper did not go to press until some six years later. In the interim, Frederic Cuvier, son of the famous French zoologist Georges Cuvier, published the first written description of the red panda. As this latter account, which went to press in 1825, was the first published report of an animal new to science, it was Cuvier, not Hardwicke, who had the honour of giving the panda its scientific name. The name he chose was *Ailurus fulgens* meaning shining or fire-coloured cat (*"Ailurus, à cause de sa ressemblance extérieure avec le Chat, et pour nom spécifique celui de Fulgens, à cause du brilliant de ses couleurs"*, *Ailurus* because of its external resemblance to the cat and *fulgens* because of its brilliant colours) [6]. Cuvier's description of the panda was based on a few remains (skin, paws, incomplete jaw bones and teeth), which had arrived in Europe, together with a general description of its appearance provided by his son-in-law, Alfred du Vaucel. Even from this limited material, Cuvier realized that he was dealing with a unique species that was not represented in the collections of the Paris Natural History Museum.

Hardwicke also provided a detailed account of the red panda describing its coat as "a beautiful fulvous brown colour which on the back becomes lighter and assumes a golden hue" [3]. He also deduced that, due to its striking peculiarities, it must belong to a new genus. In particular, he referred to the singular structure of its teeth. Hardwicke reported that red pandas lived near rivers and mountain torrents, that they spent much of their time in trees and that they fed on birds and small mammals. He also noted that they were referred to by the local people as *Wha* or *Chitwa* because of their vocalizations. Although Hardwicke's paper was not published until 1827, two years after Cuvier's publication, the then secretary of the prestigious Linnean Society felt that Hardwicke's description was too important to be omitted from the Transactions of the Society.

However, the story around the discovery of the red panda does not end there; it has been proposed that the first European to see a red panda in the wild did so before either

Hardwicke or Cuvier published their descriptions. A Dane, the botanist Nathaniel Wallich, who was director of the East India Company's Botanical Garden in Calcutta, made a number of collecting trips to the mountainous regions of Nepal in the early 19th century. He is known to have brought back a number of animals from this region for both General Hardwicke and Alfred du Vaucel. Indeed, as Morris and Morris have suggested, although there is no record of the event, it could well be that Wallich in fact provided the red panda specimens used by Cuvier and/or Hardwicke [1].

For many years, the red panda remained known only from skins. It was some 20 years before another English naturalist eventually took an interest in this creature. Brian Houghton Hodgson was a retired Indian civil servant who moved to the high mountains of Sikkim where he lived for 13 years. He was a well-known naturalist who published nearly 130 papers on the mammals and birds of the Himalayan region in addition to numerous papers on physical geography, ethnography and ethnology of the region. In 1847, Hodgson published his work on the red panda in the *Journal of the Asiatic Society of Bengal* [7]. This paper tells us much about the red panda or, as he refers to it, the Wah. Hodgson was also a good artist and two of his images of the red panda are presented in Figures 1.3 and 1.4. Much of the information that Hodgson published on the red panda was, until comparatively recently, all we knew about this species in the wild, much of it

FIGURE 1.3 Hodgson's illustration of a red panda. *(Reprinted courtesy of the Zoological Society of London)*

FIGURE 1.4 Hodgson's original sketch of a sleeping red panda. *(Reprinted courtesy of the Zoological Society of London)*

still remains undisputed. He reported that red pandas could be found in the northern parts of the central mountain region and all the forested areas of the "juxta nivean" or Cachar region between the altitudes of 7000 and 13 000 feet (2500 and 4500 metres) and emphasized that they did not live in the snows. He knew that wahs were arboreal, although they descended to the ground to feed. He said no other quadruped could surpass the wah as a tree-climber and observed how they would "climb steadily and firmly, upwards and downwards without any necessity for turning back on themselves", i.e. head first. He also noted that they were plantigrade and moved awkwardly "but without special embarrassment" when on the ground and, when they needed to travel faster, they moved with a series of bounds.

Hodgson knew a lot about the red panda's biology. He described their diet as consisting of fruits, tuberous roots, thick bamboo sprouts, acorns, beechmast and eggs. Actually, Hodgson believed that eggs were the only animal material eaten by red pandas and he doubted that they would eat any meat despite being assured to the contrary. Interestingly, he also reported that he had never seen a wah using its hands when eating despite the fact that red pandas regularly hold bamboo in their forepaws to feed using their false thumb to help them get a good grip of the stems. Hodgson also observed that wahs love to eat milk and ghee (the clarified butter used in Indian cuisine) and reported that they would raid remote dairies and cowsheds to steal these delicacies. One may wonder if this report is the source for including milk in so many zoo diets for red pandas. Hodgson also realized that red pandas were crepuscular rather than truly nocturnal and said they slept in the daytime "curled like dogs or cats with the tail over the eyes to exclude light". In his view, red pandas were monogamous and bred only once each year. He noted that the young remained with their family until the next birth at which time the mother would drive them away, that births occurred in spring or early summer, that the average litter size was two with one cub usually being bigger than the other and that this size difference was not related to gender. Interestingly, Hodgson noted that the wah he was discussing originated from Assam and Sikkim, and that he thought it was different to the panda which had been described by Cuvier, which he supposed had originated from Bhutan. Hodgson, therefore, provided a full description of his wha with the suggested name of *Ailurus ochraceus* or the Nepalese *Ailurus* that he describes as being a "deep ochreous red" with jet-black belly, ears, limbs and tail tip.

In fact, the existence of a second form of wah or panda was confirmed in 1902 in a publication by Oldfield Thomas who described and named this Szechuan panda on the basis of a specimen donated to the Natural History Museum of London by Mr F.W. Styan [8]. Thomas reported a difference in colour between this red panda and the one described by Cuvier but said that he suspected that the colours were variable, even in the Himalayan form. The main difference Thomas found between the two forms was one of size; he reported, for example, that the skull of the Chinese form was significantly larger than that of the Himalayan form. In this publication, Thomas states that he does not consider the Chinese specimen to belong to a separate species as he felt that there might be various intervening forms that had not been found. However, by 1922, Thomas seems to have revised his opinion. In a second publication on the Chinese red panda, this time based on specimens originating from northern Yunnan and north-eastern Burma, he reports that, while the examples originating from these regions are similar to each other, there are clear differences to those originating from the Himalayas. He therefore concluded that the Chinese and Himalayan forms of the red panda should be categorized as clear separate species; he therefore refers to *Ailurus styani* in this publication [9].

There is one last early red panda publication which needs to be referred to here, namely that of Cheminaud who, in his 1942 hunting memoirs, refers to the occurrence of the red panda in northern Laos. The species in question was apparently called *mi kham* (golden bear) by the local people and "ours de bamboo" (bamboo bear) by colonial hunters and its description does indeed bear some similarity to the red panda; a cat-sized animal, golden red in colour with a white face and black tail and a gentle disposition. However, this part of Laos is far outside of the normal range of the red panda and the climatic conditions of the area are very different from those normally associated with the species, therefore, one is led to suppose that Cheminaud was probably mistaken. However, his report was taken seriously and has resulted in Laos being included in the red panda's range in many important publications including those dealing with biodiversity in Laos, WWF's webpage on the red panda and the IUCN Red Data Book despite the fact that no one else has ever reported seeing a red panda in Laos since Cheminaud's time. There are other reports of red pandas occurring in unexpected climatic zones, and these reports, together with that of Cheminaud, are evaluated and discussed in Chapter 24. Cheminaud's report may well have been a case of simple misidentification resulting from the inadequate knowledge of the time. We would not expect a similar mistake to occur today when we have much more information available on the red panda and on the indigenous fauna of most countries. Nevertheless, during the collection of data for the IUCN Action Plan for Procyonids and Ailurids, a Burmese naturalist reported that red pandas were commonly seen in his country and accompanied his statement with the photograph of a palm civet [10].

The red panda's history in zoos begins some 40 years after its discovery. The first one to be seen outside of its natural range arrived at London Zoo on 22nd May 1869, the same year that Père David discovered the giant panda and sent its skin to the Museum of Natural History in Paris. This wah was the sole survivor of three animals that had been presented to the zoo by Dr H. Simpson. It is surprising that even this single animal managed to survive, as all three animals were apparently in a poor condition before they even embarked on the arduous sea voyage from India to Great Britain. A letter from Dr J. Anderson of the Indian museum in Calcutta informed the Zoological Society that the pandas were suffering so severely from the heat and that he felt it would be unlikely for any of them to survive the journey to England despite the steps he had taken to mitigate their situation [1].

On arrival at the zoo, the last surviving panda was given into the care of Abraham Bartlett, the superintendent of the zoo. Bartlett recorded that he "found the animal in a very exhausted condition, not able to stand, and so weak that it could with difficulty crawl from one end of its cage to the other. It was suffering from frequent discharges of frothy, slimy faecal matter. This filth had so completely covered and matted its fur that its appearance and smell were most offensive". The feeding instructions that came with the panda said it should be given milk, a little rice and grass each day. It is interesting to note that milk, or a milk-based gruel, remained a standard part of red panda zoo diets until well into the 1980s, indeed, it was given to a quartet of red pandas arriving at Melaka Zoo, Malaysia, as recently as 2002.

Bartlett felt that the suggested diet was not adequate and set out to find what the animal would accept; sweetened beef tea and porridge mixed with eggs became the order of the day. The red panda's condition improved, its coat started to re-grow and its beauty gradually became apparent. As it became stronger it was allowed into the garden where it was also seen eating rose shoots, unripe apples and pears. Unfortunately, although it ate well, this first zoo red panda was not destined for a long life, it died suddenly during the night of 12 December 1869, a little over 6 months after its arrival. The next red panda did not arrive in London Zoo for another 7 years [11,12].

After its death, the body of the panda was sent to William Henry Flower, the conservator of the museum. Flower's detailed and accurate description of red panda anatomy was published in 1870 and has remained the standard reference on the topic [13]. It was in this publication that Flower noted his uncertainty regarding the classification of the red panda. Although he noticed some similarities with the raccoon, he suggested the panda was distinct enough to merit its own family. How this family was related to the rest of the carnivore order was not certain, but Flower thought some light might be thrown on the topic by the then recently discovered giant panda, a view shared by Lankaster who considered the two pandas to be closely related [14]. However, not all authors shared this opinion, for example Edward Blyth classified the red panda with the American raccoons [15].

These discussions on red panda phylogeny have continued for a century and relationships have been suggested between the two red panda (sub)species and the giant panda as well as with the procyonids and the ursids. Is the red panda the only Old World member of the raccoon family, is it an aberrant bear, is it a mustellid or does is belong to a separate family, either on its own or with the giant pandas? This debate will be continued elsewhere in this volume, however, the reader could be forgiven for wondering whether it matters in the greater scheme of things if we consider the red panda a raccoon, a bear, or simply a panda. In many ways the answer to this question is no, after all we classify animals in order to simplify our understanding of them which, logically, should have no practical consequences for the species concerned. Unfortunately, this is not quite true in today's world; a world of limited time and reduced budgets where conservation priorities are often founded on the concept of taxonomic uniqueness. This strategy accords monospecific families a very high level of priority. So from this standpoint it does matter whether we consider the red panda to be yet another procyonid (albeit a very specialized and beautiful one) or to be the sole representative of the family Ailuridae. At a very practical level classification can matter too. One very renowned scientist refused to collaborate in the production of the IUCN Status Survey and Conservation Action Plan for Procyonids (later Procyonids and Ailurids) because he did not consider the red panda to be a procyonid and therefore refused to contribute [10]. The classification of the red panda is a debate which needs to be settled and it is to be hoped that Groves' chapter (Chapter 7) and recommendations later in this volume will provide a basis for this.

Our knowledge of the red panda in the wild is still fairly limited. One of the reasons for this is undoubtedly its rarity; even in Hodgson's time the red panda was described as being scarce or uncommon [7]. Even today, despite the attention given to the species by virtue of the IUCN Action Plan, the WWF eastern Himalaya programme, CITES Appendix 1 listing and its inclusion in the Zoological Society of London's EDGE of extinction programme, information remains woefully inadequate. In 2007, a meeting was held in Sikkim to determine whether a Red Panda PHVA (Population and Habitat Viability Assessment) workshop was a viable option. This meeting concluded that there were insufficient data available to provide the basis for such a workshop, even one just dealing with a limited part of the range of one subspecies. Fortunately, the Sikkim meeting led to action and, since that time, several field studies have been conducted with the result that there is now a plan to hold the first Red Panda PHVA in the autumn of 2010. A sample of the fieldwork recently undertaken is presented elsewhere in this volume. This chapter together with another presenting a summary of the fieldwork conducted in China, provide a good overview of our current knowledge on the status and distribution of the red panda. Happily, even in the absence of adequate field data, there are a number of conservation initiatives already in place which benefit the red panda across parts of its range as discussed by Wei Fuwen and Brian Williams in Chapters 21 and 22. However, these are still limited in scope and extent.

A viable zoo population of red pandas could play an important role in ensuring the red panda's survival. It could function either as a backup population for reintroduction or restocking in the wild, as a research resource for learning more about red panda biology or provide a cute flagship for engendering public sympathy and fund-raising for Himalayan conservation. Unfortunately, the zoo population is not yet up to this challenge despite the fact that our knowledge of the red panda in captivity is fairly extensive. There have been a number of publications dealing with captive management, nutrition and behaviour (see, for example, Red Panda Biology [16]). Husbandry and management guidelines [17] have been written informing zoos on best practice for caring for red pandas and captive breeding masterplans [18,19] have been produced which inform them on the best way to manage the zoo population of red pandas for posterity. Yet, despite the general availability of this information, the captive management of red pandas is still far from optimum; care in zoos leaves much to be desired. Kati Loeffler's chapter (Chapter 13) presents an overview of current husbandry practice and indicates areas in need of extra attention; in particular, her separate report on the care of red pandas kept in captivity in China is particularly disturbing. Furthermore, the status of the current captive population would indicate that either masterplan recommendations were not sufficient or that they have not been adequately implemented. The improvements in the zoo population which resulted after the initiation of regional breeding programmes and the publication of the husbandry and management guidelines seem to have worn off; the numbers of red pandas in the two largest regional populations are again declining and the healthy population which was developing in India has crashed. These problems will be discussed in more detail later, however, it suffices here to say that the future of the red panda in captivity, as in the wild, is far from assured at the present time. Unless these trends are reversed our healthy captive population could be reduced to small groups of inbred individuals.

The publication of "Red Panda" not only presents an overview of the current state of our knowledge about this intriguing species but it is also intended to bring the red panda out of obscurity and into the spotlight of public attention. The topics discussed in this volume cover red panda anatomy, reproductive physiology and behaviour, ecology, nutrition, veterinary care and pathology in addition to the subjects outlined above. It also examines the red panda's role in the regional economy and in culture, both traditional and modern and, surprisingly, it finds that this little-known species is fast on its way to becoming a star of the worldwide web and that this animal, once regarded as of no commercial value, is today unfortunately becoming a delicacy in some restaurants in China.

Finally, it discusses what the future could hold for the red panda, what is necessary to ensure its survival and whether we can afford to allow it to become extinct.

There is one fact on which almost all discussions of the red panda, both early descriptions and current observations, agree and that is that the red panda is a creature of great beauty and charm, both cute and playful with a docile, non-aggressive disposition. Its gentle nature and lack of speed or cunning have made, and continue to make, it susceptible to capture. Today, our lack of knowledge adds to its vulnerability; our ignorance of the red panda's needs may mean that we are permitting excessive destruction of its habitat at the same time as not implementing adequate measures for its protection. It is to be hoped that this volume, by demonstrating the uniqueness and beauty of the red panda, its importance to our environment and its place in our culture and traditions, will stimulate and energize people into taking the steps necessary to ensure that the shining cat bear of the Himalayas is with us to stay.

References

[1] R. Morris, D. Morris, Men and Pandas, Hutchinson & Co., London, UK, 1966.
[2] C. Catton, Pandas, Christopher Helm Ltd., Bromley, Kent, UK, 1990.

[3] T. Hardwicke, Description of a new Genus of the class Mammalia, from the Himalayan chain of hills between Nepaul and the snowy mountains, Read 6 November 1821, Trans. Linn. Soc. 15 (7) (1827) 161–165.
[4] A. Sowerby, C. de, Pandas or Cat-bears, China J. 17 (6) (1932) 296–299.
[5] E.P. Gee, The Wild Animals of India, Collins, London, UK, 1964.
[6] F. Cuvier, Histoire naturelle des mammifères, avec des figures originales, colorées, desinées d'apres des animaux vivants. Paris, (1825) 1–3.
[7] B.H. Hodgson, On the Cat-toed subplantigrades of the sub-Himalayas, J. Asiatic Soc. Bengal 16 (1847) 1113–1129.
[8] O. Thomas, On the Panda of Sze-Chuan, Ann. Mag. Nat. Hist. 10 (1902) 251–252.
[9] O. Thomas, On Mammals from the Yunnan Highlands collected by Mr George Forrest and presented to the British Museum by Col. Stephenson R. Clarke, DSO, 1922.
[10] A.R. Glatston, The Red Panda, Olingos, Coatis, Raccoons and their Relatives, IUCN, Gland, Switzerland, 1994.
[11] A.D. Bartlett, Remarks on the habits of the Panda (*Aelurus fulgens*) in captivity, Proc. Zool. Soc. London (1870) 269–772.
[12] A.D. Bartlett, Wild Beasts in the Zoo, Chapman and Hall, London, UK, 1900.
[13] W.H. Flower, On the anatomy of *Aelurus fulgens*, Fr. Cuv. Proc. Zool. Soc. London (1870) 754–769.
[14] E.R. Lankaster, On the affinities of *Aeluropus melanoleucus*, Trans. Linn. Soc. London, Zool 8 (6) (1901) 163.
[15] R. Lydekker, The Game Animals of India, Burma, Malaya and Tibet, Rowland Ward, London, UK, 1924.
[16] A.R. Glatston (Ed.), Red Panda Biology, SBP Academic Publishing, The Hague, Netherlands, 1989.
[17] A.R. Glatston, Husbandry and management guidelines, in: A.R. Glatston (Ed.), The Red or Lesser Panda Studbook No. 7, Stichting Koninklijke Rotterdamse, Diergaarde, Rotterdam, Netherlands, 1993, pp. 37–66.
[18] A.R. Glatston, F.P.G. Princee, A Global Masterplan for the Captive Breeding of the Red Panda. Part 1, *Ailurus fulgens fulgens*, Royal Rotterdam Zoological and Botanical Gardens, The Netherlands, 1993.
[19] A.R. Glatston, K. Leus, Global Captive Breeding Masterplan for The Red or Lesser Panda *Ailurus fulgens fulgens* and *Ailurus fulgens styani*, Royal Rotterdam Zoological and Botanical Gardens, Rotterdam, The Netherlands, 2005.

CHAPTER

2

People and Red Pandas: The Red Panda's Role in Economy and Culture

Angela R. Glatston[1] *and Axel Gebauer*[2]

[1]Royal Rotterdam Zoological and Botanical Gardens, Rotterdam, The Netherlands
[2]Görlitz Zoo, Görlitz, Germany

OUTLINE

The red panda is an exceptionally attractive species. Cuvier, who published the first account of the red panda in 1825, described it as: "Most handsome mammal in existence" [1] and Sowerby [2], another early reporter who saw it in the Calcutta Zoo, referred to it as "a delightful little creature". Yet, despite its striking appearance, the red panda seems to have had only limited impact on the culture, traditions or economy of the Himalayan region. In fact, the 1994 IUCN SSC Action Plan for Procyonids and Ailurids [3] concluded that, although the red panda is acknowledged as the state animal of Sikkim and its image was also used to promote the first International Tea Festival in Darjeeling, it does not form part of the culture or folklore of any of its range states. We now know that this statement is not completely true, nevertheless, reference to the red panda in folktales or rituals remains sparse. To an outsider it is difficult to understand why such a notably beautiful creature has managed to make so little impression on the traditions of its homelands. On the other hand, considering that the red panda's habitat is relatively inaccessible and that these animals are essentially nocturnal or crepuscular in their habits, this lack of impact

Red Panda. DOI: 10.1016/B978-1-4377-7813-7.00002-1

becomes easier to appreciate. However, as mentioned above, further study has indicated that the Action Plan was not entirely correct in its conclusions; the red panda does indeed have a place in the culture of its range states, albeit a limited one. What perhaps is more surprising is the effect that the red panda seems to be having on modern culture, particularly outside of its range states. Apparently, mass travel and the Internet have at last brought the red panda to the attention of the general public and today its image and its name do indeed play a significant role in culture, art and merchandising around the world, which is surprising given that most people claim that they are unaware of the red panda's existence.

THE RED PANDA IN CULTURE AND TRADITION OF THE RANGE STATES

The red panda is clearly familiar to the local peoples of the Himalayas. Hodgson [4], the first Westerner to study the red panda in the wild, provides us with numerous local names for the red panda (see Box 2.1) and other authors [5–7] have since supplied us with several more originating from other parts of the red panda's range. The species was also clearly well known in its range states long before Hardwicke presented his paper to the Linnean Society [8]; Roberts reports that a red panda was depicted in a Chinese pen

BOX 2.1

WHAT'S IN A NAME; VARIOUS LOCAL NAMES FOR THE RED PANDA

The red panda is known by a variety of names in English of which the red panda or the lesser panda are the most familiar. However, a number of other names have been used in the literature such as: (red) bear-cat, cat bear, cloud bear, bright panda, common panda, fire cat, fire fox, red cat, brilliant cat, fox bear, Himalayan raccoon.

The following local names have also been reported:

Area/Dialect	Names used
Limbu	Walsar, Jho, Kye, Wáh, Oá, Yé, Nigála pónya
Shotia	Wahe
Lepcha	Sa Nums, Sankam, Saknam, Thóng-wah, Thó-kyé
Bhotias	Wakdonka, Woker, Oakdnka U'któnka
Sikkim	Workar
Burma	Kyaung-wun
Kachin	Ru
China	Chu-chieh-liang, Xia xong mao
Bhutan	Achhu Dongkar (Dzongkha)
Central Nepal	Hobrey

and ink scroll illustrating a hunting scene [9]. This scroll dates back to the 13th century Chou dynasty. Nevertheless, this apparent familiarity has not led to a key role for the red panda in the culture and traditions of the peoples of the Himalayas. For example, Majupuria [10], in his book on the sacred and symbolic animals of Nepal, discusses many Himalayan species, all of which have symbolism in Nepalese traditional culture, but the red panda is not among them. However, Michael Oppitz [11], a Swiss anthropologist, reports that the ramma or shamans of the Northern Magar tribe in the Dhaulagiri Region of Western Nepal use the skin and fur of the red panda in their ritual dress. These people consider the red panda to be a protective animal which guards the wearer against the attacks of aggressive spirits; for this reason, its body is hung on the shaman's back when he undergoes a dangerous ritual in the course of his healing seances. The adjacent Bhuji Khola Kami people apparently have similar beliefs (Oppitz, personal communication).

The belief that red pandas and/or their fur are good luck talismans can be found in other parts of the red panda's range, for example, red panda tails were formerly used as good luck charms by some of the tribal people of Arunachal Pradesh (Dwaipayan Banerjee, personal communication) and it is considered an omen of good luck for the Yi people of Yunnan if a bridegroom wears a red panda fur hat during his wedding ceremony. Other examples of the red panda as a harbinger of good fortune can be found in the beliefs of some of the tribal peoples of Bhutan, convictions which persist right up until the present day in some parts of the country. For example, the people living in the Gasa and upper Paro regions of western Bhutan maintain that it is a good omen if you see a panda when you are travelling on a business trip; it guarantees that the trip will be successful. Similarly, the people of the Bumthang and Sengore regions of central Bhutan consider red pandas to be the reincarnation of Buddhist monks, apparently because their fur is a similar colour to that of the monks' robes. As a result of this, these people will not harm red pandas (Sangay Dorij, personal communication). Unfortunately, these convictions do not occur throughout the country; some villagers in the lower areas of Jigme Dorji National Park are of the opinion that if the red panda, or *"Yaem Dongkar"*, howls at night it is an omen that somebody from their community will die. However, red pandas cannot really be said to howl, so clearly these people are confusing the red panda with another species, probably the jackal or the Indian red fox but, nevertheless, their conviction means that, if they see a red panda during the day time, they will try to kill it or chase it away.

The Action Plan for Procyonids and Ailurids [3] also reported that red pandas have never played a significant role in the trade or economy of the range states. Again, it is surprising that such a beautifully coloured, thick, warm fur has never had any commercial value in the fur trade. On more detailed examination, it is clear that the early reports on this topic contradict each other. Some reports indeed indicate that red panda fur was considered of such poor quality that only the tail had any value and that as a duster [12]. However, other sources would seem to disagree; some authors do report that red panda skins were sold in local markets [3,13]. Indeed, some even indicate that a substantial trade in red panda fur once existed; Roy Chapman Andrews [14] made the following report in Camps and Trails in China in 1916–1917:

> ... in Ta-li Fu (Dali) and Hsia-kuan (Xiaguan) are important fur markets and we spent some time investigating the shops. One important find was the panda (*Aelurus fulgens*). The panda is an aberrant member

of the raccoon family but looks rather like a fox; in fact the Chinese call it the "fire fox" because of its beautiful, red fur. Pandas were supposed to be exceedingly rare and we could hardly believe it possible when we saw dozens of coats made from their skins hanging in the fur shops

and

Li-chiang (Lijiang) is a fur market of considerable importance for the Tibetans bring down vast quantities of skins for sale and trade. Lambs, goats, foxes, cats, civets, *pandas*, and flying squirrels hang in the shops and there are dozens of fur dressers who do really excellent tanning.

Red panda fur coats are no longer seen, however, the red panda fur hats have remained in use, particularly in China. In the past, red panda fur hats were fairly common (see Figure 21.6) throughout much of its range, they were worn both in Bhutan (Sangay Dorij, personal communication) and in the Singalila region of the Indian-Nepalese border and, in China, they formed part of the traditional uniform of Naxi or Muli soldiers of Szechuan. There are two photographs dating from 1924–25 taken by Joseph Rock during his travels in China that show these soldiers in their traditional uniform and red panda fur hat [15]. Indeed, it is in the Yunnan region of China where people wearing red panda fur hats can still be seen today. There are several references and images on the Internet of tribal peoples wearing red panda skin hats. Two of these show a man and a woman belonging to a Tibetan minority group from the Yunnan Province. The photograph was taken in the main square of the old town of Lijiang as recently as the spring of 2006. The photographer (Peter Oxford, personal communication) who posted these images from 2006 said it was still common to see the people of various tribes, particularly those belonging to the Yi tribe from the Jade Dragon Mountain area north of Lijiang, with red panda skin hats. Indeed, a man wearing such a hat came into view during a recent BBC television documentary series about the Himalayas. This particular scene was also shot in Lijiang.

Today, the sale of red panda pelts still continues illegally. The Action Plan for Procyonids and Ailurids has a photograph of a woman selling a red panda skin in a market. More recently, a report in the USA Today On Deadline blog read as follows:

According to the Beijing News, Sun Shiqun, 60, told police she paid $5000 for the 3-foot-long pelt. Police said tests confirmed that the skin came from a panda.

The Associated Press reported Sun tried to "pass off the pelt of a red panda, a smaller animal, as a giant panda" [16].

Other incidents that have come to the attention of the authors are the confiscation, in 2005, of red panda furs from a Nepalese tradesman in the Gola Pass in the Kanchenjunga Conservation Area, Nepal (D. Chapagain to R. Melisch of TRAFFIC, personal communication, September 2006). Recently, an employee of Flora and Fauna International reported seeing red panda carcasses in the homes of villagers in Eastern Myanmar. Apparently, these villagers regularly hunt red pandas and one of these hunters allowed the FFI representative to accompany him while he caught a red panda with his hands (Frank Momberg, personal communication).

Although red panda fur does have some uses, the same is not true of its meat. Hodgson specifically stated that red panda meat is not eaten and, until recently, there were very few reports of red panda meat being eaten apart from some anecdotal reports of its consumption in Arunachal Pradesh. Unfortunately, this situation has changed in recent months with reports now emerging from China of red panda meat being served in restaurants. A visitor to Zhongshan City in the Pearl River Delta near Hong Kong, who wishes to remain anonymous, reported seeing red pandas housed in tiny cages at the back of restaurants, waiting to be butchered. It is to be hoped for the future of this rare species, that this new trend will be very short lived.

In addition to these uses for dead red pandas, living specimens also seem to have had a limited role as gifts and/or pets. Hodgson [4], the author of the earliest reports on the red panda, saw the potential of red pandas as pets and suggested that, due to their gentle disposition and lack of smell, they would make good "pets for ladies". This recommendation was never taken up, however, the tribal peoples of Arunachal Pradesh do traditionally keep red pandas as pets (Forest Officer, personal communication) while the Northern Magar people of Western Nepal also use them as gifts (Oppitz, personal communication). Gee [17] in his book on Indian wildlife reports that red pandas were the favourite pets of the former Indian prime minister Jawaharlal Nehru. Indeed, some footage of Nehru and his pet red pandas can be seen in the documentary film, Cherub of the Mist [18]. Even today the studbook keeper still receives sporadic reports of live red pandas being offered for sale, presumably as pets, in the markets of China. The habit of keeping red pandas as household pets may have spread to other countries in recent times; there are two or three videos on YouTube that show red pandas in Japanese homes. The animals in question seem to be treated like spoiled pets. It is possible that the subjects of these videos could be zoo animals that are being hand-reared in the home of a zookeeper, nevertheless, to the casual viewer they appear to be charming pets.

Red pandas are not just presented as individual gifts, they have also been used as state gifts on occasion even though they are somewhat less prestigious than their giant namesake in this context. The international studbook records show that, in 1984, a pair of red pandas was presented as a state gift to Spain by the government of Nepal. Around the same period, a Chinese delegation visiting Australia donated red pandas to both the Sydney and Melbourne zoos.

THE RED PANDA IN TODAY'S RANGE STATES

Today, the use of the red panda's image is becoming more prevalent within its range states than was formerly the case. As mentioned earlier, the red panda is currently the state animal of Sikkim where its face appears on the coat of arms of the Sikkim Forest Department (Figure 2.1). Its picture was also used in the advertisements around the 1991 International Tea Festival. Since that time it has also leant its name to a local black tea, which is apparently particularly popular with tourists (Figure 2.2). The red panda's likeness also adorns the label of Red Panda beer, a beverage which is brewed by a local brewery in Bumthang, Bhutan (Figure 2.3). When the owner of the brewery was asked why he

FIGURE 2.1 Gate into Sikkim showing the coat of arms of the Forest Department.

FIGURE 2.2 Red panda tea.

FIGURE 2.3 Red panda beer label.

had named his beer after the red panda he replied that while many beers use animal names which evoke power, this was not the image that he sought for his brew. His felt his beer looked beautiful in the glass, had a soft rather than an astringent taste and was intended to savour rather than to inebriate. Therefore, he named it after an animal which, in his words, "is a special living thing and a symbol for beauty, fellowship and perhaps also a little wisdom (intelligence)".

The red panda's image also appears on postage stamps and coins in its range states. Bhutan, India (Figure 2.4) and China have all produced red panda stamps while Nepal has struck a 50 rupee coin depicting the species. It is interesting to note that several countries outside of the red panda's natural distribution have also depicted it on their stamps, for example, Mongolia, New Zealand and the Czech Republic, in addition, a special series of commemorative red panda stamps has also been produced in the USA. Indeed, it is outside of the range states where the red panda's name and image seem to have caught the public imagination turning this little-known species into a cultural icon.

FIGURE 2.4 Indian postage stamp depicting a red panda.

RED PANDAS IN WESTERN CULTURE

The soft and cute-looking red panda is clearly an animal whose form lends itself well to the production of plush toys; an assortment of cuddly red panda toys is available around the world, although not the extensive selection that is available for its giant namesake. In addition, the red panda's attractive image is used to merchandise T-shirts, postcards, posters, coasters, embroidery patterns, commemorative coins and many other items (Figure 2.5) as well as in promotional context (Figure 2.6), while its name is applied in many contexts varying from popular music groups to computer games and books to trading companies etc. However, it is on the Internet where the red panda's name and likeness seem to have had the greatest impact.

Red Pandas in the Movies

As unlikely as it may seem, red panda characters have played significant roles in four full-length movies. The most important of these are two culturally significant anime (Japanese animation) films: The *Tale of the White Serpent* and The *Jungle Book Shonen Mowgli*. The former of these was particularly noteworthy as it was the first colour anime, feature film. It was released in 1958 and was honoured at the Venice Children's Film Festival the following year. It is essentially an adaptation of the Song dynasty folktale "Madame White Snake" which tells the story of Xu-Xian, a young boy, who owns a pet snake which magically transforms into a beautiful princess. The princess and boy fall in love only to be separated by a local monk. The story ends happily with the princess giving up her magical powers and remaining in human form to prove that her love for Xu-Xian is genuine. In the film, Xu-Xian has two pets, Panda (a giant panda) and Mimi (a red panda), both of whom help Xu-Xian in his quest to find the princess. When this film was later released in the USA the name was changed to *Panda and the Magic Serpent* and the red panda character was transformed into a domestic cat.

FIGURE 2.5 Commemorative coin from Görlitz Zoo.

FIGURE 2.6 Red panda promoting recycling. *(Reprinted courtesy of Call2recycle)*

The second anime film, the *Jungle Book Shonen Mowgli* or *Janguru Bukku Shōnen Mōguri*, is an adaptation of the original Jungle Book story. It was broadcast in the early 1990s by a national television company in India where it became a popular series. It was even translated into Arabic and became a hit with Arab viewers as well. There was of course no red panda character in Kipling's original *Jungle Book*, however, the Shonen Mowgli film introduced viewers to Kichi, a red panda who is befriended by Mowgli, after humans have killed her parents, and joins him in many of his adventures (Figure 2.7).

Red panda characters have also had major roles in two western animation films; the *Bamboo Bears* and, more recently, *Kung-fu Panda*. The first of these was a 1990s European series which told the story of a group of animal eco-warriors: Slo-Lee, a giant panda, Bamboo-Lee, a red panda and Dah-Lin a female bamboo rat, who save the world from various ecological disasters. This television series was released in the early 1990s and has apparently recently been launched again in DVD form by Allumination Filmworks. Finally, 2008 saw the launch of the major Dreamworks production, *Kung Fu Panda*, a mainstream movie, in which the characters are voiced by well-known Hollywood screen actors. The red panda character, Master Shifu, is a kung fu guru who has to train Po, the giant panda, so he can fulfil the role of Dragon warrior and rid the Valley of Peace of the evil snow leopard

FIGURE 2.7 Board in the Padmaja Naidu Himalayan Zoo showing Kichi.

Tai Lung. Many people do not seem to be initially aware that Master Shifu, voiced over by actor Dustin Hofman, is indeed a red panda and this has given rise to various Internet discussions on the subject which in turn is raising awareness of the red panda with both movie-goers and Internet users worldwide.

There is one further animated film depicting a red panda that deserves a mention and that is *Barbie the Island Princess* produced by the toy company Mattel. In this film, the Barbie central character has several animal friends sharing her adventures including Sagi, the red panda. Accompanying this film is the Island (or Jungle) Princess line of Barbie doll toys, each of which has its own red panda pet included in the packaging. A second spin-off is a children's game based around this storyline where the Barbie doll character is shipwrecked on a desert island and reared by a red panda. There are undoubtedly more movies and videos to be found featuring red panda characters but the preceding list includes the major offerings to date.

In addition to these commercial films, red pandas, both real and cartoon, are depicted in a considerable number of YouTube videos; more than 3000 titles come up in response to typing red panda into the search window. Some of these are clips from the movies discussed previously, others are about acrobats or musicians who have named themselves after the red panda but, by far the majority of these are the offerings of ordinary people who are showing their appreciation for red pandas. The sheer number of these videos attests to the appeal of the red panda to people far and wide. Some show zoo visits in which the red panda is central, while others show particularly endearing behaviours such as mother pandas with their cubs, play behaviour or "kissing" (red pandas grooming each other's mouth). Red pandas standing on their hind legs and "waving" or young red pandas clumsily moving through their enclosures are particularly popular. Other contributions are simply series of red panda photographs put to evocative music and yet others show red pandas interacting with people, for example, zoo staff catching escaped red pandas; people feeding red pandas by hand or sitting with a panda on their lap. There are even films depicting the antics of red pandas in the home, for example, the very popular "Panda Attack".

Red Pandas in Other Art Forms

The image of the red panda is not confined to videos. There are numerous photographs of red pandas on the Internet available both to view and to purchase. Flickr, the Internet photograph sharing site, has a huge selection of red panda pictures. These range from the classic nature shot to more artistic representations and to the downright cute. The red panda is also a popular subject for artists and images abound varying from traditional pen and ink drawings to watercolours and oil paintings. In addition, there are also a number of artists who have used the red panda as the basis of a more imaginative fantasy genre of art depicting red pandas as warriors, princesses, magicians or monks (Figure 2.8). Finally, there are people who use red panda avatars in various Internet communities, blogs and games. In fact, recently, a specific red panda game has been produced, *Bipo Mystery of the Red Panda*, which follows the adventures of Bipo the red panda trying to find his grandfather.

FIGURE 2.8 Martin Hsu's image of the red panda as a Buddhist monk. *(Reprinted courtesy of Martin Hsu)*

It is not only the red panda's cute appearance that is popular; its name is also evocative. There are plays and books which have no relationship with the animal but which use its name in their title. For example, there is a Red Panda who is the masked hero of the Decoder Ring theatre company's popular podcast, *The Red Panda Adventures*. This light-hearted series follows the adventures of the Red Panda, "Canada's greatest superhero", and the Flying Squirrel, his trusty sidekick, as they protect the citizens of Toronto from villains ranging from gangsters to the supernatural forces of darkness. There is absolutely nothing about this hero that remotely resembles his animal namesake but the name is catchy. The red panda name also appears in the title of the eighth volume of the American comic book series, *Of Queen & Country*. Published in 2007, it is called *Operation Red Panda*. The series itself follows the adventures of Tara Chace, an operative of the Special Operations Section of SIS, known as the Minders. Again there is nothing about the story of *Operation Red Panda* that is remotely connected with the species in question, it is just making use of the evocative panda name. On the other hand, Andrea Siegal's novel, *Like the Red Panda*, may not be directly related to the animal itself but, as one reviewer says, "it is a novel which artfully evokes the despair, even the hopelessness, of an endearing character". It is perhaps in the hopelessness and charm of the main character that we may find a real link between the title of the book and the red panda species.

The use of the red panda's name extends far beyond the world of films, art, novels and radio plays. Searching Google for websites using the term red panda yields a plethora of personal websites and blogs either dedicated to or using the name of this relatively unknown species. There are red panda companies and entertainers including a record production company, a trading company, a popular music band, a troop of Chinese acrobats, a jewellery producer, a chain of Chinese restaurants, a games company and a publishing house to name but a few. However, undoubtedly the best known and most widespread use of one of the red panda's names is that by Mozilla who have named their popular web browser, Firefox. Most Firefox users are probably unaware that the Firefox is a red panda and, it is true that the logo does not help this, as it seems to represent a fox rather than a panda. Nevertheless, Mozilla acknowledges that the red panda is the namesake of their browser and have sponsored a number of red panda projects as a result of this link. Interestingly, a new, real-time, discovery engine add-on has recently appeared for use with Firefox, and its name is – Red Panda.

CONCLUSION

When a species has a high cultural value it adds an extra dimension to our motivation to preserve it. Likewise, if it has a high economic value our urge to exploit it to extinction may be mitigated by the necessity of leaving an important resource available to our descendants. These are both considerations which are taken into account in the preparation of IUCN species Action Plans. According to the IUCN Action Plan for Procyonids and Ailurids, the red panda confers no particular cultural or economic benefits to the people of its range states so the cultural dimension cannot be said to add an extra impetus to its conservation. In this chapter, we have demonstrated that this lack of cultural and economic impact is not entirely true and that the red panda does indeed have a role, albeit a limited one, in the traditional practices of the local peoples. With a few exceptions, it generally seems that, in these cultures, the red panda is associated with protection and good luck. At the same time, in economic terms, its fur and meat remain of limited use and so at present its economic value luckily remains negligible.

The red panda may only have a limited role in the traditions of the range states but it does seem to have gained a considerable, and increasing, cultural role in the western world, particularly in the virtual world of the Internet. It is unclear how such a comparatively unknown species has gained such a foothold in our lives without our being aware of it. Undoubtedly, television nature documentaries and the advent of the Internet have brought the image of this little-known, attractive animal into our homes and made more of us aware of its beauty and endearing character while, at the same time, not increasing our awareness of the actual animal species. As a result, a whole section of the general public has taken the red panda to its collective heart.

At the beginning of this chapter, the question was raised why such a striking animal as the red panda appears so rarely in the local culture and traditions of its range states. Here, at the end of the chapter, we have a second, more vexing question to answer: why have the red panda's name and likeness now become so popular in the western, or at least, the virtual world? Perhaps it is not the actual answers to either of these questions that are really important but rather the fact that the beauty and charm of this little-known species is gradually winning the hearts and minds of people around the world and that this admiration may provide an opportunity to raise more support for its conservation in the long term.

References

[1] R. Morris, D. Morris, Men and Pandas, Hutchinson & Co., London, UK, 1966.

[2] A. Sowerby, C. De, The pandas or cat-bear, China J. 17 (6) (1932) 296–299.

[3] A.R. Glatston, The Red Panda, Olingos, Coatis, Raccoons and their Relatives. Status Survey and Conservation Action Plan, IUCN, Gland, Switzerland, 1994, (compiler).

[4] B.H. Hodgson, On the cat-toed subplantigrades of the sub-Himalayas, J. Asiatic Soc. Bengal 16 (1847) 1113–1129.

[5] C. Catton, Pandas, Christopher Helm Ltd., Bromley, Kent, UK, 1990.

[6] G.S., Gurung, Reconciling biodiversity conservation priorities with livelihood needs in kangchenjunga conservation area, Nepal, Dissertation Universität Zürich, (2006) 203 pp.

[7] R. Lydekker, The Game Animals of India, Burma, Malaya and Tibet, Rowland Ward, London, UK, 1924.

[8] M.G. Hardwicke, Description of a new genus of the class mammalia, from the Himalaya chasin of hills between Nepaul and the snowy mountains, Trans. Linnean Soc. London XV (1827) 161–165.

[9] M. Roberts, The fire cat, ZooGoer March/April (1992) 13–18.

[10] T.C. Majupuria, Sacred and Symbolic Animals of Nepal, Sahayogi Press, Kathmandu, Nepal, 1977.

[11] M. Oppitz, Schamanen im Blinden Land. Film, 2nd edition on DVD from 2008, Völkerkundemuseum der Universität, Zürich, Switzerland, 1980.

[12] R. Mell, Beiträge zur Fauna sinica. I. Die Vertebraten südchinas; Feldlisten und Feldnoten der Säuger, Vögel, Reptilien, Batrachier, Arch. Naturgesch. 88 (A, 10) (1922) 1–134.

[13] D. MacKlintock, Red Pandas, a Natural History, Charles Schreiber's Sons, New York, USA, 1988.

[14] R.C. Andrews, Y.B. Andrews, Camps and Trails in China, 1916–1917: A Narrative of Exploration, Adventure and Sport in Little Known China, Appleton, New York, USA, 1918.

[15] J., Rock, National Geographic Magazine. Photographs are available from the Havard Image Library (http://via.harvard.edu:748) but the images are not easy to find. The images in question can be found at: http://

drjosephrock.blogspot.com/2005_02_27_drjosephrock_archive.html and the 2005-02-13_drjosephrock_archive, 1925.

[16] Unfortunately Mike Carney's post is no longer available on the Internet, however, the incident was referred to in the Shanghai Daily Newspaper, March 08, 2007.

[17] E.P. Gee, The Wild Life of India, Collins, London, UK, 1964.

[18] N. Bedi, R. Bedi, Cherub of the Mist. Movie, Bedi Films/Visuals, India, 2006.

CHAPTER

3

Evolution of the Family Ailuridae: Origins and Old-World Fossil Record

Manuel J. Salesa[1], *Stéphane Peigné*[2], *Mauricio Antón*[1] *and Jorge Morales*[1]

[1]**Departamento de Paleobiología, Museo Nacional de Ciencias Naturales-CSIC, Madrid, Spain**
[2]**Muséum national d'Histoire naturelle, Département Histoire de la Terre, USM 0203 – UMR 7207 CNRS/MNHN/UPMC Centre de Rechercha Sur la Paléobiodiversité et les Paléoenvironnements, Paris, France**

OUTLINE

INTRODUCTION

The red panda family, the Ailuridae, is unusual because its only extant representative is a derived form, far removed from the group's morphotypical condition. Effectively, the extreme adaptations of the red panda skull and dentition for a vegetarian diet imply that the study of the fossil record is essential if we are to gain an understanding of the original biology and ecology of the ailurids. Fossil ailurids have remained frustratingly rare and fragmentary findings for many decades, but recent years have seen a dramatic improvement of their record. This has been especially the case with the recent finds of dramatically complete remains of the simocyonine *Simocyon batalleri* from Batallones-1 in Madrid, Spain, and of the ailurine *Pristinailurus bristoli* from Gray Fossil Site in Tennessee, USA. Taking into account the information provided by these and other recent findings, it is now

Red Panda. DOI: 10.1016/B978-1-4377-7813-7.00003-3

possible to gain a more complete picture than ever before of the true place of ailurids within the Carnivora, and of the evolutionary processes that led to the remarkable convergences between the only living ailurid and the unrelated giant panda. In this chapter, we present the early evolution of the ailurids, its Old-World fossil record, and the implications of the palaeontological evidence for the position of the ailurids among other carnivoran families.

ORIGIN OF THE AILURIDAE

The earliest ancestors of the red panda within the order Carnivora were the miacids, which lived in the Late Palaeocene and Eocene of Eurasia and North America, between approximately 55 and 34 million years ago [1]. The miacids, traditionally included as a family within the Carnivora, but now widely considered as a paraphyletic group, were at the root of the Caniformia, one of the two suborders within the order Carnivora, which includes the dog, bear, weasel and raccoon families besides the Phocoidea, a clade equivalent to the traditional Pinnipedia -phocids, otariids and odobenids [2]. With their small size and weasel-like appearance, miacids were a widespread and successful group, well adapted to life in the lush forests of the Eocene [3]. Late in the Eocene, they gave rise to two separate groups, the hesperocyonines or earliest Canids on one side, and the earliest Ursida, represented by the American genus *Mustelavus* [2]. *Mustelavus* in turn was near the dichotomy that gave rise to two large groups of Caniformia: the ailurids and procyonids on one hand, and the mustelids on the other. The European fossil record of basal Mustelida includes a series of genera that share the presence of a small suprameatal fossa and a reduced M1 parastyle [2]. Among these, the genus *Amphictis*, discussed below, appears closest to the origins of the family Ailuridae.

THE PRIMITIVE AILURIDS

Interest in the evolutionary history of the family Ailuridae has recently increased due to the new discoveries of highly interesting fossils which have filled a traditional gap in the knowledge of the anatomy and biomechanics of the group, allowing a better understanding of the development of the ailurid model. Nevertheless, the earliest representatives of the family are still poorly known, due to the relative scarcity of their fossils. Furthermore, studies on these fossils have classically focused on fragmentary skulls and mandibles, and paid little attention to the rest of the skeleton, which was virtually unknown for these early forms [4–9].

Ailurids likely originated in Europe during the Late Oligocene–Early Miocene (i.e., between 25 and 18 Ma), with the genus *Amphictis* Pomel, 1853 [10,11] (Figure 3.1). This animal, with a skull length of 10 cm, was probably very similar in size to that of an extant red panda (*Ailurus fulgens*), although only the skull and dentition are relatively well known. The dentition of *Amphictis* was that of a generalized carnivore, with simple and pointed premolars, relatively trenchant carnassials (P4 and m1) and M1, M2 and m2 with grinding surfaces (Figure 3.2). While in many ways (dentition, basicranium) the morphology of

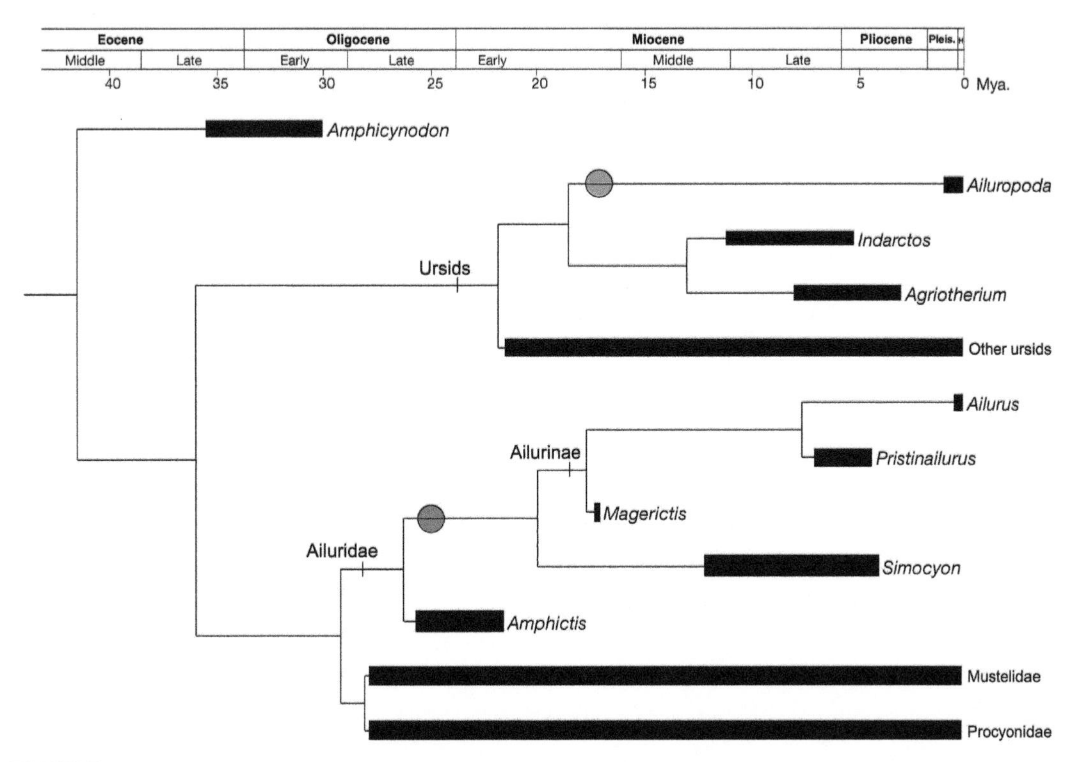

FIGURE 3.1 Schematic tree summarizing the phylogenetic relationships and temporal ranges of the members of Ailuridae and other related families of the Carnivora, highlighting the independent double appearance of the false-thumb in giant pandas and ailurids (coloured circles) *(taken from [12])*.

Amphictis remains primitive, the elongation of m2, which also displays a completely separated protoconid and metaconid, and the presence of lateral grooves on the canines, support the placement of *Amphictis* into the Ailuridae. The dentition of the individual species of this genus resembles in some ways that of extant canids or mustelids, and points towards a diet probably based on small vertebrates, invertebrates, eggs and fruit. Even if the relationship of *Amphictis* to the red panda lineage remains to be confirmed by additional material, the placement of this genus in the Ailuridae is reinforced by the close phylogenetic relationships between *Amphictis* and two later genera more securely placed within the family, *Alopecocyon* and especially *Simocyon* (see below for the latter).

Alopecocyon is known from fragmentary material from the middle Miocene of Europe, Asia, and North America (under the name *Actiocyon*, a probable junior synonym of *Alopecocyon* [10,13]) and therefore documents the early expansion of the family throughout the northern continents. The species of *Alopecocyon* (Figure 3.3) differ from those of *Amphictis* in having second molars enlarged relative to the first molars; both genera share the structure of the m2 and the presence of a lateral groove on the canines with more advanced ailurids. The close relationship between *Alopecocyon* and *Simocyon* has long been recognized [5,6,14–16] and is based on dental evidence, but postcranial and basicranial material could

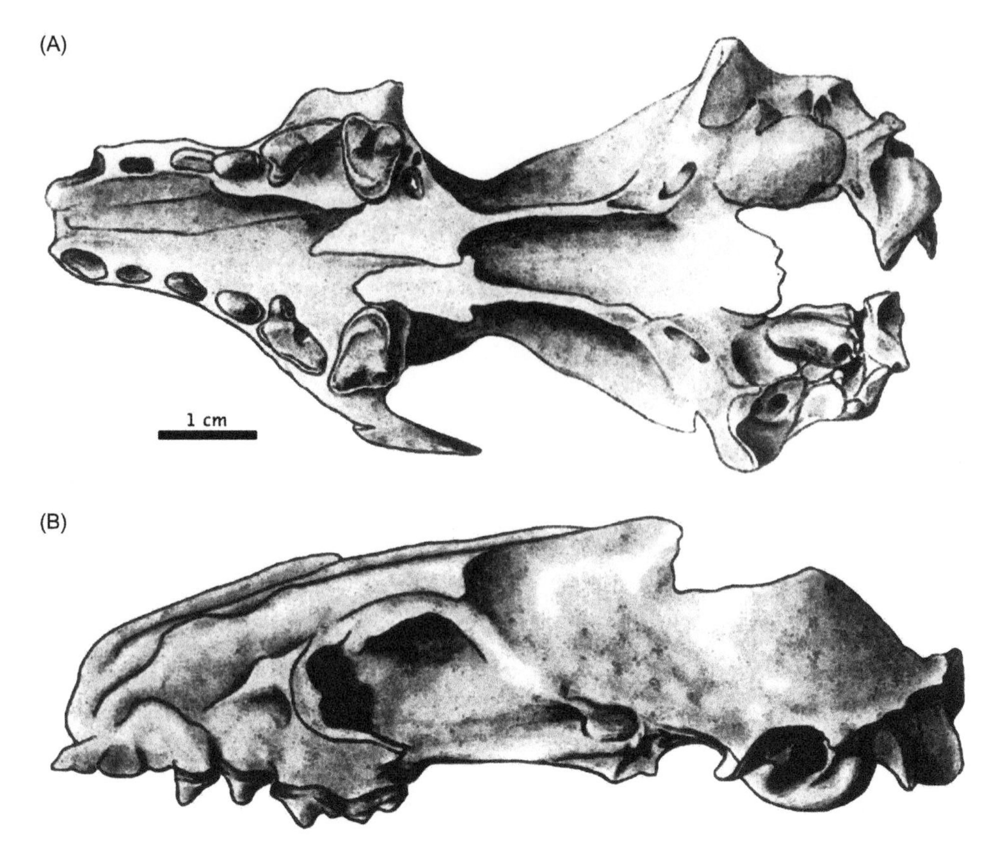

FIGURE 3.2 Skull of *Amphictis ambiguus* (PFRA-28) from the Phosphorites du Quercy (Late Oligocene, France); (A) ventral view; (B) lateral view *(image modified from [7])*.

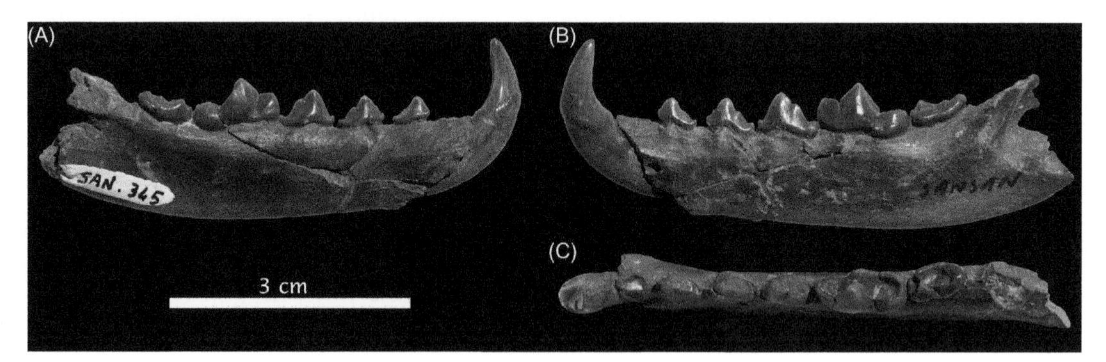

FIGURE 3.3 Fragment of left hemimandible of *Alopecocyon goeriachensis* (SAN-345) from the locality of Sansan (Middle Miocene, France); (A) lingual (internal) view; (B) labial (external) view; (C) occlusal view *(photos by Guillaume Fleury and Yves Laurent, from the Museum of Toulouse)*.

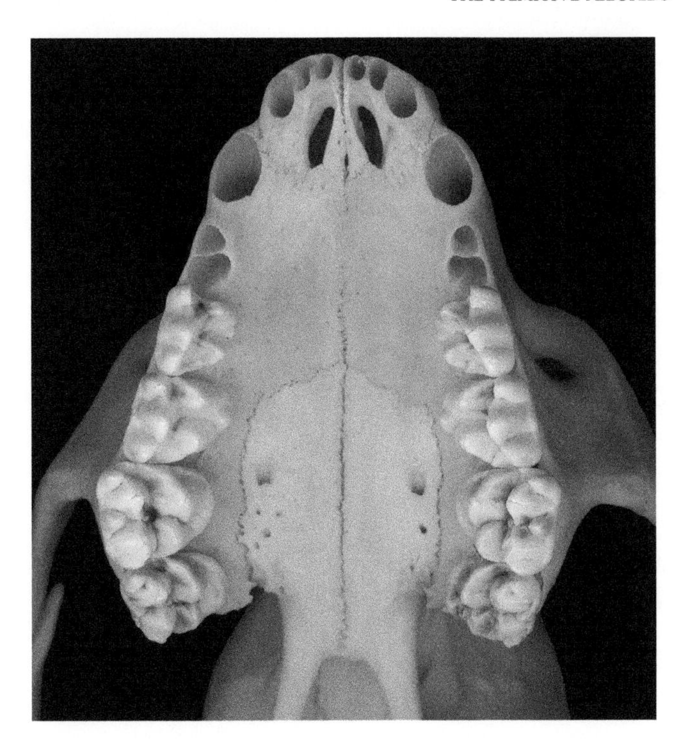

FIGURE 3.4 Detailed view of the upper cheek teeth of *Ailurus fulgens*, showing their highly specialized morphology to be a member of the Carnivora *(photo by M.J. Salesa)*

confirm this relationship. Thus, the early ailurids (*Amphictis* and *Alopecocyon*) did not exhibit the specialized dental morphology toward herbivory seen in the more recent members of the subfamily Ailurinae: *Pristinailurus*, *Parailurus*, and *Ailurus*. These species, being adapted to a diet mostly composed of plant material, share a completely different dental pattern, with occlusal surfaces composed of several blunted cusps [17], producing a model more similar to that of a pig than to any other carnivore (Figure 3.4). This cusp pattern produces a grinding surface, very efficient at processing tough vegetal material, such as bamboo leaves, which are the primary constituent of the diet of the extant red panda.

Another ailurid very close to *Amphictis* is *Magerictis imperialensis*, from the Middle Miocene. This species was described on the basis of an isolated m2 found in the fossil site of Estación Imperial (Madrid, Spain) in 1991 [18]. Although it was the only fossil of this species retrieved from the excavation, its morphology clearly fitted with the ailurine m2 morphotype, that is, exhibiting a very elongated crown, with several low cuspids and, more specifically, a metaconid and a protoconid completely separated (more clearly so than in, e.g., *Amphictis*) by a longitudinal valley that continues the full length of the tooth [18] (Figure 3.5). This animal probably inhabited the tropical forests that existed in Europe at this time but, unfortunately, until its postcrania is found, little else can be said about its paleobiology.

One of the best represented fossil ailurids is the puma-sized *Simocyon* (Figure 3.6), which is known from the Middle and Late Miocene of North America [10,19] and Europe [11,20,21], and from the Middle Miocene to Early Pliocene of China [22,23]. Its craniodental anatomy is very well known thanks to the very complete cranial and dental material

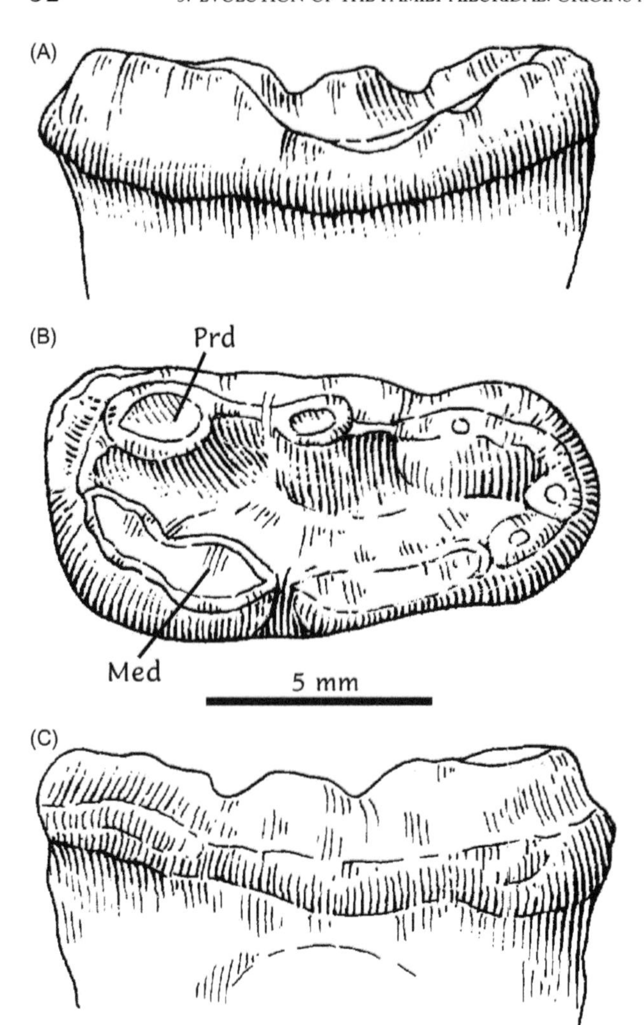

FIGURE 3.5 Detailed view of the right lower second molar of *Magerictis imperialensis* from the locality of Estación Imperial (Middle Miocene, Madrid, Spain), showing the marked separation between two of the main cuspids of the teeth: protoconid (Prd) and metaconid (Med); (A), lingual (internal) view; (B), occlusal view; (C), labial (external) view *(image modified from [18])*.

found in China and, more recently, in Spain [11,22,24], which clearly placed *Simocyon* within the family Ailuridae [11,21,22]. These exceptional fossils have allowed us to track the dental evolution in this genus. The older and more primitive species *Simocyon batalleri* from the Vallesian (Late Miocene, 11.1 to 8.7 Million years ago) had a complete set of teeth, with relatively small canines, four premolars and two molars in both maxilla and mandible, whereas the younger and more derived species *Simocyon primigenius* from the Turolian (Late Miocene, 8.7 to 5.3 Million years ago) had reduced the number of premolars to just one, yet having the same number of molars as *S. batalleri* (Figures 3.7 and 3.8). Nevertheless, this reduction is not related to an adaptation to a herbivorous diet, and both species share the same dental morphology, not very different from earlier members of the family, the genera *Amphictis* and *Alopecocyon*, e.g., pointed premolars and more or less trenchant carnassials (the teeth that carnivores use for processing the food, basically

FIGURE 3.6 Skull (above) and life appearance (below) of *Simocyon batalleri* from the Spanish locality of Batallones-1 *(artwork by M. Antón).*

cutting fragments of flesh from the bone). These three genera are grouped into the subfamily Simocyoninae in some previous works [10,22]. Thus, *Simocyon* was probably a generalized carnivore, which would prey upon small vertebrates but could also exploit fruits, seeds, eggs, and carrion.

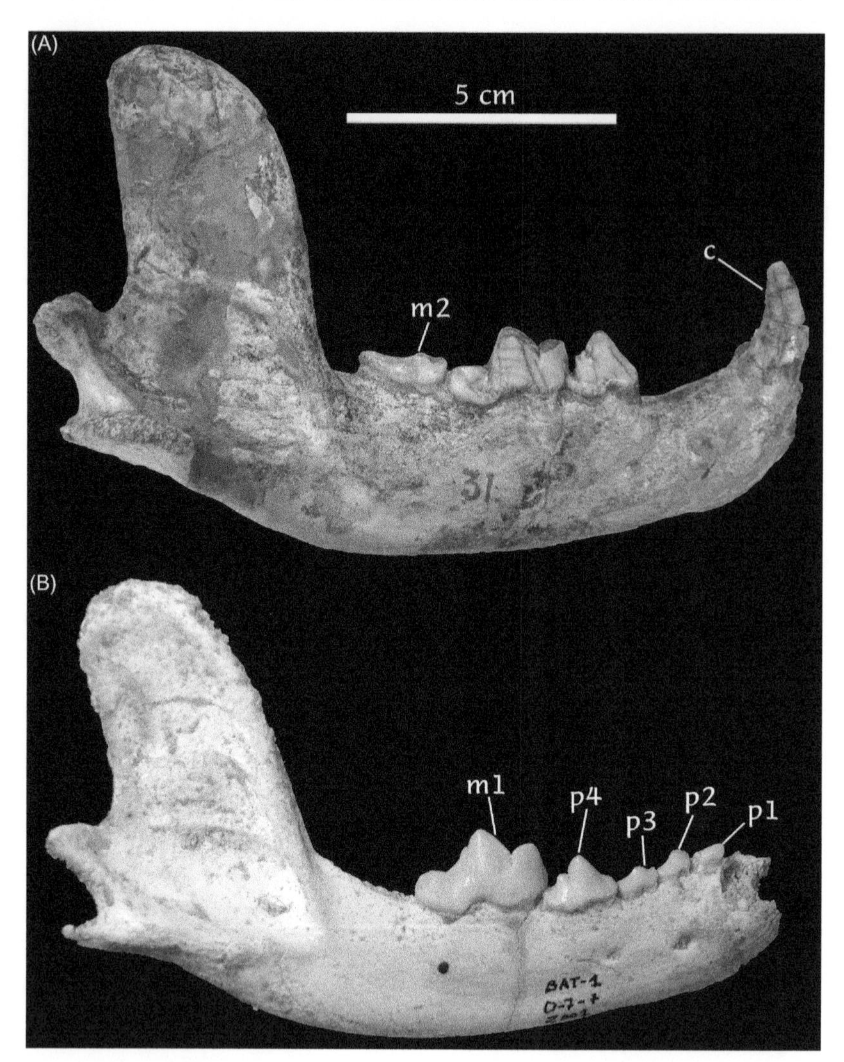

FIGURE 3.7 Comparison of the mandibular and dental morphology of two species of *Simocyon*: (A) MNHN−PIK 3020, right hemimandible of *Simocyon primigenius* from Pikermi (Greece), with lower canine (c), p4, m1 and m2. See the relatively larger p4 in *S. primigenius*, and the absence of three of the four lower premolars. (B) BAT1-D7-7-2001, right hemimandible of *S. batalleri* from Batallones-1 (Madrid, Spain), with four lower premolars (p1 to p4) and the first lower molar (m1), the lower carnassial tooth.

Unlike the dental remains, the postcranial fossils of *Simocyon* were almost unknown, with only a few published elements attributed to the genus [25,26], until the recent discovery of a rich sample of bones of at least two individuals of *Simocyon batalleri* in the Spanish locality of Batallones-1. Together these specimens represent the most complete postcranial material ever found for the genus, and have filled this "postcranial gap", yielding significant information on the proportions, palaeoecology, anatomy and biomechanics of this fossil ailurid [12,27].

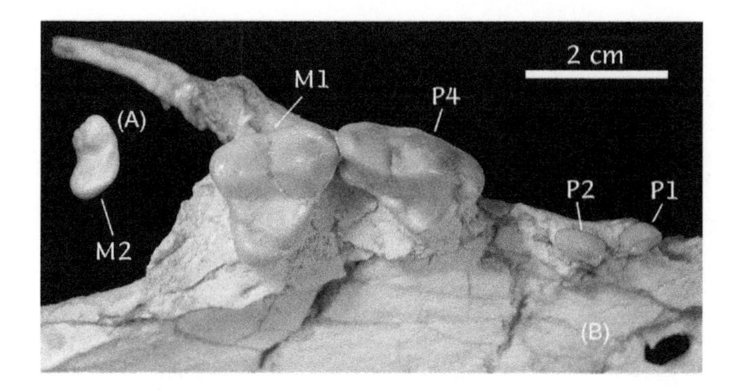

FIGURE 3.8 Upper dentition of *Simocyon batalleri* from Batallones-1 (Madrid, Spain). (A) B-3235, left upper second molar (M2) in occlusal view. (B) B-3620, upper right toothrow in occlusal view, showing the first, second and fourth upper premolars (P1, P2 and P4; the latter is the upper carnassial tooth), and the upper first molar (M1).

The extraordinary preservation of the material from Batallones-1 (located 30 km south of Madrid, Spain) is the direct result of its geologic history, believed to have consisted in the formation of an irregular cavity in sepiolite levels, at least 12 metres deep, that later filled with greenish clay [28–30]. This cavity would have acted as a natural trap for many animals, mainly carnivores (98% of the total macro-mammalian sample in number of bones) that were probably trapped while attempting to scavenge [28,29] (Figure 3.9). In addition, when sepiolite is wet, its surface becomes slippery, which would make escape from the trap very difficult; in those conditions, even big cats were not able to climb out. Besides this, the cavity was filled with clay, which caused the bottom of the trap to be covered with wet mud. This mud hindered the movement of animals and probably caused hypothermia, contributing to their more or less rapid death [28,29,31,32] and burial.

The fossil sample from Batallones-1 also includes nearly complete skeletons of previously poorly known carnivores such as the sabre-toothed cats *Promegantereon ogygia* and *Machairodus aphanistus*, the primitive hyaenid *Protictitherium crassum*, the new species of amphicyonid *Magericyon anceps*, and *Simocyon batalleri* [11,12,27,28,31,33–35]. Concerning *S. batalleri*, the preservation of its fossils is so good that they clearly show several important details, such as articulation facets or muscular attachment areas, which make it possible to propose a palaeoecological model for this rare and specialized Miocene carnivore. Specifically, most of the post-cranial specializations present in *S. batalleri* suggest developed arboreal abilities, although given the inferred size for this animal, around 60 kg [27], and its generalized dentition, it is difficult to imagine *S. batalleri* as a herbivore tree dweller like the living red panda.

One of the most striking features possessed by *S. batalleri* is the so-called "false thumb", a wrist bone (the radial sesamoid) that was hypertrophied, partly resembling the function of an opposable thumb [12]. This trait is also present in the two extant species of pandas, the giant panda (*Ailuropoda melanoleuca*) and the red panda (Figure 3.10), which use this structure with remarkable ability, holding the bamboo branches tight enabling them easily to strip off the leaves, the main part of their diet. Although the presence of a false-thumb in both pandas has long been known (see [36] for the giant panda; [37] for the red panda), and despite its uniqueness among the Carnivora, only in a few occasions has it been used to support proposals of a close relationship between the two pandas (see e.g., [37,38]). However, recent molecular studies have shown that the giant panda is a bear (family Ursidae) and that the red panda should be considered as the single extant member of a

FIGURE 3.9 Reconstruction of the bottom of the natural trap of Batallones-1, showing the carcass of a rhinoceros surrounded by two adult individuals of the sabre-toothed felid *Machairodus aphanistus*. Several remains of other carnivores are scattered on the muddy ground. *(Artwork by M. Antón).*

monotypic family Ailuridae inside a Musteloidea clade (a group including the Ailuridae, the Mephitidae, the Procyonidae, and the Mustelidae) [39–46]. In other words, this specialized structure must have a different origin in each of the pandas (see Figure 3.1). Indeed, study of the functional morphology of the radial sesamoid of the two pandas shows clear differences in shape, relative size and even in muscular attachments (Figure 3.11) [47,48].

The discovery that *S. batalleri* had a "false thumb" was very surprising, mostly because this structure was present in an animal that did not present any trait of herbivorous adaptation in its dentition. Then, how did *S. batalleri* use its "false thumb"? When its dental

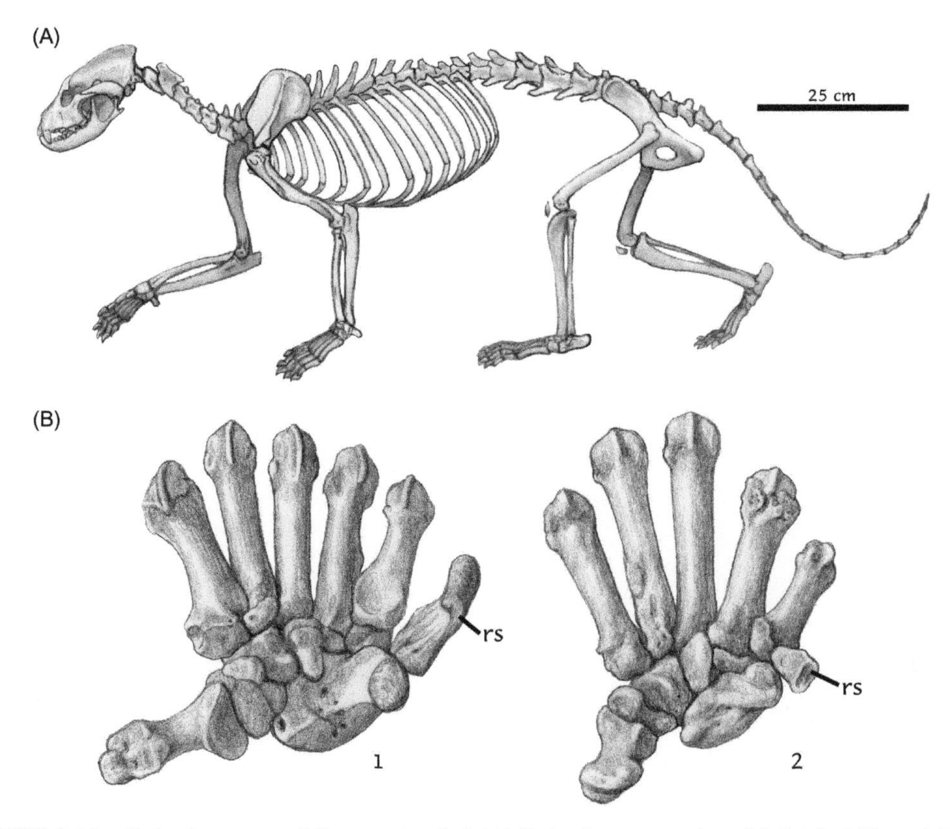

FIGURE 3.10 Skeletal anatomy of *Simocyon batalleri*: (A) Skeletal reconstruction of *S. batalleri*. The pelvis, femora, tibiae, fibulae, sacrum, and caudal vertebrae are not known and have been reconstructed on the basis of the homologous pieces of *A. fulgens*. (B) Articulated right carpus and metacarpus in palmar view, showing the position of the "false thumb" or radial sesamoid (rs) in the giant panda (*Ailuropoda melanoleuca*) (Left; 1) and *S. batalleri* (Right; 2) (not to scale). (*Artwork by M. Antón*) (*taken from [12]*).

morphology and postcranial anatomy are considered together, it is more likely that this animal used its arboreal abilities as a way to escape from the larger predators of the Late Miocene, such as amphicyonids or sabre-toothed cats, which are abundantly recorded at Batallones-1. All were strongly built carnivores, probably aggressive towards other competing predators, and a close encounter with them could have been very dangerous for a medium-sized and relatively vulnerable animal such as *S. batalleri*. A deep look into the anatomy of this animal indicates the presence of strong muscles in the shoulder, forearm, and lumbar region, which produced the necessary force to propel its body vertically fast enough to escape from these encounters [27]. Within this context the "false thumb" acquires its functionality, acting as a kind of pincer that allowed *S. batalleri* to reach relatively thin branches, where it could escape from the large predators or even to predate in turn upon small vertebrates, such as lizards, birds or rodents. Thus, the adaptations of this ailurid for climbing trees would have both dietary and ethological components producing a rare fossil carnivore (Figure 3.12) with no clear extant analogue [12,27].

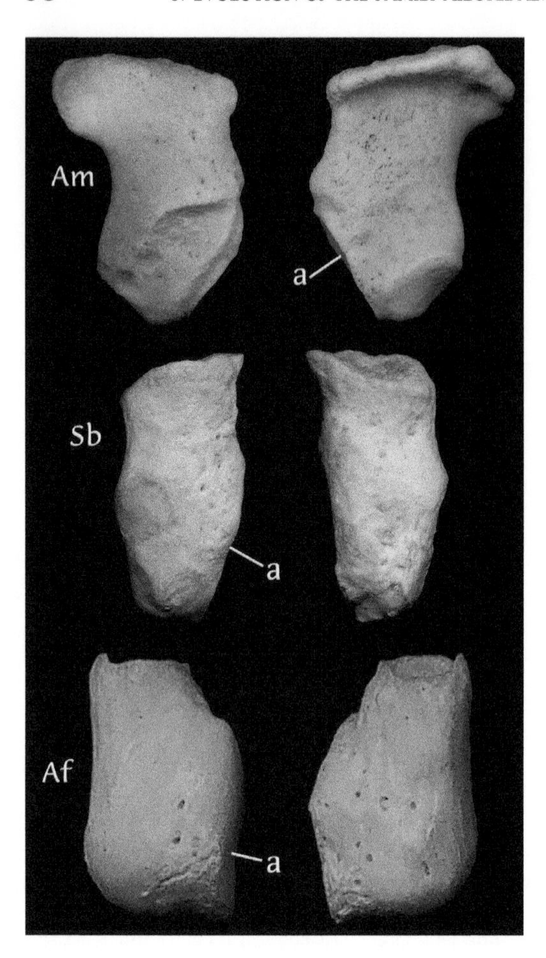

FIGURE 3.11 Comparisons of left "false thumbs" or radial sesamoids of the three species of panda, represented at the same size (internal face at left, external face at right; a, articulation facet with the scapholunar, a carpal bone): Am, giant panda (*Ailuropoda melanoleuca*); Sb, *Simocyon batalleri*; Af, red panda (*Ailurus fulgens*). The main morphological differences between the "false thumb" of the giant panda and that of the red pandas can be observed: in the former this bone is compressed and has a folded tip, whereas in the latter it is cylindrical, with a concave tip *(taken from [12])*.

The presence of a "false thumb" in two different lineages of carnivores constitutes one of the most remarkable examples of convergence (the development of a similar structure from different origins) among all vertebrates. Both panda lineages developed a similar structure, but from completely different ancestors: in the case of the giant panda, this structure would have evolved from terrestrial, bear-like ancestors that primarily used the "false thumb" as an aid in feeding, whereas the red pandas hypertrophied the radial sesamoid in an arboreal context, and only in the more recent forms, such as the genus *Ailurus*, adopted a secondary function in feeding, when they evolved towards herbivory [12,27,48].

CONCLUSIONS

The ailurids are a family of carnivorans where the derived condition of the only extant form makes it especially important to study the fossil record in order to understand the

FIGURE 3.12 Reconstruction of the life appearance of *Simocyon batalleri*, reconstructed with a similar colour pattern to that of its extant relative, the red panda *(artwork by M. Antón)*.

original morphotype and ecology of the whole family. The Middle Miocene red panda relatives found in Madrid (Spain) illustrate the early stages in the evolution of this group with a generalized dentition showing little specialization for hypo-carnivory. On the other hand, the puma-sized *Simocyon batalleri*, from the Late Miocene of Madrid, exemplifies a very different ecomorph within the ailurid family: a large, hyper-carnivore animal with developed climbing abilities and terrestrial foraging habits. This type of predator has no extant analogue, being more similar to a pantherine cat, than to the small, vegetarian modern red panda. Fossil ailurids have traditionally been rare findings but recent discoveries are filling the gaps of our knowledge of the anatomy and adaptations of one of the most unusual families of the Carnivora.

ACKNOWLEDGEMENTS

We thank Guillaume Fleury and Yves Laurent (Museum of Toulouse, France) for the photos of the hemimandible of *Alopecocyon goeriachensis* from Sansan, illustrated in Figure 3.5.

References

[1] G.D. Wesley-Hunt, J.J. Flynn, Phylogeny of the Carnivora: basal relationships among the Carnivoramorphans, and assessment of the position of 'Miacoidea' relative to Carnivora, J. Syst. Palaeontol. 3 (2005) 1–28.
[2] X. Wang, M.C. McKenna, D. Dashzeveg, *Amphicticeps* and *Amphicynodon* (Arctoidea, Carnivora) from Hsanda Gol Formation, Central Mongolia and Phylogeny of Basal Arctoids with Comments on Zoogeography, Am. Mus. Novit. 3983 (2005) 1–57.
[3] L.D. Martin, in: J.L. Gittleman (Ed.), Carnivore Behavior, Ecology, and Evolution, Cornell University Press, New York, USA, 1989, pp. 536–568.
[4] J. Viret, Les Faunes de Mammifères de l'Oligocène supérieur de la Limagne Bourbonnaise, Ann. Univ. Lyon, S. I. Sci. Med. 47 (1929) 1–328.
[5] G. de Beaumont, Remarques préliminaires sur le genre *Amphictis* Pomel (Carnivore), Bull. Soc. Vaud. Sci. Nat. 350 (73) (1976) 171–180.
[6] G. de. Beaumont, Qu'est-ce que le *Plesictis leobensis* Redlich (Mammifère, Carnivore)?, Arch. Sci. 3 (2) (1982) 143–152.
[7] E. Cirot, L. de Bonis, Le Crâne d'*Amphictis ambiguus* (Carnivora, Mammalia): son importance pour la compréhension de la phylogénie des mustéloïdes, C. R. Acad. Sci. Paris 316 (1993) 1327–1333.
[8] E.P.J. Heizmann, M. Morlo, Amphictis schlosseri n. sp. – eine neue Carnivoren-Art (Mammalia) aus dem Unter-Miozän von Südwestdeutschland. Stuttg. Beit. Natur. Ser. B (Geol. Paläontol.) 216 (1994) 1–25.
[9] E. Cirot, M. Wolsan, Late Oligocene Amphictids (Mammalia: Carnivora) from La Milloque, Aquitaine Basin, France, Geobios 28 (1994) 757–767.
[10] J.A. Baskin, in: C.M Janis, K.M Scott, L.L Jacobs (Eds.), Evolution of Tertiary Mammals of North America. Volume 1: Terrestrial Carnivores, Ungulates, and Ungulatelike Mammals, Cambridge University Press, Cambridge, UK, 1998, pp. 144–151.
[11] S. Peigné, M.J. Salesa, M. Antón, J. Morales, Ailurid carnivoran mammal *Simocyon* from the late Miocene of Spain and the systematics of the genus, Acta Paleontol. Pol. 50 (2) (2005) 219–238.
[12] M.J. Salesa, M. Antón, S. Peigné, J. Morales, Evidence of a false thumb in a fossil carnivore clarifies the evolution of pandas, Proc. Natl. Acad. Sci. USA 103 (2006) 379–382.
[13] S.D. Webb, The Pliocene Canidae of Florida, Bull. Flo. State Mus. 14 (4) (1969) 273–308.
[14] J. Viret, Catalogue critique de la faune des mammifères miocènes de La Grive Saint-Alban (Isère). Première partie. Chiroptères, Carnivores, Edentés Pholidotes, Nouv. Arch. Mus. Hist. nat. Lyon 3 (1951) 1–104.
[15] E. Thenius, Zur Herkunft der Simocyoniden (Canidae, Mammalia), Sitz. Österr. Akad. Wiss. Math. Natur. Kla., Abt. 1 (158) (1949) 799–810.
[16] G. de Beaumont, Essai sur la position taxonomique des genres *Alopecocyon* Viret et *Simocyon* Wagner (Carnivora), Eclog. Geol. Helv. 57 (1964) 829–836.
[17] I. Sasagawa, K. Takahashi, T. Sakumoto, H. Nagamori, H. Yabe, I. Kobayashi, Discovery of the extinct red panda *Parailurus* (Mammalia, Carnivora) in Japan, J. Vert. Paleontol. 23 (4) (2003) 895–900.
[18] L. Ginsburg, J. Morales, D. Soria, E. Herráez, Découverte d'une forme ancestrale du Petit Panda dans le Miocène moyen de Madrid (Espagne), C. R. Acad. Sci. Paris 325 (1997) 447–451.
[19] A.R. Tedrow, J.A. Baskin, S.F. Robinson, in: D.D. Gillette (Ed.), Vertebrate Paleontology in Utah, Utah Geological Survey, Salt Lake City, USA, 1999, pp. 487–493.
[20] S. Fraile, B. Pérez, I. de Miguel, J. Morales, in: J.P. Calvo, J. Morales (Eds.), Avances en el conocimiento del Terciario Ibérico, Departamento de Petrología y Geoquímica, UCM and Museo Nacional de Ciencias Naturales-CSIC, Madrid, Spain, 1997, pp. 77–80.
[21] L. Ginsburg, in: G.E. Rössner, K. Heissig (Eds.), The Miocene Land Mammals of Europe, Verlag Dr Friedlich Pfeil, Munich, Germany, 1999, pp. 109–148.
[22] X. Wang, New cranial material of *Simocyon* from China, and its implications for phylogenetic relationships to the red panda (Ailurus), J. Vert. Paleontol. 17 (1997) 184–198.
[23] X. Wang, J. Ye, J. Meng, W. Wu, L. Liu, S. Bi, Carnivora from Middle Miocene of Northern Junggar Basin, Xinjiang autonomous region, China, Vert. PalAsiat. 36 (1998) 218–243.
[24] O. Zdansky, Jungtertiäre carnivoren Chinas, Palaeontol. Sin. ser. C 2 (1924) 1–149.

[25] J. Kaup, Vier neue Arten urweltlicher Raubthiere, welche im zoologischen Museum zu Darmstadt aufbewahrt werden, Arch. Miner. Geogn. Bergbau- Hüttenkd 5 (1832) 150–158.
[26] Gaudry, A. (1862). Animaux fossiles et Géologie de l'Attique. F. Savy Editeur, Paris, France.
[27] M.J. Salesa, M. Antón, S. Peigné, J. Morales, Functional anatomy and biomechanics of the postcranial skeleton of *Simocyon batalleri* (Viret, 1929) (Carnivora, Ailuridae) from the Late Miocene of Spain, Zool. J. Linn. Soc. 152 (2008) 593–621.
[28] J. Morales, L. Alcalá, L. Amezua, et al., in: J. Morales, M. Nieto, L. Amezua, et al. (Eds.), Patrimonio Paleontológico de la Comunidad de Madrid, Comunidad de Madrid, Madrid, Spain, 2000, pp. 179–190.
[29] J. Morales, L. Alcalá, M.A. Álvarez-Sierra, et al., Paleontología del sistema de yacimientos de mamíferos miocenos del Cerro de los Batallones, Cuenca de Madrid, Geogaceta 35 (2004) 139–142.
[30] M. Pozo, J.P. Calvo, P.G. Silva, J. Morales, P. Peláez-Campomanes, M. Nieto, Geología del sistema de yacimientos de mamíferos miocenos del Cerro de los Batallones, Cuenca de Madrid, Geogaceta 35 (2004) 143–146.
[31] M. Antón, J. Morales, in: J. Morales, M. Nieto, L. Amezua, et al. (Eds.), Patrimonio Paleontológico de la Comunidad de Madrid, Comunidad de Madrid, Madrid, Spain, 2000, pp. 190–201.
[32] M.J. Salesa, M. Antón, A. Turner, J. Morales, Inferred behaviour and ecology of the primitive sabre-toothed cat *Paramachairodus ogygia* (Felidae, Machairodontinae) from the Late Miocene of Spain, J. Zool. 268 (2006) 243–254.
[33] M. Antón, M.J. Salesa, J. Morales, A. Turner, First known complete skulls of the scimitar-toothed cat *Machairodus aphanistus* (Felidae, Carnívora) from the Spanish late Miocene site of Batallones-1, J. Vert. Paleontol. 24 (2004) 957–969.
[34] S. Peigné, M.J. Salesa, M. Antón, J. Morales, A new Amphicyonine (Carnivora: Amphicyonidae) from the Late Miocene of Batallones-1 (Madrid, Spain), Palaeontology 51 (2008) 943–965.
[35] M.J. Salesa, M. Antón, A. Turner, J. Morales, Aspects of the functional morphology in the cranial and cervical skeleton of the sabre-toothed cat *Paramachairodus ogygia* (Kaup, 1832) (Felidae, Machairodontinae) from the Late Miocene of Spain: Implications for the origins of the machairodont killing bite, Zool. J. Linn. Soc. 144 (2005) 363–377.
[36] P. Gervais, De l'*Ursus melanoleucus* de l'Abbé Arman David, J. Zool. 4 (1875) 79–87.
[37] Lankester, E.R. (1901). On the affinities of *Aeluropus melanoleucus*, A. Milne-Edwards. Trans. Linn. Soc. London, 2nd ser. 8, Zool., part 6, 163–172.
[38] K.S. Bardenfleth, On the systematic position of *Aeluropus melanoleucus*, Mindeskrift anledn. hundredaaret for Japetus Steenstrups Fødsel udg. natur. 1 (17) (1914) 1–15.
[39] I. Delisle, C. Strobeck, A phylogeny of the Caniformia (order Carnivora) based on 12 complete protein-coding mitochondrial genes, Mol. Phylogenet. Evol. 37 (2005) 192–201.
[40] X. Domingo-Roura, F. López-Giráldez, M. Saeki, J. Marmi, Phylogenetic inference and comparative evolution of a complex microsatellite and its flanking region in carnivores, Genet. Res. 85 (2005) 223–233.
[41] J.J. Flynn, M.A. Nedbal, Phylogeny of the Carnivora (Mammalia): Congruent vs incompatibility among multiple data sets, Mol. Phylogenet. Evol. 9 (1998) 414–426.
[42] J.J. Flynn, J.A. Finarelli, S. Zehr, J. Hsu, M.A. Nedbal, Molecular Phylogeny of the Carnivora (Mammalia): Assessing the Impact of Increased Sampling on Resolving Enigmatic Relationships, Syst. Biol. 54 (2005) 317–337.
[43] J.J. Flynn, M.A. Nedbal, J.W. Dragoo, R.L. Honeycutt, Whence the Red Panda? Mol. Phylogenet. Evol. 17 (2000) 190–199.
[44] T.L. Fulton, C. Strobeck, Molecular phylogeny of the Arctoidea (Carnivora): Effect of missing data on supertree and supermatrix analyses of multiple gene data sets, Mol. Phylogenet. Evol. 41 (2006) 165–181.
[45] T.L. Fulton, C. Strobeck, Novel phylogeny of the raccoon family (Procyonidae: Carnivora) based on nuclear and mitochondrial DNA evidence, Mol. Phylogenet. Evol. 43 (2007) 1171–1177.
[46] J.J. Sato, M. Wolsan, H. Suzuki, et al., Evidence from nuclear DNA sequences sheds light on the phylogenetic relationships of Pinnipedia: Single origin with affinity to Musteloidea, Zool. Sci. 23 (2006) 125–146.
[47] D.D. Davis, The giant panda. A morphological study of evolutionary mechanisms, Fieldiana: Zool. Mem. 3 (1964) 1–339.
[48] M. Antón, M.J. Salesa, J.F. Pastor, S. Peigné, J. Morales, Implications of the functional anatomy of the hand and forearm of *Ailurus fulgens* (Carnivora, Ailuridae) for the evolution of the 'false-thumb' in pandas, J. Anat. 209 (2006) 757–764.

CHAPTER

4

Advanced Members of the Ailuridae (Lesser or Red Pandas – Subfamily Ailurinae)

Steven C. Wallace

Department of Geosciences and Don Sundquist Center of Excellence in Paleontology, East Tennessee State University, Johnson City, Tennessee, USA

OUTLINE

INTRODUCTION

When trying to interpret the fossil record of the family Ailuridae (red pandas and their relatives), our view is of course biased by the living red panda (*Ailurus fulgens*), an arboreal (meaning that it spends most of its time in the trees) herbivore that specializes in eating bamboo [1]. Though highly modified for its particular lifestyle, the evolutionary pathway taken by this unique little hypocarnivore (an animal that is technically a member of the order Carnivora, yet is a herbivore) was actually more complicated than one would expect and, like many other organisms, was not a straight line. Unfortunately, members

Red Panda. DOI: 10.1016/B978-1-4377-7813-7.00004-5

of the Ailuridae are rare in the fossil record, as is the case of many members of the Carnivora. Moreover, some taxa are small and therefore even less likely to be preserved as fossils. As a result, little material is available for study and interpretation.

Consequently, the status of the Ailuridae is still somewhat controversial (e.g. [2–16]). However, there are a few clear statements that can be made. All carnivores can be divided into two main groups: the "dog" group, which contains taxa like true dogs, bears, bear dogs (now extinct), seals, sea lions, walruses, raccoons, skunks, red pandas and weasels; and the "cat" group, which contains taxa like true cats, nimravids (extinct cat-like animals), hyenas, viverrids (genets and civets) and herpestids (mongooses). How red pandas fit within the "dog" group is not quite as clear, but they are indeed nested nicely within it. Based on fossil evidence, previous authors [5] have argued that the Ailuridae should be more closely tied to the Procyonidae (the raccoons and their relatives), while others [17,18] suggested a position at or near the base of the clade containing the procyonids and the Mustelidae (weasels). More recent studies have strongly suggested a closer tie to either the Mephitidae (skunks) or the pair Procyonidae + Mustelidae within a larger Musteloid clade (e.g. [6–12,15,16]). Though it has been difficult to figure out the exact internal relationships within this Musteloid clade, there is consistency in that it typically contains ailurids, "weasels", skunks and raccoons, but lacks the bears (Ursidae).

Wherever you place the Ailuridae within a larger context, it is clear that there are two main lineages within the group. The primitive ailurids, which were discussed in the previous chapter, consisted primarily of larger, less-specialized species found in Europe, Asia, and North America; and the Ailurinae (lesser or red pandas), which were/are more specialized, typically smaller and have a similar distribution. Though the early ailurids are typically carnivorous, with a tendency towards hypercarnivory (eating exclusively meat like a lion or a polar bear), the ailurines exhibit a trend towards hypocarnivory (eating mostly or only vegetation). There are several questionable fossil specimens in the literature that likely represent members of either of these two groups (e.g., [19]), however, they will not be discussed here due to their ambiguous nature.

Though fragmentary, true red pandas (subfamily Ailurinae) first appear during the Miocene. One candidate for the earliest example is the middle Miocene *Magerictis imperialensis* from Spain [20]. Though clearly more primitive than any other ailurine, exhibiting little complexity and remaining fairly low crowned, the m2 (lower second molar) does have an elongated talonid with well-differentiated cusps like all other ailurines. Unfortunately, it is only represented by this single tooth, so its exact relationship within the Ailuridae is uncertain [20,21]. The first confirmed member of the Ailurinae, the primitive *Pristinailurus bristoli* from the late Miocene of eastern North America, was originally described based on a single M1 (upper first molar) [22], but is now known from several additional specimens including a nearly complete skeleton [23,24], and will therefore be addressed in detail below.

By the early Pliocene, ailurines appear in both Europe and North America in the form of the genus *Parailurus* (see summaries in [21,22,25]; Chapter 5; and dental discussion below). The oldest records come from the east coast of England (Figure 4.1) in the form of several dental fragments, which were originally placed within the living genus *Ailurus* [26,27]. Subsequent recovery of cranial material from Romania highlighted the many differences and justified erection of the genus *Parailurus* [28]. Additional European records

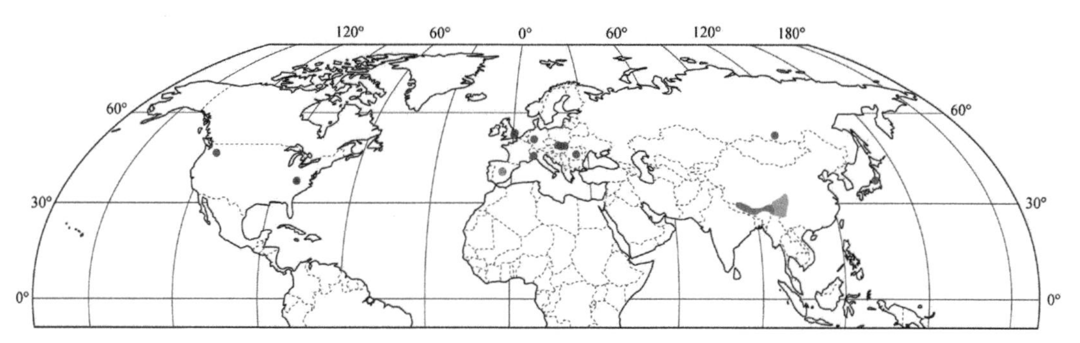

FIGURE 4.1 Map showing the distribution of the Ailurinae. Fossil localities are dots: *Parailurus* = red, *Pristinailurus bristoli* = blue, and *Magerictis* = yellow. The modern distribution of *Ailurus fulgens* is in green with two subspecies indicated (*A. f. fulgens* = dark green and *A. f. styani* = light green). *(Modified from [1,21]).*

and several from parts of Asia, including Russia [21] and even Japan [29], show that the genus was quite widespread by the middle to late Pliocene (see Figure 4.1) and that it included at least three species – *P. anglicus* and its junior synonym *"P. hungaricus"* (see Chapter 5), *P. baikalicus* and an unnamed species from Slovakia [30,31] (see Chapter 5). Though mostly known from cranial fragments (teeth and jaws), it is clear that these forms were larger than *Pristinailurus* (except perhaps the undescribed Včeláre specimen, which appears to be roughly equal in size [30]), likely reaching sizes ranging from 15 to 20 kg. Several additional isolated teeth have not been assigned to any of these three taxa and may well represent additional species (e.g., [19,25,29,33]).

For nearly 100 years *Parailurus* was thought to be an exclusively Old World taxon. However, the first New World member of the genus (an unnamed species) is represented by only a single M1 (Figure 4.2A) from the early Blancan (early Pliocene) Taunton Local Fauna of Washington State [25]. Though fossils continue to be recovered from this site, no additional *Parailurus* material has surfaced. Fortunately, the tooth does offer some insight into the species as it is highly derived, containing many enlarged and isolated accessory cusps [22,25]. Though it is difficult to say with certainty exactly what it ate, it is clear that the Washington species was highly specialized for grinding tough vegetation. In fact, the molar is more specialized than the equivalent tooth of the living *Ailurus*, and likely represents a unique species within *Parailurus* [22]. Additional material is needed, however, before the species can be fully described and interpreted.

The second New World ailurine to be described (*Pristinailurus bristoli* [22]) was recovered from the Gray Fossil Site, a prolific late Miocene deposit in eastern Tennessee, which continues to yield additional specimens of this unique carnivoran. Although the first specimen, a canine (ETMNH 359), was discovered in 2002, its significance was not recognized until the recovery of an M1 (ETMNH 360, Figure 4.2B) in the fall of 2003, which exhibited the diagnostic inflated and isolated metaconule, alerting researchers to the presence of an ailurid. Only after this second discovery was the canine further prepared to reveal the lateral grooves consistent with other members of the Ailuridae and with some procyonids

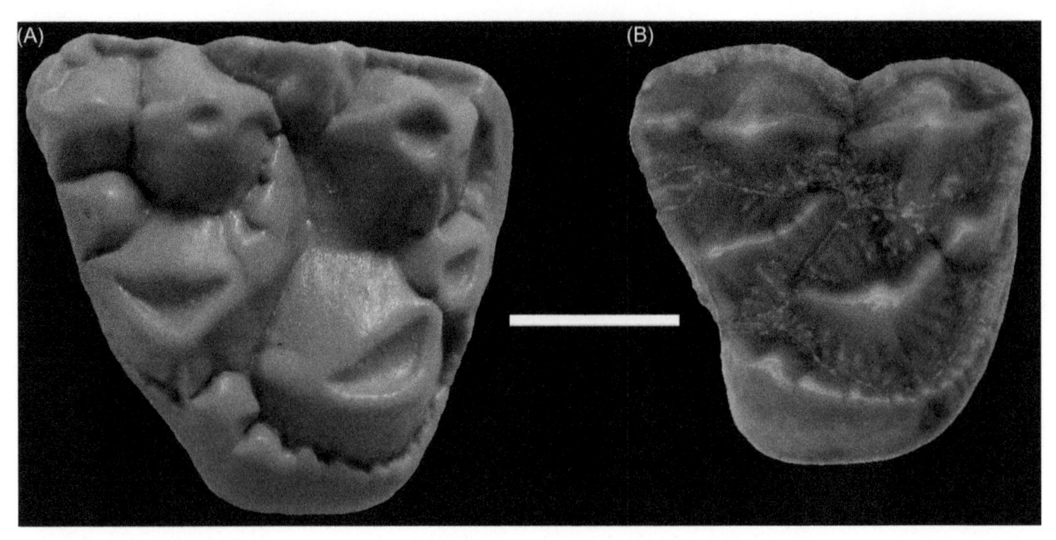

FIGURE 4.2 (A) Cast of right upper first molar of *Parailurus* sp. (LACM 10808) from Washington [25]. (B) Right upper first molar of *Pristinailurus bristoli* (Holotype – ETMNH 360) from the Gray Fossil Site in eastern Tennessee [22]. Scale bar is 5 mm.

[5,32,33]. The unique morphology, yet simple cusp pattern of the molar, warranted the erection of both a new genus and species [22]. Additionally, by recovering the first upper molar of *Pristinailurus*, direct comparison with all species of *Parailurus* was possible.

Although larger than the living *Ailurus* and slightly smaller than *Parailurus* [21,22], the overall proportions of the M1 of *Pristinailurus bristoli* more closely match European *Parailurus*, than the Washington specimen. Additionally, the number, overall development, and isolation of the accessory cusps are distinctly less advanced on *Pristinailurus bristoli* than *Parailurus* (e.g., the former lacks a mesostyle and only exhibits a rudimentary parastyle) (Figures 4.2 and 4.3). Moreover, the accessory cusp located between the metaconule and the metastyle is poorly developed on *Pristinailurus bristoli*. Both *Parailurus* and *Pristinailurus bristoli* share a well-developed metastyle and protoconule, however, the latter is further developed in all species of *Parailurus*. *Pristinailurus bristoli* also exhibits an enlarged hypocone by enlargement of the posterolingual corner of the lingual cingulum (a cingulum is a small shelf of enamel which can occur along any edge of a tooth). Recent finds show that the hypocone is even more exaggerated on the second upper molar (Figure 4.4). This is in stark contrast to the significantly reduced cingulum, with a discrete hypocone, in all *Parailurus*. Though unique, the single molar offered little information about the diet or life history of the animal [21,22].

Fortunately, additional material was discovered fairly quickly, in the form of a right lower dentary (ETMNH 596) including most of the teeth (Figure 4.5) in early 2006 [23] and, in the fall of 2007, a nearly complete skeleton (ETMNH 3596) was discovered eroding near the surface [24]. It should be mentioned that less than 1% of the fossil deposit has been excavated, so it seems likely that many additional individuals will be recovered over

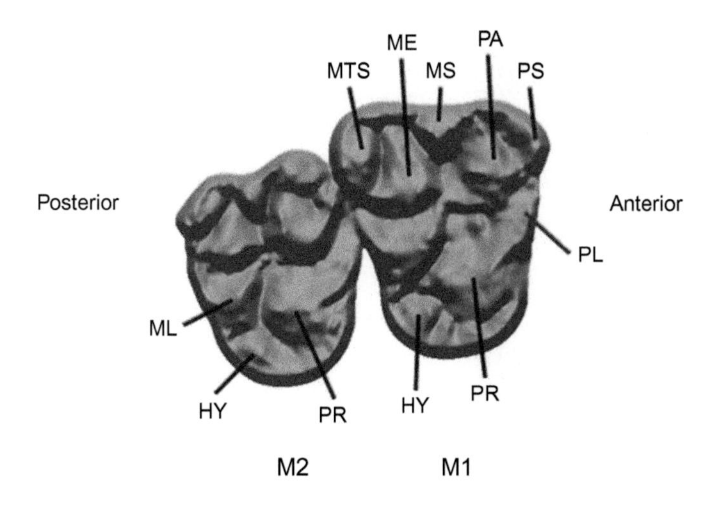

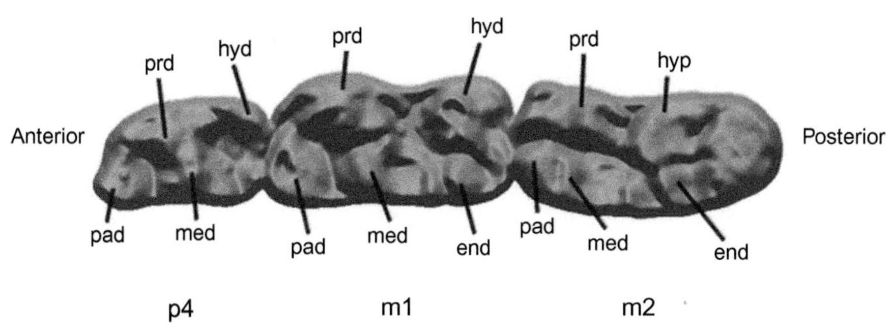

FIGURE 4.3 Generalized dental morphology of a hypothetical ailurine (based on *Ailurus fulgens*). M1 and M2 are upper right molars one and two respectively; p4 is the lower right fourth premolar; and m1 and m2 are lower right molars one and two respectively. Abbreviations for upper teeth: HY, hypocone; ME, metacone; ML, metaconule; MS, mesostyle; MTS, metastyle; PA, paracone; PL, protoconule; PR, protocone; and PS, parastyle. Abbreviations for lower teeth: end, entoconid; hyd, hypoconid; med metaconid; pad, paraconid; and prd, protoconid.

the next few decades. Though excavation continues on the new skeleton (ETMNH 3596), the individual already represents the most complete ailurid ever recovered and includes: a complete skull and dentaries (Figure 4.6); both front limbs (lacking only two phalanges and two claws); all seven cervical, twelve thoracic, four lumbar, and six caudal vertebrae; part of the sacrum; thirteen ribs; six sternebrae; most of the left innominate; both femora; the left patella; the distal two-thirds of the right tibia; the distal one-third of the left tibia; the distal end of the left fibula; both astragali; both calcanea; the right navicular; all of the right metatarsals (proximal 1/2 only of MTIII); and four phalanges (hind feet) (Figure 4.7). Though strikingly similar to the living red panda, the skeletal remains exhibit a mosaic of both primitive and derived characters.

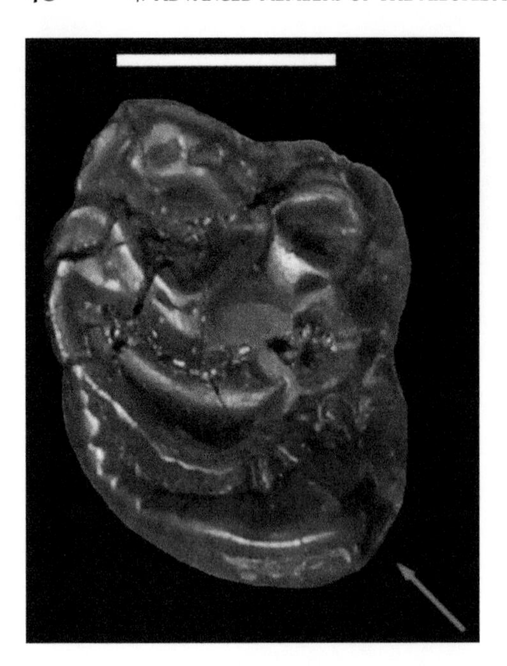

FIGURE 4.4 Left upper second molar of *Pristinailurus bristoli* (ETMNH 3596) from the Gray Fossil Site in eastern Tennessee highlighting (red arrow) the enlarged hypocone. Scale bar is 5 mm.

THE FIRST FOSSIL AILURINE SKELETON

Although this new skeleton of *Pristinailurs* is currently the most complete ailurine ever found, only a few isolated post-cranial elements of *Parailurus* [35] exist for direct comparison. Therefore, cranial comparisons will include *Parailurus*, whereas those of the post-crania will mostly focus on *Ailurus* and *Simocyon*. Detailed description and cladistic analysis of the new specimen are under preparation, and therefore are only briefly covered here. Lastly, recent studies on the myology of the living red panda [36,37], will allow a more thorough interpretation of functional aspects of the skeleton of *Pristinailurus*.

Dentary (Modified from [23])

Though the two specimens (ETMNH 596 and 3596 – see Figure 4.5) exhibit some individual variation, they are identical in all major features. Overall, the dentary of *Pristinailurus* is strikingly similar to *Parailurus* and *Ailurus*. Specifically, the angular process is enlarged, and clearly extends past the mandibular condyle as in *Parailurus* and *Ailurus*. The anterior border of the coronoid process is more procumbent than in *Simocyon*, but not quite as much as in *Parailurus* or *Ailurus*. Incisors are relatively larger than in *Ailurus*. Like the upper, the lower canine exhibits the typical ailurid lateral groove. All premolars except p1 are double rooted as in *Parailurus* and *Ailurus*; however, p1–p3 are small and only exhibit a single primary cusp as in *Simocyon*. The p4 is also small, but does contain several cusps; however, like *Simocyon*, but unlike *Parailurus* and *Ailurus*, it lacks a paraconid. Lower molars are nearly identical to *Ailurus*, including all accessory cusps and the enlarged talonids, but seem to be far less specialized than *Parailurus*. Interestingly, the m1 exhibits a slight labial cingulum, which is more developed than on the corresponding molar in *Ailurus*, but not as developed

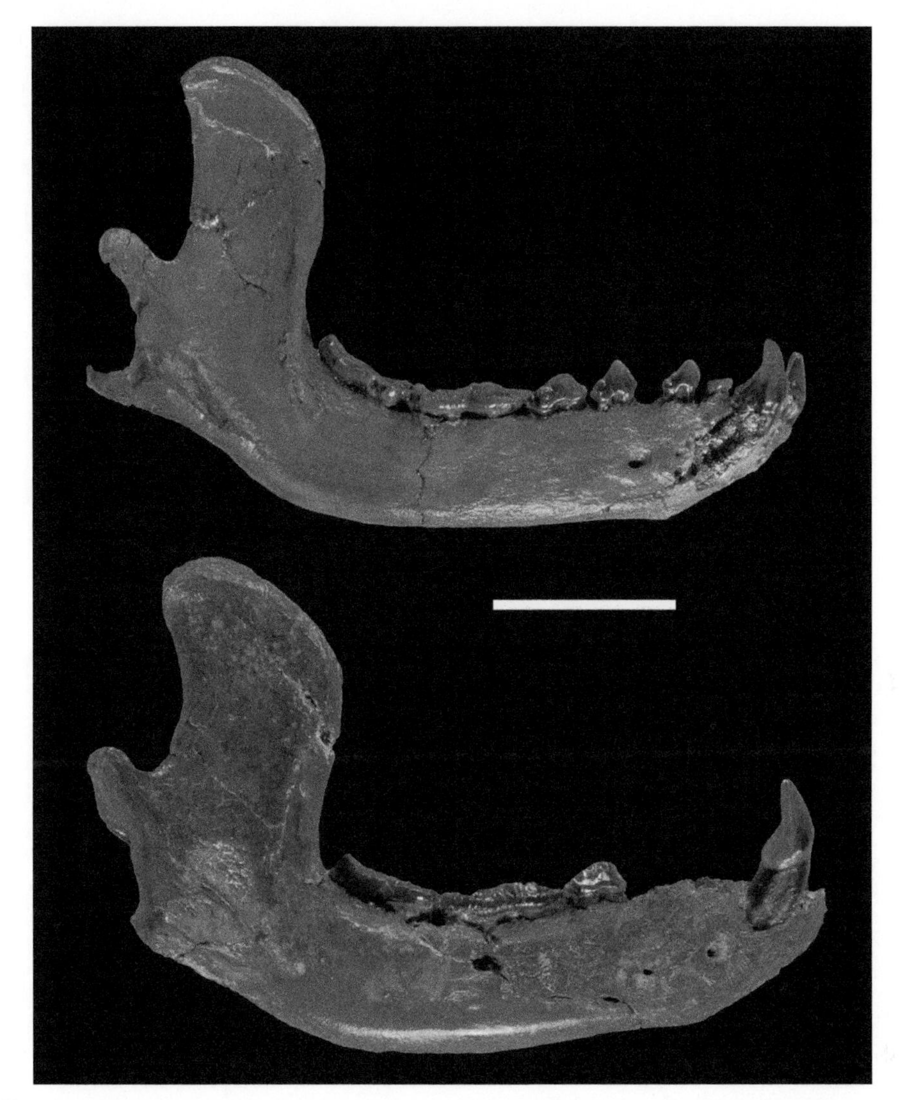

FIGURE 4.5 Right lower dentary of two different individuals of *Pristinailurus bristoli*, ETMNH 3596 (top) and ETMNH 596 (bottom), from the Gray fossil site in eastern Tennessee [23,24]. Both specimens exhibit teeth in full wear (and are therefore both adults), yet note the gracile nature of ETMNH 3596. Scale bar is 2 cm.

as those in some *Parailurus*. Wear obscures its presence or absence on m2, but the width of the tooth seems to suggest its presence. Though heavily worn, there is evidence of small crenulations/cuspids on both molars, which appear to be more developed than in *Ailurus*. Taken as a whole, the lower dentition exhibits a combination of both primitive and derived characters. Clearly, the dentary is more similar to *Ailurus* and *Parailurus* than to *Simocyon*, but (overall) is not as derived as the former.

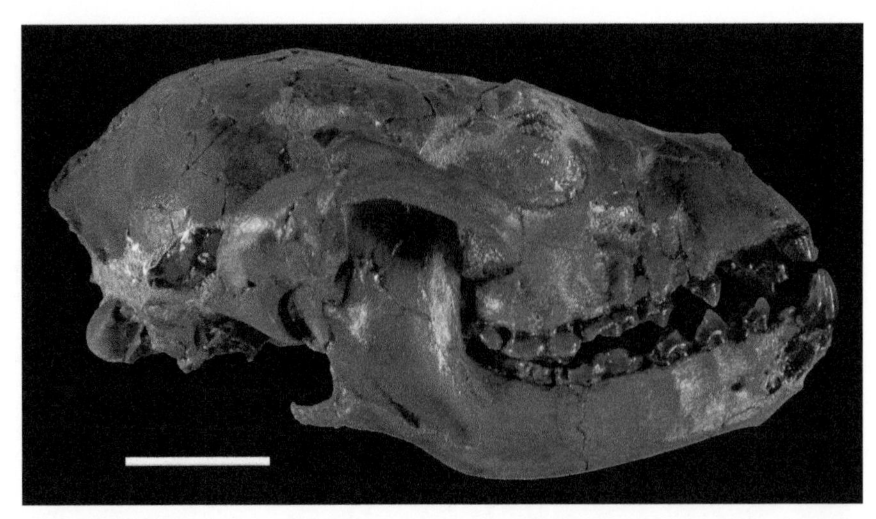

FIGURE 4.6 Skull and dentaries of *Pristinailurus bristoli* (ETMNH 3596) from the Gray fossil site in eastern Tennessee. The glassy/clear material is filler. Scale bar is 2 cm.

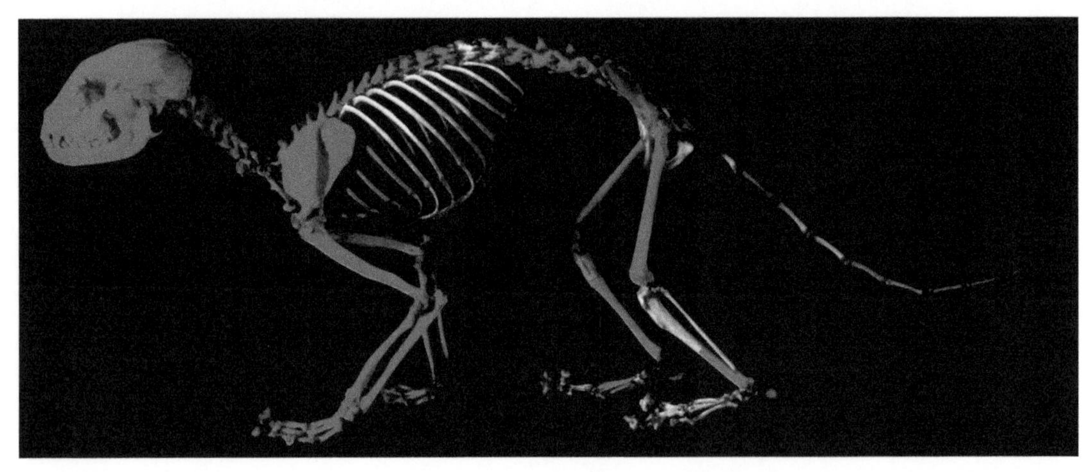

FIGURE 4.7 Skeleton of modern *Ailurus fulgens* (modified from [34]) illustrating (in red) what has been recovered thus far of this new skeleton of *Pristinailurus bristoli* (ETMNH 3596) from the Gray Fossil Site in eastern Tennessee.

Cranium

Unlike *Ailurus*, the skull of *Pristinailurus bristoli* is not as domed or inflated, lacks a sagittal crest, and the snout remains relatively long (see Figure 4.6). The upper dentition is quite interesting (see Figure 4.6) in that the canines are small for a carnivore. However, the incisors are still quite large like *Parailurus* but unlike *Ailurus*, a feature which is somewhat exaggerated by the small canines (see Figure 4.6). As with the lowers, the upper premolars remain simple and lack the accessory cusps of more derived taxa. In addition, the upper fourth premolar remains longer than it is wide, as in most *Parailurus*, but unlike *Ailurus*, which instead exhibits a very molariform fourth premolar (being wider than long). Like the upper first molar, the upper second is simple and lacks many of the accessory cusps seen in more advanced taxa, however, it retains a well-developed and expanded hypocone (see Figure 4.4). Except for the Včeláre specimen (Kundrát, personal communication), *Parailurus* typically exhibits a reduced hypocone, and the cusp is variably reduced in *Ailurus* (however, it should be noted that regardless of its size in *Ailurus*, like the other cusps on the teeth of this derived taxon, the hypocone is higher crowned and is separated from other cusps by a distinct valley).

Post-Crania and Size

One of the most obvious features exhibited by the new skeleton is the larger size of *Pristinailurus bristoli* compared to the living *Ailurus* (though direct comparison of post-cranial elements is not possible, dental measurements strongly suggest that *Pristinailurus bristoli* was smaller than most *Parailurus*). Though roughly the same length, the forelimb of this new specimen is more robust than in *Ailurus* (Figure 4.8). Moreover, both the femur and the tibia are 10–15% longer than those of *Ailurus* (Figure 4.8), resulting in significantly longer hindlimbs relative to the forelimbs, and further supporting the larger size of *Pristinailurus bristoli*. Based on the diameter of the humerus of this new skeleton, and its smaller size when compared to other elements (of the same taxon) from the Gray site, *Pristinailurus bristoli* likely weighed between 8 and 10 kg (mass equations are from [38] following [39]), while *Ailurus* typically averages closer to 5 kg [1]. Though larger, individuals of the subspecies *Ailurus fulgens styani* have been reported from zoos with equivalent weights to that estimated for *Pristinailurus bristoli*, so further sampling may shed light on the significance of the size differences between the two taxa. Regardless, the proportional differences suggest a different life habit than the living form.

Radial Sesamoid

Like the European *Simocyon batalleri*, and as predicted by Salesa et al. [40], *Pristinailurus bristoli* possesses a modified radial sesamoid, or "false thumb" (Figure 4.9), which it likely used for climbing. If *Simocyon* is indeed a basal ailurid, one would expect all (or at least most) of its descendants (particularly the advanced ailurines) to retain this unique adaptation; however, recovery of the "false thumb" of *Pristinailurus bristoli* reveals a striking oddity within the ailurines. Though present, the radial sesamoid is nearly identical in size to that of *Ailurus* (if not even slightly smaller), an animal nearly half the weight (Figure 4.9). This smaller relative size, taken in conjunction with the differing proportions of the front and hindlimbs, seems to suggest that *Pristinailurus bristoli* was utilizing the "false thumb" for climbing to a lesser extent, perhaps spending more time on the ground. This is further supported by a brachial index (length of radius relative to the length of the humerus) of roughly 84.5 for *Pristinailurus bristoli*, which is closer to that of *Simocyon* [40] and other semi-arboreal mammals, than to that of the arboreal *Ailurus* (78.5 – within the collections at ETSU). Moreover, following Heinrich and Houde [41], the combined humerus/femur

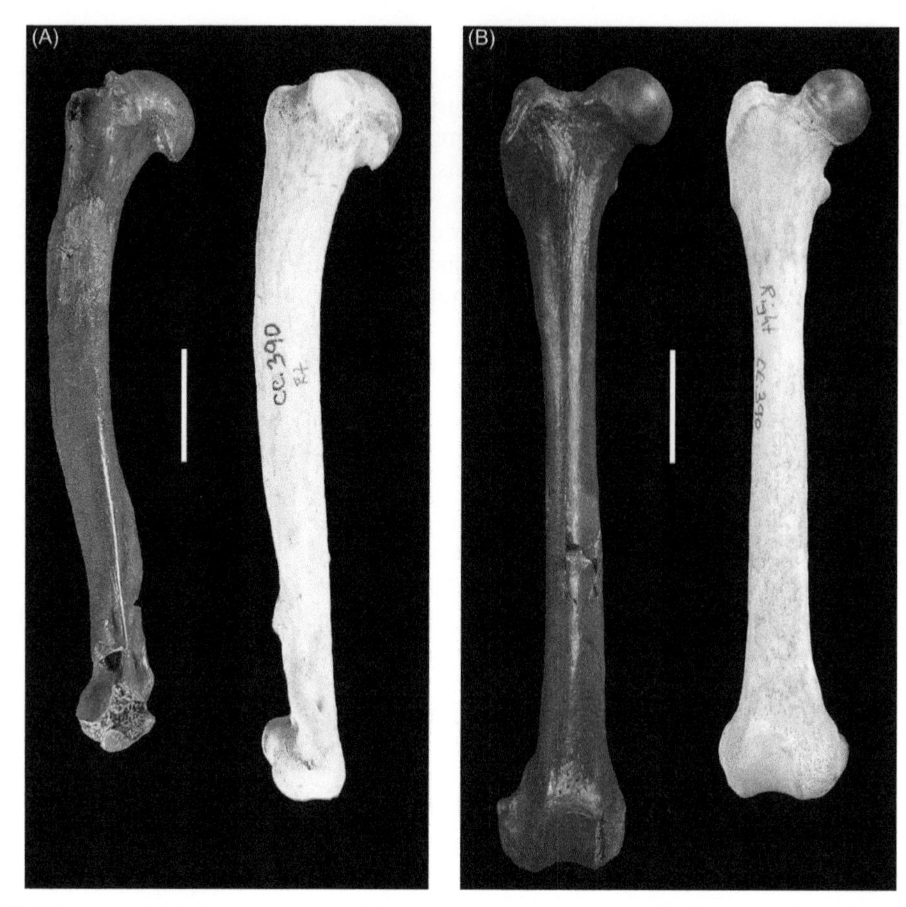

FIGURE 4.8 Major limb elements of *Pristinailurus bristoli* (dark) compared to the modern *Ailurus fulgens* (white). (A) Right humerus of both species, lateral view. (B) Right femur of both species, dorsal view. Scale bar is 2 cm.

and humerus/radius ratios are more similar to those of scansorial to terrestrial, rather than arboreal, carnivores. However, none of this should be surprising considering the dentition of *Pristinailurus*, which suggests a more versatile diet.

RELATIONSHIPS WITHIN THE AILURINAE

The divergence of the Ailuridae + Mephitidae from other Musteloids (Procyonidae + Mustelidae) is estimated to have occurred just over 33 Ma (million years ago) [10,42]. However, probable members of the subfamily Ailurinae do not appear in the fossil record until 16–17 Ma in the form of *Magerictis* from Spain [20]. Unfortunately, because no additional material has been described, little else can be said about its relationship to the more advanced ailurines [20,21]. *Pristinailurus bristoli* on the other hand (now that the full

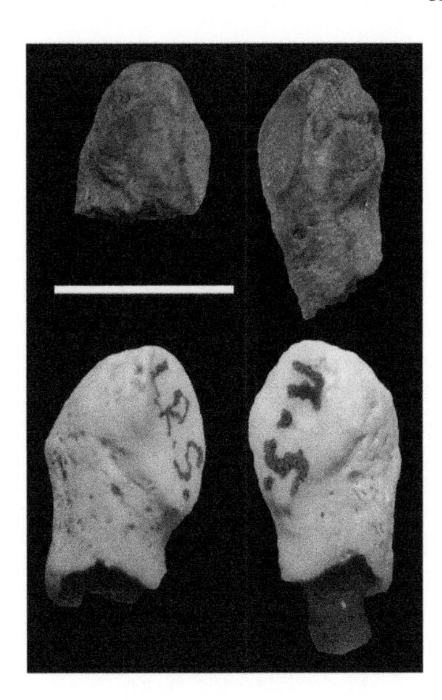

FIGURE 4.9 Radial sesamoids of *Pristinailurus bristoli* (top, ETMNH 3596) compared to those of *Ailurus fulgens* (bottom, NVPL 390). Scale bar is 5 mm.

dentition is known, see above), appears to be (dentally) more similar to *Parailurus* than to other members of the clade (e.g., *Ailurus* or the more primitive *Magerictis* and/or *Simocyon*), as suggested by Sotnikova [21]. Aside from the development of cingula and small crenulations/cuspids, the lower molars of *Pristinailurus* are nearly identical to those of *Ailurus* and *Parailurus*, yet the uppers are far less derived. All premolars are present, but remain small and simple. However, those of *Parailurus* are intermediate between *Ailurus* and *Pristinailurus* (mainly in becoming molariform through the addition of accessory cusps and cuspids).

Therefore, *Pristinailurus bristoli* does appear to represent a very primitive member of the Ailurinae (Figure 4.10), however, there are several trends within the clade. Specifically:

1. There seems to be a group, comprised of the primitive *Pristinailurus bristoli* and the more derived *Parailurus* (Figure 4.10), that possess premolars which are only slightly modified, with a tendency towards amplification of the molars. Specifically, *Pristinailurus bristoli* retains enlarged hypocones on both upper molars, cusps which become reduced on the molars, but more important on the premolars, of later forms [21,45]. Moreover, *Parailurus* tends to add complexity to the molars through the addition of numerous accessory cusps and cuspids resulting in an overall reduction in the size of the hypocone in at least some forms. The lower molars of *Pristinailurus bristoli* appear to exhibit enlarged cingula and several accessory cusps (wear obscures the extent and detail of these additional cusps) relative to *Ailurus*, while *Parailurus* again selected the addition of many accessory cusps (even occurring on expanded and heavily crenulated cingula in some forms with many small cuspids). Within described *Parailurus*, the Washington specimen seems to be the most derived (though more material is needed to be certain), exhibiting a large surface area for an M1, but also the highest complexity, size, and isolation of accessory cusps. The crenulations/cuspids

mentioned above are actually common on the teeth of all ailurines, but are particularly exaggerated within *Pristinailurus bristoli* and the species of *Parailurus*, and are not typical of *Ailurus*. Through selection, these crenulations could have been the "seeds" for many of the additional isolated accessory cusps (particularly many of the stylids). Essentially, certain crenulations became enlarged and isolated at the expense of the others. In total, these trends suggest a more omnivorous diet, with molars for grinding vegetation, yet functional premolars, canines and incisors for eating meat (particularly evident in *Pristinailurus bristoli*).

2. The living *Ailurus* on the other hand, though exhibiting moderate complexity in the molars, seems to follow the route of selecting molariform premolars; likely in response to its highly specialized diet of eating almost exclusively bamboo (though they are also known to occasionally eat invertebrates, as well as small vertebrates [1,46,47]. In addition, *Ailurus* exhibits fewer crenulations/cuspids, but instead exhibits larger, isolated and more hypsodont (high crowned) primary cusps. Although tied firmly within the Ailurinae, these morphological differences seem to suggest that *Ailurus* split off early from the rest of the clade (see Figure 4.10), pursuing a slightly different ecologic path (that of a true hypocarnivore). Though older records are currently lacking, future finds will likely clarify the early history of *Ailurus* and its divergence from *Pristinailurus bristoli* and *Parailurus*.

Within *Parailurus*, the exact relationships remain difficult to interpret because of the fragmentary nature of most of the material and the overall lack of precise dating of the specimens (e.g., *Parailurus anglicus* has been recovered from both the Red Crag of England, which has been dated to roughly 2.6 Ma [48, 49], and the Capeni locality of Romania, which has been dated to ~3.7 Ma [50]). Fortunately, the discovery of a new species [21], represented by several teeth (in addition to M1), has provided the opportunity for an interpretation broader than, yet similar to, that provided by Wallace and Wang [22]. Specifically, Sotnikova [21] points out that the new species *Parailurus baikalicus* seems to be

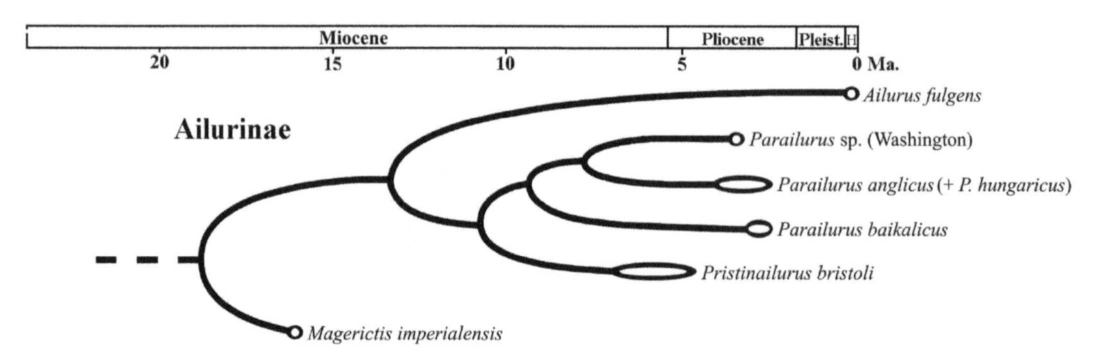

FIGURE 4.10 Schematic phylogeny of the group. The approximate geologic ages of the individual taxa are indicated by the open circles. Please note that this is not a cladogram, so the branches are only intended to show relative relationships and timing of divergences. A true cladistical analysis for this group is underway, but is not included here. Geologic boundaries follow Tedford et al. [43] and Bell et al. [44]. Ma = Million years ago.

the most primitive member of the genus and likely aligns close to *Pristinailurus bristoli*. The slightly smaller *Parailurus anglicus*, though potentially older geologically (see Figure 4.10), is more derived. The two largest forms, *Parailurus* sp. from Washington, and the mid- to late-Pliocene "*Parailurus hungaricus*" (a junior synonym of *Parailurus anglicus*, see Chapter 5), represent the most derived members of the genus described to date. There has been prior debate over the validity of "*Parailurus hungaricus*", with some authors placing it within *Parailurus anglicus* (e.g., [51], see Chapter 5). Indeed, significant morphological, temporal, and geographical overlap between the two species suggests that they more likely represent different sexes (or perhaps subspecies) of the same taxon (see Chapter 5). Concerning the Washington specimen, only the M1 exists, so a detailed analysis is not possible at this time. What can be stated, however, is that though one would expect the two North America ailurines to be closely related, the separation in time, taken in conjunction with the overall primitive nature of *Pristinailurus bristoli* relative to the advanced Washington specimen, clearly shows that the two species are not closely related. The simplest explanation of this difference is that the species are the result of two separate and distinct immigration events (from ancestral "stock"), as suggested by Wallace and Wang [22]. Finally, though the Japanese *Parailurus* spp. [29] is only represented by a single tooth, a P4 (upper fourth premolar), its greater width to length ratio (the tooth is much wider than it is long) implies that the species is a more advanced member of the genus (possessing a premolar which is becoming molariform; similar to that seen in *Ailurus*), likely falling at least above *Parailurus anglicus* (which still retains a P4 that is longer than it is wide). It should also be noted that the Japanese tooth exhibits many small crenulations/cuspids, which is typical of the genus [29].

PALEOGEOGRAPHY

Eurasian Ailurines

Analysis of the distribution of fossil ailurines in Europe and Asia (see Figure 4.1) shows a clear correlation with the distribution of temperate forests through time. Moreover, declines in the number and diversity of broad-leaf deciduous forest and woody taxa in the mid- to high-latitudes of Europe, western North America and Alaska from the middle Miocene through the Pliocene [52] meant that northern ecosystems were becoming stressed. Fortunately, mammals are uniquely proficient at simply following their most favourable habitat should it alter or shift. Thus, these changing forests could have encouraged various species to seek out more favourable habitats, resulting in not only the wide distribution of this group, but also its diversity in size and morphology. In other words, the climatically driven stresses on the ecosystem may have actually heavily influenced the evolution of the group as a whole.

A cursory look at the distribution of Eurasian fossil ailurines (see Figure 4.1) however, reveals an obvious "east Asian hole" (specifically China and much of Russia) in the record. This seems odd considering that conditions (i.e., temperature and vegetation) in southern and eastern China during the late Miocene may have been very similar to that of Europe [53] at that time. Moreover, based on the distribution of the living *Ailurus*, one would expect to find fossil remains in (at least) China, let alone eastern Russia. Even if the current distribution represents a recent migration event, one would still expect at least some fossils to be preserved along the route taken. However, no material has been described from China to date. Taken in conjunction with the current distribution of *Ailurus* (see Figure 4.1), new ailurid material from China (or eastern Russia) could shed new light on the potential early offshoot of *Ailurus* from the rest of the clade, and will be significant to our understanding of the early diversification of this group.

North American Ailurines

Although the land exposed between Siberia and Alaska during low sea level (effectively creating the land bridge known as Beringia) has been suggested to have acted as a filter to mammalian dispersals [54], it provided the most likely dispersal route for North American ailurines. In fact, by the end of the Miocene, global cooling (relative to the much warmer early Cenozoic) meant that Beringia was covered by mostly deciduous forest [55–57]. Moreover, much of the Arctic would have been covered in what has been interpreted as cool temperate forest, similar to that covering much of eastern North America today [55]. In addition, temperatures across Beringia at that time were 2–4 degrees Celsius (°C) warmer than today [57], and recent evidence from southern Alaska has shown the persistence of mixed northern hardwood and warm-temperate mesophytic forests in that region during the late Miocene to early Pliocene [58]. Such a combination of conditions would have provided easy passage for *Pristinailurus* (or its ancestors) across Siberia into Alaska, and then across Canada and down into the southern Appalachians.

During the Pliocene, additional cooling resulted in the dispersal of more conifers with fewer deciduous trees across Beringia [59]. Though sea levels varied throughout the Pliocene, several low stands [60] – prior to some of the high stands that would have blocked passage [61,62] – provided the opportunity, which likely allowed for the second immigration into North America (*Parailurus* sp. from Washington) via this green corridor. A short warming event at the end of the Miocene through the early Pliocene [61], could also have helped create a favourable "window" for the migration event. Interestingly, the conditions of Beringia might have been similar to that of much of Europe and Asia at that time [61,62], which could explain the Washington specimen's closer affiliation with the European *Parailurus*, rather than to *Pristinailurus*.

CONCLUSIONS/SUMMARY

Aside from a few well-represented taxa such as *Simocyon batalleri* and *Pristinailurus bristoli*, the fossil record of the Ailuridae consists of mostly isolated teeth and fragmentary specimens. However, these two taxa are beginning to shed new light on this enigmatic group of specialized carnivorans. In particular, it appears that, as suggested by Salesa et al. [40], climbing evolved early within the group (most likely as a mechanism for eluding the larger predators of the time), and that highly specialized dentition was only later selected.

However, the selection of the various dental characters appears to have been mosaic, with different teeth specializing at different rates (and with different focal zones). In addition, some forms specialized in hypercarnivory, while others were becoming adapted to hypocarnivory to varying degrees. In fact, it is quite likely that *Simocyon*'s move to the trees may have directly led to the selection of features which allowed this group to exploit flowers and/or fruit (in addition to eating meat), and therefore hypocarnivory may have been a direct result of this change in lifestyle. Moreover, it appears that although both living "pandas" use their so-called "false thumb" for feeding, each lineage acquired this unique trait via different pathways – one for feeding and one for climbing – making this one of the most fascinating examples of convergent evolution [40]. Lastly, as demonstrated by *Pristinailurus bristoli*, not every member of this group remained arboreal, which may (in part) have reduced its ability to compete with other more generalized omnivores, and ultimately to the extinction of nearly every member of this group.

Favourable passageways, in combination with increasing ecological stresses, during the late Miocene and into the Pliocene, may have encouraged not only the diversification of this group as a whole, but also its wide distribution. However, the various species (both fossil and living) seem to be restricted to the northern temperate forests. Specifically, the oldest known fossil ailurines are from the middle Miocene of southwestern Europe (Spain). By the late Miocene to early Pliocene they appear to quickly diversify and spread throughout Eurasia, and even include two distinct genera which reach North America. Morphologic characteristics of the teeth within this group suggest that the living *Ailurus* is part of a lineage which split off early from other fossil ailurines. Because the range of the living *Ailurus* includes parts of southern China, which currently lacks a fossil record of this unique group, any fossil ailurine (or ailurid in general) material from there, or even portions of eastern Russia, would be significant to our understanding of this separation.

References

[1] M.S. Roberts, J.L. Gittleman, Ailurus fulgens, Mammalian Species 222 (1984) 1–8.

[2] J.J. Flynn, M.A. Nedbal, Phylogeny of the Carnivora (Mammalia): congruence vs incompatibility among multiple data sets, Molec. Phylogenet. Evol. 9 (1998) 414–426.

[3] J.J. Flynn, J.A. Finarelli, S. Zehr, J. Hsu, M.A. Nedbal, Molecular phylogeny of the Carnivora (Mammalia): assessing the impact of increased sampling on resolving enigmatic relationships, System. Biol. 54 (2005) 317–337.

[4] J.J. Flynn, M.A. Nedbal, J.W. Dragoo, R.L. Honeycutt, Whence the red panda? Molec. Phylogenet. Evol. 17 (2000) 190–199.

[5] X. Wang, New cranial material of Simocyon from China, and its implications for phylogenetic relationships to the red panda (Ailurus), J. Vert. Paleontol. 17 (1997) 184–198.

[6] L. Yu, Q. Li, O.A. Ryder, Y. Zhang, Phylogenetic relationships within mammalian order Carnivora indicated by sequences of two nuclear DNA genes, Molec. Phylogenet. Evol. 33 (2004) 694–705.

[7] T.L. Fulton, C. Strobeck, Molecular phylogeny of the Arctoidea (Carnivora): effect of missing data on supertree and supermatrix analyses of multiple gene data sets, Molec. Phylogenet. Evol. 41 (2006) 165–181.

[8] J.J. Sato, M. Wolsan, H. Suzuki, et al., Evidence from nuclear DNA sequences sheds light on the phylogenetic relationships of Pinnipedia: single origin with affinity to Musteloidea, Zool. Sci. 23 (2006) 125–146.

[9] T.L. Fulton, C. Strobeck, Novel phylogeny of the raccoon family (Procyonidae: Carnivora) based on nuclear and mitochondrial DNA evidence, Molec. Phylogenet. Evol. 43 (2007) 1171–1177.

[10] T. Yonezawa, M. Nikaido, N. Kohno, Y. Fukumoto, N. Okada, M. Hasegawa, Molecular phylogenetic study on the origin and evolution of Mustelidae, Gene 396 (2007) 1–12.
[11] J.A. Finarelli, A total evidence phylogeny of the Arctoidea (Carnivore: Mammalia): relationships among basal taxa, J. Mammal. Evol. 15 (2008) 231–259.
[12] L. Yu, J. Liu, P. Luan, et al., New insights into the evolution of intronic sequences of the β-fibrinogen gene and their application in reconstructing Mustelid phylogeny, Zool. Sci. 25 (2008) 662–672.
[13] C. Ledje, U. Arnason, Phylogenetic analyses of complete cytochrome b genes of the order Carnivora with particular emphasis on the Caniformia, J. Molec. Evol. 42 (1996) 135–144.
[14] C. Ledje, U. Arnason, Phylogenetic relationships within caniform carnivores based on analyses of the mitochondrial 12S rRNA gene, J. Molec. Evol. 43 (1996) 641–649.
[15] I. Delisle, C. Strobeck, A phylogeny of the Caniformia (order Carnivora) based on 12 complete protein-coding mitochondrial genes, Molec. Phylogenet. Evol. 37 (2005) 192–201.
[16] U. Arnason, A. Gullberg, A. Janke, M. Kullberg, Mitogenomic analysis of caniform relationships, Molec. Phylogenet. Evol. 45 (2007) 863–874.
[17] N. Schmidt-Kittler, Zur Stammesgeschichte der marderverwandten Raubtiergruppen (Musteloidea, Carnivora), Eclog. Geol. Helvet. 74 (1981) 753–801.
[18] M. Wolsan, Phylogeny and classification of early European Mustelida (Mammalia: Carnivora), Acta Theriol. 38 (1993) 345–384.
[19] L. Ginsburg, O. Maridet, P. Mein, Un Ailurinae (Mammalia, Carnivora, Ailuridae) dans le Miocène moyen de Four (Isère, France), Geodiversitas 23 (2001) 81–85.
[20] L. Ginsburg, J. Morales, D. Soria, E. Herráez, Découverte d'une forme ancestrale du Petit Panda dans le Miocène moyen de Madrid (Espagne), Comptes. Rend. Acad. Sci. Paris 325 (1997) 447–451.
[21] M.V. Sotnikova, A new species of lesser panda Parailurus (Mammalia, Carnivora) from the pliocene of Transbaikalia (Russia) and some aspects of Ailurine phylogeny, Paleontol. Zh. 1 (2008) 92–102.
[22] S.C. Wallace, X. Wang, Two new carnivores from an unusual late Tertiary forest biota in eastern North America, Nature 431 (2004) 556–559.
[23] S.C. Wallace, X. Wang, First mandible and lower dentition of Pristinailurus bristoli, with comments of life history and phylogeny, J. Vert. Paleontol. 27 (2007) 162A.
[24] S.C. Wallace, First ailurine post crania: a nearly complete skeleton of Pristinailurus bristoli, Southeast. Assoc. Vert. Paleont. Proc. 1 (2008) 27.
[25] R.H. Tedford, E.P. Gustafson, First North American record of the extinct panda Parailurus, Nature 265 (1977) 621–623.
[26] W.B. Dawkins, On Ailurus anglicus, a new Carnivore from red Crag, Q. J. Geol. Soc. 44 (1888) 228–231.
[27] E.F. Newton, On some new mammals from red and Norwich Crags, Q. J. Geol. Soc. 46 (1890) 444–453.
[28] M. Schlosser, Parailurus anglicus and Ursus böckhi, aus den Ligniten von Baróth-Köpecz, Comitat Háromezèk in Ungarn, Mittheil.m Jahrbuch Königl. Ungarisch. Geol. Anst. 13 (1899) 66–95.
[29] I. Sasagawa, K. Takahashi, T. Sakumoto, H. Nagamori, H. Yabe, I. Kobayashi, Discovery of the extinct red panda Parailurus (Mammalia, Carnivora) in Japan, J. Vert. Paleontol. 23 (4) (2003) 895–900.
[30] M. Kundrát, A morphological resolution to the enigma of lesser panda phylogeny, J. Morphol. 232 (3) (1997) 282.
[31] M. Kundrát, New dental remains of an extinct lesser panda – Morphotype or new species? J. Vert. Paleontol. 17 (1997) 58A.
[32] G. de Beaumont, Essai sur la position taxonomique des genres Alopecocyon Viret et Simocyon Wagner (Carnivora), Eclog. Geol. Helvet. 57 (1964) 829–836.
[33] D.M. Decker, W.C. Wozencraft, Phylogenetic analysis of recent procyonid genera, J. Mammal. 72 (1991) 42–55.
[34] J-B. de Panafieu, P. Gries, L. Asher, Evolution, Seven Stories Press, New York, 2007.
[35] O. Fejfar, M. Sabol, Pliocene carnivores (Carnivora, Mammalia) from Ivanovce and Hajnacka (Slovakia), Cour. Forschungsinst. Senckenberg 246 (2004) 15–53.
[36] R.E. Fisher, B. Adrian, C. Elrod, M. Hicks, The phylogeny of the red panda (Ailurus fulgens): evidence from the hind limb, J. Anat. 213 (2008) 607–628.

[37] M. Antón, M.J. Salesa, J.F. Pastor, S. Peigné, J. Morales, Implications of the functional anatomy of the hand and forearm of Ailurus fulgens (Carnivora, Ailuridae) for the evolution of the 'false-thumb' in pandas, J. Anat. 209 (2006) 757–764.
[38] P. Christiansen, J.M. Harris, Body size of Smilodon (Mammalia: Felidae), J. Morphol. 266 (2005) 369–384.
[39] M.J. Salesa, S. Peigné, M. Antón, J. Morales, Functional anatomy and biomechanics of the postcranial skeleton of Simocyon batalleri (Viret, 1929) (Carnivora, Ailuridae) from the late Miocene of Spain, Zool. J. Linnean Soc. 152 (2008) 593–621.
[40] M.J. Salesa, M. Antón, S. Peigné, J. Morales, Evidence of a false thumb in a fossil carnivore clarifies the evolution of pandas, Proc. Natl. Acad. Sci. USA 103 (2006) 379–382.
[41] R.E. Heinrich, P. Houde, Postcranial anatomy of Viverravus (Mammalia, Carnivora) and implications for substrate use in basal Carnivora, J. Vert. Paleontol. 26 (2006) 422–435.
[42] M. Wolsan, Fossil-based minimum divergence dates for the major clades of musteloid carnivorans, in: G.E. Rössner, K. Heissig, (Eds.), Abstracts of Plenary, Symposium, Poster and Oral Papers Presented at Ninth International Mammalogical Congress (IMC 9), 2005, pp. 372–373.
[43] R.H. Tedford, L.B. Albright III, A.D. Barnosky, et al., Mammalian biochronology of the Arikareean through Hemphillian Interval (Late Oligocene through early Pliocene Epochs), in: M.O. Woodburne (Ed.), Late Cretaceous and Cenozoic Mammals of North America, Columbia University Press, 2004, pp. 167–231.
[44] C.J. Bell, E.L. Lundelius Jr., A.D. Barnosky, et al., The Blancan, Irvingtonian, and Rancholabrean Mammal Ages, in: M.O. Woodburne (Ed.), Late Cretaceous and Cenozoic Mammals of North America, Columbia University Press, 2004, pp. 232–314.
[45] L. Ginsburg, In The Miocene Land Mammals of Europe, Verlag Dr Friedlich Pfeil, Munich, 1999, pp. 109–148
[46] B.H. Hodgson, On the cat-toed subplantigrades of the sub-Himalayas, J. Asiatic Soc. 16 (1847) 1113–1129.
[47] A. Sowerby, C. de, The pandas or cat-bears, China J. Sci. Arts 17 (1932) 296–299.
[48] B.M. Funnell, Plio-pleistocene palaeogeography of the southern North Sea basin (3.75–0.60 Ma), Quat. Sci. Rev. 15 (1996) 391–405.
[49] A.A. McMillan, A provisional quaternary and neogene lithostratigraphical framework for Great Britain, Netherlands J. Geosci. 84 (2005) 87–107.
[50] C. Radulescu, P-M. Samson, A. Petculescu, E. Stiuca, Pliocene large mammals of Romania, Coloq. Paleontol., Vol. Extraord. 1 (2003) 549–558.
[51] M. Morlo, M. Kundrat, The first carnivoran fauna from the Ruscinium (Early Pliocene, MN 15) of Germany, Palaontol. Zeits. 75 (2001) 163–187.
[52] J.A. Wolfe, Relations of environmental change to angiosperm evolution during the late Cretaceous and Tertiary, in: K. Iwatsuki, P.H. Raven (Eds.), Evolution and Diversification of Land Plants, Springer-Verlag, Tokyo, Japan, 1997, pp. 269–290.
[53] L. Francois, M. Ghislain, D. Otto, A. Micheels, Late Miocene vegetation reconstruction with the CARAIB model, Palaeogeo. Palaeoclimatol. Palaeoecol. 238 (2006) 302–320.
[54] L.J. Flynn, R.H. Tedford, Z.X. Qiu, Enrichment and stability in the Pliocene mammalian faunas of northern China, Paleobiology 17 (1991) 246–265.
[55] C.J. Williams, E.K. Mendell, J. Murphy, W.M. Court, A.H. Johnson, S.L. Richter, Paleoenvironmental reconstruction of a middle Miocene forest from the western Canadian Arctic, Palaeogeo. Palaeoclimatol. Palaeoecol. 261 (2008) 160–176.
[56] J.A. Wolfe, Distribution of major vegetational types during the Tertiary, in: E.T. Sundquist, W.A. Broecker (Eds.), The Carbon Cycle and Atmospheric CO_2; Natural Variations, Archean to Present, Geophysical Monograph, 32, 1985, pp. 357–375.
[57] J.A. Wolfe, An analysis of Neogene climates in Beringia, Palaeogeo. Palaeoclimatol. Palaeoecol. 108 (1994) 207–216.
[58] L.M. Reinink-Smith, E.B. Leopold, Warm climate in the late Miocene of the south coast of Alaska and the occurrence of Podocarpaceae pollen, Palynology 29 (2005) 205–262.
[59] H. Dowsett, R. Thompson, J. Barron, et al., Joint investigations of the middle Pliocene climate I: PRISM paleoenvironmental reconstructions, Glob. Planet Change 9 (1994) 169–195.

[60] G.S. Dwyer, M.A. Chandler, Mid Pliocene sea level and continental ice volume based on coupled benthic Mg/Ca paleotemperatures and oxygen isotopes. Philosoph. Transact. Roy. Soc. A (In Press – online article DOI 10.1098/rsta.2008.0222) 2008.

[61] L. Marincovich Jr., E.M. Brouwers, D.M. Hopkins, M.C. McKenna, Late Mesozoic and Cenozoic paleogeographic history of the Arctic Ocean basin, based on shallow-water marine faunas and terrestrial vertebrates, in: A. Grantz, L. Johnson, J.F. Sweeney (Eds.), The Arctic Ocean Region. The Geology of North America, Geological Society of America, 1990, pp. 403–426.

[62] M. Volker, T. Utescher, D.L. Dilcher, Cenozoic continental climate evolution of Central Europe, Proc. Natl. Acad. Sci. USA 102 (2005) 14964–14969.

CHAPTER

5

Phenotypic and Geographic Diversity of the Lesser Panda *Parailurus*

Martin Kundrát

Centre of Excellence for Integrative Research of the Earth's Geosphere, Geological Institute, Slovak Academy of Sciences, Banská Bystrica, Slovak Republic

OUTLINE

INTRODUCTION

The living red panda, *Ailurus fulgens*, represents a terminal relic of the once fluorishing carnivoran family of lesser pandas, the Ailuridae. Their ancestors and other extinct close relatives had successfuly spread around the northern hemisphere during the Miocene and Pliocene age. Today, the lesser panda leads a hidden life high in the refugial territories of alpine bamboo forests in China, Myanmar, India, Bhutan, and Nepal [1,2].

The red panda has a massive head, large flattened teeth and well-developed jaw muscles that provide the grinding power necessary for its high-fibre vegetarian diet. It exists primarily (80–95%) by consuming the leaves of arrow bamboo ["eater of bamboo" = "poonya" (original local name) has been anglicized to "panda"], however, it is not adapted to a strict herbivorous diet [3]. The panda feeds occasionally also on the meat of birds, rodents, and lizards, but pregnant females in particular show an increased carnivorous appetite [4].

Red Panda. DOI: 10.1016/B978-1-4377-7813-7.00005-7

Although the first known written record of the red panda comes from a scroll of the 13th century Chinese Dynasty of Chou, it was not discovered by Europeans until 1821 by T. Hardwicke, and by America in 1897 by F.W. Styan [3]. In 1825, the red panda was officially introduced to the scientific world by Geoffroy Saint-Hilaire and Cuvier, who described the herbivorous tree-dwelling animal and gave it the generic name *Ailurus* [5] based on its superficial likeness to that of the domestic cat [6]. It was not the domestic cat finally, but the giant panda (*Ailuropoda melanoleuca*) and the raccoon (*Procyon lotor*), whose morphological similarities led to confusion in identifying the closest extant relative of the red panda. Neither morphological nor molecular character analyses provided a consistent solution of the unresolved bear–panda–raccoon phylogenetic trichotomy, and varied among placements of the red panda.

For more than a century, biologists and paleontologists debated over the controversial views whether the red panda:

1. should be placed within the bears – Ursidae [7–10]
2. is closely related to *Ailuropoda* with uncertainty about their relationship within Arctoidea [11,12]
3. is a member of the panda clade (with *Ailuropoda*) that is sister taxon to Ursidae [13]
4. is related to raccoons – Procyonidae [14–22]
5. should be placed within musteloids – Musteloidea including procyonids plus some or all mustelids [23–31]
6. or should be defined as a monotypic lineage of uncertain phylogenetic affinities within Arctoidea [32,33].

Recent studies suggest that considerable progress has been made in the solution of the riddle of the red panda's ancestry. The current prevailing consensus has emerged as placement of the red panda to a close relationship with Procyonidae (original idea: nearly all 19th century taxonomists usually assigned the red panda to a representative of an ancient lineage of procyonids, an Asiatic raccoon [34–38]) or Mephitidae (a novel interpretation), with both now included within the musteloid clade that has a sister group relationship to Ursidae (original idea) and/or Pinnipedia (a novel interpretation). These studies demonstrated that the red panda represents a unique example of a surviving descendant of the musteloid stem that diverged early after the stem lineage radiated from the ancient phylogenetic split of basal arctoids. Based on an estimated date of African ape–human divergence at 35 million years ago, O'Brien et al. [17] concluded that the ancesors of the modern ursids and procyonids (and hence musteloids) split into two separate lineages between 30 and 50 million years ago. The divergence time of the lesser panda lineage from the Musteloidea stem has been suggested to have happened 28 Ma [39], 30 Ma [26], to 31–37 Ma [6].

The evolutionary history of the lesser panda's predecessors was very unclear until recently due to the fragmentary nature of fossil specimens. It has been accepted (but see [24]) that Ailuridae (and Ailurinae *sensu stricto*) include the recent *Ailurus fulgens* and extinct lesser pandas represented by a single genus *Parailurus* from the Pliocene of Europe [40], Asia [41,42], and North America [43]. Although *Amphictis ambiguus* from the late Oligocene of Quercy (France) may be more closely related to the Simocyoninae and perhaps Ailurinae [44], both subfamilies are currently included in the Ailuridae. It has been shown that the poor fossil record prevented the recognition of further Miocene ailurids first hand. Although *Magerictis imperialis* from the middle Miocene of Spain was described as an ailurid [45,46], it was later assigned to Ailurinae [47]. Only recent findings allowed us to reveal the ailurid identity of other taxa such as *Protursus* and *Simocyon*, all already known to specialists. *Protursus simpsoni* was originally labelled as an ursid [48], but recently was replaced into Ailuridae [49]. The *Simocyon* genus [50] from the middle and late Miocene of Europe (*Simocyon batalleri* [49,51], *Simocyon diaphorus* [52]) and North America (*Simocyon primigenius* [44]), and from the middle Miocene to early Pliocene of China (*Simocyon primigenius* [22]) was mostly considered a canid, but the whole Simocyoninae subfamily had a complicated taxonomic history (see [22]). The subfamily was later relocated to Procyonidae [44,49,53,54] and finally nested within Ailuridae [47]. The most recent contribution that considerably expanded our knowledge about fossil ailurines came from the late Miocene of eastern North America and was given the name *Pristinailurus bristoli* [55–57] (see also Chapter 4). The Ailurinae subfamily (lesser pandas *sensu stricto* or true red pandas) include to this date: *Ailurus* (Pleistocene [58,59] through recent of Asia only), *Parailurus* (Pliocene of Euroasia and North America), and *Pristinailurus* (Miocene of Euroasia and North America). It remains open to further investigation if the ailurines were derived from symocyoinines or share the common ancestor.

Order Carnivora (Bowdich, 1821)
Suborder Caniformia (Kretzoi, 1943)
Infraorder Arctoidea (Flower, 1869)
Superfamily Musteloidea (Fischer von Waldheim, 1817)
Family Ailuridae (Gray, 1843)
Problematica
Amphictis ambiguus (Gervais, 1876)
Incertae sedis
Magerictis imperialis (Ginsburg et al., 1997)
Protursus simpsoni (Crusafont and Kurtén, 1976)
Subfamily Simocyoninae (Dawkins, 1868)
Simocyon primigenius (Roth and Wagner, 1854)
Simocyon diaphorus (Kaup, 1832)
Simocyon batalleri (Viret, 1929)
Subfamily Ailurinae (Flower, 1869)
Ailurus fulgens (Geoffroy Saint-Hilaire and Cuvier, 1825)
Parailurus anglicus (Dawkins, 1888)
Parailurus baikalicus (Sotnikova, 2008)
Pristinailurus bristoli (Wallace and Wang, 2004)

Apparently, the least known period of the ailurine evolution is the period of the *Parailurus* pandas, despite their first remains being known from the end of the 19th century [36,60,61], and the specimens are known from 13 localities (Figure 5.1) of three continents (see Appendix). In this chapter, I will address two issues associated with the genus:

1. validity of the European *Parailurus* species
2. discovery of a skull of the extinct lesser panda in Slovakia.

Finally, I will provide a comprehensive review of all European and non-European specimens assigned to the genus *Parailurus*.

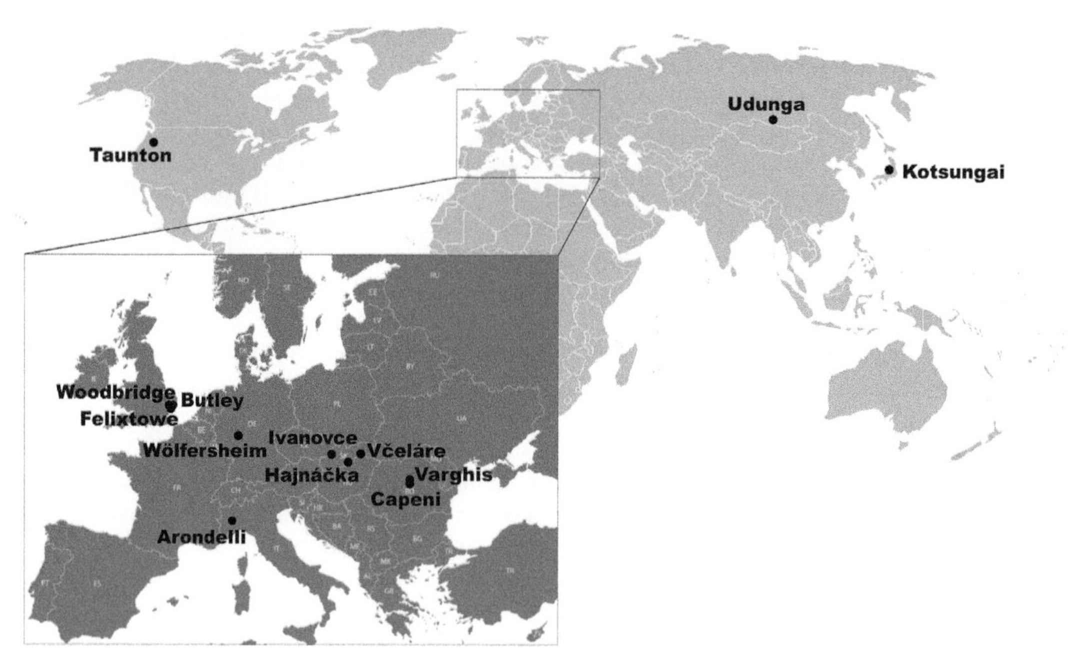

FIGURE 5.1 World map showing the occurrence of fossil remains of the lesser panda *Parailurus*.

TAXONOMIC REDUNDANCY OF *PARAILURUS HUNGARICUS*

The original placement of the first fossil ailurine in Europe, *Ailurus anglicus* [36], into the living genus *Ailurus* was soon questioned by more complete material found at the Baróth-Köpecz locality [61]. Having found almost complete sets of lower and upper dentition, as well as the craniofacial part of the skull (although considerably damaged), Schlosser noticed a morphological disparity in cranio-dental patterns (elongated snout, zygomatic process and infraorbital foramen shifted posteriorly, maxillary margin gradual between the maxillary tuber and choanic wall) that provided him with reasons to replace the *anglicus* species into a new extinct genus of lesser pandas, *Parailurus* [61]. The new ailurine genus has been widely accepted until it was recently considered a synonym of *Ailurus* [24]. Wolsan synonymized the *Parailurus* with *Ailurus* in order to maintain the monophyly of the latter taxon and make his taxonomic analysis more informative by reducing the number of monotypic genera. The next findings of the Pliocene lesser pandas showed that the *Parailurus* [61] is far from being a monotypic genus and includes now at least two valid species, *P. anglicus* [36] and *P. baikalicus* [42], a controversial species of *P. hungaricus*, and two potential candidates for a new species, from the Včeláre [62] and Taunton [43] localities. The new specimens have also offered a significant set of diagnostic dental characters to further sustain the taxonomic validity of Schlosser's *Parailurus* [42,63]. However, there is no such agreement for the intrageneric taxonomy. Recently, a controversy over the validity of the *hungaricus* species has occurred.

In 1915, the Hungarian paleontologist Kormos visited the Pliocene volcanic beds around the Hajnáčka village near Fil'akovo. He found a fragment of the right maxilla with the first and second molar *in situ* and considered this find as proof of an animal that differs to those found in the Red Crag beds and Baróth-Köpecz lignite mines. He erected a new species of extinct lesser panda in Europe: *Parailurus hungaricus* [64]. As this name first appeared in publication without a detailed diagnosis [65], in such circumstances *P. hungaricus* was considered a nomen nudum. A detailed description of the new species was not accomplished until 1935, and published together with description of new specimens of *P. anglicus* found at Baróth-Köpecz [64]. The Hajnáčka locality yielded other specimens in the middle of the twentieth century. Fejfar [66] assigned this material to the Kormos' species *Parailurus hungaricus*. Recently, the numerous specimens, found by Tobien, more than fifty years ago at the Wöfersheim locality, have been brought to light [63]. The occlusal patterns found in the first upper and lower molars of the Wölfersheim's lesser panda have combined features known in the teeth of both *P. anglicus* and *P. hungaricus* [63]. Despite minor variabilities in the size of the crown, the cross-similarities in dental morphologies of both the taxa cannot further endorse the Hajnáčka-based species of *Parailurus*. Therefore, Morlo and Kundrát [63] suggested *P. hungaricus* to be considered a junior synonym of *P. anglicus*. This suggestion has been oppossed by Fejfar and Sabol [67]. Although they provided a thorough list of any possible differences found in the occlusal morphology of the first molars in the "two" species, these are established only on the Fejfar's specimens from Hajnáčka, and most of the differential diagnosis is due to reasons of size. It is common that the bigger teeth will have proportionately higher crowns, more massive cusps, and broader cingula (see description of the Arondelli lesser panda below) – all the features representing major reasons why Fejfar and Sabol [67] found the validity of *P. hungaricus* further justified. Although one must admit that recognizing a new species is a subjective matter, there are several apparent faults in the approach employed by Fejfar and Sabol [67] to evaluate and interpret the differences they observed.

First, Fejfar and Sabol [67] have exclusively applied a one-sided view to comprehend phenotypic variability: the view of disparity, rather than making a balance through weighing the differences against similarities as well. Second, they focused on making a list of any tiny difference and found the list satisfactory to support the validity of *P. hungaricus*. However, such a list of different characters can be done for any well-established recent species, including *Ailurus fulgens*. What really matters is to recognize how relevant conclusions for taxonomic status are made without weighing inside a larger set of dental morphologies of *Parailurus* and *Ailurus*. Third, they have overlooked that most of their listed morphological differences can be explained as alometric modulations. This is apparent through linearity of plotted L/W dimensions of the upper first molar in the European adult specimens of *Parailurus* [63]. The same linear trend and variability of dimensional area is found in the recent species *Ailurus fulgens* that includes two subspecies *A.f. fulgens* and *A.f. styani* [1]. Fourth, Fejar and Sabol [67] underestimated the intraspecific molar polymorphism in the living panda, although distinct phenotypes of some of the characters (e.g., lingual cingulum in M1, protocone, lingual shelf in m1) they use to distinguish between the taxa are evident even in their small sample ($n = 4$) of *Ailurus*. The *Ailurus* specimens in the Field Museum of Natural History and the American Museum of Natural History give a much better perspective of occlusal plasticity within the recent genus (personal observations; n = 33) that should be considered to weigh their arguments (Figure 5.2). Fifth, their character argumentation based on such properties as more versus less developed, increased versus less increased, massive versus moderate, steeper versus rounded, loses its integrity when the sample is enlarged for the *hungaricus* and *anglicus* specimens from other localities.

The problem with the taxonomic validity of *P. hungaricus* is partly the result of the fragmentary nature of its fossil record, as only a handful of dental specimens are known for the Hajnáčka lesser panda. For this reason, neither the character variability of the Hajnáčka population nor the ontogenic modulation, including sexual dimorphism, can be grasped. My comments do not exclude a possibility of adaptive radiation of the Pliocene lesser pandas in Europe. Rather they offer serious reasons to question validity of the taxon that shares more similarities than differences with another, primary taxon.

Although critical doubts on the validity of *P. hungaricus* have been demonstrated (see below), these character ambiguities can also be considered by others as "considerable distinctions in morphology of the first upper and lower molars in pandas from Hajnáčka and Wölfersheim" [42]. In particular, the designation of *P. baikalicus* demonstrates character differences that clearly distinguished the Udunga specimen from the European specimens. However, when demonstrating morphological disparity between the Udunga and European specimens, Sotnikova [42] made differential analysis of her specimen exclusively

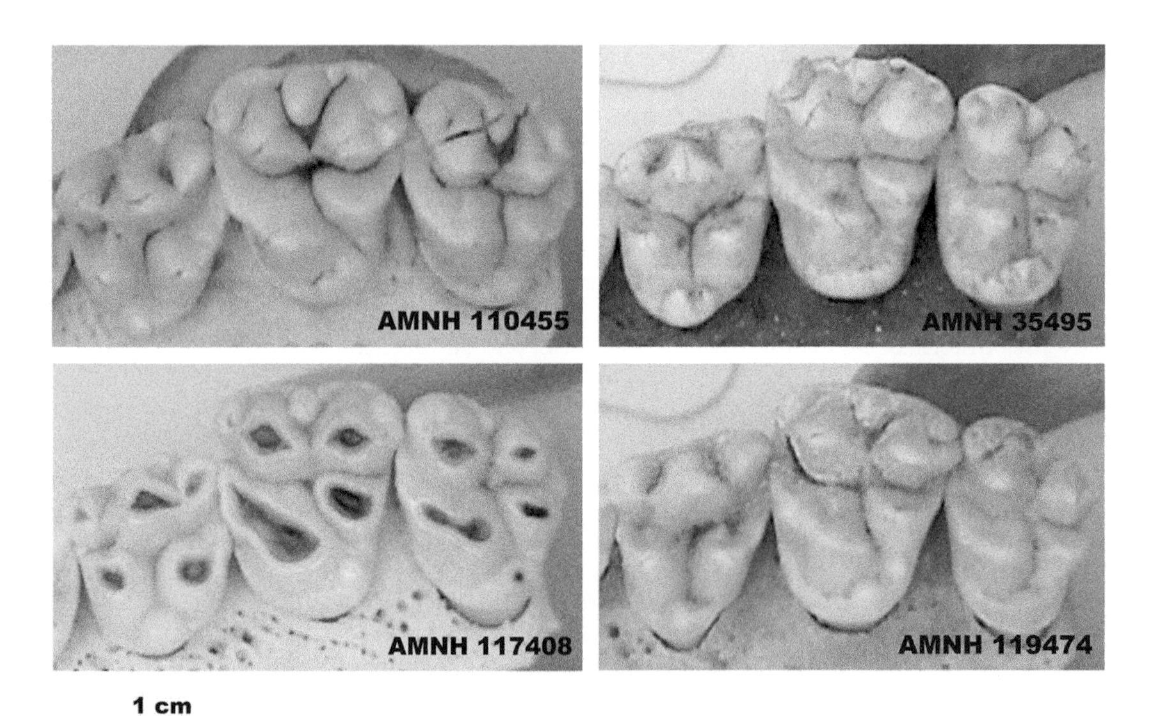

FIGURE 5.2 Occlusal polymorphism in the first upper molar of the living red panda *Ailurus fulgens*, collections of the American Museum of Natural History.

against characters of the European *Parailurus* as it would be a monotypic genus. She did not provide taxonomic separation of *P. baikalicus* through individual dianostics from the two "valid" species of *P. anglicus* and *P. hungaricus*, respectively. In addition, she accepted the reinstatement of *P. hungaricus*, and further claimed that all specimens of *P. anglicus* belong to the early Pliocene and *P. hungaricus* to the middle Pliocene. This is misleading and forces us to see the issue in a simple and incorrect way. The Red Crag localities, where the three specimens of *Parailurus anglicus* are known from, have been known for a long time to have an age of 3.0 to >2.6 Ma [68], which makes *P. anglicus* even younger than those from Hajnáčka. Therefore, *P. hungaricus* cannot represent a successive species replacement of *P. anglicus* (contra Sotnikova [42]). Moreover, the paleomagnetic data [69] indicate that the Hajnáčka fossil fauna may have had an older first occurrence than expected (up to 3.5 Ma), and therefore partly overlapped with the latest Ruscinian fossil faunas including lesser pandas diagnosed as *P. anglicus*.

In conclusion, I would like to point out that:

1. distinct apomorphies (e.g., comparable to those of *Parailurus baikalicus* [42]) to support *Parailurus hungaricus* as the valid species of the Pliocene lesser pandas in Europe are absent
2. the differences listed by Fejfar and Sabol [67] are within range of intraspecific modularity extrapolated from conditions in *Ailurus fulgens*, and most of them are cross-present in other European specimens assigned to *Parailurus anglicus*
3. a similar set of differences found between the Hajnáčka and Wölfersheim specimens can be provided for specimens from Wölfersheim versus Baróth-Kopecz or Wölfersheim versus Red Crag, all included in *P. anglicus*
4. being bigger does not justify the presence of a new species; furthermore it cannot be ruled out that the massiveness of some of the occlusal morphologies observed in the Hajnáčka specimens is due to either a higher ontogenic age of individuals (supported by considerably worn cusps), sexual dimorphism, and/or a dietary adaptation to local environmental conditions (warm and humid forest as suggested by Fejfar and Sabol [67])
5. the Hajnáčka lesser pandas were more or less coevals of the other Pliocene lesser pandas attributable to the primary species of *Parailurus*: *P. anglicus*, and hence, to grant again the validity of *P. hungaricus* through arguments that this is a taxonomic replacement of the primary species cannot be accepted. New characters such as evident innovations, absent in all other *Parailurus* specimens, must be present in order to bring on a consensus on the taxonomic integrity of *Parailurus hungaricus*. So far, the continued use of *P. hungaricus* in scientific literature can only be looked at as the use of a descriptive name of no taxonomic value for specimens found at the Hajnáčka locality.

THE EXTINCT LESSER PANDA FROM THE VČELÁRE LOCALITY

The evolution of the European lesser pandas is for some reason more frequently recorded in Slovakia than in other countries and includes specimens from three localities: Hajnáčka [64,66], Ivanovce [67,70] and Včeláre [40,62,71–74]. It is interesting that these localities yielded remains of extinct ailurines from different geological ages, starting with the oldest specimen from Ivanovce – MN 15b [75], followed by the specimens from Hajnáčka – MN 16a (+ latest 15b) [69], and finished with the most recent findings from Včeláre – MN 16b/17-Q2 [76].

Despite numerous other localities, the paleontological record of the *Parailurus* lesser panda remains restricted mostly to the isolated teeth, occasionally with post-cranial elements [66,74], and rarely cranial fragments [42,61]. Therefore, it is not surprising that more complete skeletal remains are needed to get an idea of how *Parailurus* might look. I am pleased to present here preliminary information concerning the discovery of the nearly complete skull of an extinct lesser panda. In 1992, I led an exploration of the Včeláre locality known mostly for findings of numerous assemblages of fossil micromammals [76]. The Včeláre locality represents a large active limestone quarry (Figure 5.3) located approximately 200 m above the small village of Včeláre near Košice, the district of Eastern Slovakia (Figure 5.4). The Včeláre quarry spreads across the top part of the easternmost edge of the Dolný Vrch Plateau of the Slovakian Karst, at an altitude of 420–450 metres above sea level (coordinates: 48°33′N latitude, and 20°49′E longitude). The approximately 1.8 km long southern face of the quarry passes along the Slovak–Hungarian border. Bones and teeth of various vertebrate groups [76–78] were found in the karst fissures filled with ochrous clayed sediments or plastic and lithified terra rossa. Despite the fact that all fossil sites are concentrated in the eastern section of the quarry, in particular at the upper and the uppermost working floor, the fauna they contain differs considerably in age, from the latest Pliocene to early Pleistocene.

FIGURE 5.3 The limestone quarry near the Včeláre village, Slovakia; arrow points to the place where the fossil ailurine skull was found.

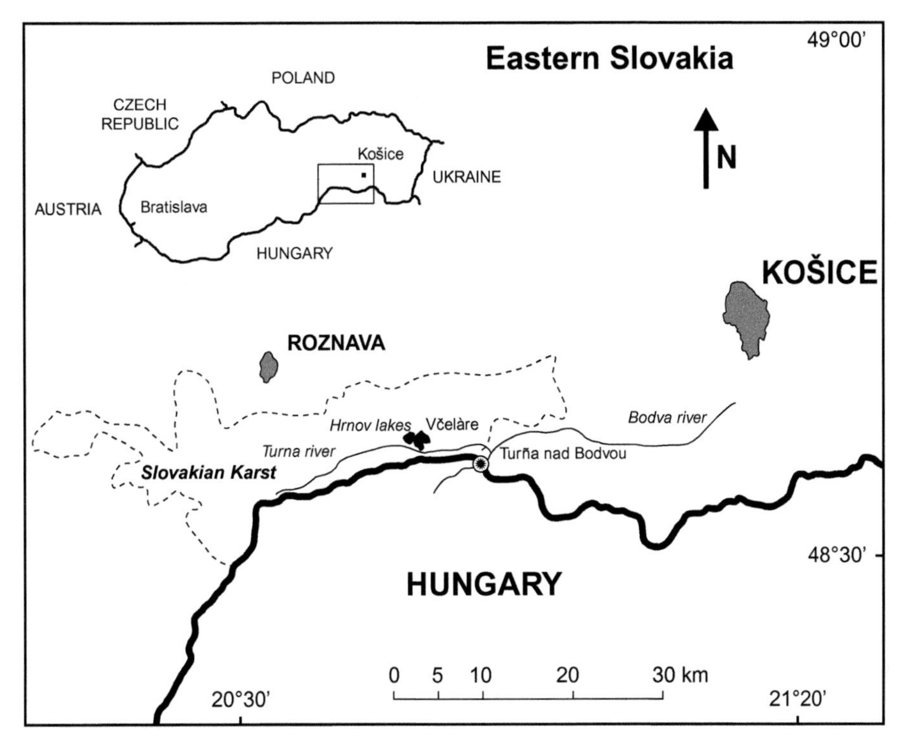

FIGURE 5.4 Location of the Včeláre locality that yielded several specimens of extinct lesser panda.

FIGURE 5.5 The lithified block of terra rossa with exposed skulls of a fossil ailurine and a carnivoran from the Včeláre quarry, Slovakia.

During the field trip undertaken in the late fall, two carnivoran skulls (Figure 5.5) partly exposed from a separated lithified and sintered block of dark terra rossa were discovered on November 11, 1992. It turned out that one of them represented the first complete skull of the lesser panda [71], and probably the only skull of a Pliocene lesser panda in Europe; other ailurine cranial remains of the Miocene/early Pliocene lesser panda *Pristinailurus* were discovered in North America recently (Wallace, personal communication). The other carnivoran skull is preserved in fact only by the basicranial and occipital region, and is now under systematic determination.

Due to the allochtonous character of the finding, direct dating of the Včeláre panda cannot be assessed. Perhaps, it is the skull fragment that might be of some help to elucidate the age of the specimen. Apart from it, I assume that the specimen belongs to the macrofauna from the Včeláre 2 site, found by Holec (personal communication), described partly (Proboscidea) in 1985 [78], and in part (Carnivora) in 2008 [74]. The Včeláre panda shows significant dental innovations in comparison to the specimens from Hajnáčka and Butley that are the closest chronological relatives to it. It is also a much smaller specimen, reaching the adult size of the living red panda.

The Včeláre fauna is principally much younger than faunas of other panda-yielding localities, and sheds light on faunistic succession events at the very end of the Pliocene and Pleistocene [76]. The fossil vertebrate assemblages found at the Včeláre locality came from different fissure or cavern-like systems formed inside the Middle Triassic (Ladinian) Wetterstein limstones. The latest fauna of the Včeláre locality, referred to as Včeláre 2 site, shows some similarities to that in Hajnáčka at the generic level: *Anancus arvernensis*, *Tapirus* spp., *Macaca* spp., and *Parailurus* spp. However, at least two of these genera, *Parailurus* and *Anancus*, represent a different morphotype of lesser panda [72] and a progressive morphotype of the mastodon (*sensu* Holec) [78].

First Holec [78] and recently Sabol et al. [74] suggested that the ancient Včeláre environment consisted of a forest combined with open grassland biotope. If so, then the Včeláre biotope changed significantly in comparison to that of the warm jungle forest in Hajnáčka [66]. A high representation of batrachofauna and the occurrence of the semiaquatic form of *Micromys* show that the Včeláre 2 fauna indeed inhabited humid swamp foliaceous forest with discontinuities of steppe-like areas in the frequently humid but moderate climatic zone [76,77]. Consequences of the next environmental changes include a further retreat of the forest and succession of the mesophile ekotop at the Plio–Pleistocene boundary (MN 17-Q 1). This is evident in particular by appearance of *Apodemus* (*Karstomys*) cf. *A. mystacinus* (the characteristic element of the extremely xerothermic regions with shrub vegetation of the Mediterranean type) [76]. As to macromammal representatives, only *Macaca* spp., present also in the older Holec fauna, is recorded at the beginning of Pleistocene, while new forms including *Ochotona* spp., *Hypolagus* spp., *Mustela* cf. *M. nivalis*, *Ursus* cf. *U. meditteraneus*, canids, *Equus* cf. *E. stenonis*, and cervids became more evident (see Horáček) [76]. In my opinion, the invasion of new competitors and predators, as well as the retreat of hiding forest spots, rather than climatic (temperature/humidity) changes themselves, might have built up the uncrossable barrier for further survival of lesser pandas in Europe at the end of Pliocene. Although it cannot be ruled out that the lesser pandas might have survived to the Plesitocene age at local European refuges, I suggest a more cautious estimation of a biochron for the Včeláre panda, which would probably fall within the interval of 2.0–2.5 Ma (late MN 17 preferred).

The fossil ailurine from Včeláre shows some phenotypic similarities with other specimens of *Parailurus*, based on the characteristic dental configuration of highly diagnostic first molars. In many aspects the specimen is the most complete preservation of extinct European ailurines and has become the key source of information necessary for our understanding of the terminal episode of the lesser pandas in Europe. The specimen consists of several separate remains (Figure 5.6) as follows:

- nearly complete skull with the right upper first and second molar *in situ*, and the right upper third premolar with a damaged occlusal area
- nearly complete right mandible with a row of more or less intact teeth except the second incisive preserved by root only, the third incisive with a damaged occlusal area, canine partly eroded, the first premolar with damaged occlusal area, and the second molar having the lingual part of the crown broken
- fragment of the left upper third premolar
- fragments of the left upper first and second molar with incomplete crowns
- fragment of the left zygomatic process of the temporal bone
- fragment of the left retroarticular process with mandibular fossa preserved
- fragment of the left coronoid process

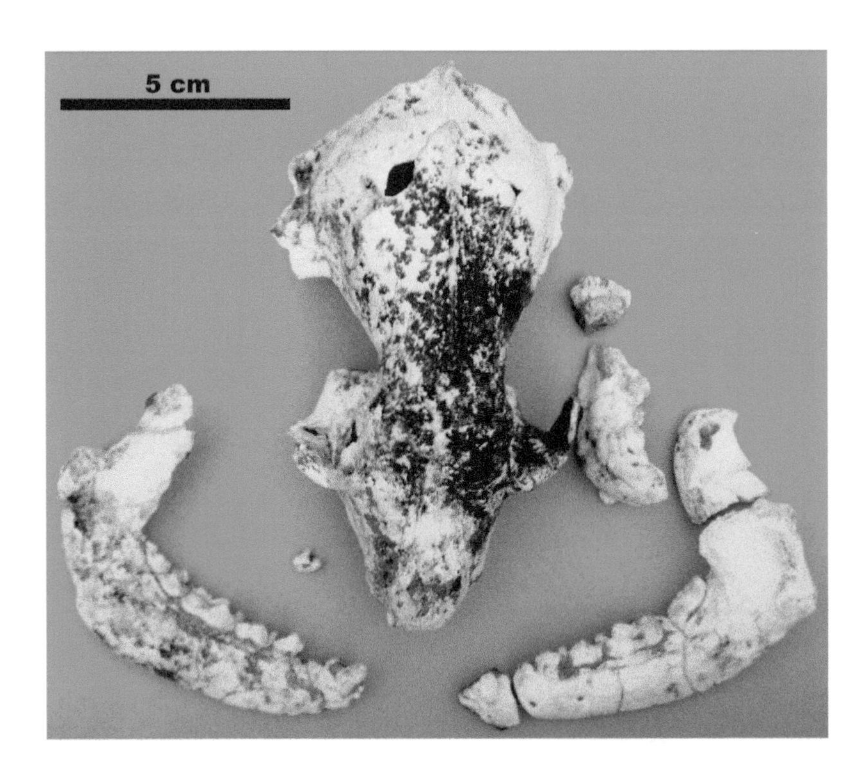

FIGURE 5.6 The cranial remains of undescribed specimen of fossil ailurine from Včeláre, Slovakia.

- fragment of the incisive part of the left mandible with roots of all three incisives *in situ*, and with almost complete canine and the first premolar
- fragment of the left mandible showing incomplete crowns of the second and third premolar, the fourth premolar with damaged buccal part of the crown, the first molar with posterior part of the trigonid crushed, and the second molar with incomplete anterior part of the crown.

The Včeláre lesser panda is of particular interest for three reasons:

1. it is preserved as a nearly complete skull, the first one reported for any fossil ailurinae in Europe
2. it is also the youngest specimen among European extinct lesser pandas; and finally
3. it is the smallest representative of the fossil ailurids known so far.

A comprehensive analysis of the Včeláre specimen, including designation of a new species, will be published by the author soon.

APPENDIX

European Specimens of the Extinct Lesser Panda *Parailurus*

The Felixtowe specimen – 1888

Parailurus anglicus (Dawkins, 1888)
Synonym: *Ailurus anglicus* Dawkins, 1888
Holotype: YORYM: YM509 – fragment of the posterior part of the right mandible with the second molar *in situ* (Figure 5.7)
Institute: Yorkshire Museum, York, UK
Source: Dawkins, W. B. (1888) Quart. J. Geol. Soc. London 44:228–231(Pl. X, Figs. 1–4) [36]
Newton, E. T. (1891) Mem. Geol. Sur. UK, pp. 13–14 (Pl. I, Figs. 17a,b) [79]

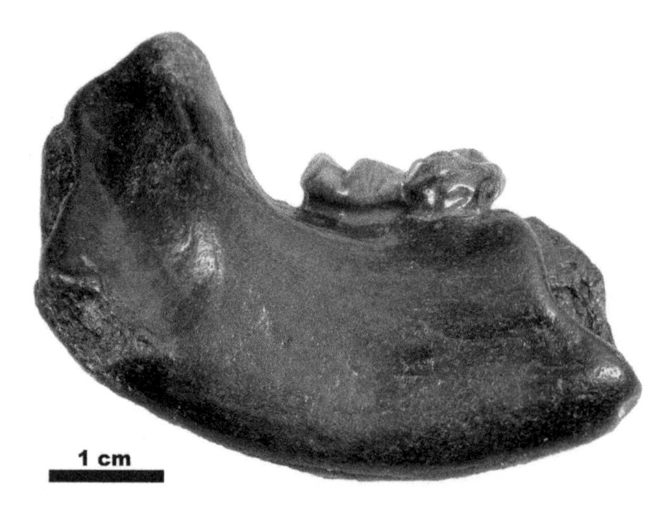

FIGURE 5.7 The holotype of *Parailurus anglicus* (Dawkins, 1888) from Felixstowe, UK; YORYM: YM509 (*image courtesy of Yorkshire Museum Trust*).

Locality: Red Crag Nodule-bed near Felixstowe, UK

Chronologic range: MN 16, late Pliocene [80]

Remarks: Based on the material available, Dawkins found only minor differences between the Felixstowe lesser panda and the living panda. He considered such characters as (1) "the cusps of the lower second molar as a whole are larger and blunter"; and (2) "the dental foramen situated further back" are not of more than specific value and therefore he designated the second species of the genus *Ailurus*: *A. anglicus,* rather than erecting a new genus for the extinct panda. However, in comments on the findings (p. 231) [36], Lyddeker said that "the fossil seemed to indicate a genus closely related to *Ailurus*", while Seeley pointed out that "the mode of grouping of the denticles in the recent and fossil types was different".

The Woodbridge specimen – 1891

Parailurus anglicus (Dawkins, 1888)

Synonym: *Ailurus anglicus* Dawkins, 1888

Referred specimen: fragment of a mandible with no teeth was reported to be in the Red Crag collection of the York Museum; the specimen found near Woodbridge [79]

Institute: Yorkshire Museum, York, UK (not confirmed)

Source: Newton ET (1891) Mem. Geol. Sur. UK, p. 13 [79]

Locality: Red Crag Nodule-bed near Woodbridge, UK

Chronologic range: MN 16, late Pliocene [80]

The Butley specimen – 1890

Parailurus anglicus (Dawkins, 1888)

Synonym: *Ailurus anglicus* Dawkins, 1888

Referred specimen: BGS GSM 3924 – the crown of the left upper first molar (Figure 5.8)

Institute: British Geological Survey, Nottingham, UK

Source: Newton ET (1890) Quart. J. Geol. Soc., London 46:444–543 (Pl. XVIII, Figs. 9a,b) [60]

Newton ET (1891) Mem. Geol. Sur. UK, pp. 13–14 [Pl. I, Figs. 18a,b [79]

Locality: Red Crag Nodule-bed at the locality of Butley near Boyton (a lower part of the Red Crag Formation, Suffolk bone bed), UK

Chronologic range: MN 16, late Pliocene [80]; isolated Red Crag localities in East England were suggested to have an age comparable to the Red Crag at Walton-on-the-Naze, 3.0 to >2.6 Ma [68] which corroborates the previous conclusion by Balson [80]

The Capeni (Baróth-Köpecz) specimens – 1899, 1901

Parailurus anglicus (Dawkins, 1888)

Parailurus new genus Schlosser, 1899

Genotype (1899; these remains belonged to three individuals at least): considerably damaged craniofacial part of skull (Figure 5.9A) including maxillary teeth (from the third premolar through the second molar) and both mandibles having an almost completely preserved row of teeth (the left mandible should be referred to as number o/531-532 *sensu* Kormos, 1935) [64]

FIGURE 5.8 Specimen of *Parailurus anglicus* from Butley, UK; BGS GSM 3924 (*image courtesy of Mr P. Shepherd*).

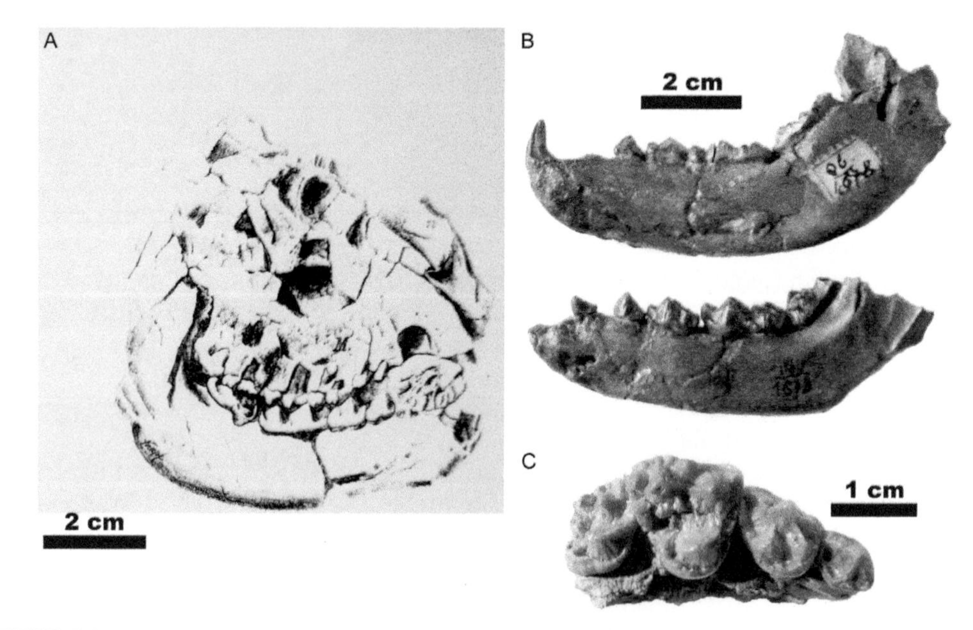

FIGURE 5.9 The Schlosser's *Parailurus* genotype (A) and Kormos' specimen (B,C) of *Parailurus anglicus* from Baróth-Köpecz (Capeni), Romania; Ob-1578 (*image courtesy of L. Kordos*).

fragment of another left lower mandible with the fourth premolar *in situ*
isolated left fourth lower premolar
isolated left lower canine

Institute: unknown

Source: Schlosser M (1899) Mitth. A.D. Jahrb. D. Kgl. Ung. Geol. Anst. Bd. XIII 2:67–95 (Pl. X, Fig. 1; Pl. XI, Figs. 2–5,7,8,10,11) [61]

Referred specimens (1901; these remains belonged to four individual at least): Ob-1578 – a pair of the lower mandibles with complete rows of teeth without incisors (Figure 5.9B)

Ob-1580 – an incomplete lower left mandible with the first and second molar *in situ* (Figure 5.9C)
Ob-1581+1597 – fragment of the right maxilla with the third and fourth premolars and the first and second molars *in situ;* Ob-1597 (the third premolar) +1581 (the third premolar through the second molar)
isolated left upper second incisive
isolated left upper third incisive
Ob-1584 – isolated right upper canine
Ob-1588 – isolated right lower fourth premolar
Ob-1589 – isolated right lower first molar
Ob-1590 – isolated right lower second molar
Ob-1599 – isolated left upper second molar

Institute: Geological Museum of Hungary, Budapest, Hungary

Source: Kormos T (1935) Mitt. A. D. Jahrb. D. Kgl. Ung. Geol. Anst. Bd. XXX 2:1–40 (Pl. I, Figs. 2–7; Pl. II, Figs. 1–5) [64]

Locality: the lignite mine (coal bed III) Köpecz near Baróth in the former district of Háromszék, Hungary, today Capeni (Baraolt Basin), Transylvania, Romania

Chronologic range: MN 15, late Ruscinian (Csarnotian age in Romania), early Pliocene [43]; MN 15b, late Ruscinian, early late Siennsian (Dacian Basin Stage), early Pliocene; the late Gilbert magnetic chron, approx. 3.7 Ma [81]

The Hajnáčka specimens – 1935, 1964

Parailurus anglicus (Dawkins, 1888)

Synonym: *Parailurus hungaricus* Kormos, 1935

Referred specimen (1935): Ob-3267 – fragment of the right maxilla with first and second molar *in situ* (Figure 5.10)

Institute: Geological Museum of Hungary, Budapest, Hungary

Source: Kormos T (1935) Mitt. A. D. Jahrb. D. Kgl. Ung. Geol. Anst. Bd. XXX 2:1-40 (Pl. I, Figs. 8,9) [64]

Kormos T (1917) Állattani Közlemények 16:137–138 [65]

Referred specimens (1964): SNM-MNH Z-26381 (SÚÚG OF 65372) – the right upper first molar (probe 9/56) (Figure 5.11A)

SNM-MNH Z-26380 (former SÚÚG OF 65371) – the left lower first molar (probe 8/56) (Figure 5.11B)
SNM-MNH Z-26382 (former SÚÚG OF 65373) – a lingual fragment of the right upper first molar (probe 9/56)
SNM-MNH Z-26383/2 – fragment of the trochanter major and distal part of right femur (probe 3/56)

FIGURE 5.10 The Kormos' "*hungaricus* holotype", now assigned to *Parailurus anglicus* from Hajnáčka, Slovakia; Ob-3267 (*images courtesy of L. Kordos*).

SNM-MNH Z-26383/1 – right tibia (probe 3/56)
SNM-MNH Z-26383/5 – a distal part of the right fibula (probe 3/56)
SNM-MNH Z-26383/3 – right astragalus (probe 3/56)
SNM-MNH Z-26383/4 – right calcaneus (probe 3/56)
SNM-MNH Z-26384 – left calcaneus (probe 8/56)

Institute: Museum of Natural History, The Slovak National Museum, Bratislava, Slovakia

Source: Fejfar O (1964) Roz. Ústřed. Úst. Geol., Praha 30:55–59 (Figs. 36a-e) [66]

Fejfar O, Sabol M (2004) Cour. Forsch. Inst. Senckenberg 246:15–53 (Figs. 5 (1–4,9,12,13); Pl. 7, Figs. 1–3,7,9,11,12–23; Pl. 8, Figs. 1–13) [67]

Locality: uncemented, partly redeposited lapili tuffs and lacustrine tuffitic sediments at the locality of Kostná dolina in Hajnáčka near Fil'akovo, Slovakia

Chronologic range: MN 16a, early Villányian, late Pliocene [75]; however, Lindsay et al. [69] suggested the early Gauss magnetic chron, 3.0–3.5 Ma, for the locality, which implies that Hajnáčka fauna might be older than previously suggested: MN16/15

The Wölfersheim specimens – 1952, 2001

Parailurus anglicus (Dawkins, 1888)

Referred specimen: SMF 2000/223 (former Wö 138) – isolated left upper third incisive with complete root and crown, the latter abraded on the lingual side

SMF 2000/224 (former Wö 136) – isolated left upper canine with mostly incomplete root, and abraded crown, broken across the top and damaged at the base
SMF 2000/225 (former Wö 239/1) – isolated right lower canine with an almost complete crown abraded on the lingual side
SMF 2000/226 (former Wö 239/2) – isolated right lower second premolar
SMF 2000/227 (former Wö 239/3) – isolated left lower second premolar

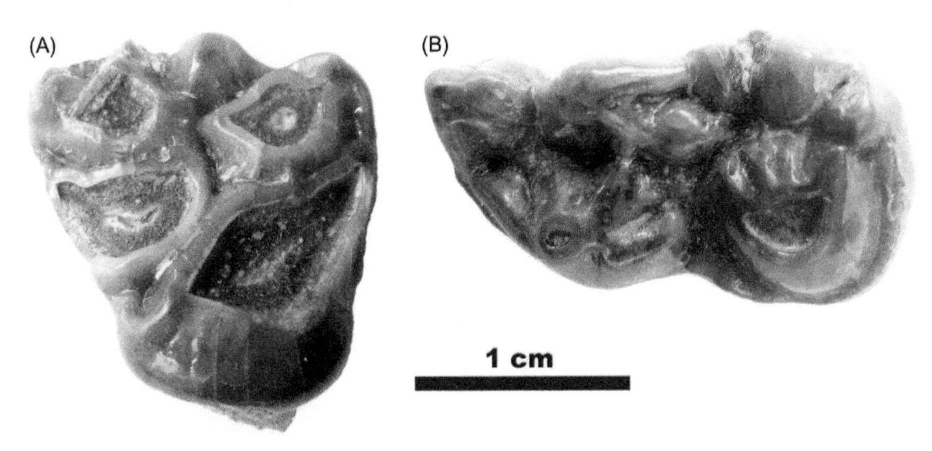

FIGURE 5.11 Specimens of *Parailurus anglicus* from Hajnáčka, Slovakia; SNM-MNH Z-26381 (A) and 26380 (B), (*images courtesy of A. Ďurišová and D. Pakozdyová*).

SMF 2000/228 (former Wö 239/6) – isolated right upper second premolar with fragmentary posterior root present
SMF 2000/229 (former Wö 239/5) – isolated left lower fourth premolar
SMF 2000/230 (former Wö 239/4) – isolated left upper fourth premolar
SMF 2000/231 (former Wö 148) – isolated left upper second premolar
SMF 2000/232 (former Wö 301) – left mandible fragment showing the posterior part of the canine alveolus, the anterior and posterior roots of the second premolar inside alveoli, the third premolar *in situ*, considerably abraded fourth premolar *in situ*, anterior part of the posterior alveolus of the first molar; included in SMF 2000/232 is an isolated fragment of the left lower first molar with heavily abraded crown lacking the antero-lingual part of the trigonid and the postero-labial part of the talonid
SMF 2000/233 (former Wö 144) – isolated right lower first molar (Figure 5.12A)
SMF 2000/234 – isolated postero-labial part of the crown of the right upper first molar with the apex of the metacone slightly abraded
SMF 2000/235 (former Wö 219) – isolated right upper first molar with intact crown (except a slight apical abrasion of the labial cusps) and two labial roots present (Figure 5.12B)
SMF 2000/236 (former Wö 142) – isolated intact crown of the deciduous left upper fourth premolar with protoconule and metaconule slightly abraded at the tip, and damaged posterior side of the metacone
SMF 2000/237 (included in former Wö 301) – isolated fragment (postero-lingual part of the talonid) of the right lower second molar
SMF 2000/238 – isolated fragment (posterior part of the talonid) of the left lower first molar with slightly abraded hypoconid
SMF 2000/239 (included in former Wö 301) – isolated postero-lingual part of the right upper second molar

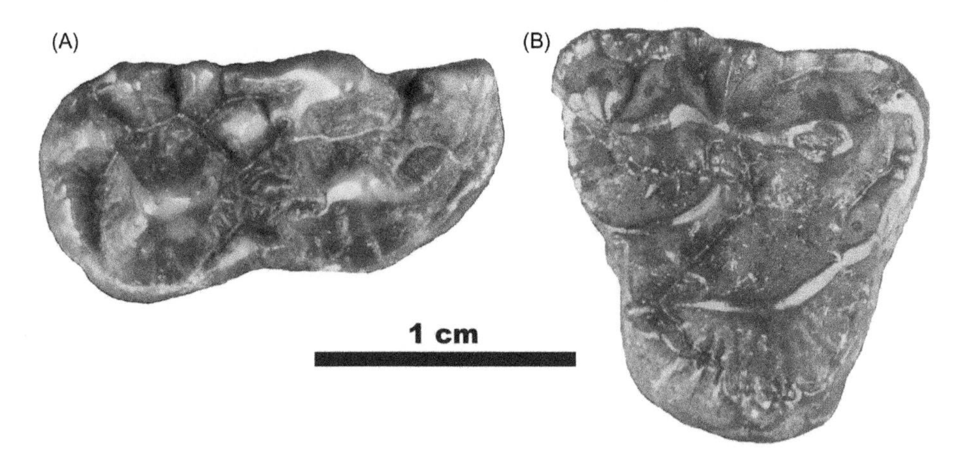

FIGURE 5.12 Specimens of *Parailurus anglicus* from Wöfersheim, Germany; SMF 2000/233 (A) and 235 (B).

Institute: Forschungsinstitut und Naturmuseum Senckenberg, Frankfurt am Main, Germany

Source: Tobien H (1953) Z. Deutsch. Geol. Ges. Bd. 104:191 [82]

Kundrát M, Morlo M (1999) J. Vert. Paleontol. 19(Suppl.3):57–58A [83]

Morlo M, Kundrát M (2001) Paläontol. Zeitschr. 72:163–187 (Figs. 15–30) [63]

Locality: Wölfersheim near Frankfurt am Main, Hesse, Germany; 50°24′30″ N and 8°51′25″E

Age: MN 15b, late Ruscinian, early Pliocene [75]

Remarks: The first reference to the extinct lesser panda from Wölfersheim was given in a faunal list by Tobien [82]. Only the generic name of *Parailurus* appeared in the report and it has long been believed that only one tooth, the first right upper molar SMF 2000/235, was unearthed in Wölfersheim. This is partly due to Tedford and Gustafson [43], who presented a drawing (without description) of this specimen only. The authors classified the Wölfersheim specimens as *Parailurus* cf. *P. anglicus*. However, re-evaluation of the Wölfersheim collection finally revealed a total of 18 specimens, which makes the site one of the most important concerning the fossil history of the ailurines.

The Ivanovce specimen – 1961

Parailurus cf. *anglicus* (Dawkins, 1888)

Referred specimen: SNM-MNH Z-26390 – distal fragment of the left lower second molar (the horizontal filling 6513) (Figure 5.13)

Institute: Museum of Natural History, The Slovak National Museum, Bratislava, Slovakia

Source: Fejfar O (1961) N. Jb. Geol. Paläont., Abh. 3:257–273 [70]

Fejfar O, Sabol M (2004) Cour. Forsch. Inst. Senckenberg 246:15–53 [67]

Locality: karst fissures at the Ivanovce village near Trenčín, Slovakia

Chronologic range: MN 15b, late Ruscinian, early Pliocene (sensu Fejfar and Heinrich) [75]

The Arondelli specimen – 1967

Parailurus cf. *anglicus* (Dawkins, 1888)

Synonym: *Parailurus* cf. *hungaricus* Kormos, 1935

FIGURE 5.13 Specimen of *Parailurus* cf. *P. anglicus* from Ivanovce, Slovakia; SNM-MNH Z-26390 (*image courtesy of A. Ďurišová and D. Pakozdyová*).

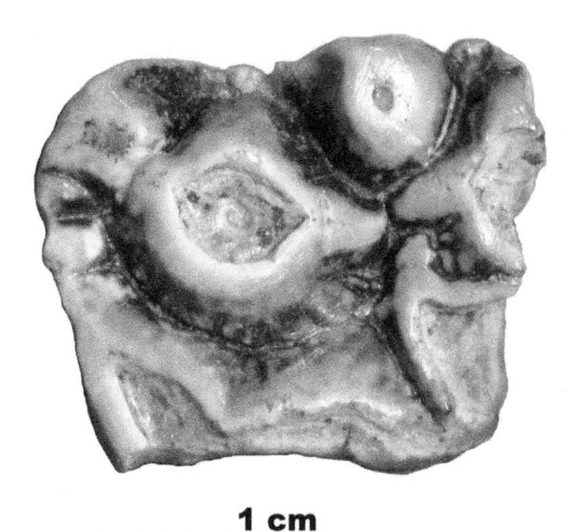

FIGURE 5.14 The Arondelli specimen of *Parailurus* cf. *P. anglicus*, Italy; UCMP 77039 (*image courtesy of P. Holroyd*).

Referred specimen: UCMP 77039 – an antero-buccal fragment of the left upper first molar (Figure 5.14)

Institute: Museum of Paleontology, The University of California, Berkeley, CA, USA

Source: Berzi A, Michaux J, Hutchison JH, Lindsay E (1967) Giorn. Geol. 35:1–4 [84]

Locality: quarry near the village of Villafranca d'Asti and on the Cascina Arondelli, Italy

Chronologic range: MN 16a, early Villányian, late Pliocene [75]; early Gauss magnetic chron, 3.5–3.0 Ma [69]

Description: The only report of the lesser panda from Arondelli [84] lacked both a description and illustration of the specimen, although it was finally diagnosed as *Parailurus* cf. *P. hungaricus*. For this reason I describe and illustrate here the Arondelli specimen for the first time, based on the cast provided to me by the California University Museum of Paleontology. The specimen represents the antero-buccal quadrant of the occlusal area of the left first upper molar. Comparing available dimensions of the specimen to the other *Parailurus* specimens, the tooth belongs to the size category of the lesser pandas found in Hajnáčka. This might have been the reason for Berzi et al. [84] to conclude that their specimen is similar to the "*hungaricus*" species. The other feature that also might have evoked the same conclusion was considerable abrasion of the cusps, like in the specimen (SNM-MNH Z-26381 found by Fejfar [66]). However, when the diagnostic occlusal features and the abrasion are considered in detail, the Arrondeli specimen represents an almost identical copy of the Butley specimen (see Figure 5.8), although slightly larger. Thus the Arondelli specimen extends the size range of *Parailurus anglicus.* The larger size makes the corresponding features massive or robust in their appearance. These features have been claimed [67] to be specific to "*Parailurus hungaricus*" exclusively (see also comments to the issue above).

The molar fragment of the Arondelli lesser panda shows preservation of three complete cusps: the paracone, the mesostyle, and the metastyle. The most striking feature that makes the Arondelli specimen different from the others is advanced enlargement of the mesostyle and parastyle in comparison to the paracone. However, both styles are variable in size in *Ailurus fulgens* as well (see below) and, therefore, are not significant for the specific taxonomy. The external cingulum, having a cuspule formed on it, is constrained by buccal ridges projected from the parastyle and mesostyle; the parastyle ridge is more pronounced than in the other European ailurines. In contrast to the Butley and Wölfersheim specimens, a cross-shaped ridge that connects the paracone and the metacone, as well as the mesostyle and the metaconule is incomplete in the Arondelli specimen, similar to that of the Baróth-Köpecz and Hajnáčka specimens. The grooves that separate the cusps are well expressed in the findings from Arondelli, Butley, and Wölfersheim, while these are mostly obliterated in the specimens from Baróth-Köpecz and Hajnáčka. The protoconule continues antero-buccally as a shell-like cingulum in the Arondelli specimen; a character found also in the Butley specimen and the Kormos' specimen from Hajnáčka. Finally, the Arrondeli specimen shows a larger depression surrounded by the protocone, the paracone, and the metaconule. A similar character is known in the Butley specimen, and less expressed in the Fejfar's specimen.

In conclusion, the Arondelli specimen provides new arguments against acceptance of the Hajnáčka lesser pandas forming the new species, *Parailurus hungaricus.* In contrast to the original paper, I have found that the Arondelli lesser panda shares considerable details with the specimen of *Parailurus anglicus* from Butley regardless of the size. Furthermore, it combines features of all known specimens, making it impossible to find unambiguous characters that can distinguish between the two different species of European ailurines from Pliocene.

The Včeláre specimens –1996, 2008

cf. *Parailurus* spp.

Referred specimens (1996): nearly complete skull (see above)

Institute: Centre of Excellence for Integrative Research of the Earth's Geosphere, Geological Institute, Slovak Academy of Sciences, Slovakia

Source: Kundrát M (1996) Nature Carpatica 37:211–213 [71]

Kundrát M (1997) J. Morphol. 232(3):282 [72]

Kundrát M (1997) J. Vertebr. Paleontol. 17(3):58A [62]

Kundrát M (1998) Abstracts, Euro-American Mammal Congress: Challenges in Holarctic Mammalogy, Santiago de Compostela, Spain, p. 345 [73]

Parailurus spp.

Referred specimens (2008): SNM-MNH Z-26703/1 – a right maxillary fragment with the unerupted third and fourth premolar

SNM-MNH Z-26703/2 – fragment of the right upper first molar
SNM/MNH Z-26704/1 – fragment of the left dentary with the damaged first molar
SNM-MNH Z-26704/2 – fragment of the lower left fourth premolar
SNM-MNH Z-26705/1 – fragment of the right mandible with the second molar
SNM-MNH Z- 26705/2 – fragment of the right dentary with the third premolar (damaged), fourth premolar, and first molar (damaged)

Institute: Museum of Natural History, The Slovak National Museum, Bratislava, Slovakia

Source: Sabol M, Holec P, Wagner J (2008) Paleontol. J. 42:531–543 [74]

Locality: karst fissue system in the limestone quarry at the Včeláre village near Košice, Slovakia; 48°33″ N and 20°49′ E

Chronologic range: MN 16b/17-Q2, early Villányian-Biaharian, late Pliocene–Pleistocene [76]

The Varghis specimen – 2003

Parailurus anglicus (Dawkins, 1888)

Referred specimens: unknown

Institute: unknown

Source: Radulescu C, Samson PE, Petculescu A, Sitiucă E (2003) Coloq. Paleo.1:549–558 [81]

Locality: the lignite mine (coal bed III) near Capeni in the Baraolt Basin, Transylvania, Romania

Chronologic range: MN 15b, late Ruscinian, early late Siennsian (Dacian Basin Stage). Early Pliocene; the late Gilbert magnetic chron, approx. 3.6 Ma; the Varghis locality is slightly younger than the Capeni locality [81]

Out of Europe Specimens of the Extinct Lesser Panda *Parailurus*

The Taunton specimen – 1977

Parailurus spp.

Referred specimen: LACM 10808 – isolated right upper first molar (Figure 5.15)

Institute: The Los Angeles County Museum, Adams County, Los Angeles, USA

Source: Tedford RH, Gustafson EP (1977) Nature 265:621–623 (Fig. 1) [43]

Locality: fluviatile gravels and crossbedded sand of the Ringold Formation near Taunton, WA, USA; 46 41′ N and 119°21′ W

Chronologic range: MN 16a, middle Blancan (= early Villányian), late Pliocene; late Gauss magnetic chron 3.0 Ma [85] alternatively early Blancan, 2.7–4.9 Ma [86]

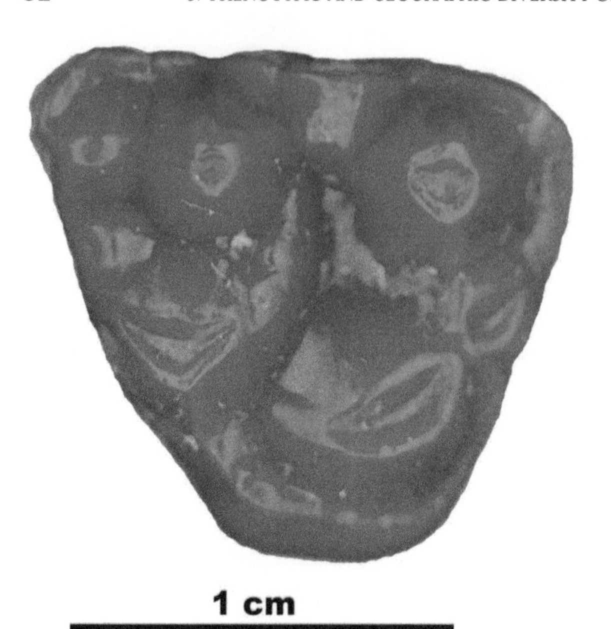

FIGURE 5.15 The Taunton specimen of *Parailurus* sp. from USA; the cast of LACM 10808 (*image courtesy of H. Tedford*).

The Kotsungai specimen – 2003

Paraïlurus spp.

Referred specimen: NSGR-V-11 – isolated right upper fourth premolar (Figure 5.16)

Institute: Department of Geology, Faculty of Science, Niigata University, Japan

Source: Sasagawa I, Takahashi K, Sakumoto T, Nagamori H, Zabe H, Kobazashi I (2003) J. Vert. Paleontol. 23:895–900 (Figs. 2,3a) [41]

Locality: lenticular conglomerate bed of the Ushigakubi Formation at the Kotsungai locality near Tochio, Niigata Prefecture, Japan; 37°29′ N and 138°59′ E

Chronological range: MN 15b, late Ruscinian, early Pliocene, 3.0–4.0 Ma [41,87]

The Udunga specimen – 2008

Parailurus baikalicus Sotnikova, 2008

Synonym: *Ailurus* spp.

Holotype: BF GIN 962/58 – fragmentary left maxilla with the posterior root (in alveolus) of the third premolar, and the intact fourth premolar, first and second molar *in situ* (Figure 5.17)

anterior part of the jugal

Institute: Geological Institute of the Siberian Branch of the Russian Academy of Sciences, Ulan-Ude, Russia

Source: Sotnikova MV (2008) Paleontol. J. 42:90–99 (Fig. 1a,b) [42]

Kalmykov NP, Shabunova VV (2005) Gos. Univ., Tomsk, Russia, pp. 293–295 (Fig. 3) [88]

Maschenko EN (1994) Priroda 11:64–70 [89]

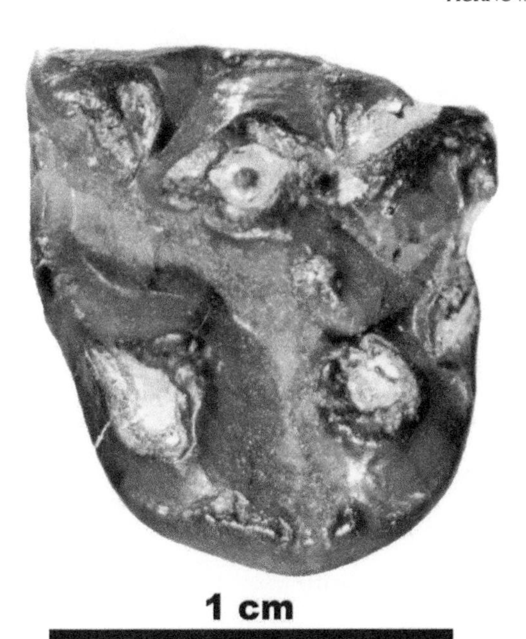

FIGURE 5.16 The Kotsungai specimen of *Parailurus* sp. from Japan; NSGR-V-11 (*image courtesy of I. Sasagawa*).

FIGURE 5.17 The holotype of *Parailurus baikalicus* from Udunga, Transbaikal, Russia; BF GIN 962/58, Sotnikova, 2008.

Locality: red beds of the Chikoi Formation, Udunga locality in the valley of the Temnik River (left tributary of the Selenga River), Transbaikalia, Russia

Chronologic range: MN 15b, late Ruscinian, middle Pliocene [90]

ACKNOWLEDGEMENTS

I thank Dr Angela Glatston for inviting me to contribute to her book on the lesser pandas, Dr Steven Wallace for his valuable comments, as well as the following individuals and institutions for their generous donations of the *Parailurus* images: RNDr. Ďurišová (Museum of Natural History, The Slovak National Museum); Dr Pat Holroyd (Museum of Paleontology, University of California); Prof. László Kordos (Geological Museum of Hungary); Prof. Ichiro Sasagawa (School of Dentistry at Niigata, The Nippon Dental University); Mr Paul Shepherd (British Geological Survey, Kingsley Dunham Centre); and York Museums Trust (Yorkshire Museum).

References

[1] M.S. Roberts, J.L. Gittleman, *Ailurus fulgens*, Mammal Species 222 (1984) 1–8.
[2] A.R. Glatston, Red Panda Biology, SPB Academic Publishing B.V., The Hague, The Netherlands, 1989.
[3] G.B. Schaller, The Last Panda. The University of Chicago Press, Chicago, USA, 1993.
[4] M.C.K. Bleijenberg, in: A.R Glatston (Ed.), The Red or Lesser Panda Studbook, vol. 3, The Royal Rotterdam Zoological and Botanical Gardens, Rotterdam, The Netherlands, 1984, pp. 23–36.
[5] E. Geoffroy Saint-Hilaire, G. Cuvier, Histoire Naturelle des Mammifères, avec des Figures Originales, Coloriées, Dessinées d'après des Animaux Vivans, Livraison L. Published under the authority of the administration of Muséum d'Histoire Naturelle, by E. Geoffroy Saint-Hilaire and G. Cuvier, Chez A. Belin, Paris, France, 1825.
[6] J.J. Flynn, M.A. Nedbal, J.W. Dragoo, R.L. Honeycutt, Whence the Red Panda? Molec. Phylogenet. Evol. 17 (2000) 190–199.
[7] V.M. Sarich, The giant panda in a bear, Nature 245 (1973) 218–220.
[8] W.C. Wozencraft, in: J. Gittleman (Ed.), Carnivore Behavior, Ecology, and Evolution, The Cornell University Press, Ithaca, USA, 1989, pp. 495–535.
[9] A.R. Wyss, J.J. Flynn, in: F. Szalay, M. Novacek, M. McKenna (Eds.), Mammal Phylogeny: Placentals, Springer-Verlag, New York, USA, 1993, pp. 32–52.
[10] P.B. Vrana, M.C. Milinkovitch, J.R. Powell, W.C. Wheeler, Higher level relationships of the arctoid Carnivora based on sequence data and total evidence, Molec. Phylogenet. Evol. 3 (1994) 47–58.
[11] É.-L. Trouessart, Catalogus mammalium tam viventium quam fossilium, R. Friedläder, Berolini, 1898.
[12] W. Segall, The auditory region of the arctoid carnivores, Zool. Ser. Field Museum Nat. Hist. 29 (1934) 33–59.
[13] L. Ginsburg, Sur la position systématique du petit panda, *Ailurus fulgens* (Carnivora, Mammalia), Géobios, Mém. Spéc. 6 (1982) 247–258.
[14] R.I. Pocock, The external characters and classification of the Procyonidae, Proc. Zool. Soc. London (1921) 389–422.
[15] W.K. Gregory, On the phylogenetic relationships of the giant panda (Ailuropoda) to other arctoid Carnivora, Am. Museum Novit. 878 (1936) 1–29.
[16] E. Thenius, Zur systematischen und phylogenetischen Stellung des Bambusbären: *Ailuropoda melanoleuca* David (Carnivora, Mammalia), Zeits. Saugetierk 44 (1979) 286–305.
[17] S.J. O'Brien, W.G. Nash, D.E. Wildt, M.E. Bush, R.E. Benveniste, A molecular solution to the riddle of the giant panda's phylogeny, Nature 317 (1985) 140–144.
[18] D. Goldman, P.R. Giri, S.J. O'Brien, Molecular genetic distance estimates among the Ursidae as indicated by one- and two-dimensional protein electrophoresis, Evolution 43 (1989) 282–295.
[19] R.K. Wayne, R.E. Benveniste, D.N. Janczewski, S.J. O'Brien, in: J. Gittleman (Ed.), Carnivore Behavior, Ecology, and Evolution, The Cornell University Press, Ithaca, USA, 1989, pp. 465–494.
[20] J.P. Slattery, S.J. O'Brien, Molecular phylogeny of the red panda (*Ailurus fulgens*), J. Hered. 86 (1995) 413–422.
[21] J.J. Flynn, N.A. Neff, R.H. Tedford, in: M Benton (Ed.), The Phylogeny and Classification of the Tetrapods. Volume 2, Mammals, Clarendon Press, Oxford, UK, 1988, pp. 73–116.
[22] X. Wang, New cranial material of *Simocyon* from China, and its implications for phylogenetic relationship to the red panda (*Ailurus*), J. Vert. Paleontol. 17 (1997) 184–198.
[23] N. Schmidt-Kittler, Zur Stammesgeschichte der marderverwandten Raubtiegruppen (Musteloidea, Carnivora), Eclog. Geolog. Helvet. 74 (1981) 753–801.

[24] M. Wolsan, Phylogeny and classification of early European Mustelida (Mammalia: Carnivora), Acta Theriol. 4 (1993) 345–384.

[25] J.J. Flynn, M.A. Nedbal, Phylogeny of the Carnivora (Mammalia): Congruence vs incompatibility among multiple data sets, Molec. Phylogenet. Evol. 9 (1998) 414–426.

[26] O.R.P. Bininda-Emonds, J.L. Gittleman, A. Purvis, Building large trees by combining phylogenetic information: a complete phylogeny of the extant Carnivora (Mammalia), Biol. Rev. 74 (1999) 143–175.

[27] J.J. Flynn, J.A. Finarelli, S. Zehr, J. Hsu, M.A. Nedbal, Molecular phylogeny of the Carnivora (Mammalia): assessing the impact of increased sampling on resolving enigmatic relationships, System. Biol. 54 (2005) 317–337.

[28] T.L. Fulton, C. Strobeck, Molecular phylogeny of the Arctoidea (Carnivora): effect of missing data on supertree and supermatrix analyses of multiple gene data sets, Molec. Phylogenet. Evol. 41 (2006) 165–181.

[29] T.L. Fulton, C. Strobeck, Novel phylogeny of the raccoon family (Procyonidae: Carnivora) based on nuclear and mitochondrial DNA evidence, Molec. Phylogenet. Evol. 43 (2007) 1171–1177.

[30] T. Yonezawa, M. Nikaido, N. Kohno, Y. Fukumoto, N. Okada, M. Hasegawa, Molecular phylogenetic study on the origin and evolution of Mustelidae, Gene 396 (2007) 1–12.

[31] J.A. Finarelli, A total evidence phylogeny of the Arctoidea (Carnivore: Mammalia): relationships among basal taxa, J. Mammal. Evol. 15 (2008) 231–259.

[32] C. Ledje, Ú. Árnason, Phylogenetic analyses of complete cytochrome b genes of the order Carnivora with particular emphasis on the Caniformia, J. Molec. Evol. 42 (1996) 135–144.

[33] C. Ledje, Ú. Árnason, Phylogenetic relationships within caniform carnivores based on analyses of the mitochondrial 12S rRNA gene, J. Molec. Evol. 43 (1996) 641–649.

[34] H.M.D. de Blainville, Ostéographie ou description iconographique comparée du squelette et du système dentaire des mammifères récents et fossiles pour servir de base à la Zoologie et à la Géologie. 4 (1841) 1–888.

[35] W.H. Flower, On the value of the characters of the base of the cranium in the classification of the order Carnivora, and on the systematic position of *Bassaris* and other disputed forms, Proc. Zool. Soc. London (1869) 4–37.

[36] W.B. Dawkins, On *Ailurus anglicus*, a new carnivore from Red Crag, Q. J. Geol. Soc. London 44 (1888) 228–231.

[37] W.H. Flower, R. Lydekker, An Introduction to the Study of Mammals, Living and Extinct, Verlag Black, London, UK, 1891.

[38] H. Winge, Jordfundne og nulevende rovdyr (Carnivora) fra Lagoa Santa, etc, E. Museo Lundii, Copenhagen 2 (4) (1895–96) 1–130.

[39] S.J. O'Brien, The ancestry of the giant panda, Sci. Am. 11 (1987) 102–107.

[40] M. Kundrát, Pliocene lesser panda world distribution, Abstracts, Euro-American Mammal Congress: Challenges in Holarctic Mammalogy, Santiago de Compostela, Spain, 1998, pp. 354–355.

[41] I. Sasagawa, K. Takahashi, T. Sakumoto, H. Nagamori, H. Zabe, I. Kobazashi, Discovery of the extinct red panda *Parailurus* (Mammalia, Carnivora) in Japan, J. Vert. Paleontol. 23 (2003) 895–900.

[42] M.V. Sotnikova, A new species of lesser panda *Parailurus* (Mammalia, Carnivora) from the Pliocene of Transbaikalia (Russia) and some aspects of Ailurinae phylogeny, Paleontol. J. 42 (2008) 92–102.

[43] R.H. Tedford, E.P. Gustafson, First North American record of the extinct panda *Parailurus*, Nature 265 (1977) 621–623.

[44] J.A. Baskin, in: C.M Janis, K.M Scott, L.L Jacobs (Eds.), Evolution of Tertiary Mammals of North America, Volume 1: Terrestrial Carnivores, Ungulates, and Ungulatelike Mammals, Cambridge University Press, Cambridge, UK, 1998, pp. 144–151.

[45] L. Ginsburg, J. Morales, D. Soria, E. Herráez, Découverte d'une forme ancestrale du Petit Panda dans le Miocène moyen de Madrid (Espagne), Comptes Rend. Acad. Sci. Paris 325 (1997) 447–451.

[46] L. Ginsburg, O. Maridet, P. Mein, Un Ailurinae (Mammalia, Carnivora, Ailuridae) dans le Miocène moyen de Four (Isère, France), Geodiversitas 23 (2001) 81–85.

[47] M.J. Salesa, M. Antón, S. Peigné, J. Morales, Evidence of a false thumb in a fossil carnivore clarifies the evolution of pandas, Proc. Natl. Acad. Sci. USA 103 (2006) 379–382.

[48] P. Crusafont, B. Kurtén, Bears and bear-dogs from the Vallesian of the Vallés-Penedés Basin, Spain. Acta Zool Fen. 144 (1976) 1–29.

[49] S. Peigné, M.J. Salesa, M. Antón, J. Morales, Ailurid carnivoran mammal *Simocyon* from the late Miocene of Spain and the systematics of the genus, Acta Palaeontol. Pol. 50 (2005) 219–238.
[50] A. Wagner, Geschichte der Urwelt, mit besonderer Berücksichtingung der Menschenrassen und des mosaischen Schöpfungsberichtes, Pt. 2. Leipzig, Germany, 1858.
[51] M.J. Salesa, M. Antón, S. Peigné, J. Morales, Functional anatomy and biomechanics of the postcranial skeleton of *Simocyon batalleri* (Viret, 1929) (Carnivora, Ailuridae) from the Late Miocene of Spain, Zool. J. Linnean Soc. London, 152 (2008) 593–621.
[52] O. Kullmer, M. Morlo, J. Sommer, et al., The second specimen of *Simocyon diaphorus* (Kaup, 1832) (Mammalia, Carnivora, Ailuridae) from the type-locality Eppelsheim (Early Late Miocene, Germany), J. Vert. Paleontol. 28 (2008) 928–932.
[53] G. de Beaumont, Essai sur la position taxonomique des genres *Alopecocyon* Viret et *Simocyon* Wagner (Carnivora), Eclog. Geolog. Helvet. 57 (1964) 829–836.
[54] G. de Beaumont, Remarques preliminaries sur le genre *Amphictis* Pomel (Carnivora), Bull. Soc. Vaudoise Sci. Nat. 73 (1976) 171–180.
[55] S.C. Wallace, X. Wang, Two new carnivores from an unusual late Tertiary forest biota in eastern North America, Nature 431 (2004) 556–559.
[56] S.C. Wallace, X. Wang, First mandible and lower dentition of *Pristinailurus bristoli*, with comments of life history and phylogeny, J. Vert. Paleontol. 27 (2007) 162.
[57] S.C. Wallace, First ailurine post crania: a nearly complete skeleton of *Pristinailurus bristoli*, Southeast. Assoc. Vert. Paleontol. Proc. 1 (2008) 27.
[58] H.D. Khalke, On the complex of *Stegodon-Ailuropoda* fauna of southern China and the chronological position of *Gigantopithecus blacki* von Koenigswald, Vert. Palasiat. 5 (2) (1961) 83–108.
[59] D.Z. Chen, G. Qi, Human remains and mammals accompanied in Xi Chuo, Yunnan, Vert. Palasiat. 16 (1978) 35–46.
[60] E.T. Newton, On some new mammals from the Red and Norwich Crags, Q. J. Geol. Soc. London 46 (1890) 444–453.
[61] M. Schlosser, *Parailurus anglicus* und *Ursus Böckhi* aus dem Ligniten von Baróth-Köpecz Comitat Háromszék in Ungarn, Mitteil. Jahrbuch r König. Ungarisch. Geologisch. Anst. 13 (1899) 1–40.
[62] M. Kundrát, New dental remains of an extinct lesser panda – morphotype or new species? J. Vert. Paleontol. 17 (Suppl. 3) (1997) 58A.
[63] M. Morlo, M. Kundrát, The first carnivoran fauna the Ruscinian (Early Pliocene, MN 15) of Germany, Paläontol. Zeits. 72 (2001) 163–187.
[64] T. Kormos, Beiträge zur Kenntnis der Gattung *Parailurus*, Mitteil. Jahrbuch r König. Ungarisch. Geologisch. Anst. 2 (1935) 1–40.
[65] T. Kormos, Macskamedvék a magyar pliocaenben, Állatt. Közlemén. 16 (1917) 137–138.
[66] O. Fejfar, The Lower-Villafranchien vertebrates from Hajnáčka near Fiľakovo (Slowakei, ČSSR), Rozp. Ústřed. Ústavu Geol. 30 (1964) 55–59.
[67] O. Fejfar, M. Sabol, Pliocene carnivores (Carnivora, Mammalia) from Ivanovce and Hajnáčka (Slovakia), Cour. Forschungsinst. Senckenberg 246 (2004) 15–53.
[68] M.J. Head, Pollen and dinoflagellates from the Red Crag at Walton-on-the-Naze, Essex: evidence for a mild climatic phase during the early late Pliocene of eastern England, Geol. Mag. 135 (1998) 803–817.
[69] E.H. Lindsay, N.D. Opdyke, O. Fejfar, Correlation of selected Late Cenozoic European mammal fauna with the magnetic polarity time scale, Palaeogeog. Palaeoclimatol. Palaeoecol. 133 (1997) 206–226.
[70] O. Fejfar, Die plio-pleistozänen Wirbeltierfaunen von Hajnáčka und Ivanovce (Slowakei) ČSR, Neu. Jahrbuch Geol. Paläontol. Abhandl. 3 (1961) 257–273.
[71] M. Kundrát, The first record of the extinct lesser panda *Parailurus* on Eastern Slovakia, Nat. Carpat. 37 (1996) 211–214.
[72] M. Kundrát, A morphological resolution to the enigma of lesser panda phylogeny, J. Morphol. 232 (3) (1997) 282.
[73] M. Kundrát, The first *Parailurus* skull from Slovakia: a new light on the late phase of lesser panda evolution, Abstracts, Euro-American Mammal Congress: Challenges in Holarctic Mammalogy, Santiago de Compostela, Spain, 1998, 345.

[74] M. Sabol, P. Holec, J. Vagner, Late Pliocene Carnivores from Včeláre 2 (Southeastern Slovakia), Paleontol. J. 45 (2008) 531–543.
[75] O. Fejfar, W.D. Heinrich, Biostratigraphic subdivision of the European Late Cenozoic based on muroid rodents (Mammalia), Mem. Soc. Geol. Ital. 31 (1986) 185–190.
[76] I. Horáček, Survey of the fossil vertebrate localities Včeláre 1-7, Časop. Mineral. Geol. Praha 4 (1985) 353–366.
[77] M. Hodrová, Amphibian of Pliocene and Pleistocene Včeláre localities (Slovakia), Časop. Mineral. Geol. Praha 2 (1983) 145–161.
[78] P. Holec, Finds of *Mastodon* (Proboscidea, Mammalia) relics in Neogene and Quaternary sediments of Slovakia (ČSSR), Západ. Karpat. Sér. Paleontol. 10 (1985) 13–53.
[79] E.T. Newton, The vertebrata of the Pliocene deposits of Britain, Mem. Geol. Surv. UK, 1891, 13–14.
[80] P.S. Balson, The Neogene of East Anglia – a field excursion report, Tert. Res. 11 (1990) 179–189.
[81] C. Radulescu, P.-E. Samson, A. Petculescu, E. Sitiucă, Pliocene large mammals of Romania, Coloq. Paleontol. 1 (2003) 549–558.
[82] H. Tobien, Die oberpliozäne Säugerfauna von Wölfersheim-Wetterau, Zeits. Deutsch. Geol. Gesells. 104 (1953) 191.
[83] M. Kundrát, M. Morlo, The extinct lesser panda and other carnivorans from the Pliocene of Wölfersheim, Germany, J. Vert. Paleontol. 19 (Suppl. 3) (1999) 57–58A.
[84] A. Berzi, J. Michaux, J.H. Hutchison and E. Lindsay. The Arondelli local fauna, an assemblage of small vertebrates from the Villafranchian stage near Villafrance d'Asti, Italy, Committee Mediterranean Neogene Stratigraphy Proceedings, IV Session, Bologna, 1967, Giornae di Geologia (2) XXXV, fasc.1, 1967, pp. 1–4.
[85] J.A. White, N.H. Morgan, The Leporidae (Mammalia, Lagomorpha) from the Blancan (Pliocene) Taunton local fauna of Washington, J. Vert. Paleontol. 15 (1995) 366–374.
[86] R.S. White Jr., G.S. Morgan, in: R.D. McCord (Ed.), Vertebrate Paleontology of Arizona, Mesa Southwest Museum Bull., 11, 2005, 117–133 + Appendix 1 and 2.
[87] I. Kobayashi, in: R. Tuschi (Ed.), Pacific Neogene Events, University of Tokyo Press, Tokyo, Japan, 1990, pp. 77–83.
[88] N.P. Kalmykov, V.V. Shabunova, Evolution of Life on the Earth, Tomskij Gosudarstvennyj Universitet, Tomsk, Russia (2005) 293–295.
[89] E.N. Maschenko, The northernmost primate in Asia, Priroda 11 (1994) 64–70.
[90] M.V. Sotnikova, N.P. Kalmykov, Paleogeography and Biostratigraphy of the Pliocene and Anthropogene, Geological Institute, Russian Academy of Sciences, Moscow, Russia, 1991, pp. 146–159.

CHAPTER

6

Red Panda Anatomy

Rebecca E. Fisher

Department of Basic Medical Sciences, University of Arizona, College of Medicine-Phoenix, Phoenix, Arizona; School of Life Sciences, Arizona State University, Tempe, Arizona, USA

OUTLINE

HISTORY OF ANATOMICAL STUDY

The red panda was first introduced to western science by Major-General Hardwicke, in a paper read to the Linnean Society in 1821; however, 5 years passed before this account was published [1]. In the interim, F.G. Cuvier presented the first description of the red panda, naming it *Ailurus fulgens* [2]. Most of the early accounts [1–3] focused on the external appearance of *Ailurus*, but descriptions of the cranium and dentition soon followed [4–6]. In fact, in 1902, Thomas proposed a new subspecies, *Ailurus fulgens styani*, based on a number of cranial features, as well as minor differences in coat colour (e.g., white face mask less pronounced) [7]. The cranial features distinguishing the new subspecies included its larger overall size, an inflated frontal bone, due to more extensive frontal sinuses, and prominent zygomatic arches and coronoid processes on the mandible [7]. Today, two subspecies are generally recognized, including *Ailurus fulgens fulgens*, which has a broader range, and *Ailurus fulgens styani*, which is restricted to China's Sichuan and Yunnan provinces, northern Burma, and Tibet.

Red Panda. DOI: 10.1016/B978-1-4377-7813-7.00006-9

While many authors focused on craniodental features, Lankester [6] and Bardenfleth [8] also provided descriptions of the post-cranial skeleton, while Mivart [4], Flower [9] and Hodgson [10,11] described aspects of the viscera. Flower's account, based on the dissection of a young adult male, was by far the most comprehensive, and still represents the primary source of information on red panda visceral anatomy. Additional studies emerged in the 1920s. Based on the analysis of several males and females, Pocock documented the morphology of the ears, rhinarium, facial vibrissae, plantar pads, and external genitalia [12]. Four years later, Carlsson published a survey of red panda anatomy, including notes on many organ systems [13]. In Davis' seminal volume on the anatomy of the giant panda, he cites Carlsson [13] and Flower [9] as authorities on red panda anatomy. However, Davis also refers to his own "partial dissections" of *Ailurus* [14]. Recently, the musculoskeletal system of the red panda has been investigated by Fisher [15,16] (Figure 6.1), while a number of authors have focused specifically on the radial sesamoid and its associated muscles [17–19]. These recent investigations have largely been prompted by a desire to interpret

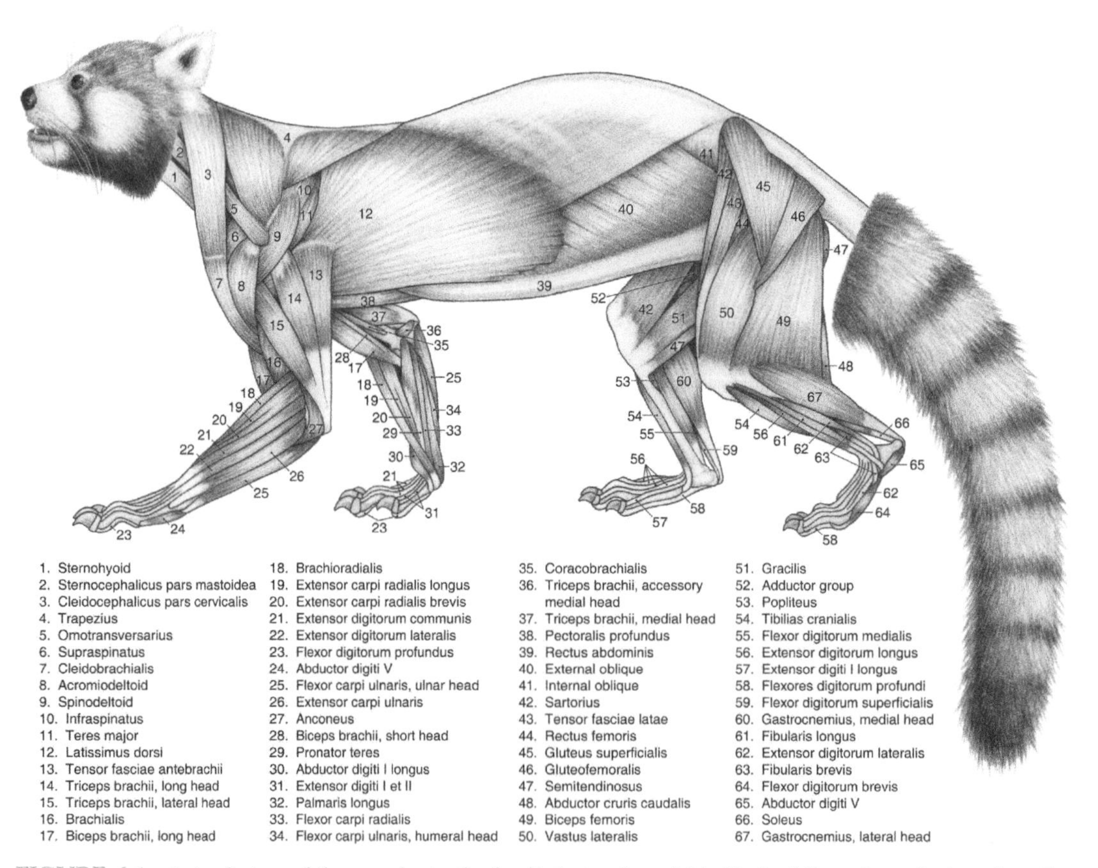

FIGURE 6.1 Lateral view of the muscles in the forelimb, trunk, and hindlimb of the red panda, based on the dissection of four specimens by Fisher and colleagues [15,16].

newly discovered fossil ailurids [20–23]. What follows is a comprehensive review of what is currently known about red panda anatomy.

EXTERNAL ANATOMY

The coat pattern of the red panda is striking. Its white face is marked by red patches coursing from the lateral angle of the eyes to the corners of the mouth (see Figure 6.1). As noted above, *Ailurus fulgens styani* is characterized by a more restricted white face mask, compared to *Ailurus fulgens fulgens* [7]. The facial vibrissae of the red panda, located in the buccal, mandibular, and submental regions, are moderate in length [12,13]. The erect, triangular-shaped ears are covered with white hair ventrally, apart from a red patch in the centre (see Figure 6.1). In marked contrast, the dorsal aspects of the ears and torso are covered with brilliant red or orange-brown hair, while the ventral aspects of the torso and limbs are black. Of course, one of the most distinctive features of the red panda is its striped, bushy tail (see Figure 6.1). The length of the tail reflects its use as an organ of balance during arboreal locomotion.

Interestingly, the soles of the red panda's paws are also covered with hair. This feature likely represents an adaptation for walking on snow and other cold substrates, as in the Tibetan sand fox and polar bears [24]. In contrast to the rest of the limb, the hair on the paws is white. These hairs largely conceal the plantar pads, which are extremely reduced in size. In fact, the metatarsal pads are absent altogether in the red panda [12]. The soles of the feet are also the site of small glands that secrete a colourless and odourless fluid of uncertain function [25]. Finally, the paws are equipped with sharp, curved, semiretractable claws (see Figures 6.1 and 6.2). Red pandas utilize these claws extensively while climbing.

Scent-marking anal glands are present in most carnivores, and red pandas possess a pair that empty into the distal aspect of the rectum [9,24]. The fluid secreted by these glands is quite pungent. Pocock noted that the glands open into a slight invagination of skin, representing a rudimentary anal pouch [12]. However, this pouch is by no means as extensively developed as those found in herpestids and hyaenids [12,24].

In terms of the overall body proportions, the head of the red panda is small, lacking a pronounced rostrum, and the limbs are roughly equal in length (see Figures 6.1 and 6.2). Data from both wild-caught and captive red pandas indicate that the length of the head and body ranges from 510 to 635 millimetres, while the tail ranges from 280 to 485 millimetres [25–27]. Red pandas have seven cervical, 14 thoracic, six lumbar, three sacral and up to 19 coccygeal vertebrae (see Figure 6.2) [14,25]. Adult weights have been reported for captive red pandas, and vary from 3.7 to 6.2 kilograms in males and 4.2 to 6 kilograms in females, indicating there is no sexual dimorphism in size [25,28].

SKULL AND DENTITION

The red panda possesses 36 to 38 teeth and has the following dental formula: i 3/3, c 1/1, p 3/3-4, m 2/2 (Figure 6.3). The first upper premolar is absent, while the first lower premolar is absent or vestigial. The incisors and canines are characterized by low crown heights, while the premolars and molars are distinguished by a number of accessory cusps, creating

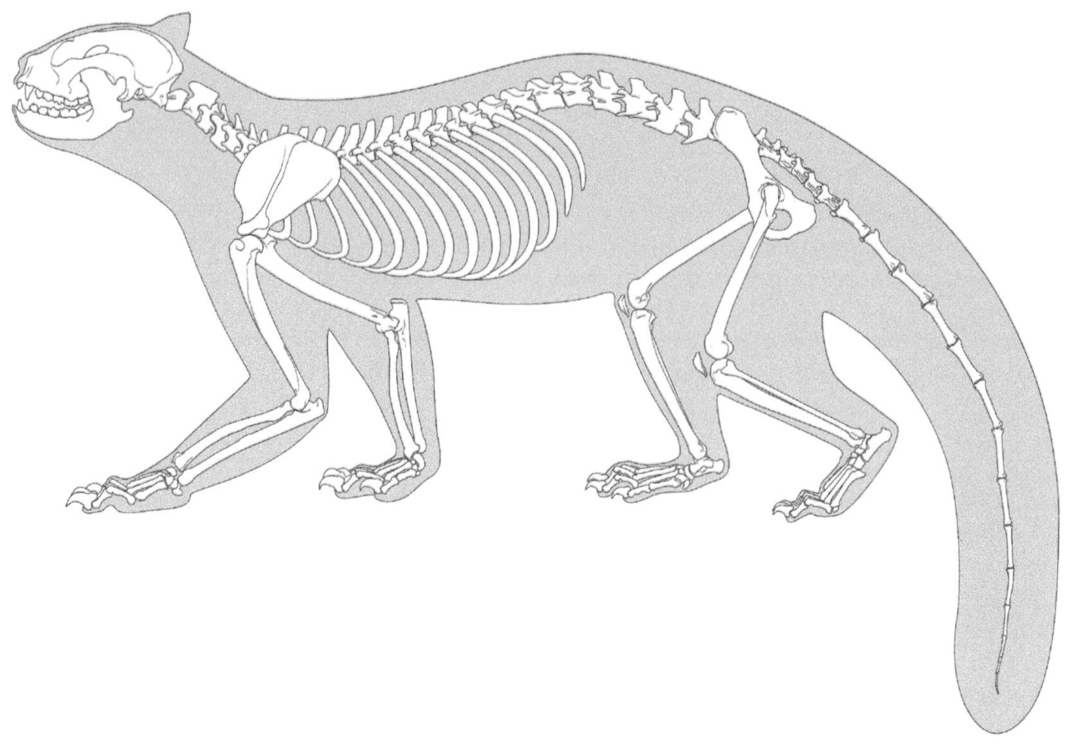

FIGURE 6.2 The skeleton of the red panda. Note the postscapular fossa on the caudal border of the scapula, the entepicondylar foramen of the distal humerus, and the moderately developed greater trochanter of the femur.

an elaborate crown pattern that facilitates the effective mastication of bamboo. In contrast to the shearing crests typical of most carnivores, red panda premolars have low, rounded cusps. In particular, the third upper premolar has a well-developed paracone and hypocone, while the fourth upper premolar (the carnassial), is characterized by five cusps. The fourth upper premolar and the upper molars are particularly robust, with their widths exceeding their length (Figure 6.3).

Compared to other carnivores, the red panda's mandible is more robust than expected based on body size; however, the mandible of the giant panda is even more robust, reflecting its dependence on the hardest part of the bamboo plant, the stems [29]. In the red panda, features of note include the large mandibular condyles, the tall mandibular rami, and robust coronoid processes (see Figures 6.2 and 6.3). The last two features reflect the impressive size of the chewing muscles, including the masseter and temporalis, which attach to the rami and coronoid (Figure 6.4). Another interesting trait related to the red panda mandible is its symphysis. Characterized by a fibrocartilaginous pad, relatively low rugosities on the adjoining symphyseal plates, and an area of smooth articulation anterosuperiorly, the symphysis is capable of moderate movement [30]. This type of symphysis most likely represents the primitive condition in carnivores [30].

The red panda's cranium has prominent and widely flared zygomatic arches [31,32]. The wide arches create an enlarged temporal fossa, housing the temporalis muscle (see

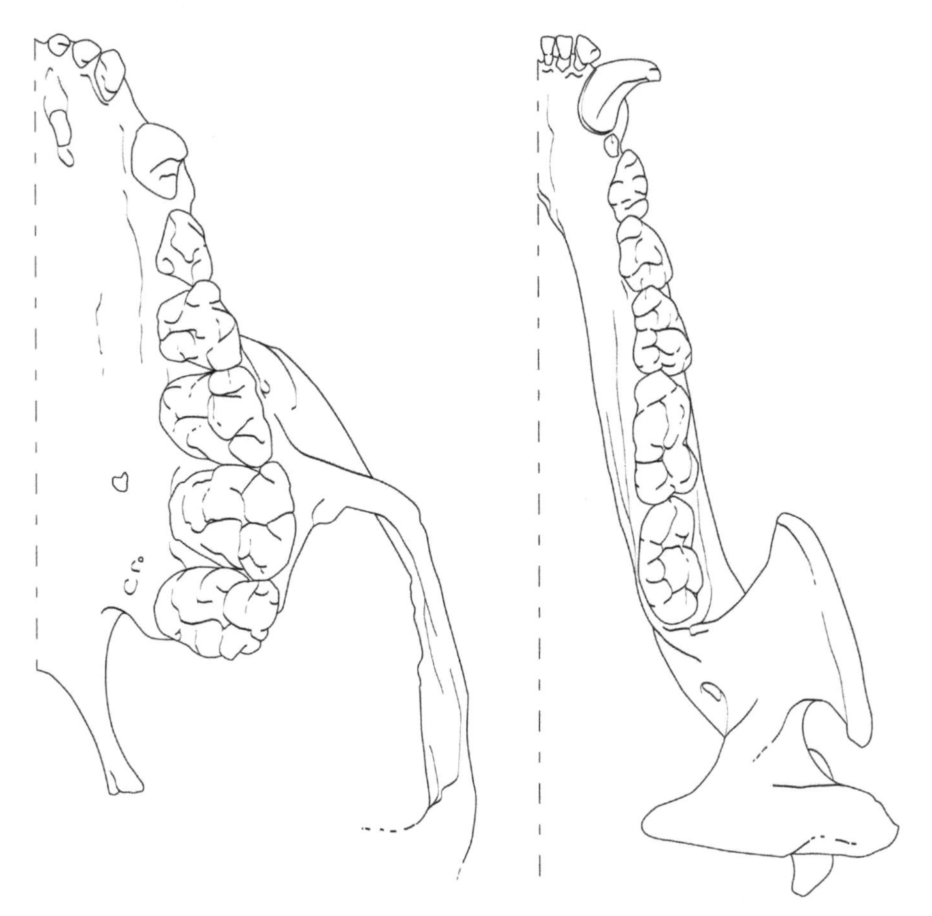

FIGURE 6.3 Occlusal views of the maxillary and mandibular dentition in the red panda.

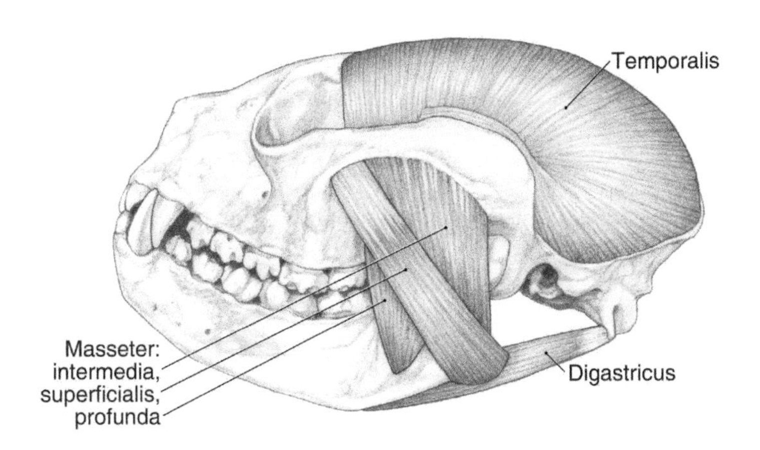

FIGURE 6.4 Lateral view of the skull and chewing muscles in the red panda.

Figure 6.4). In addition, the zygomatic arches provide the site of origin of the masseter muscles (see Figure 6.4). As noted above, both of these chewing muscles are well developed in the red panda. However, it is interesting to note that although the temporalis muscle is pronounced, the sagittal crest is low (see Figures 6.2 and 6.4).

Other distinctive features of the red panda cranium include diminutive postorbital processes, palatine bones that extend past the level of upper molars, a robust and anteriorly recurved postglenoid process, and the presence of an alisphenoid canal [5,6,9,25]. The alisphenoid canal conducts the external carotid artery and is present in *Ailurus*, canids and ursids (except the giant panda), but it is lacking in procyonids and mustelids [9,24,33]. In addition, the auditory bulla of the red panda is largely formed by the tympanic, while the rostral and caudal entotympanics form a minor portion of the medial wall [34]. As a result, the middle ear cavity is small in *Ailurus*. They lack the hypertrophied caudal entotympanic and the invasion of the middle ear cavity into the mastoid as seen in other carnivores. Consequently, red pandas may have less auditory sensitivity [34].

NECK AND THORACIC VISCERA

The larynx of the red panda resembles that of the procyonids, in that the thyroid, cricoid, arytenoid, and corniculate cartilages are reduced, while the cuneiform cartilages are absent [9,14]. The larynx gives rise to a trachea consisting of 38 cartilaginous rings, and this in turn bifurcates to form the main stem bronchi. In the red panda, the right main bronchus is shorter and more robust than the left [9].

The bronchi continue to divide as they enter the lungs. The right lung consists of four divisions in *Ailurus*, including superior, inferior, middle, and azygous lobes. This arrangement is found in all carnivores apart from the pinnipeds [9,13,14,35]. In contrast, the left lung in *Ailurus* is composed of a superior and inferior lobe, separated by a horizontal fissure [9,13,14,35]. This mirrors the condition found in procyonids, mustelids, and ursids, while the rest of the carnivores have three lobes in the left lung [25].

The aortic arch of the red panda was originally described by Flower [9] and his findings were confirmed by Davis [14]. The left subclavian arises independently from the arch, while the common carotids and right subclavian share a common trunk of origin. The right subclavian is the first to branch off so that the carotids briefly share a common trunk. This branching pattern is variably present in procyonids and ursids [14]. The common carotids eventually give rise to the external and internal carotid branches in the neck. In *Ailurus*, the external carotid artery supplies the maxilla, mandible, dura, and the orbit, while the internal carotid and vertebral arteries supply the brain [36]. Interestingly, red pandas are similar to ursids in lacking anastomoses between their extracranial and intracranial circulations [36].

ORAL CAVITY AND ABDOMINAL VISCERA

The salivary glands are large in red pandas, particularly the parotid glands, which open into the oral cavity opposite the third premolars [9,13]. Enlarged salivary glands are common in omnivorous carnivores, and this is especially true in herbivorous mammals [14]. The red panda's tongue is moderate in length, measuring three inches (760 mm) long in the male dissected by Flower [9]. Its surface is marked by a V-shaped row of 11 circumvallate papillae, laterally arranged foliate papillae, and fungiform papillae scattered throughout [9,13,14]. Forming the superior aspect of the oral cavity, the palatal ridges are low and irregular in shape, while posterior to the level of the molars, the palate is smooth and lacks an uvula [9].

Despite the fact that they are specialized herbivores, red pandas have a simple digestive tract and lack the microbial symbionts that facilitate the digestion of plant material [37]. The digestive tract is relatively short in length, measuring 4.2 times the body length, and the stomach consists of a single chamber with a muscular, thick-walled pylorus [9–11,13,25]. The diameter of the gastrointestinal tract gradually diminishes from the duodenum to the rectum, where it enlarges once again. Thus, there is no clear distinction between the ileum and the colon, and the red panda lacks a caecum [9]. In addition, the small intestine has only a few Peyer's patches, and these are confined to the middle of the small intestine [9,13].

The liver consists of three main divisions, the left, middle, and right lobes. However, the visceral surface of the right lobe also gives rise to Spigelian and caudate lobes [9]. The caudate lobe is particularly well developed in *Ailurus*. In addition, the visceral surface of the left lobe features an accessory lobule near the transverse fissure, also present in the giant panda [9,14]. The gall bladder sits in a cleft on the visceral surface of the middle lobe of the liver and is unremarkable in appearance. Situated to the left of the stomach, the spleen is elongate, lacking a notch or fissure [9]. On the posterior abdominal wall, the right kidney sits at a slightly more superior level than the left, and both lack lobules on their surface [9]. Finally, the suprarenal glands are situated at the superior poles of each kidney [9].

PELVIC AND PERINEAL VISCERA

The glands associated with the reproductive system are reduced or absent in the red panda. Similar to ursids, *Ailurus* is characterized by a small prostate gland; in fact, in the giant panda, the gland is absent altogether [14,24]. In addition, the bulbourethral, or Cowper's, glands are absent, as in all caniforms [9,14]. Furthermore, the glands of the ductus deferens and ampullae are absent in red pandas, mirroring the condition found in canids and procyonids [14].

The testes form rounded protuberances in the perineum, but a scrotum is not present [9]. The penis and its associated baculum are relatively small in the red panda [9,10,12]. The baculum measures 23 millimetres, which is one-third the size of the baculum in *Potos*, despite the smaller body size of the kinkajou [12]. The penis of the red panda mirrors the primitive condition in carnivores, also seen in giant pandas, felids, and some viverrids [14]. In these taxa, the corpus is principally composed of cavernous tissue and there is a relatively small baculum. In contrast, in more derived carnivores, the erectile tissue of the penis has largely been replaced by an osseous baculum.

LIMBS

In *Ailurus*, the clavicle is reduced to a tendo clavicularis located between the cleidocephalicus pars cervicis and the cleidobrachialis (see Figure 6.1). Within the forelimb proper, the scapula features a postscapular fossa, the site of origin for part of the subscapularis muscle. This fossa is located on the caudal border of the scapula, adjacent to the infraspinous fossa (see Figure 6.2). The postscapular fossa is particularly well developed in ursids and it is also present, although in a more reduced form, in procyonids [38]. In carnivores, the chief function of the subscapularis muscle is to fix the glenohumeral, or shoulder, joint. Fixing the shoulder joint is particularly important when retracting the forelimb against resistance, as in climbing. Davis argued that the postscapular fossa was especially large in bears, as their large body size puts them at a mechanical disadvantage, requiring a more robust subscapularis to achieve skillful climbing [38].

In the red panda, the humerus is characterized by a prominent entepicondylar foramen (see Figure 6.2). This foramen is absent in ursids and canids [37]. In *Ailurus*, the entepicondylar foramen transmits the median nerve; however, the brachial artery does not travel through this foramen [39]. This arrangement is also found in procyonids; however, in mustelids and feliforms, the artery and the nerve travel through the foramen [39].

A feature that has received a great deal of scrutiny is the enlarged radial sesamoid of the forepaw (Figure 6.5). The radial sesamoid of the giant panda, and to a lesser extent, the red panda, is believed to increase forepaw dexterity [40]. In fact, both pandas utilize this structure when grasping bamboo stems. However, myological differences may be even more important in determining dexterity, especially if the osteology of the manus is conserved throughout evolution [40]. A radial sesamoid is not uncommon in carnivores. A small sesamoid is present on the radial side of the carpus in the caniforms and the felids, and the male kinkajou is known to use its enlarged radial sesamoid to stimulate the female during mating [24]. However, the radial sesamoid is most developed in the pandas.

In giant pandas, the radial sesamoid articulates with the fused scapholunar and varies in its degree of contact with the first metacarpal. It can be drawn into opposition with the first digit and forms a pseudo-thumb. In the red panda, the radial sesamoid is smaller, includes a distal cartilaginous cap, and articulates with the scapholunar only [18,19]. According to Endo and colleagues, the radial sesamoid of the red panda acts as a supportive ridge while grasping bamboo [18]. Antón and colleagues challenged this interpretation, arguing that the sesamoid and its muscular attachments effect supination of the forearm and adduction of the palm, adaptations most likely linked to efficient locomotion along thin branches [19].

Four muscles attach to the radial sesamoid, including opponens digiti I, abductor digiti I brevis, abductor digiti I longus and, to a variable extent, palmaris longus (see Figure 6.5) [16,19]. The bellies of the opponens digiti I and abductor digiti I brevis are fused in the red panda; together, these muscle fibres flex the first digit (see Figure 6.5) [16,19]. The belly of abductor digiti I longus originates in the forearm, and its tendon inserts on the base of the first metacarpal as well as the radial sesamoid (see Figure 6.5) [16,19]. The muscle belly of palmaris longus arises from the medial epicondyle of the humerus, and its tendon may insert on the radial sesamoid as well as the palmar fascia. The attachment of palmaris

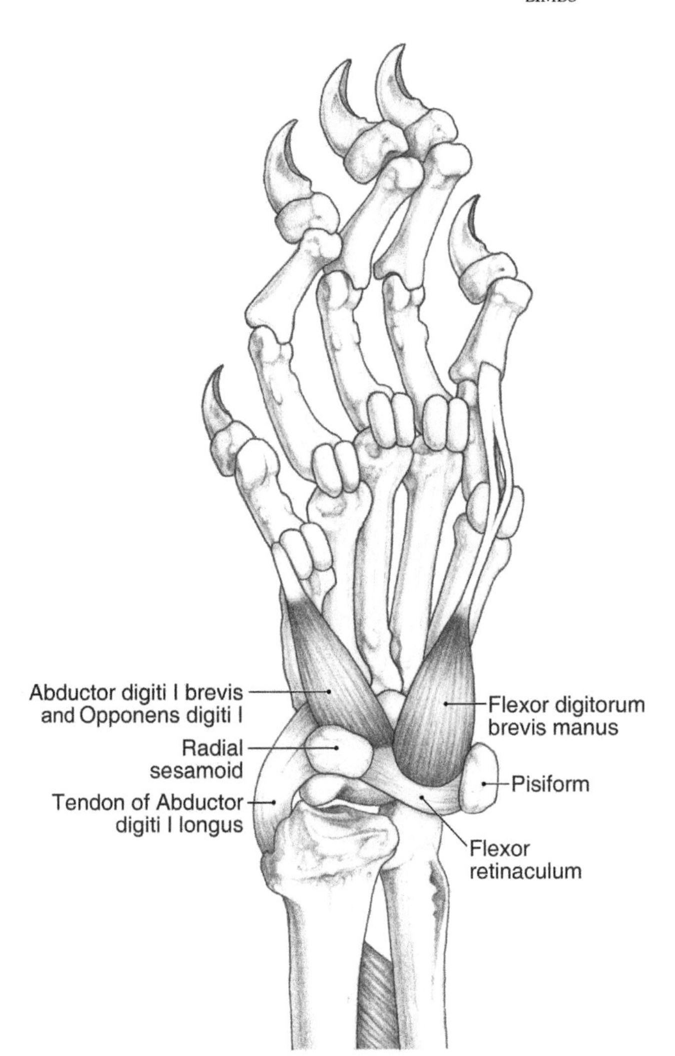

FIGURE 6.5 The radial sesamoid apparatus in the forepaw of the red panda.

longus to the sesamoid varies in *Ailurus* [16]. Antón and colleagues observed this insertion in two red pandas, but Endo and co-authors failed to document this attachment in four specimens [18,19]. Finally, Fisher and colleagues observed a radial sesamoid insertion in three out of eight limbs dissected [16].

Other features of interest in the red panda forelimb include the soft tissue origin of flexor digitorum superficialis, the variable presence of biceps brachii caput breve, and the lack of a rhomboid capitis [16]. In *Ailurus*, the superficial flexor arises in the distal forearm, from the surface of flexor digitorum profundus, rather than from the medial epicondyle of the humerus; this condition is likely to be derived, and is also seen in *Ailuropoda*, mustelids and some procyonids [16]. Another feature that is likely to be derived in *Ailurus*, ursids, and procyonids is the variable presence of a short head of biceps brachii; however, the red panda's lack of a rhomboid capitis is most distinctive, as this condition has only been documented in *Ailurus* and some ursids [16].

The bones of the hindlimb demonstrate fewer specializations compared to those of the forelimb. The femur has a straight shaft, similar to ursids and procyonids, and its greater trochanter, the site of insertion for the gluteal muscles, is moderately developed (see Figure 6.2) [14]. In the leg, the fibula and tibia are joined by synovial joints distally and proximally. This arrangement is primitive for carnivores and allows the fibula to rotate about its long axis, permitting the foot to adjust to uneven substrates, as in climbing [24].

The muscles of the hindlimb are also highly conserved, featuring numerous primitive retentions (see Figure 6.1) [15]. In particular, red pandas retain a number of muscles and tendons related to their weight-bearing hallux, including a robust deep flexor tendon to digit I, the presence of adductor digiti I, and a pair of flexor breves profundi for the hallux. Other retentions include the presence of four lumbricals, as well as gluteofemoralis and soleus. However, red pandas do exhibit a few derived features in the hindlimb. They resemble ursids and canids in lacking a caudal belly of semitendinosus [15]. In addition, their semimembranosus muscle is composed of up to three bellies while the adductor compartment consists of numerous subdivisions; these features are also observed in procyonids [15].

Unfortunately, little is known about the vasculature of the limbs. Davis surveyed the blood vessels of the forelimb and concluded that red pandas resemble procyonids and ursids [39]. These groups share a number of derived features, such as a distal bifurcation of the common median artery, and independent origins of the dorsal and recurrent interosseous arteries [39]. In addition, as noted above, the brachial artery bypasses the entepicondylar foramen in these groups.

ACKNOWLEDGEMENTS

R. Fisher is a Research Collaborator in the Division of Mammals at the National Museum of Natural History, Smithsonian Institution. She would like to thank Jim Mead and Richard Thorington (National Museum of Natural History) for granting permission to work on four red panda specimens. She would also like to thank Linda Gordon, Helen Kafka, and Dave Schmidt (National Museum of Natural History) for facilitating the shipment of these specimens to Arizona. Special thanks to Charley Potter and John Ososky (National Museum of Natural History) for their continued support, and to Kerr Whitfield (University of Arizona College of Medicine-Phoenix in Partnership with Arizona State University) and Autumn Ervin (Arizona State University) for article translation. Original artwork (Figures 6.1–6.5) by Brent Adrian.

References

[1] M.-G. Hardwicke, Description of a new genus of the class mammalia, from the Himalaya Chain of Hills between Nepaul [sic] and the Snowy Mountains, Transact. Linnean Soc. London 15 (1827) 161–165.

[2] Cuvier, F. (1825). Histoire naturelle des Mammiferes, avec des figures originales, colorees, desinees d'ares des animaux vivants. Paris.

[3] A.D. Bartlett, Remarks on the habits of the panda (*Aelurus fulgens*) in captivity, Proc. Zool. Soc. London (1870) 769–772.
[4] S.G. Mivart, On the anatomy, classification, and distribution of the Arctoidea, Proc. Zool. Soc. London (1885) 340–404.
[5] W.H. Flower, R. Lydekker, An Introduction to the Study of Mammals Living and Extinct, Arno Press, New York, USA, 1891.
[6] E.R. Lankester, On the affinities of *Aeluropus melanoleucus*, A. Milne-Edwards, Transact. Linnean Soc. London, 2nd Ser. Zool. 8 (1901) 163–171.
[7] O. Thomas, On the pandas of Sze-chuen, Ann. Mag. Nat. Hist. 10 (ser. 7) (1902) 251–252.
[8] K.S. Bardenfleth, On the systematic position of *Aeluropus melanoleucus*, Mindersk Jactus Steenstrup 17 (1913) 1–15.
[9] W.H. Flower, On the anatomy of *Aelurus fulgens*, Proc. Zool. Soc. London (1870) 752–769.
[10] B.H. Hodgson, On the cat-toed subplantigrades of the sub-Himalayas, J. Asiatic Soc. Bengal 16 (2) (1847) 1113–1129.
[11] B.H. Hodgson, Addendum on the anatomy of *Ailurus*, J. Asiatic Soc. Bengal 17 (1848) 573–575.
[12] R.I. Pocock, The external characters and classification of the Procyonidae, Proc. Zool. Soc. London (1921) 389–422.
[13] A. Carlsson, Uber *Ailurus fulgens*, Acta Zool. 6 (1925) 269–305.
[14] D.D. Davis, The Giant Panda: A Morphological Study of Evolutionary Mechanisms, Chicago Natural History Museum, Chicago, USA, 1964.
[15] R.E. Fisher, B. Adrian, C. Elrod, M. Hicks, The phylogeny of the red panda (*Ailurus fulgens*): evidence from the hindlimb, J. Anat. 213 (2008) 607–635.
[16] R.E. Fisher, B. Adrian, M. Barton, J. Holmgren, The phylogeny of the red panda (*Ailurus fulgens*): evidence from the forelimb, J. Anat. 215 (2009) 611–637.
[17] T. Inaba, K.W. Takahashi, An anatomical study of the pseudothumb and skeleton on the red panda (*Ailurus fulgens*), Jap. J. Zoo. Wildlife Med. 1 (2) (1996) 87–92.
[18] H. Endo, M. Sasaki, H. Kogiku, M. Yamamoto, K. Arishima, Radial sesamoid bone as a part of the manipulation system in the lesser panda (*Ailurus fulgens*), Ann. Anat. 183 (2) (2001) 181–184.
[19] M. Antón, M.J. Salesa, J. Pastor, S. Peigne, J. Morales, Implications for the functional anatomy of the hand and forearm of *Ailurus fulgens* (Carnivora, Ailuridae) for the evolution of the 'false-thumb' in pandas, J. Anat. 209 (2006) 757–764.
[20] S. Peigne, M.J. Salesa, M. Antón, J. Morales, Ailurid carnivoran mammal *Simocyon* from the late Miocene of Spain and the systematics of the genus, Acta Palaeontol. 50 (2) (2005) 219–238.
[21] M. Salesa, M. Antón, S. Peigne, J. Morales, Evidence of a false thumb in a fossil carnivore clarifies the evolution of pandas, Proc. Natl. Acad. Sci. USA 103 (2) (2006) 379–382.
[22] M.J. Salesa, M. Antón, S. Peigne, J. Morales, Functional anatomy and biomechanics of the postcranial skeleton of *Simocyon batalleri* (Viret, 1929) (Carnivora, Ailuridae) from the Late Miocene of Spain, Zool. J. Linnean Soc. 152 (2008) 593–621.
[23] S.C. Wallace, X. Wang, Two new carnivores from an unusual late Tertiary forest biota in eastern North America, Nature 431 (2004) 556–559.
[24] R.F. Ewer, The Carnivores, second edn., Cornell University Press, Ithaca, New York, USA, 1998.
[25] M.S. Roberts, J.L. Gittleman, *Ailurus fulgens*, Mammalian Species 222 (1984) 1–8.
[26] R.M. Nowak (Ed.), Walker's Mammals of the World, sixth edn., Johns Hopkins Press, Baltimore, USA, 1999.
[27] R.I. Pocock, The fauna of British India, including Ceylon and Burma: Mammalia, Volume II, Taylor and Francis, London, UK, 1941.
[28] M.S. Roberts, D.S. Kessler, Reproduction in red pandas, *Ailurus fulgens* (Carnivora: Ailuropodidae), J. Zool. London 188 (1989) 235–249.
[29] S. Zhang, R. Pan, M. Li, C. Oxnard, F. Wei, Mandible of the giant panda (*Ailuropoda melanoleuca*) compared with other Chinese carnivores: functional adaptation, Biol. J. Linnean Soc. 92 (2007) 449–456.
[30] R. Scapino, Morphological investigation into functions of the jaw symphysis in carnivorans, J. Morphol. 167 (1981) 339–375.
[31] W.K. Gregory, On the phylogenetic relationships of the giant panda (*Ailuropoda*) to other arctoid carnivora, Am. Museum Novit. 878 (1936) 1–29.

[32] L.B. Radinsky, Evolution of the skull shape in carnivores. I. Representative modern carnivores, Biol. J. Linnean Soc. 15 (1981) 369–388.
[33] J.J. Flynn, N.A. Neff, R.H. Tedford, The Phylogeny and Classification of the Tetrapods. Vol. 2 (Mammals) in: M.J Benton (Ed.), Clarendon Press, Oxford, UK, 1988, pp. 73–115.
[34] R.M. Hunt, The auditory bulla in carnivora: an anatomical basis for reappraisal of carnivore evolution, J. Morphol. 143 (1974) 21–76.
[35] E. Goppert, Kelkopf und trachea. Handbuch vergleichend. Anat, Wirbelti. 3 (1937) 797–866.
[36] J. Bugge, The cephalic arterial system in carnivores, with special reference to the systematic classification, Acta Anat. 101 (1) (1978) 45–61.
[37] Z. Zhang, F. Wei, M. Li, B. Zhang, X. Liu, J. Hu, Microhabitat separation during winter among sympatric giant pandas, red pandas, and tufted deer: the effects of diet, body size, and energy metabolism, Can. J. Zool. 82 (2004) 1451–1458.
[38] D.D. Davis, The shoulder architecture of bears and other carnivores, Fieldana Zool. 31 (34) (1949) 285–305.
[39] D.D. Davis, The arteries of the forearm in carnivores, Zool. 1 Ser. 27 (1941) 137–227.
[40] A.N. Iwaniuk, S.M. Pellis, I.Q. Whishaw, Are long digits correlated with high forepaw dexterity? A comparative test in terrestrial carnivores (Carnivora), Can. J. Zool. 79 (2001) 900–906.

CHAPTER

7

The Taxonomy and Phylogeny of *Ailurus*

Colin Groves

School of Archaeology & Anthropology, Australian National University, Canberra, Australia

OUTLINE

INTRODUCTION

In the first half of the 20th century most authors, as represented by such influential figures as Allen [1], Simpson [2], and Ellerman and Morrison-Scott [3], classified the red panda (*Ailurus*) as a member of the family Procyonidae (raccoons and their relatives, a predominantly Neotropical group). Only Pocock [4,5] separated it into a family of its own, Ailuridae and, for his pains, he was roundly condemned by Allen [1]: "this course has the disadvantage of obscuring the evident affinity between the Asiatic and American genera" [of Procyonidae] (p. 313) – overlooking the fact that, evidently, to Pocock, the affinity had not been especially evident!

As well as whether the red panda is a Procyonid or not, there was another burning question: is it related to the giant panda (*Ailuropoda*)? Most influential authors [2,3] thought

Red Panda. DOI: 10.1016/B978-1-4377-7813-7.00007-0

so, allocating the giant panda to the same subfamily, Ailurinae, of the Procyonidae; a few, however, awarded the giant panda a family of its own: Ailuropodidae [1,5].

How many species and/or subspecies exist of the red panda has also been somewhat controversial. *Ailurus fulgens* was originally described [6] as being from the "East Indies", but the specimen was almost certainly from Nepal where the collector, Duvaucel, is known to have collected mammals [5]. Thomas [7] described a new subspecies *Ailurus fulgens styani* from "Yang-liu-pa, N.W.Sze-chuen" (= Yangliuba, Shaanxi, 32°07′N, 109°05′E), distinguishing it from "*A. fulgens*" (correctly, *A. fulgens fulgens*) of the Himalayas by skull characters – also questionably by characters of the pelage, stating curiously that "I doubt if any of these differences will prove constant, as the Himalayan specimens vary in all of them". Later, the same author [8] raised his Chinese subspecies to species rank as *Ailurus styani*, having in the meantime studied three further specimens, from Li-kiang Range (= Lijiang Range, approx. 27°20′N, 100°E), Yunnan, which conformed to the original characters "especially those of the skull". Allen [1] was less than impressed: "The differences are hardly those separating species" (p. 316). Pocock [5] took issue with Thomas's revised opinion on more substantive grounds: "The characters... vary, however, individually in both races, so that they intergrade and negative Thomas's opinion in 1922 that specific status might be granted to *styani*" (p. 259).

By mid-century, therefore, there were two outstanding issues concerning the taxonomy of the red panda: its relationships to other Carnivora, and its alpha taxonomy (i.e., one species or two). A considerable amount of evidence has accumulated since then on the first question, and it is possible to come to a definitive conclusion. On the second question, the evidence is less abundant, but a tentative answer is still possible. A third line of research of panda evolution, which has emerged since mid-century, the fossil record, is considered in detail by Manual Salesa and Steve Wallace (Chapters 3 and 4).

RELATIONSHIPS TO OTHER CARNIVORA

The turning point in understanding pandas was the anatomical dissection of a giant panda (*Ailuropoda melanoleuca*) [9]. The resulting monograph, although astounding in its thoroughness and detail, is slightly annoying because the author announced at the very beginning that it had become clearer and clearer to him as his work had progressed that the giant panda is a bear, and his text reflects that conclusion to the extent that comparisons with procyonids and the red panda are sporadic at best. He seemed to regard *Ailurus* as a member of the Procyonidae, though he did note that the two pandas share specializations in the forefoot, the male genitalia, and the masticatory system. So upsetting and controversial were his conclusions about the giant panda, however, that subsequent papers on pandas for some years seemed concerned almost entirely with the question of whether he was right or not about *Ailuropoda* – the red panda seeming to be in the main just an afterthought.

The earliest molecular study of panda relationships [10], an immunological study, corroborated Davis's anatomical conclusion that the giant panda is a bear, but placed the red panda as sister to the now enlarged Ursidae, with the raccoon as sister to both.

The first subsequent attempt to elucidate the affinities of the red panda itself [11] carefully considered the dental and cranial similarities to, and differences from, the giant panda, other bears, and the Procyonidae; *Ailurus* was allocated to a separate family, Ailuridae (and this monograph was one of the first, since Pocock [5], to do so). On balance, the author of this study favoured a relationship with the Ursidae, placing the Procyonidae in a separate clade with the Mustelidae. As an interesting footnote: at that time, it was widely believed that the pinnipeds were diphyletic, and the author [11] assigned the Otariidae to the ursid/ailurid clade and the Phocidae to the procyonid/mustelid clade.

A major molecular study [12], using DNA hybridization, electrophoretic distance (more than 50 isozyme loci), immunological distance (serum proteins), and G-banded chromosomes, again confirmed the giant panda as a bear, but reached different conclusions about the red panda, linking it either to raccoons (hybridization, isozymes), or to bears (immunological distance). The study's authors also put a date on the separation of giant panda from (other) bears, 15–20 million years ago (Ma). A similar study on a greater range of species, in a book on Carnivora [13], linked red panda more decisively to raccoons; using an estimated fossil divergence time of about 40 Ma for all caniform families excluding Canidae, they calculated that red panda had separated from procyonids as early as 38 Ma. This contrasted somewhat with morphological assessments in the same book [14] that the red panda belongs in the ursid clade, or at the very least should be placed "incertae sedis" [15].

With one exception [16], all subsequent molecular analyses have used DNA sequencing.

A study using cytochrome *b*, 12S rRNA, and two tRNA sections [17] found that red panda is actually sister to a clade containing both bears (including giant panda) and procyonids, and concluded that it is best to place the red panda in a family of its own.

A subsequent very elaborate piece of research [18] combined 12S and 16S, cytochrome *b*, and morphology. Irrespective of whether transitions and transversions were equally weighted or transversions were given two to four times the weight of transitions, red panda was sister to bears plus pinnipeds; procyonids and mustelids came next, with skunks the sister to them all. The authors of this paper, too, therefore favoured placing the red panda in a family of its own.

The following year, a study using 12S rRNA [19] placed *Ailurus* well within the procyonids using maximum parsimony and maximum likelihood, but outside them using neighbour joining. Further analysis of 592 fibroblast proteins by electrophoresis again found the red panda forming a clade with the raccoon – but this was in this case the only procyonid studied.

An analysis of the whole of the suborder Caniformia using 12s rRNA [20] could not resolve the phylogeny using parsimony, except that mustelids and procyonids formed a clade: pinnipeds, the mustelid-procyonid clade, *Ailurus*, skunks, bears, and dogs formed an unresolved comb. Neighbour joining linked skunks with bears; majority-rule linked skunks with the mustelid-procyonid clade.

A further attempt to trace the evolution of the Caniformia [21] sequenced intron I of the transthyretin gene, and then concatenated it with cytochrome *b* and partial 12S rRNA. The Feliformia were taken as outgroup. The phylogeny was well resolved – (dogs (bears (pinnipeds (red panda (procyonids, mustelids))))) – but the skunks were unfortunately not included in this analysis. When morphology was added to the three molecular sequences, the pinnipeds became basal to the non-dog clade, but the tree reverted to being more compatible with the molecular data when fossil evidence was taken into account. A later study by the same group [22] used the same sequences plus 16S rRNA, this time including the skunks, and substituting a few related species where the "target" species were not available. In this case, maximum parsimony joined *Ailurus* to the skunks, but the bootstrap value was only 54%, and maximum likelihood did not replicate this clade. The same group again, five years later [23], sequenced yet more genes, including the mitochondrial ND2 and nuclear IRBP and TBG, obtaining long sequences, mostly over 1000 bp. A Bayesian tree, with all 1.00 posterior probabilities, produced much the same phylogeny as before, except that the skunks came out as sister to the mustelid plus procyonid clade, with red panda sister to all of them: (pinnipeds (red panda (skunks (procyonids, mustelids)))).

A new approach to chromsome studies [16] attempted to reconstruct the karyotype evolution of carnivores. Red panda appeared related to mustelids by this method.

Sequencing of a microsatellite locus, Mel08 [24], found that skunks were basal to procyonids plus mustelids, and red panda was basal to the entire clade, but with only low bootstrap support (55%) (Table 7.1).

TABLE 7.1 History of phylogenetic placements of red panda (*Ailurus*) from the 1970s on

Date	Sources	Method	Overall phylogeny	Position of *Ailurus*	Notes
1. NOT BASED ON SEQUENCING					
1973	[10]	ID	U, P	Sister to U	No other groups studied
1982	[11]	Morphology	U(M,P)	Sister to U	pinn diphyletic
1985	[12]	DNA-DNA, electrophoresis, ID, G-banding	U,P	Sister to P	No other groups studied
1989	[13]	DNA-DNA, electrophoresis, ID, G-banding	sk(M,P,U, pinn)	Sister to P	
1989	[14]	Morphology	U(P(M,sk))	Related to U	pinns diphyletic
2002	[16]	Chromosomes	-	Related to M	Few groups studied
2. USING DNA SEQUENCING					
1993	[17]	cyt.*b*, 12S, tRNA	U,P	Basal to both	No other groups studied
1994	[18]	12S, 16S, cyt.*b*, morphology	sk(M,P)(U, pinn)	Sister to U + pinn	P paraphyletic
1995	[19]	12S and electrophoresis	U,P	With P	No other groups studied
1996	[20]	12S	pinn(M,P) (sk) (U)	Part of basal comb	
1998	[21]	TR-I-1, cyt.*b*, 12S, morphology	U(pinn (M,P))	Basal to M,P	sk not tested
2000	[22]	TR-I-1, cyt.*b*, 12S, 16S	pinn(M,P) (sk)	Weakly with sk	U not tested
2005	[23]	TR-I-1, cyt.*b*, 12S, 16S, ND2, IRBP, TBG	U(pinn(sk (M,P)))	Sister to sk + M,P	
2005	[24]	Mel08	pinn(sk (M,P))	Sister to sk + M,P	

U = Ursidae, M = Mustelidae, P = Procyonidae, pinn = pinnipeds, sk = skunks.

The history of phylogenetic analysis of the red panda is one that is fairly familiar in mammalian evolutionary studies. Early morphological studies, once old assumptions were cast aside and cladistic thinking began to predominate, pulled *Ailurus* away from the Procyonidae, where it had been lodged for so long more by tradition than anything else, and indicated a rather isolated phylogenetic position for it. The initial focus on the relationships of the giant panda (*Ailuropoda*) was both a help and a hindrance to the understanding of the relationships of the red panda. Early molecular studies, mainly by immunology and electrophoresis, gave a variety of results, as did early DNA sequencing. Once one other traditional assumption, that the Mustelidae are monophyletic, was tested and found erroneous, the consequent extraction from that family of the skunks (nowadays awarded a separate family, Mephitidae) in turn assisted a better understanding of the relationships of the red panda.

The recent simultaneous analysis of many taxa, using many independent DNA sequences, both mitochondrial and nuclear [23], is convincing; it is consistent with previous studies, given that most of these earlier studies made assumptions (such as mustelid monophyly) or lacked certain key taxa. It is also consistent with another recent analysis [24] using quite a different portion of the genome. In summary: after the Caniformia split off from the Feliformia, the ancestors of the Canidae separated, followed by the Ursidae, followed by the pinnipeds, followed by the Ailuridae, leaving a clade consisting of Mephitidae, Mustelidae, and Procyonidae, from which the Mephitidae split off first [25].

ALPHA TAXONOMY

Red pandas are found in three Chinese provinces (Sichuan, Yunnan, Tibet), in the mountains of northernmost Burma and the northern rim of India as far west as Sikkim, and in Bhutan and Nepal. A survey [26] of all the regions in China where pandas had been reported found that they had disappeared from several areas because of deforestation and hunting for fur; the authors of the survey considered that the Nujiang (Salween) River is the likely boundary between the two subspecies *Ailurus fulgens styani* to the east (Sichuan and almost all of Yunnan) and *A. f. fulgens* to the west (Tibet, and a tiny portion of Yunnan, bordering on Tibet and Burma). Here, I will try to approach the question of whether the differences between the described taxa are real, and if so at what taxonomic level they should be separated and what we can conclude about their respective distributions.

Early 20th century authors were in agreement that the Chinese red panda is distinct from that of the Himalayas, but most did not think that it should rank as a separate species [1,3,5], referring to them respectively as *Ailurus fulgens styani* (Chinese) and *A. f. fulgens* (Himalayan). The most meticulous of these older authors, Pocock [5], compared some Burmese skins with those from the Himalayas and China, and concluded that there were no marked differences from the Chinese subspecies. He listed the differences between the two subspecies as follows:

- the winter coat is longer in the Chinese, up to 70 mm as opposed to 40–50 mm in the Himalayan panda

- Himalayan pandas are deep red in colour, often lightened (at least on the rump and hind part of the back) by ochraceous tips to the hairs, whereas the Chinese/Burmese form has more black in the pelage
- In the Chinese/Burmese panda, the dark tail stripes are more distinct, sometimes quite black, the face is redder, with less white on it
- The skull of the Chinese and Burmese panda is a little bigger, with larger, more robust teeth; this is functionally associated with stouter, more widely spreading zygomatic arches, as indeed had been noted by the original describer of *styani* [7]
- Finally, the frontals of Chinese/Burmese skulls are definitely inflated with air cells, making the forehead strongly curved and the muzzle slope steeper, contrasting with the Himalayan skulls in which the forehead is not noticeably inflated and the dorsal profile is evenly, gently curved. This means that in Chinese skulls the brain case is more swollen anteriorly, so that from above it appears more parallel-sided instead of evenly contracting forwards like Himalayan skulls [7].

In order to explore the nature of geographic variation in red pandas, I collected measurements of skulls, including some from the literature and compared photographs of specimens of known origin, both museum skins and living animals. In what follows, and in order not to pre-empt the status of the Chinese and Himalayan taxa, I will refer to them as simply *fulgens* and *styani* respectively, without indicating whether they are species, or subspecies of a single species. "Chinese" in this instance refers only to Sichuan and Yunnan, keeping "Tibet" as a separate category; and generally I refer to Burmese pandas as *styani*, giving them special consideration where necessary.

I have also been fortunate to see photos of pandas of known origin from zoos in Greenville, Knoxville, Potter Park, Binder Park, and Cincinnati, taken by Lucy Dueck. These were formerly published on a website, "Conservation Genetics of Red Pandas", but this website is unfortunately no longer available.

1. Cranial Comparisons

I took measurements of skulls of *Ailurus* in the Natural History Museum (London), and found an additional specimen in the Zoologisches Staatssammlung (Munich). Many, but not all, of the NHM scales were also measured by Pocock [5]; some of the specimens had come into the collection subsequent to Pocock's study, whereas others had simply been omitted from his list. Measurements of skulls from Chinese territory have been published by Zhang [26], but I did not include these in the analysis for two reasons:

- First, because of slight uncertainty that the measurements were comparable. This is probably not in fact a major consideration, given that my own measurements in the NHM are very close to those published by Pocock [5] on the same specimens, and there is presumably a limited number of ways of taking comparable measurements.
- Secondly, and more important, because of the lack of age control. As will be seen (below), only skulls in which the basilar suture is fused can be considered fully grown.

The skulls in question [27] lack information on the fusion of this suture, with one exception, number T070, which was illustrated in a different publication [28] and does seem to have the basilar suture fused.

Nonetheless, some of these skulls are of great interest, being from Tibet, and I will refer to them where necessary.

The measurements were as follows:

1. Greatest skull length
2. Condylobasal length
3. Bizygomatic width
4. Interorbital width
5. Postorbital width
6. Maxillary breadth (at canine alveoli)
7. Maxillary toothrow length (including canine)
8. Mandibular length (greatest length, parallel to alveolar margin)
9. Basion to bregma height
10. Anterior skull height, from highest point on frontal above one orbit (usually the left) to surface of palate perpendicularly below. This measurement was specially designed, after visual inspection of the material, to test the describer's [7,8] contention that *styani* can always be distinguished by the extremely convex forehead.

I entered them into an SPSS data file, and constructed several univariate and bivariate diagrams. In order to obtain a reasonable sample size for putative *A. f. styani*, I combined the Burmese sample with those from China (Sichuan, Yunnan). There being only two samples (Chinese and Himalayan, corresponding to the two described taxa, *A. f. fulgens* and *A. f. styani*), I did not run any multivariate analyses.

No differences between the sexes could be found.

Figure 7.1 gives the boxplot for greatest skull length. It is to be noted, first of all, that young-adult skulls ("ya") are much smaller than those which are fully adult (except for the existence of a very small but fully adult *fulgens* skull); this is heuristically important, because all too frequently in mammalogy it is the eruption of the last molars which is taken to mark maturity, but in mammals such as red pandas this is clearly not the case (Figure 7.1). Age for age, *styani* are very much larger than *fulgens*, with no overlaps. It is noteworthy that Tibetan skull T070 [27,28] is very small, falling in the range of *fulgens*, whereas the other Tibetan skulls [27] are much larger, within the range of *styani*. Notably, T070 is from the Zhangmu district, immediately to the north of Nepal, whereas the other skulls may be from south-eastern Tibet.

There is also a big size difference in zygomatic breadth (Figure 7.2), and again there is an unusually small adult *fulgens* (a different specimen from Figure 7.1).

Figure 7.3 shows maxillary toothrow length and, in this case, there is not much difference between the two named taxa or between young-adult and fully adult. A specimen of *styani* from Burma has an unusually short toothrow. The other measurements listed above likewise show only average differences between the two taxa.

Figure 7.4 is a bivariate plot of measurements 9 (babr) and 10 (palfron), and Figure 7.5 plots palfron against greatest skull length. Thomas [7,8] was correct: there is

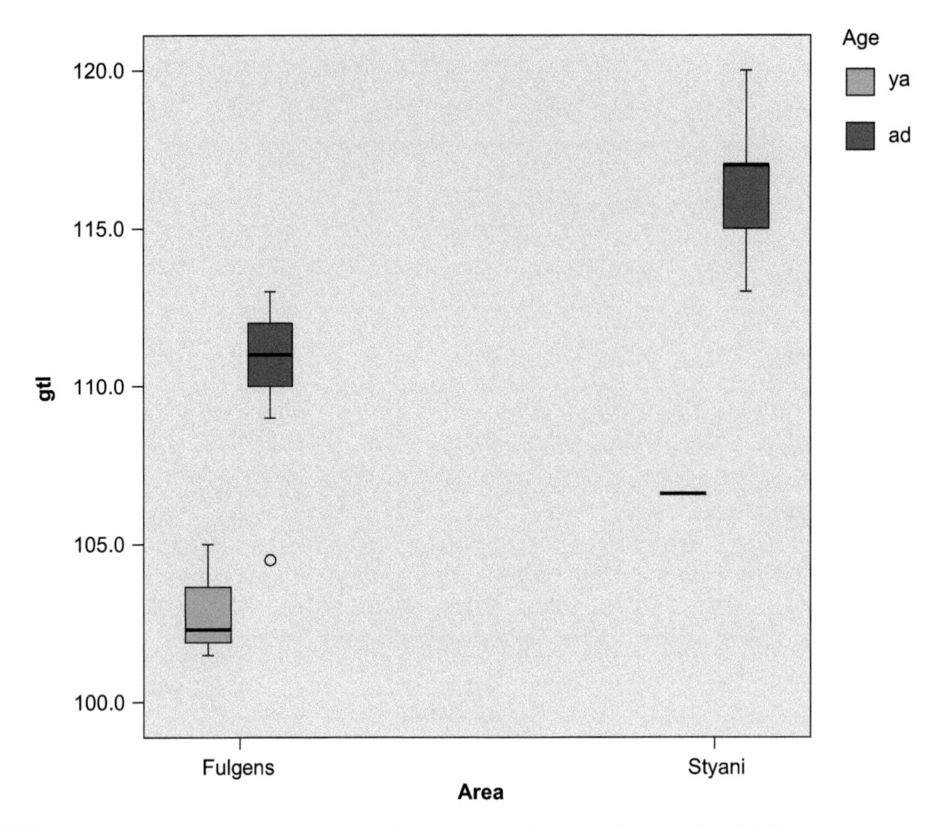

FIGURE 7.1 Medians, quartiles and ranges of Greatest Skull Length for samples of *Ailurus*.

a large and consistent difference between the two taxa. Visually, the frontals medial to the superior orbital margins are swollen and convex, only the midline being slightly concave. I did not take x-rays, but I would expect to find inflated frontal sinuses. It is especially striking that, though both young adults and full adults are depicted in these two figures, there is still no overlap: the swollen lateral frontals have developed well before subadulthood. The striking difference between the two taxa is shown in Figure 7.6, in which two photographs, chosen at random, represent *fulgens* and *styani* respectively.

To sum up: there are striking and apparently consistent differences between Sichuan/Yunnan/Burma (*styani*) and Himalayan (*fulgens*) skulls. As earlier authors [5,7,8] agreed, there is a consistent visual difference (the steep, inflated frontals of *styani*), and – when age is taken into account – *styani* is consistently larger. Table 7.2 illustrates this in another way, giving means, standard deviations, and observed ranges for adult specimens of the two taxa.

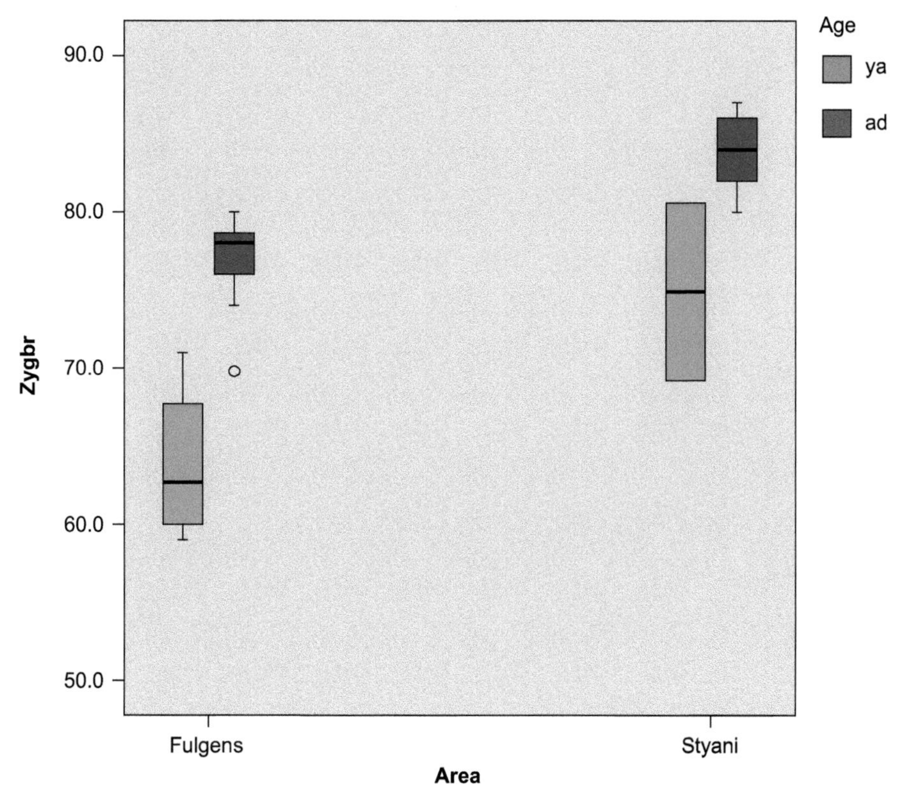

FIGURE 7.2 Medians, quartiles and ranges of Zygomatic Width for samples of *Ailurus*.

It is possible that skulls from Southeastern Tibet may be different again, although they seem much more like *styani* (contra some recent suggestions [29,30]); but the skull from Zhangmu, further west in Tibet, is more comparable to *fulgens*, as it is small in size and the figured skull from the region [28] has the relatively flat frontal type.

2. External Features

External measurements are given in the same publications as craniodental measurements, and I have plotted some of these, with all due caution, given the small sample sizes and absence of age control in the case of the Chinese sources. The evidence suggests that head and body length in Chinese pandas is subject to sexual dimorphism, which it is not in Himalayan pandas (Figure 7.7), but extra caution in this case is due because no such difference between the sexes is appreciable in the skull. The tail is longer in *styani* than in

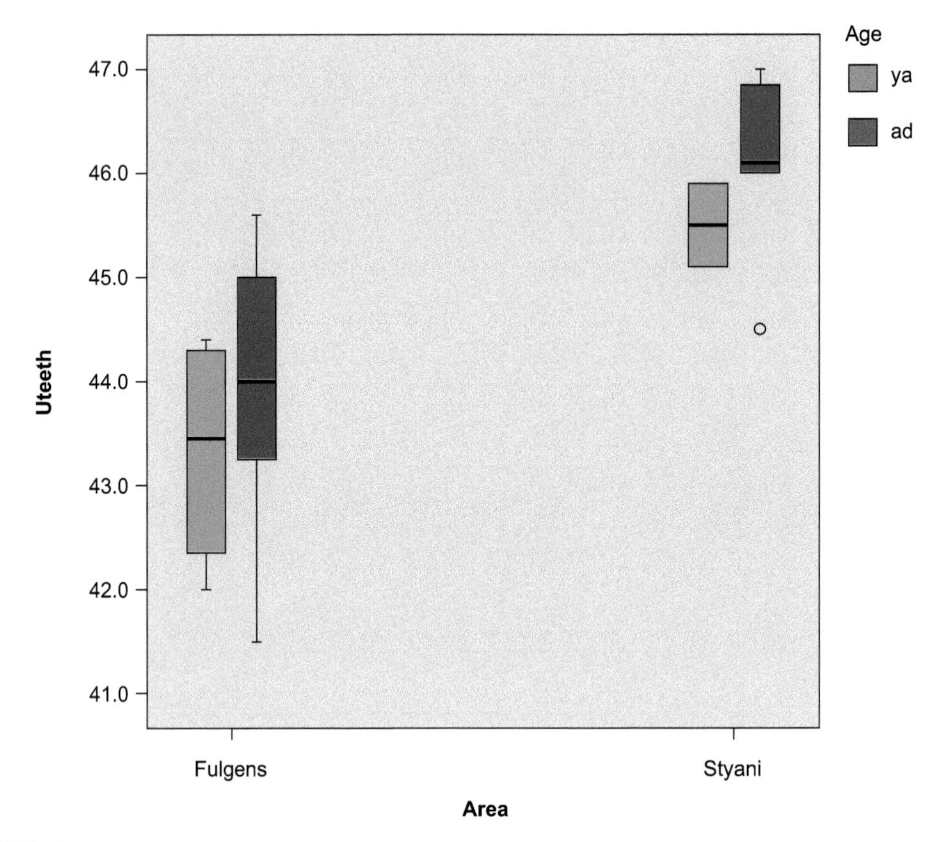

FIGURE 7.3 Medians, quartiles and ranges of Maxillary Toothrow Length (including canine) for samples of *Ailurus*.

fulgens, with very little overlap, although Tibetan specimens are short-tailed like *fulgens* (Figure 7.8).

Dr David Happold kindly photographed the skins in the Natural History Museum, London, on my behalf; there are 15 adequately localized specimens in all, including types (Figures 7.9). I have used these in conjunction with photos from websites depicting pandas of presumably known origin, mainly from captive breeding groups (*fulgens*: Greenville, Knoxville, Potter Park; *styani*: Binder Park, Cincinnati). The following differences emerge:

1. Dorsum: *fulgens* skins are reddish, from very pale to deep red; *styani* are clear red to deep maroon, overlapping slightly with darker *fulgens* specimens, while being on average more brilliant red. As was already noted over 60 years ago [5], skins from Burma fit comfortably within the range of those from China.
2. Black on arms: in *fulgens* the black zone reaches anywhere from just above the elbow to halfway up the shoulder; in *styani* well up onto the shoulder, sometimes to the top. Again, there is an overlap, but it is slight.

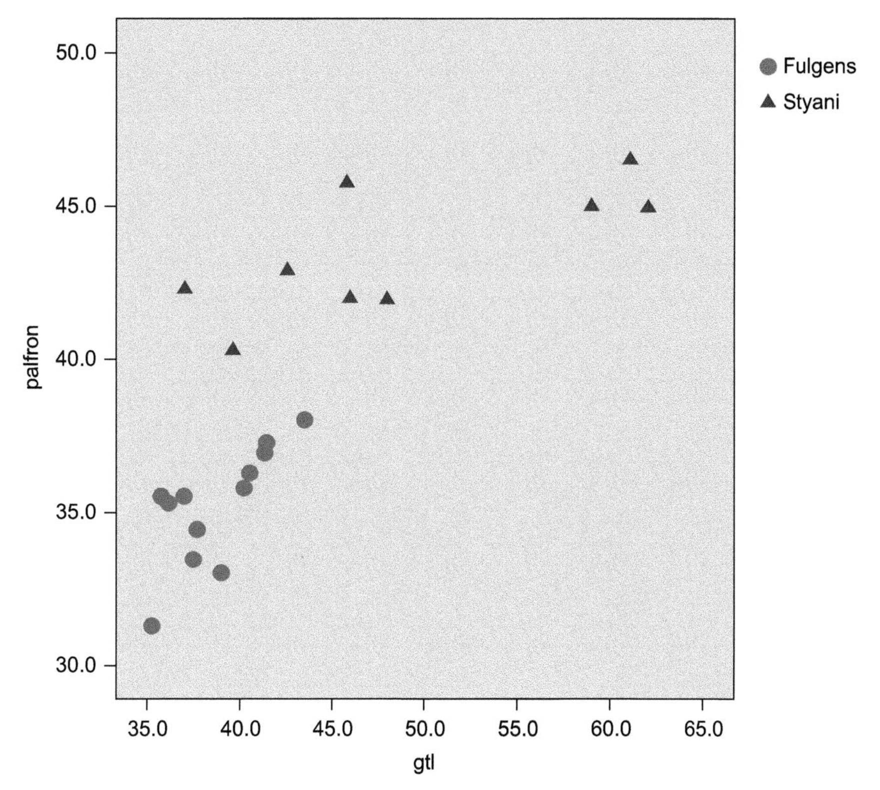

FIGURE 7.4 Anterior cranial height (palfron) plotted against posterior (babr) (all ages except infants).

3. **Black on legs:** *fulgens* – above stifle to halfway up haunch; *styani* – always halfway up haunch.
4. **Tail rings:** in *fulgens* the dark rings are dark reddish and the pale ones a somewhat lighter reddish; in *styani* there is more contrast, the dark rings being dark red and the pale ones gingery to whitish.

   There are, therefore, good differences at least on average; the difference in the tail rings is the best candidate for a consistent difference, and more detailed study on larger series would be necessary to test this. More striking differences can be seen in the facial pattern (Figure 7.10), as was first noted in the original description of *styani* [7].
5. **Forehead:** *fulgens* vary from pale whitish-red through gingery to deep maroon; *styani* from pale gingery or red to dark red. On average, therefore, *styani* seem darker, like the dorsal colour as a whole, but the difference is not consistent.
6. **Upper eye strip** (the short dark strip joining the crown to the supero-external corner of the eye): in *fulgens* this may be absent or distinctly marked but occasionally is clear, thick; in *styani* it is always present, varying from pale to dark, and usually thick. Accordingly, the pattern is less variable in *styani*, but occasionally an individual of *fulgens* has an upper eye strip that is nearly as thick and dark as the average *styani*.

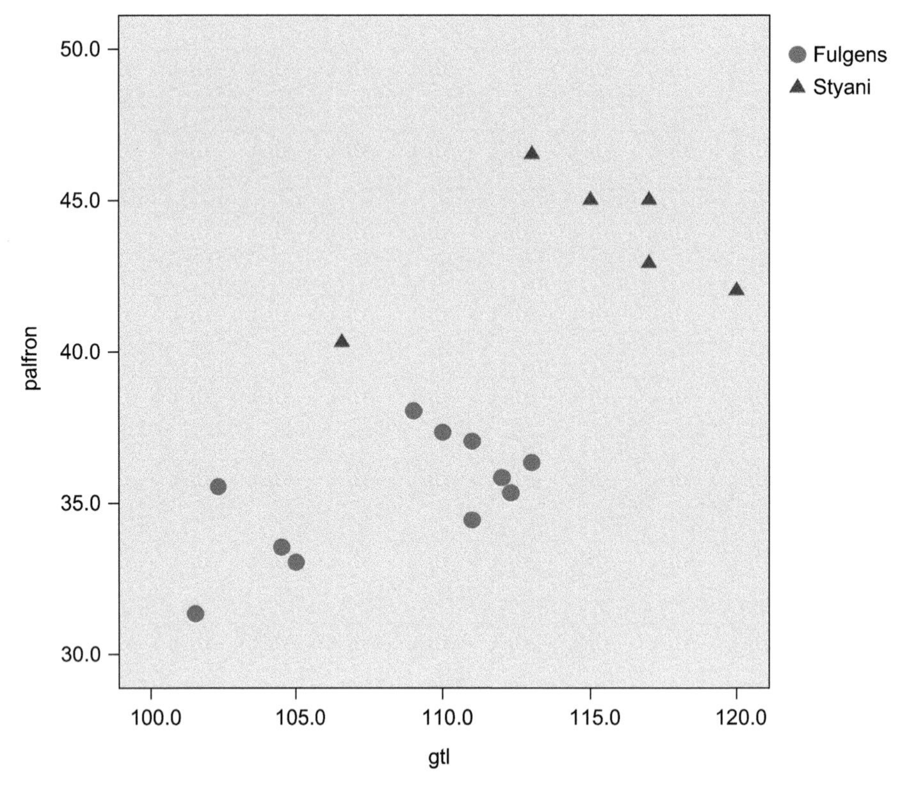

FIGURE 7.5 Anterior cranial height (palfron) plotted against greatest length (all ages except infants).

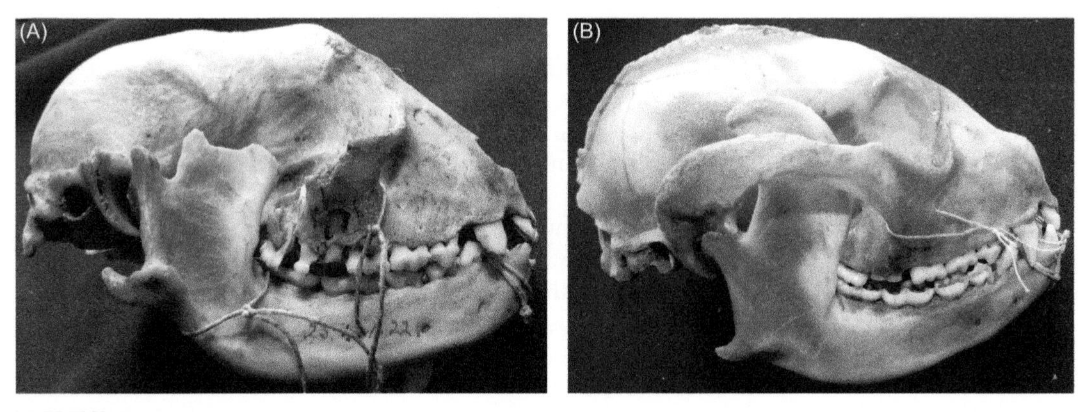

FIGURE 7.6 Lateral views of two skulls, to show strong frontal convexity in *A. styani*. (A) BM 23.1.23.1, from Chumbi valley (*A. fulgens*); (B) BM 23.4.1.22, from Lijiang Range, Yunnan (*A. styani*). Both are adult females.

7. Lower eye strip (the long strip joining the full width of the lower eyelid to the external corner of upper lip, curving around the convexity of the muzzle): in *fulgens* this varies from thin and dark to thick but gingery, often hardly appreciably darker than the crown; in *styani* it is very thick, dark (often nearly black), and

TABLE 7.2 Means and standard deviations for *Ailurus* samples.

Area		gtl	zygbr	maxbr	mandl	babr	palfron
fulgens	Mean	110.311	76.827	24.523	78.730	39.788	35.956
	N	9	11	13	10	8	9
	Std. Deviation	2.5167	2.8845	.9876	1.3483	2.4439	1.4187
	Minimum	104.5	69.8	22.8	77.0	36.2	33.5
	Maximum	113.0	80.0	26.0	81.0	43.5	38.0
styani	Mean	116.400	83.833	26.086	82.800	48.550	44.214
	N	5	6	7	5	6	7
	Std. Deviation	2.6077	2.5626	1.5994	1.4832	9.4686	1.7920
	Minimum	113.0	80.0	24.2	81.0	37.0	42.0
	Maximum	120.0	87.0	29.0	85.0	61.0	46.5

Sources: see text.

always noticeably darker than the crown. This difference is consistent on present evidence.

8. Interorbital strip (a short thick median strip going down from the crown, and touching the inner corner of the eye; together with the upper eye strips, it cuts off a white spot, of varying size, above each eye): in *fulgens* this is usually very pale, continuing the tone of the crown itself; in *styani* it is very thick and dark, especially bilaterally, often making the white spots very small. This difference is consistent on present evidence and, together with the other differences in the facial pattern (characters 5–7), produce the impression of a predominantly white facial mask in *fulgens* and a dark one in *styani*.
9. Ear: this appears to be a consistent difference. The ears in all red pandas are dark-furred externally, white inside, and white and shaggy on the rim; those of *fulgens* are relatively thinly haired inside, so appearing black centrally because of the skin showing through, and the rim hair is relatively short; the ears of *styani* are always thickly haired inside so may appear entirely white, and the rim hair is long with a long downwardly directed tuft at the base of the outer margin.

3. Molecular Genetics

The earliest study of the genetics of different red panda groups [31] analysed 25 loci by electrophoresis in 22 zoo specimens of known origin. Twenty-three of the loci were found to be monomorphic; the G6PDH locus was polymorphic, having three alleles, one unique to each subspecies, the third occurring in both; finally the TF1 locus had a fixed difference – a fast allele (A) in *fulgens*, a slow allele (B) in *styani*. On this basis, a minimum divergence time between the two taxa could be calculated as 160 000 years, and a maximum of 2 million years.

Study of the mitochondrial D-loop in pandas from Sichuan and Yunnan [29] found considerable diversity, with 25 different haplotypes, nine being unique to each provincial sample. There was greater diversity in Sichuan than in Yunnan, and the authors suggested

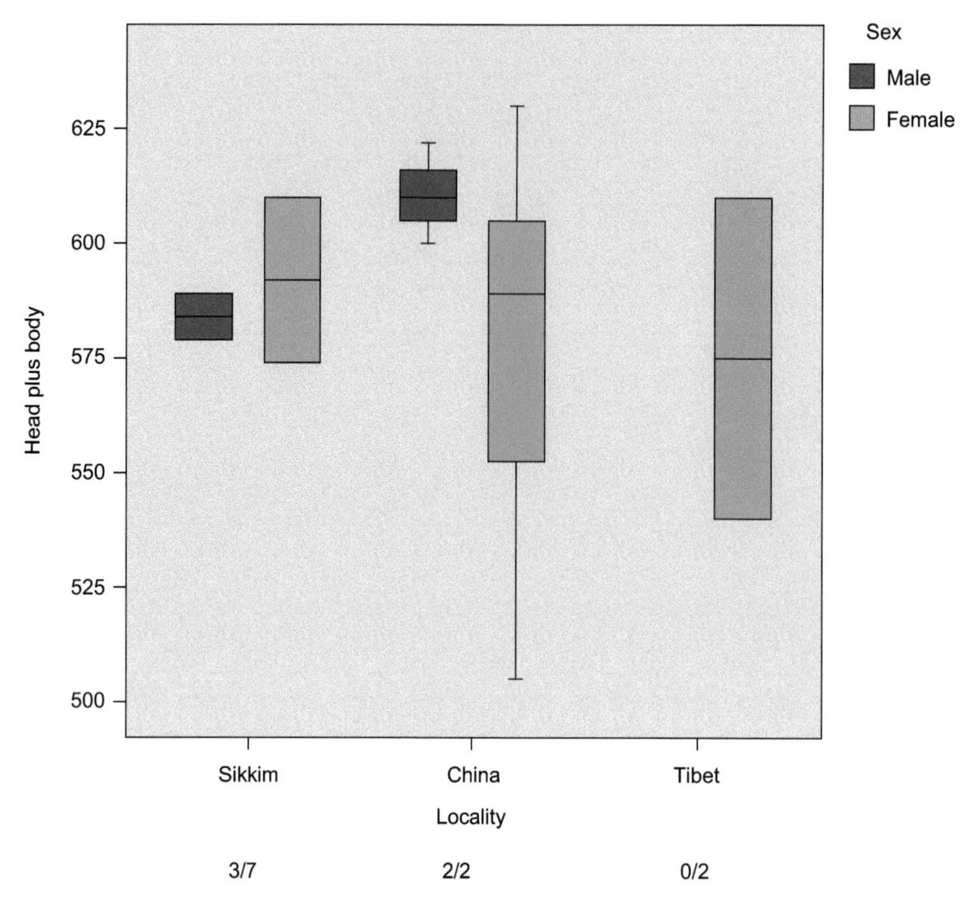

FIGURE 7.7 Medians, quartiles and ranges of Head plus Body length for samples of *Ailurus*.

that the population had expanded in a north–south direction. It is interesting that, of the Yunnan sample, the authors of the study referred five (from Gongshan) to *A. f. fulgens*, the remaining 17 (from Lushui) to *A. f. styani*, although they did not refer to taxonomic divisions subsequently in the paper.

In the same year, a DNA fingerprinting study [32] contrasted samples identified as *fulgens* from "Tibet and northwest of Yunnan" (n = 9) with samples denoted *styani* from Sichuan and elsewhere in Yunnan (n = 18). There was one DNA band unique to each taxon; two captive-born hybrids possessed both bands. In accord with an earlier proposal [26], the Nujiang was regarded as the boundary; most of the Tibetan sample was not of securely known origin, except for two individuals which, interestingly, were from Gaoligongshan (Wei Fuwen, personal communication).

The most detailed of the recent genetic studies [30] sequenced cytochrome *b* and the control region (CR) of mtDNA in pandas from Sichuan (n = 43), Yunnan (35), Tibet (3), and Burma (7), incorporating some data that had been previously reported [29]. There was

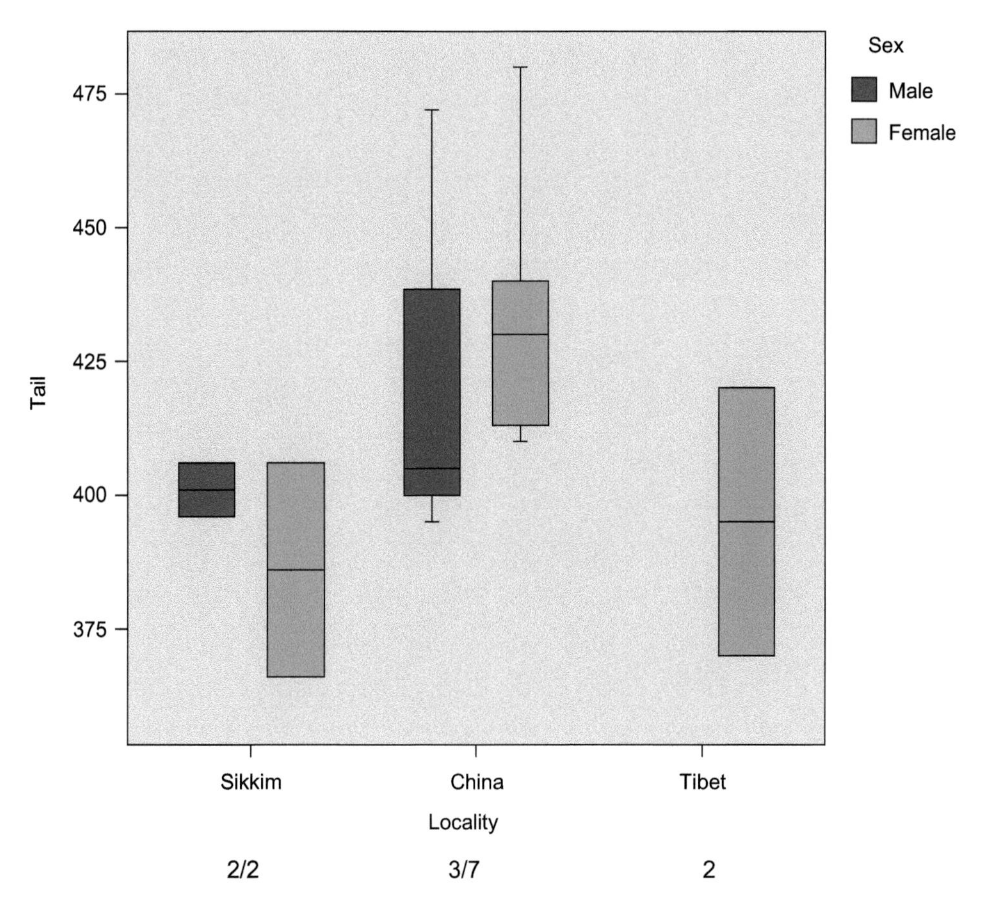

FIGURE 7.8 Medians, quartiles and ranges of Tail length for samples of *Ailurus*.

significant geographic variation, which the authors, following earlier suggestions [26,27,32] divided into *styani* and *fulgens*, the boundary between them being drawn at or around the Nujiang (Salween) River. Classified as *styani* were all populations from Sichuan and those from Yunnan to the east of the Nujiang; classified as *fulgens* were those from Yunnan to the west of the Nujiang (Lushui and Gongshan) and all those from Burma and from Tibet (Changdu region in the east, and Zhangmu just north of the border with Nepal – the same region from which a panda skull, previously reported as *fulgens* [28], had been figured). What we can see from Table 2 of the paper [30] is that cyt *b* haplotype C is widely shared among samples (all but the single specimen from Changdu), but the other haplotypes of the two regions are distributed as follows:

- *styani*: cyt *b* haplotypes A and B are unique to the taxon, the other haplotypes shared with other samples; in CR, most haplotypes are unique to the taxon, but one (haplotype 12) shared with Lushui
- Burma: cyt *b*, haplotype G shared with Lushui and with *styani*, haplotypes H, I, and J unique to the sample; CR, haplotype 11 shared with Lushui, all others unique to the

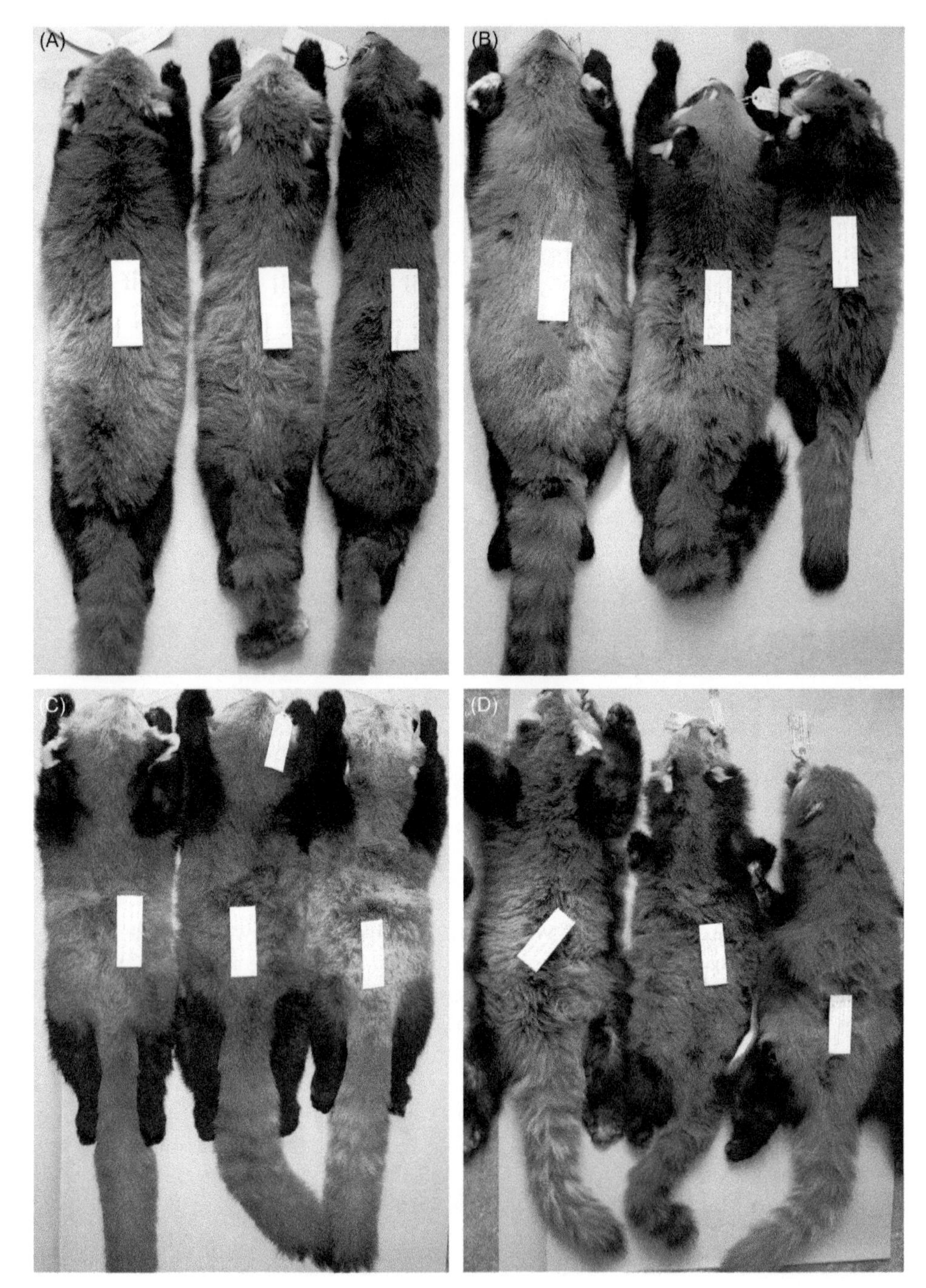

FIGURE 7.9 Skins of *Ailurus*, Natural History Museum, London. (A) *A. fulgens*, Nepal: 58.6.24.79, 39.717, 1990.64; (B) *A. styani*, Lichiang Range: 22.9.1.39, 22.9.1.38, 23.4.1.21; (C) *A. fulgens*, Chuntang, Sikkim: 15.9.1.98, 15.9.1.97, 39.468; (D) *A. styani*, Burma: 39.501 (Chimeli Pass), 50.581 (Taron valley), 34.11.3.1 (240 km N of Myitkyina). *(Photos courtesy of DCD Happold.)*

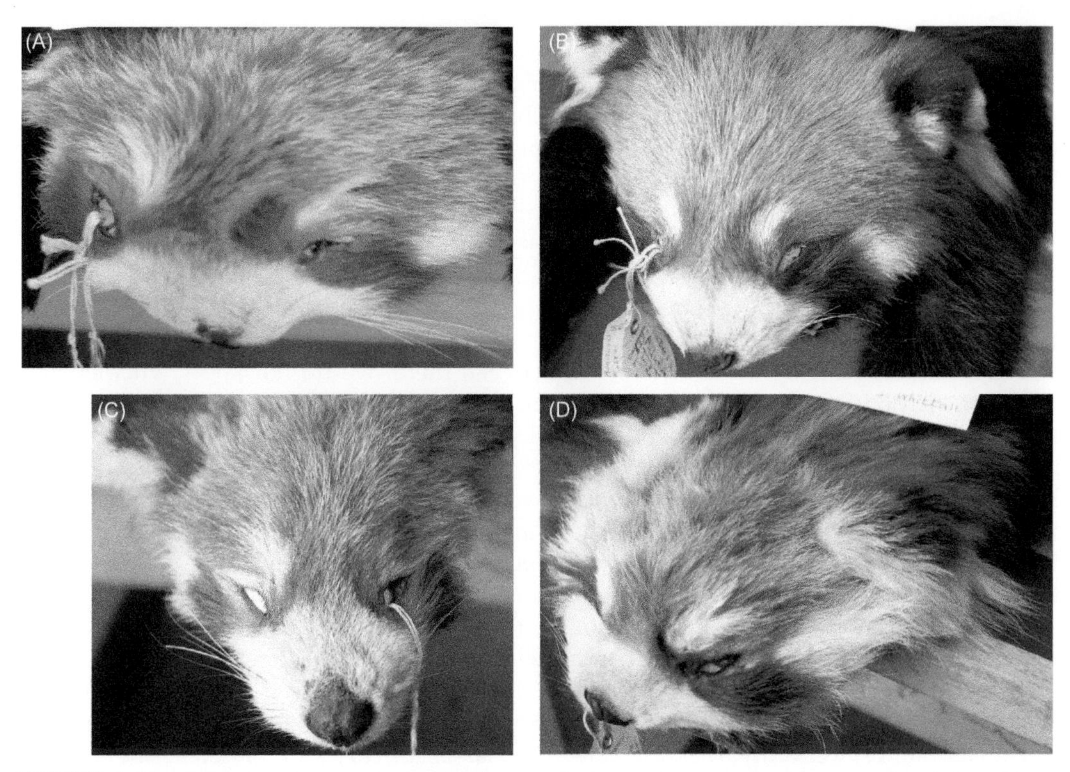

FIGURE 7.10 Close-ups of faces of selected skins in the Natural History Museum, London. (A) *A. fulgens*, Chuntang, Sikkim, 15.9.1.9.67; (B) *A. styani*, Lichiang Ra., 22.9.1.38; (C) *A. fulgens*, Nepal, 58.6.24.79; (D) *A. styani*, 240 km N of Myitkyina, 34.11.3.1. *(Photos courtesy of DCD Happold.)*

sample. The mtDNA evidence, therefore, cannot be said to contradict that of skins and skulls in assigning Burmese pandas to *styani*

- Lushui, and the sole specimen from Changdu, far to the north and actually east of the upper Nujiang: cyt *b*, all haplotypes shared with other samples; CR, four haplotypes shared with other samples (one with *styani*, one each with Gongshan, Changdu, and Burma), four unique to the sample
- Gongshan: cyt *b* haplotype E shared with *styani*; CR haplotype 17 shared with Lushui, the other four unique to the sample
- Zhangmu: cyt *b*, haplotype F unique; CR, both haplotypes unique.

What we see from this is that it would probably be premature to assign the samples along the Nujiang, even on its right bank, to *fulgens* (although the relatively small sample sizes should be borne in mind), since all share haplotypes with their neighbours including *styani* but also have some unique ones (private haplotypes). Only the isolated small sample from Zhangmu shares no haplotypes with any other (the ubiquitous cyt *b* haplotype C excepted). Study of these two DNA regions in Himalayan pandas would be of the greatest interest.

4. Conclusions on Alpha Taxonomy

This survey obviously suffers from a number of drawbacks. The geographic coverage is sporadic, and too few specimens are available. For all this, there are some strong pointers, and it has become clear where the gaps are, so that future research can attempt to fill them in.

It does seem likely that Chinese and Himalayan red pandas are consistently different, and this is important. I have elsewhere [33] argued that the "traditional" view of what a species is, the so-called Biological Species Concept whereby species are reproductively isolated from each other in a state of nature, is so widely inoperable that, in most cases, it provides no guidance at all. In the present instance, what we do know is that there are two taxa with apparently fixed differences between them; we do not know whether their ranges of distribution meet, and we are not at liberty to speculate what would happen – would they interbreed? – if they did. The Phylogenetic Species Concept regards species as evolutionary units, "diagnosably distinct", and this, unlike the Biological Species Concept, is testable on whatever evidence is to hand [33]. The evidence to hand, incomplete as it is, indicates strongly that Himalayan and Chinese red pandas are diagnosably distinct, and so constitute two distinct species, *Ailurus fulgens* and *Ailurus styani* respectively.

Burmese red pandas seem to fall into the range of *Ailurus styani*, and so (with some qualifications) do those from southeastern Tibet, whereas those from Zhangmu in southwestern Tibet seem likely to be *A. fulgens*. In all these cases, further evidence (in the form of larger samples) is needed. I am very far from proposing expeditions to collect further specimens; almost certainly the requisite material is already out there, in the form of "trophy" skulls, trade skins, and pets.

I would finally note that these conclusions are at variance with those of Miles Roberts, the only other author in recent times to test geographic variation in *Ailurus* [34]. That author concluded, on the basis of craniometric and odontometric comparisons, that there was probably too much variability to enable one to distinguish clearly between "subspecies", but stressed that more research was necessary before intermixing them in captivity. The number of specimens measured by Roberts was larger than in the present study (16 *fulgens*, 13 *styani*), but ageing criteria were not clearly given, "cranial suture closure" and "adult dentition" being mentioned but not described, and one suspects the inadvertent inclusion of subadults in the comparisons; I have noted (above) how much cranial growth occurs between the "ya" and "adult" stages, as determined on the strict basis of basilar suture closure. Notably, in Robert's study, it is dental metrics, which do not depend on full maturity, which show much stronger differentiation between the two "subspecies" than do cranial metrics.

CONCLUSIONS AND SUMMARY

The red panda belongs to a family, Ailuridae, all by itself, long-separated from other Carnivora. The fossil record (see Chapters 3 and 4) fully confirms this early separation. This makes the red panda, as the only living representative of a long separate evolutionary

fulgens:		*styani:*	
1 - Chumbi	(27°26'N, 88°55'E)	5 - Janyawng Bur	(..............................)
2 - Chuntang, Sikkim	(27°37'N, 88°37'E)	6 - Taron Valley	(27°42'N, 98°12'E)
3 - Chuntang, Sikkim	(27°37'N, 88°37'E)	7 - Chimeli Pass	(27°07'N, 98°35'E)
4 - Darjiling	(27°01'N, 88°16'E)	8 -	(27°20'N, 97°30'E)
		9 - Yu-lin-pu (Shaanxi)	(33°45'N, 106°44'E)
		10 - Lichiang Range	(27°30'N, 100°10'E)
		11 - Lichiang Range	(27°30'N, 100°10'E)

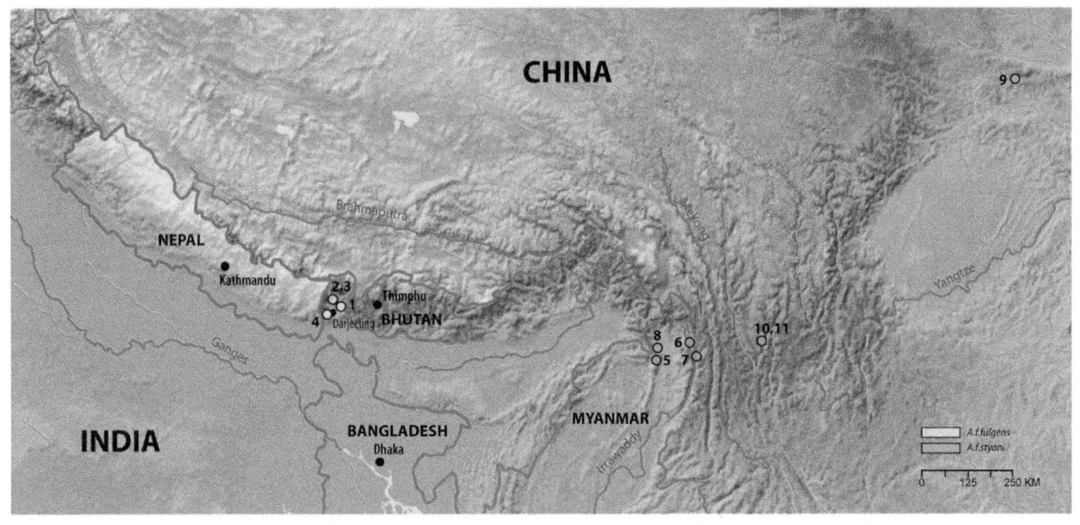

FIGURE 7.11 Map of localities of *Ailurus* from Museum specimens: Nepal, Sikkim, Darjeeling, and Chumbi Valley (*Ailurus fulgens*); Myanmar (Burma), Yunnan, and Sichuan (*Ailurus styani*). (*Map produced by Leen Zuydgeest, Rotterdam Zoo.*)

branch, extremely significant biologically and gives it high conservation value. It emphasizes that intense efforts need to be focused on its survival prospects.

Living red pandas are genetically diverse, and as far as present evidence goes belong to two distinct species: *Ailurus fulgens* from the Himalayas and perhaps Zhangmu, and *Ailurus styani* from Yunnan, Sichuan, Burma, and perhaps eastern Tibet. Localities of museum specimens are shown in Figure 7.11. The two species may be briefly differentiated as follows, differences which on the available evidence seem to be consistent (i.e., diagnostic) being marked with an asterisk:

Ailurus fulgens F. Cuvier, 1825

Synonyms:
Ailurus fulgens F.Cuvier, 1825:3 [35]
Ailurus ochraceus Hodgson, 1847:1127 [36]
Ailurus refulgens Milne-Edwards, 1874:380 [37]

*Size small, with *relatively broader muzzle and smaller teeth. *Dorsal outline of frontal bone follows an even curve backward from facial skeleton to parietals. Tail comparatively short. Fur varying from very pale to quite deep reddish. Limbs black, this tone reaching sometimes only to just above elbow and stifle, sometimes to halfway up shoulder and haunch. *Tail with alternating dark and light reddish rings. *Face white; forehead usually pale reddish but varying to maroon, usually sending a short dark strip to the external corner of each eye (but this may be absent), and a usually pale red median strip which touches internal corner of each eye. *A red or gingery strip, of varying width, from eye around curve of muzzle to corner of mouth. Ears relatively thinly haired inside and *with short hairs on rim. *TF1 locus has a fast allele (A). Differences from *A. styani* in the mitochondrial control region and cytochrome *b* gene are probable, but remain to be corroborated by studies of Himalayan samples.

Concerning the type locality, the type description [35] states as follows:

> Cet animal est originaire des Indes orientales, mais nous ignorons quelle est le partie des ces contrées à laquelle il appartient plus particulièrement; ses dépouilles et un trait général de ses formes nous ont été envoyés par A. Duvaucel.
>
> [This animal is a native of the East Indies, but we do not know to which part of these regions it particularly belongs; its skins and a general impression of its appearance have been sent to us by A. Duvaucel.]

Alfred Duvaucel died in India in 1824, the year preceding Frédéric Cuvier's description. He had been largely based in Calcutta, and one assumes that he had collected, or sent collectors, somewhere within a reasonable distance of there. The accompanying plate (Figure 7.12), at all events, depicts an animal that could only have come from the Himalayan region.

Hodgson [36], who had ample experience in Nepal and Sikkim, did not know of the name "panda", which does tend to suggest that Duvaucel had collected further east than those regions, and he thought it "probable that the Nipaulese and Sikim species may be different from the Panda, and that the latter is a species peculiar to Bhútán".

FIGURE 7.12 Plate of Panda from F. Cuvier, 1825.

Consequently, he described a new species, *Ailurus ochraceus*, from Nepal; although the description was very detailed, he at no point stated specifically what the differences were supposed to be. The syntypes are in the NHM, and like other Nepalese specimens they do not seem to differ from those from Sikkim. Accordingly, it seems probable that *A. ochraceus* is a synonym of *A. fulgens*.

The final name in the synonymy, *Ailurus refulgens*, is a *lapsus calami*. Milne-Edwards, after listing and describing mammals collected in the region of Moupin (now Baoxing), in Sichuan, remarked that other species were also known from the region [37]:

> Ainsi le Panda éclatant, ou *Ailurus refulgens* (1), qui est le seul représentant de l'un des genres les plus remarquables de l'order des Carnassiers...
>
> [Likewise, the resplendent panda, or *Ailurus refulgens* (1), which is the sole representative of one of the most remarkable genera of the carnivore order...]

The footnote reads: "(1) Fr.Cuvier, *Histoire Naturelle des Mammifères*, pl.CCCIII."

It is evident from this that Milne-Edwards had simply misread Cuvier's name, and the name *refulgens* is not an available name for the Chinese panda.

Ailurus styani Thomas, 1902

Synonyms:

Ailurus fulgens styani Thomas, 1902:251 [7]

[*Ailurus fulgens*] *refulgens* Wozencraft, 2005:628 [38]

*Size larger with large teeth. *Profile of frontal bone rises steeply above naso-frontal suture, with markedly convex postorbital processes, behind which it flattens out to form a nearly horizontal cranial vault. Tail relatively longer. Fur generally brighter coloured reddish, varying to maroon. Back of limbs reaching well up on shoulder and haunch, overlapping with *A. fulgens* but only slightly. *Tail rings more contrasted, the dark rings being dark red and the light ones whitish to gingery. Forehead usually darker red; the dark strip to the external corner of each eye is consistently present, and the median strip to the internal corner of each eye is very thick and darker than the tone of the crown. *Eye-to-mouth strip thick and dark, often nearly black, darker than crown. Ears with thick white hair inside, and *with long hairs on rim which form a long downwardly directed tuft at the outer base. *TF1 locus has a slow allele (B).

There seems no doubt that Burmese pandas belong to *A. styani*, but whether those from South-eastern Tibet do so as well needs closer study; they seem, for example, to have larger teeth [27], and Choudhury has recorded some remarkably large skins from western Arunachal Pradesh [39].

As noted above, the name *refulgens* Milne-Edwards is not available for Chinese red pandas, *pace* Wozencraft [38].

The taxonomic divisions among red pandas have been hitherto underrated. It is clear that there are two quite distinct species, and that these have to be managed as separate entities for conservation.

ACKNOWLEDGEMENTS

I thank Angela Glatston for inviting me to contribute to this valuable compendium. Thanks very much to Wei Fuwen for his rapid and detailed responses to my queries, and to Lucy Dueck for allowing me to use her photos, formerly on her website. Above all, I am deeply grateful to David Happold for taking time from his own research to photograph the red panda skins in the Natural History Museum (NHM or BMNH), London.

NOTE

On a visit to the Beijing Institute on Zoology (13–16 October, 2009), I found BIZ 21648, the skull (without skin) figured in [28] and mentioned above, and confirmed that it is a specimen of *A. fulgens* and is from Zhangmu, and is fully mature. It measures 105.4 mm in greatest length. It contrasts with four other skulls (H1662 and 29354, -5, -6), all from Baoxing, which have the characters of *A. styani* and measure respectively 120.1, 117.8, 114.0 (subadult), and 118.7 mm. The first of these is listed in [27], where its origin is given incorrectly as "Tibet". A skin with no skull, 21649, has the characters of *A. fulgens*; it is from Chang Du (31°11′N, 97°18′E), on the upper Mekong (Chiang), and two others, 18032 and -3, are labelled simply "Tibet". Although it would be advisable to study matched skins and skulls before being definitive, this does strongly suggest that the distribution map in [27], and the taxonomic allocations in [29], may be approximately correct.

References

[1] G.M. Allen, The Mammals of China and Mongolia, American Museum of Natural History, New York, USA, 1938.
[2] G.G. Simpson, The principles of classification and a classification of mammals, Bull. Am. Museum Nat. Hist. 85 (1945) 1–350.
[3] J.R. Ellerman, T.C.S. Morrison-Scott, Checklist of Palaearctic and Indian Mammals 1758 to 1946, Trustees of the British Museum, London, UK, 1951.
[4] R.I. Pocock, The external characters and classification of the Procyonidae, Proc. Zool. Soc. London (1921) 389–422.
[5] R.I. Pocock, The Fauna of British India including Ceylon and Burma. Mammalia, Volume II, Taylor and Francis, London, UK, 1941.
[6] F. Cuvier, Panda. No.50. In Histoire Naturelle des Mammifères. E. Geoffroy Saint-Hilaire and F. Cuvier (1825) 3.
[7] O. Thomas, On the panda of Sze-chuen, Ann. Mag. Nat.l Hist. 10 (1902) 251–252.
[8] O. Thomas, On mammals from the Yunnan Highlands collected by Mr George Forrest and presented to the British Museum by Col. Stephenson R. Clarke, DSO, Ann. Mag. Nat. Hist. 10 (1922) 391–397.
[9] D.D. Davis, The giant panda: a morphological study of evolutionary mechanisms, Fieldiana Zool. Mem. 3 (1964) 1–339.
[10] V.M. Sarich, The giant panda is a bear, Nature 245 (1973) 218–220.
[11] L. Ginsburg, Sur la position systématique du petit panda, *Ailurus fulgens* (Carnivora, Mammalia), Géobios, Mém. Special 6 (1982) 247–258.
[12] S.J. O'Brien, W.G. Nash, D.E. Wildt, M.E. Bush, R.E. Benveniste, A molecular solution to the riddle of the giant panda's phylogeny, Nature 317 (1985) 140–144.
[13] R.K. Wayne, R.E. Benveniste, D.N. Janczewski, S.J. O'Brien, Molecular and biochemical evolution of the Carnivora, in: J.L. Gittleman (Ed.), Carnivore Behavior, Ecology and Evolution, Cornell University Press, New York, USA, 1989, pp. 465–494.

[14] W.C. Wozencraft, The phylogeny of the recent Carnivora, in: J.L. Gittleman (Ed.), Carnivore Behavior, Ecology and Evolution, Cornell University Press, New York, USA, 1989, pp. 495–535.
[15] W.C. Wozencraft, Classification of the recent Carnivora, in: J.L. Gittleman (Ed.), Carnivore Behavior, Ecology and Evolution, Cornell University Press, New York, USA, 1989, pp. 569–593.
[16] W. Nie, J. Wang, P.C.M. O'Brien, et al., The genome phylogeny of domestic cat, red panda and five mustelid species revealed by comparative chromosome painting and G-banding, Chrom. Res. 10 (2002) 209–222.
[17] Y-P. Zhang, O.A. Ryder, Mitochondrial DNA sequence evolution in the Arctoidea, Proc. Natl. Acad. Sci. USA 90 (1993) 9557–9561.
[18] P.B. Vrana, M.C. Milinkovitch, J.R. Powell, W.C. Wheeler, Higher level relationships of the arctoid Carnivora based on sequence data and "total evidence", Mol. Phylogenet. Evol. 3 (1994) 47–58.
[19] J. Pecon Slattery, S.J. O'Brien, Molecular phylogeny of the red panda (*Ailurus fulgens*), J. Hered. 86 (1995) 413–422.
[20] C. Ledje, U. Arnason, Phylogenetic relationships within caniform carnivores based on analyses of the mitochondrial 12S rRNA gene, J. Molec. Evol. 43 (1996) 641–649.
[21] J.J. Flynn, M.A. Nedbal, Phylogeny of the Carnivora (Mammalia): congruence vs incompatibility among multiple data sets, Molec. Phylogenet. Evol. 9 (1998) 414–426.
[22] J.J. Flynn, M.A. Nedbal, J.W. Dragoo, R.L. Honeycutt, Whence the red panda? Molec. Phylogenet. Evol. 17 (2000) 190–199.
[23] J.J. Flynn, J.A. Finarelli, S. Zehr, J. Hsu, M.A. Nedbal, Molecular phylogeny of the Carnivora (Mammalia): assessing the impact of increased sampling on resolving enigmatic relationships, System. Biol. 54 (2005) 317–337.
[24] X. Domingo-Roura, F. López-Giráldez, M. Saeki, J. Marmi, Phylogenetic inference and comparative evolution of a complex microsatellite and its flanking regions in carnivores, Genet. Res. Cambridge 85 (2005) 223–233.
[25] J. Marmi, J.F. López-Giráldez, X. Domingo-Roura, Phylogeny, evolutionary history and taxonomy of the Mustelidae based on sequences of the cytochrome b gene and a complex repetitive flanking region, Zool. Scripta 33 (2004) 491–499.
[26] F. Wei, Z. Feng, Z. Wang, J. Hu, Current distribution, status and conservation of wild red pandas *Ailurus fulgens* in China, Biol. Conserv. 89 (1999) 285–291.
[27] M. Zhang, Procyonidae, in: Gao Yaoting (Ed.), Fauna Sinica, Mammalia. 8: Carnivora, Science Press, Beijing, China, 1987, pp. 103–110 (in Chinese).
[28] Z.-j. Feng, G.i-q. Cai, C.-l Zheng, The mammals of Xizang, Science Press, Beijing, China, 1986 (in Chinese).
[29] B. Su, Y. Fu, Y. Wang, L. Jin, R. Chakraborty, Genetic diversity and population history of the red panda (Ailurus fulgens) as inferred from mitochondrial DNA sequence variations, Molec. Biol. Evol. 18 (2001) 1070–1076.
[30] M. Li, F. Wei, B. Goossens, et al., Mitochondrial phylogeography and subspecific variation in the red panda (Ailurus fulgens): implications for conservation, Molec. Phylogenet. Evol. 36 (2005) 78–89.
[31] E.J. Gentz, Genetic divergence between the red panda subspecies *Ailurus fulgens fulgens* and *Ailurus fulgens styani*, in: A.R. Glatston (Ed.), Red Panda Biology, SPB Academic Publishing, The Hague, Netherlands, 1989, pp. 163–170.
[32] F. Wei, M. Li, S. Fang, G. Rao, Z. Feng, Subspecies recognition in red panda by DNA fingerprinting, in: A.R. Glatston (Ed.), The Red or Lesser Panda Studbook, No.11, Stichting Koninklijk Rotterdamse Diergaarde, Rotterdam, The Netherlands, 2001, pp. 5–8.
[33] C.P. Groves, Primate Taxonomy, Smithsonian Institution Press, Washington, USA, 2001.
[34] M. Roberts, On the subspecies of the red panda, *Ailurus fulgens*, in: A.R. Glatston (Ed.), The Red or Lesser Panda Studbook, No. 2, Stichting Koninklijk Rotterdamse Diergaarde, Rotterdam, The Netherlands, 2001, pp. 13–24.
[35] F. Cuvier, Panda. In Histoire Naturelle des Mammifères. E. Geoffroy St. Hilaire and F. Cuvier (1825) 3 (livraison 50).
[36] B.H. Hodgson, On the cat-toed subplantigrades of the Himalaya, J. Asiatic Soc. Bengal 16 (1847) 1113–1129.
[37] A. Milne-Edwards, Recherches pour servir á l'Histoire Naturelle des Mammifères, tome 1, G. Masson, Paris, 1874.

[38] W.C. Wozencraft, Order Carnivora, in: D.E. Wilson, D.-A.M Reeder (Eds.), Mammal Species of the World, 3rd edn, Johns Hopkins University Press, Baltimore, USA, 2005, pp. 532–628.
[39] A. Choudhury, On some large-sized red pandas *Ailurus fulgens* F. Cuvier, J. Bombay Nat. Hist. Soc. 99 (2002) 285–286.

CHAPTER

8

Reproduction of the Red Panda

Lesley E. Northrop[1] *and Nancy Czekala*[2]

[1]Reproductive Medicine Associates of New Jersey, Morristown, New Jersey, USA

[2]Papoose Conservation Wildlife Foundation, Del Mar, California, USA

OUTLINE

REPRODUCTIVE RELATION TO MUSTELIDS AND PROCYONIDAE

The first account of the red panda was described by Cuvier [1] as resembling the Procyonidae (raccoon) but gave it a species name of *Ailurus* due to its cat-like characteristics. Over the years, phylogenetically positioning the red panda in its proper family has proven to be a daunting task. Several accounts have grouped them with the Ursid (bear) family, while other accounts suggest it is more similar to the Mustelida (skunk) and Procyonidae families [2]. Looking at nucleotide sequences of three mitochondrial genes and a single intron, Mustelida and Procyonidae show the red panda to be the closest relative [2]. In a more recent study, the red panda simulates various characteristics that link the species in a mix of interrelationships with Mephitidae (stink badger), Procyonidae and Mustelidae species [3]. Even though the red panda has been linked as a sister species to other families, evolutionary lines strongly suggest it should be categorized as a single species in the Ailuridae family (see Chapter 7), showing a significant time span of separation

Red Panda. DOI: 10.1016/B978-1-4377-7813-7.00008-2

from other Carnivora families. However, linking the red panda in the Procyonidae and Mustelidea families would help clarify some of its reproductive characteristics. This includes defining its style of the ovulatory cycle and whether a delay in embryo development occurs prior to implantation. Similarities are seen in red panda reproduction when compared to the raccoon and skunk. This includes the age of sexual maturation, seasonal breeding in late winter, parturition in early summer, variation in gestational length, young born in a cavity-forming tree and weaning age. The North American raccoon show signs of a second late oestrus in females that lost a litter soon after parturition [4]. In comparison, Mustelids also show a second oestrus if the first breeding is ineffective [5]. Furthermore, female raccoons and skunks are induced ovulators [6], with skunks' ovulation occurring between 40 and 50 hours post-coitus [7]. Following fertilization in the skunk, a delay in embryo development occurs with an observable blastocyst at 11 days followed by a delay in implantation at day 19 [8]. Gestational lengths range from 59 to 77 days, with parturition occurring in May, early June [9].

As will be discussed further in this chapter, the red panda shows other similar characteristics to the raccoon and skunk, including such traits as induced ovulation and a delay in implantation. Even if the red panda is not closely linked taxonomically to these small carnivorous species, maybe they are linked reproductively. In this chapter, we will explore further where the red panda lies reproductively with respect to other members of the Carnivora.

AGE OF SEXUAL MATURATION

In animals, sexual maturity is measured by the age at which it can reproduce.

The red panda young becomes independent at approximately 8 months of age and the mother begins a new breeding season [10]. As reported by a 6-year study on a captive population on *A. f. fulgens*, males and females achieved sexual maturity at 18 to 20 months of age and gave birth between 24 and 26 months of age [11]. According to information compiled by the red panda global studbook [12] on first birth, sexual maturity averaged around 22 months of age. It still needs to be determined whether wild populations show the same age at sexual maturity as the captive population.

INTER-BIRTH INTERVAL

The inter-birth interval is a measure between two successfully bred litters. In the captive red panda populations, inter-birth intervals last 12 months [13]. However, there is a selective case in where a female showed an inter-birth interval of 7 months. In February of 1977, a female was transferred from the Rotterdam Zoo to Sydney, Australia. An unobserved mating occurred in Rotterdam prior to leaving for Australia and produced an early birth in May. The offspring only survived to 1 month of age. She mated again, with a male in Australia and gave birth a second time in December of 1977. Even with this case documented, the common inter-birth interval is 12 months, with offspring weaned at 8 months of age to allow cessation of milk letdown and begin the oestrous cycle for the following year. Whether or not this is an innate reproductive characteristic, rather than an evolutionary transition due to man, we cannot conclusively state.

OVULATION STYLE: MONO- OR POLYOESTROUS

Red pandas exhibit associated reproductive patterns; meaning phases of sexual desire are restricted to times when the gonads are active [14]. The red panda ovulatory cycle has shown indications of a single event (mono-oestrous cycle) or a multiple event (poly-oestrous cycle) over a brief breeding period. In order to understand better the reproductive parameters that the red panda most resembles, the basic biology behind a female oestrous cycle is imperative.

The actual behavioural definition of 'oestrus' is a period of sexual desire by the female. However, there are also periods of anoestrous where the female ovaries are quiescent and male attempts of mating are rejected [15]. The oestrous cycle is comprised of physiological changes that are induced by reproductive hormones in a cyclical manner. The ovarian cycle is represented by two phases that are separated by ovulation. The first phase is defined by an elevation in oestradiol (follicular phase) for follicular development followed by the second phase (luteal phase) defined by elevated progesterone that signifies ovulation and sustains pregnancy. In mammals, the ovaries begin hormone production in response to gonadotropin stimulation, lutenizing hormone (LH) and follicle stimulating hormone (FSH). In the beginning of the ovulatory cycle, FSH stimulates ovarian follicles that house and cultivate the oocytes (eggs) to produce oestradiol. Increasing levels of oestradiol aid in follicular development and trigger LH to expel egg(s) for ovulation. After the egg(s) are ovulated, the high levels of LH also contribute to transforming the collapsed follicle into a post-ovulatory follicle (i.e., a corpus luteum). The significance of the corpus luteum is to secrete progesterone to prepare the uterus for implantation of a fertilized egg(s). It is imperative that the female's oestrus behaviours are synchronized with ovulation to ensure breeding with the male. If mating does not go to fruition, the corpus luteum regresses and the female red panda will enter anoestrous until the following year when new follicle development begins.

In mammals, there are three main types of oestrous cycles; polyoestrous, dioestrous, and mono-oestrous. Female species that are polyoestrous, such as sheep, hamsters and horses, come into heat several times during the breeding season. A polyoestrous cycle can occur at a specific time of the year (spring/summer or autumn/winter) and is controlled by day length. A dioestrous cycle occurs twice a year and is commonly witnessed in dogs. Mono-oestrous species ovulate once a year, typically in the spring, allowing offspring development during the warm season [16].

Having a single oestrus makes it more crucial that hormone priming of the ovaries correlates with mating to increase the probability of a successful pregnancy. In the northern hemisphere, with the red panda living at high elevations in a temperate forest, a seasonal oestrus, whether mono- or poly, is more beneficial to surviving offspring. This allows for the birthing season to occur in early summer at a time when food is more plentiful and gives the offspring time to develop fully through the following winter. In captive populations of the northern hemisphere, females come into oestrus during late winter

(January–March), with birthing occurring in early summer (June–August), whereas captive populations of the southern hemisphere have a shift in oestrus by 6 months. The benefit of the species having oestrus occurring in the winter coupled with a seasonal environment increases the odds of successful births and healthy development for the offspring at the most favourable environmental conditions.

The earliest report of the red pandas' ovulatory cycle was based on a behavioural study on a Nepalese captive population at the National Zoo in Washington, DC [11]. Roberts and colleagues reported that mating occurred once during the year, indicating the red panda is a mono-oestrous species over a three-week breeding season. However, two females mated twice at a 15–21-day interval, indicating a polyoestrous cycle in a short window.

A more recent study on Nepalese captive red pandas in New Zealand indicates a stronger correlation to a polyoestrous cycle, depending on male interaction [16]. Female oestradiol levels measured through faecal extracts during the breeding season showed differences in peak concentration and occurrence under different mating paradigms. Two of the females were separated from their mate until the breeding season, whereas the third female was housed throughout the year with a male partner. Interestingly enough, the female that was paired showed a single ovulatory peak of oestradiol that coincided with a behavioural oestrus (Figure 8.1). Following mating, a sustained level of progesterone occurred and continued for 18 days. In contrast, the two females that were housed individually from a male until mating behaviour was apparent showed multiple peaks of oestradiol with no sustained progesterone levels and no observable mating (Figure 8.2). In this instance, it is suggested an external stimulus is required to induce ovulation. Ovulation requiring coitus as a stimulus is a feature shared in the Mustelidae family and more commonly seen in the Felidae (cats) family [17]. This is referred to as induced ovulation. Whether or not the red panda can be categorized as an induced ovulator will be discussed further below.

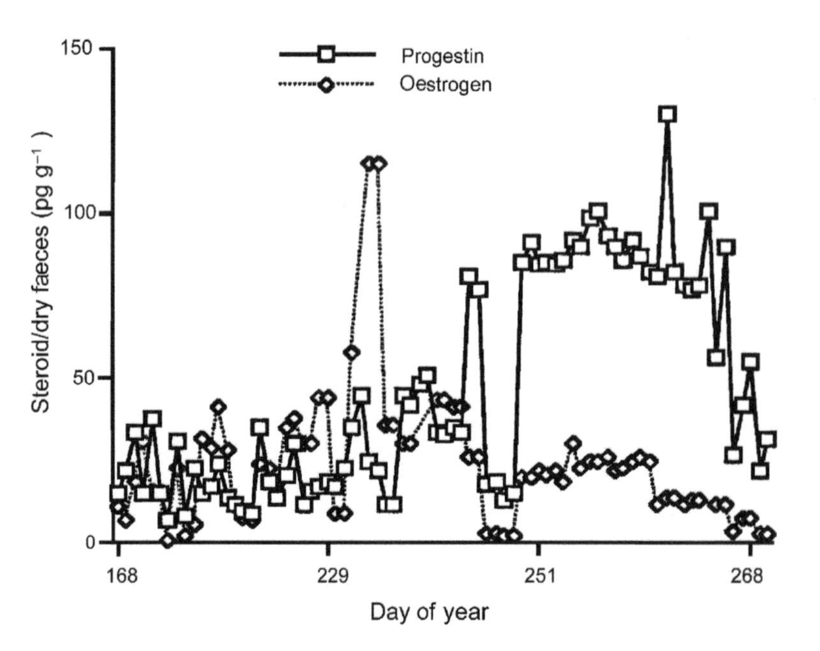

FIGURE 8.1 Oestrogen and progestin profile of continuous mate-pairing. Faecal progestin and oestrogen of a single female housed with a male and mated on days 223 and 224. Behavioural oestrus coincided with a single peak of oestrogen with a preceding rise in progestins 18 days later *(adapted from Spanner et al. [16])*.

A third study measured reproductive success by differences of husbandry management in a captive *A. f. styani* population [18]. Females that were housed with a single male throughout the year displayed two distinct peaks of oestradiol during the breeding season, with mating behaviour occurring on the day of the second peak, whereas a female housed with four males throughout the year showed erratic peaks of oestradiol over the breeding season with no mating behaviour [18]. This study suggests oestradiol is acting as the behavioural basis by driving the mating behaviour of the female.

The benefits of the red panda having multiple oestradiol peaks over a short time frame may increase the possibility of becoming pregnant. In a natural setting, pairing with a male at the exact time of ovulation is critical and this can prove to be very difficult. In solitary social animals, males compete for the attention of the female during the breeding season. Having a finite period of time to fertilize an ovum mandates the necessity for strong sexual readiness by the female. With that being said, a polyoestrous cycle would increase the chances of pregnancy by allowing the possibility of multiple matings. In contrast, having a single oestrus once a year minimizes the females' chances of bearing offspring on a consistent basis and thereby keeps the population small.

At this time, a short film on a red panda pair in the Singálila National Park of India is the only documentation of a mating courtship in the wild ("Cherubs of the Mist") [19]. However, no studies have been completed on a wild population's gonadal function to compare oestrous biology with captive populations. Studies in the captive population have shown, by behavioural and faecal hormone measurement, the red panda comes into oestrous once a year. The significance of measuring oestradiol and progesterone levels across an oestrus period would define follicular development and decipher whether stimulus is required for ovulation. Without this information, the question of whether or not the female ovulates multiple times or once over the breeding season still remains unanswered.

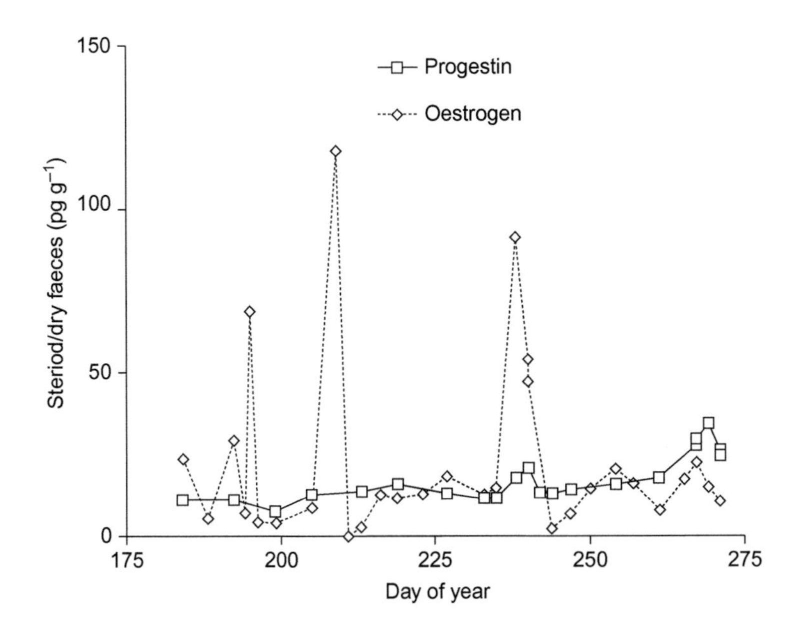

FIGURE 8.2 Oestrogen and progestin profile of mate-pairing on oestrus. Faecal progestin and oestrogen from a single female housed separate from a male showed peaks of oestrogen 14 and 27 days apart with no observed mating or rise in progestin *(adapted from Spanner et al. [16])*.

INDUCED OVULATION

Induced ovulation is a sensory mechanism that requires an external stimulus to trigger the release of a viable egg. The maturation of the follicles and sexual receptivity are under spontaneous hormonal control, but the expulsion of the egg requires vaginal stimulus. No spontaneous, steroid-induced LH surge occurs during the reproductive cycle. Usually copulation activates the preovulatory LH surge as witnessed in the cat, rabbit, raccoon, skunk, and ferret [20]. The female is behaviourally receptive during the breeding season, but still requires coitus to ovulate. The sensory stimulus of the vaginal nerve endings activate ascending pathways to the brain to initiate LH release to induce ovulation [21]. More importantly, it is the peak elevation as well as the duration of LH which determines the occurrence of ovulation. Insufficient copulatory stimuli will not induce LH release and thereby prevent ovulation. As witnessed in ferrets, multiple copulations need to take place on successive days of oestrus in order to achieve a significant peak of LH [21]. If the copulation is not successful, the mature oocyte undergoes atresia and the female must wait until the following oestrous cycle to conceive. Systematically, ovarian control over ovulation benefits the life of the mature ova by securing its full development until fertilization outcome is high.

There have been implications that the red panda shows evidence of induced ovulation. The existing evidence linking the red panda as an induced ovulator are two early papers that measured oestrogen and/or progestin profiles during the breeding season.

The first report showed females that were not housed with a male throughout the year showed elevated levels of oestradiol in accordance with male introduction [16]. This suggests that ovarian activity may be due to a male presence. Furthermore, these same females were not observed mating and showed no apparent progestin secretion, as witnessed in a mated female (see Figure 8.2) [16]. Another study also showed some characteristics of an induced ovulator by comparing the progestin profiles of mated vs non-paired females [22]. Oestradiol profiles were not measured in this study to support evidence of follicle development. However, non-mated females showed varied progestin levels. Differences were observed in four females' progestin levels. Two remained baseline throughout the entire study, whereas two other females showed similar progestin profiles to females that were mated and did not give birth (Figure 8.3) [22]. In this instance, the housing of the male and female were next to one another in separate cages, whereas others were housed far away from a male. This suggests ovarian response to olfactory stimulus, as witnessed in the musk shrew and mice [23,24]. However, if the red panda is in fact an induced ovulator [16], the absence of male stimulation or male presence should result in baseline progesterone levels in all females. Due to limited data, the question still remains unanswered. In order to determine a resolution to these questions, extensive investigations of ovulation are required. Even though many questions remain unanswered, we can propose an evolutionary hypothesis on the reproductive benefits of an induced ovulatory cycle in the red panda. More than likely, the red panda ovulatory cycle simulates characteristics of a mono-oestrous due to the seasonality of the breeding behaviours. The requirement of an external stimulus (such as mating) to activate ovarian activity for follicle release would benefit the reproductive fitness for this species. If a female in the wild does not encounter a male, her 'spontaneously' growing follicle will undergo atresia and is then

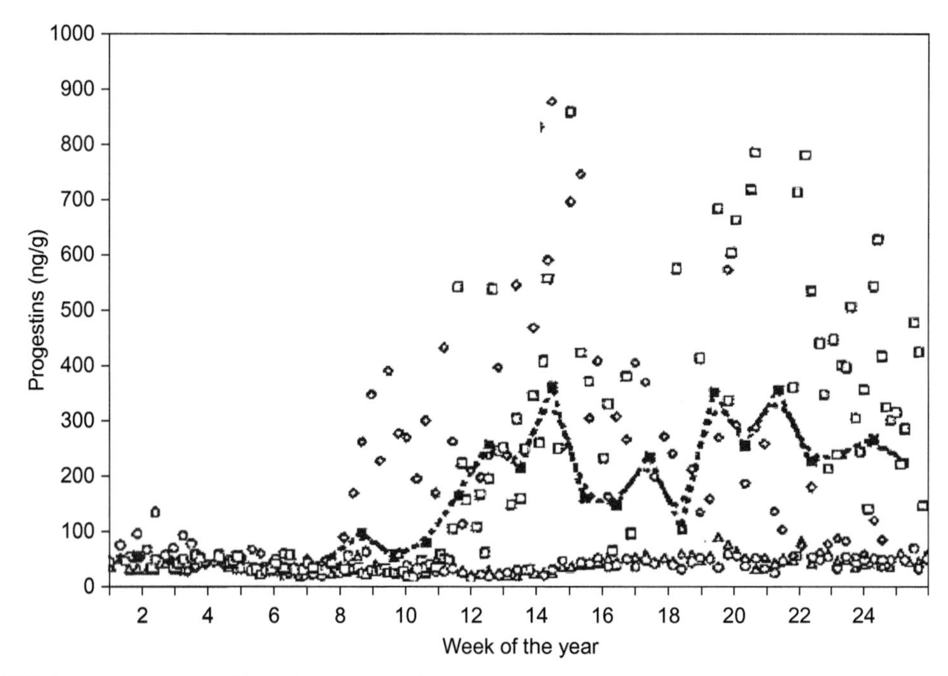

FIGURE 8.3 Progestin profiles of non-paired females during oestrus. Scattered graph representing progestin levels in individual females that were not paired with a male. The line represents the weekly average of progestins measured from four females with two of the females at baseline levels *(adapted from MacDonald et al. [22])*.

wasted. In comparison, if the female is an induced ovulator there is a mature follicle in the waiting. This allows solitary species assurance that her egg will not ovulate spontaneously, at an inconvenient time, when the male is not available. This would provide multiple opportunities to mate and ensure a fertilized ovum. As it remains, information needs be gathered to determine definitively the reproductive status of this species.

BREEDING SEASON

In all mammals, the breeding season is described as a time when the females' oestrous cycle is at a period of sexual desire by the female [15]. In the red panda, the breeding season occurs in the late winter months in captive populations of the northern (January–March) [11,18], and the southern (June–August) hemispheres [16]. It is believed that the red panda in the wild is a solitary animal by nature. Captive animals housed in contact show the male and female maintaining a steady distance from one another throughout the non-breeding season [11].

As mentioned earlier, the mammal breeding season is dependent on gonadal hormone activation. In animals, species differentiation regarding the timing of hormonally driven mating periods has been adapted by physical responses to their respective environments. Across all species, individual breeding seasons have been selected to benefit bearing young during times of beneficial food sources and environmental temperatures. Chinese populations of red panda live in mountainous ranges of the Western Sichuan Province. Temporal variations include very cold winter months, followed by mild spring and hot summer months. Whereas, in the Singélila National Park of India, birthing months occur during the monsoon season (June–August). Evolution of mating seasons has caused individuals to select, reproduce, and pass on their genes to following generations under favourable conditions. Factors that determine the timing of breeding season include ample food supply, the availability of nesting materials, and safety from predation pressures to their young at parturition. For example, in the spring months, new bamboo shoots are plentiful, which provides a hearty food source for the prospective mother and her developing fetuses. As birthing months approach, temperature conditions are more favourable in the early spring for the offspring, rather than the cold, wintry months. Also, adequate time is given to the young to develop fully through the summer to survive the next winter. The red panda young leave the nest at 8 months of age and must be able to forage for food and survive as a solitary species from then on.

MATING BEHAVIOUR: PAIRED IN CAPTIVITY, SOLITARY IN THE WILD

The red panda is sexually a solitary species during the non-breeding season and found in small groups during the breeding season [25]. Husbandry management of the red panda differs across establishments, with either pairing or non-pairing of the sexes during the year. More commonly, the male and female are paired throughout the year and show increased interactions only during the breeding season. Interactions during the non-breeding season include brief social interaction with defensive behaviours such as threat stances and high-pitched vocalizations [11]. However, at the beginning of the breeding season, the male and female show a significant increase in social interactions. These encounters show increases in scent marking rates by the male and female during the oestrus months (January to March) in the northern hemisphere [11]. Females showed increased frequencies of scent marking in comparison to the non-oestrus months [26]. The female comes into oestrus over a short period, at which time sexual behaviours are evident due to increasing ovarian oestradiol levels [27]. Behavioural observations during the mating season showed a correlation between the frequency of reproductive behaviours and the level of oestradiol secretion [18]. At the time the female comes into oestrus, she begins to vocalize chirping and heavily scent marks her territory and interactions between the male and female increase 24 hours prior to copulation [11]. The males are more active in the breeding season than the females, suggesting that the males act more as an oestrous instigator and the female is the beneficiary [26]. Besides the female showing elevated hormonal levels during the breeding season, males show increased levels of testosterone as compared to non-breeding months and testosterone concentrations correlate with their intense behaviours toward the females [28]. The male begins to follow the female intensely to be able to sniff the females' flank and ano-genital region. The female begins increased scent marking and walks around with her tail in an arched position to easily expose her genitals to the male. In accordance with these behaviours, several high-pitched vocalizations are exchanged between the two sexes which are not apparent during other periods of the year. To invite the male to mount, the female holds an arched stance in her back, with her head tilted upward and her tail pointed sharply to the side to allow proper penile penetration. This posture is commonly witnessed in other species and is defined as a lordosis posture. This positioning freezes the female in a stance to allow the male to mount and copulate. Once the male mounts the female, he engages in thrusting bouts at approximately 120 thrusts per minute including multiple intromissions. This courtship continues for several hours, usually ending within 24 hours during the breeding season [11].

To date, there are no published data of red panda mating behaviour in the wild. However, as mentioned above, the documentary film "Cherubs in the Mist" follows a reintroduced female red panda in the Singálila National Park of India [19]. During the female oestrous period, a consort relationship was witnessed between a single male and the female. Wei (see Chapter 11) reports that red panda home ranges partially overlap each other; this opens the possibility of competition among males to protect the females sharing parts of their range. This in turn increases the genetic gene pool of the species, whereas in captive populations, careful mate matching is controlled by the use of a population studbook in order to maximize genetic variation in the zoo population [12]. This allows better management of the species in setting up different pairs for the breeding season to ensure a large, captive gene pool.

PREGNANCY

As mentioned earlier, the red panda shows variability in pregnancy profiles across the population when measured in faecal progesterone metabolites and behavioural studies in captive Nepalese pandas [11,22]. An early report on reproductive status of a captive red panda population showed gestation lengths varied from 114 to 145 days after the last observable copulation [11]. As suggested by Roberts, these variations in gestation length support evidence of a delay in the developing embryo and the embryo does not implant immediately to the uterus.

Another study investigated the differences in gestation length through measuring progestin concentrations of faecal samples. The study found significant variation in gestational length under different pairing groups [22]. Females were analysed in three separate breeding groups (mated with birth, mated no birth, and non-paired groups) across different establishments. Females showed variations in progestin concentrations throughout the study period (from January to July) across the different groups. However, females that gave birth showed significantly higher levels of progestin concentrations than females that mated and did not give birth. By weeks 13–20 after mating, females that gave birth had progestin levels that were 50% higher than females that were paired and did not give birth. Overall, only one female's gestation length (98 days) was recorded in this study, since mating was not observed in the other two females that gave birth. When combining published data of pairing dates to parturition, sustained progestin levels varied from 98 to 158 days in length [11,22]. This wide range in pregnancy lengths suggests that there is a variable length delay in embryo implantation.

Delayed implantation is an evolutionary reproductive mechanism found in numerous taxa including Edentata, Mustelidae, Procyondae, Marsupialia, Rodentia, Chiroptera, and Carnivora [29]. During delayed implantation, the embryo either ceases cell division or undergoes slow cell division [30]. Several theories have been made to clarify the adaptive significance of a delayed conceptus. Some scientists suggest a delayed embryo is advantageous to animals living in extreme climatic conditions and provides species with a mechanism to give birth under optimal energy conditions [31,32]. Others suggest that delayed implantation evolved to ease the time constraints on breeding, gestation, and parturition [29]. In the red panda, there is no strong evidence to suggest a delay in embryo development. However, one indication is the various lengths of gestation observed across females (range of 98–158 days) [11,22]. The second indication is the stage of development of the young at parturition, being small in size with little hair and no eye sight, after a 3–5-month gestation. These two observations suggest an embryonic diapause due to similar development in the womb regardless of gestation length.

In order for implantation to engage effectively, hormonal preparation of the uterus is necessary. Changes in epithelial surface are driven by progesterone. In other species, oestrogen stimulates release of cytokine secretion, which further stimulates activation of the embryo. Secondly, oestrogen acts on the epithelial cells to make them responsive to trophoblast integration. In the red panda, oestrogen response to implantation has not been evaluated. The ability of a species to suspend the blastocyst *in utero* for several days or weeks allows birthing to occur under optimal nutritional conditions. For example, the skunk embryo remains in diapause for 200 days, then resumes development at a slow pace due to poor uterine conditions [33].

We cannot clearly state whether the red panda's embryo ceases development following coitus. To test this hypothesis non-invasively, several hormonal profiles can be measured through faecal metabolites to categorize the red panda with other mammals that undergo diapause. One such hormone that could allow an accurate determination is measuring prolactin levels. Prolactin already has been defined as the key hormone in seasonal control of diapause [30]. Regarding the skunk, prolactin is necessary for proper embryo development by linking environmental stimuli such as day length with ovarian control [30]. As far as the rat is concerned, prolactin is the determining hormone in maintaining the corpus luteum in early stages of pregnancy until the placenta is fully developed and takes over the role [34]. In order to define the red panda's gestation period as a diapause cycle, one would need to map out hormone profiles (oestradiol, progesterone, and prolactin) before, during, and after parturition. Included with captive animal research, field studies on the red panda's gestational period in their natural environment would allow a more accurate consensus of their reproductive parameters. Conducting tandem studies may show the wild populations have less variation in gestational length compared to the captive populations as witnessed in the giant panda [35].

PARTURITION

The red panda gives birth in the early summer months – northern hemisphere (June–August) and southern hemisphere (December–February) – to small litters of one

TABLE 8.1 Litter size for both subspecies

No. of young	*A. f. fulgens*	*A. f. styani*
1	36%	41%
2	57%	49%
3	7%	9%
4	1%	1%

Average litter size is 1.7. Data were compiled from recent 2007 studbook data by Glatston [12].

TABLE 8.2 Birth months for both subspecies

Northern hemisphere			Southern hemisphere	
Month	***A. f. fulgens***	***A. f. styani***	**Month**	***A. f. fulgens***
May	3%	3%	November	3%
June	64%	53%	December	67%
July	32%	42%	January	27%
August	1%	2%	February	3%

Data were compiled from recent Red Panda 2007 studbook by Glatston [12].

to two young, occasionally three to four in a litter, with a birth weight of 113 grams (Tables 8.1 and Table 8.2) [10]. The mother keeps her young hidden from predators in a hollow tree den (wild) or nest box (captive) where development is slow (6.9 grams/day) and the young first leave the den at 3 months of age [13]. Since red panda young become independent within 8 months of life, the mother can reproduce the following year [10].

In captivity, the survival rate varies considerably across regional populations, with ranges between 50 and 86%. Over the years, captive management has worked on providing a positive environment to help increase birth numbers. Looking at the captive *A. f. fulgens* and *A. f. styani* population mortality rate globally, the southern hemisphere (Australia/ New Zealand) show lower mortality numbers of both species compared to North American populations (Table 8.3). Unfortunately, the infant survival rate has not improved over the past 10 years.

To assess the parturition date better in the captive management setting, decreases in progestin levels could be a good indication of birth date. As witnessed in females that mated and gave birth, progestin levels did decrease as parturition neared but did not return to baseline before birth (Figure 8.4) [22]. This did not allow a good indication of parturition for animal management to prepare for birth and be able to monitor mother care at early onset. As suggested by MacDonald et al., other species show decreased levels of progestins that allow an exact prediction of parturition for captive species [22]. Unfortunately, in the red panda, such a measurement cannot be made on progestin levels alone. Instituting a better assessment of the timing of birth would allow successful

TABLE 8.3 Infant mortality in the first year of life (1996–2006)

Region	Subspecies	Male mortality	Female mortality
Global	*A. f. styani*	26%	35%
Global	*A. f. fulgens*	42%	35%
Japan	*A. f. styani*	27%	34%
Japan	*A. f. fulgens*	28%	14%
Japan	Both	27%	32%
Europe	*A. f. fulgens*	42%	33%
N. America	*A. f. styani*	23%	37%
N. America	*A. f. fulgens*	52%	48%
N. America	Both	45%	46%
Australia/NZ	Both	24%	14%

An average of infant mortality across the first year of life was calculated for a 10-year period from the Red Panda studbook by Glatston [12].

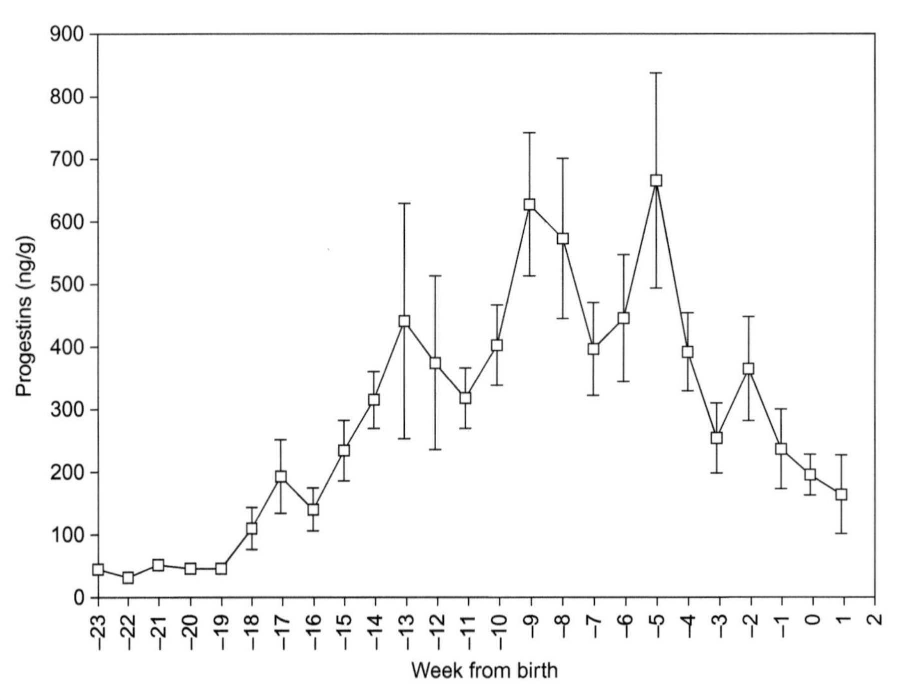

FIGURE 8.4 Average progestin profiles prior to birth. Weekly average of progestin in three females before birth. Scale = SEM *(adapted from MacDonald et al. [22]).*

management programmes to decrease infant mortality rates in captive populations. More importantly, the red panda is listed as an IUCN threatened species and listed on Appendix 1 of CITES as a highly endangered animal. Better parturition detection could allow zoo managers to prepare for birthing and decrease disturbances to the prospective mother by preventing public access to the enclosures.

There are other methods that could be implicated in timing the birth of a captive population, such as measuring levels of relaxin. Relaxin is a hormone released by the placenta and has been indicated in cervical softening [36]. Other indicators could also be explored. For example, prolactin showed elevated levels in relation to decreases in progesterone in the timing of labour in marsupials [37]. These hormone metabolites can be measured non-invasively in urine or faeces samples under a validated protocol in the laboratory. However, this type of measurement has not been reported and needs to be tested in the red panda.

In order to help keep captive populations alive and thriving, several studies need to be conducted fully to characterize the pregnancy profile of the red panda.

FUTURE GOALS OF DEFINING RED PANDA REPRODUCTION

As highlighted throughout this chapter, our understanding of red panda reproductive biology is fairly limited, leaving a wide range of areas to be further explored. Future studies need to be organized to fill in the gaps of red panda reproductive biology. There is already a motivated group of researchers and scientists working to save habitat degradation for the wild populations and conducting recent demographic population studies in Nepal. It is imperative to implement a more aggressive study on the red panda in the wild, observe them during the breeding season in their natural habitat, and collect faecal samples to take back to the laboratory to measure reproductive hormones. This should include the monitoring of breeding behaviour and tracking female behaviour after mating to parturition. Determination on whether man has altered their innate, reproductive behaviours by maintaining their environments and selecting their mate choice in captivity should also be a goal moving forward. Utilizing the current captive populations at our disposal and conducting studies may answer the key questions revolving around whether or not they are an induced ovulator, display a mono- or polyoestrous cycle, and if delayed implantation ultimately occurs.

Unlocking certain reproductive characteristics of the red panda would aid in providing an environment for the captive species that mimics their natural environment. Better monitoring of the birthing process would provide a good starting point to decrease infant fatalities further in captive environments. Even more, setting up semi-captive management environments to reintroduce animals into the wild could aid in increasing the overall population. Overall, we should strive to convert the red panda from a threatened species to a thriving, saved species.

In retrospect of several biological discoveries and scientific advancements in the 20th century, the red panda is patiently waiting to be exposed. Our main goal as conservation biologists is to protect one species at a time, allowing generations of people to be graced with this species' beautiful personality.

PHOTOGRAPHIC ILLUSTRATIONS OF MATING BEHAVIOUR

Axel Gebauer

Place: Goerlitz Zoo, Germany

Weather: Cold, −7°C, strong winds and 15 cm snow

At 08.55 in the morning of 13 January 2009 male Kelsang (born June 2007) and female Sonam (born June 2007) are seen playing under a yew tree (Figure 1). Pre-copulatory play behaviour had already been noted 5 days earlier on 8 January.

The pandas' behaviour was observed by zoo staff for about one hour. In that time the pandas were seen mating on 24 occasions. Copulation lasted between 30 seconds and 2 minutes. The animals assumed very different positions; often the female would lie on her back under the male (Figure 2), but they also mated lying side by side (Figure 3 and 4) and in a standing position (Figure 5). The female ended each bout of mating by standing up and walking away (Figure 6). The male followed immediately (Figure 7). He sometimes lost sight of the female but, as she scent marked very frequently it was not a problem for him to find her again using his sense of smell. The male scent marked as well (Figure 8) using the same places as those used by the female. As soon as the male saw the female he increased his pace and he jumped over the brook to get to her (Figure 9). They began to play again, the female lay on her back, playfully swatting the male with her forepaw (Figure 10). The male began to copulate again (Figure 11). During mating quiet twittering and squeaking calls was made by both male and female.

FIGURE 1

FIGURE 2

FIGURE 3

(Continued)

PHOTOGRAPHIC ILLUSTRATIONS OF MATING BEHAVIOUR *(Con't)*

FIGURE 4

FIGURE 5

FIGURE 6

FIGURE 7

(Continued)

PHOTOGRAPHIC ILLUSTRATIONS OF MATING BEHAVIOUR *(Con't)*

FIGURE 8

FIGURE 9

FIGURE 10

FIGURE 11

(Continued)

PHOTOGRAPHIC ILLUSTRATIONS OF MATING BEHAVIOUR *(Con't)*

FIGURE 12

Only when the keeper arrived with their morning bamboo did the pair stop to feed (Figure 12).

This mating was successful and two cubs were born on 06 June 2009, after a gestation period of 145 days.

References

[1] F. Cuvier, Panda. Histoire Naturelle des Mammifères. In E. Geoffroy St. Hilaire, F. Cuvier, (Eds.), 3 (livraison 50), 1825.

[2] J.J. Flynn, M.A. Nedbal, J.W. Dragoo, R.L. Honeycutt, Whence the Red Panda, Molec. Phylogenet. Evol. 17 (2000) 190–199.

[3] J.J. Flynn, J.A. Finarelli, S. Zehr, J. Hsu, M.A. Nedbal, Molecular phylogeny of the carnivora (Mammalis): assessing the impact of increased sampling on resolving enigmatic relationships, Systems Biol. 24 (2005) 3317–3337.

[4] S.D. Gehrt, E.K. Fritzell, Second estrus and late litters in raccoons, J. Mammal. 77 (1996) 388–393.

[5] J. Wade-Smith, M.E. Richmond, Reproduction in captive striped skunks (Mephitis mephitis), Am. Midland Natl. 100 (1978) 452–788.

[6] L.M. Llewellyn, R.K. Enders, Ovulation in the raccoon, J. Mammal. 35 (1954) 440.

[7] J. Wade-Smith, M.E. Richmond, Induced ovulation, corpus luteum development and tubal transport in the striped skunk (Mephitis mephitis), Am. J. Anat. 153 (1978) 123–142.

[8] J. Wade-Smith, Hormonal and gestational evidence for delayed implantation in the stripe skunk, Mephitis mephitis, Gen. Comp. Endocrinol. 42 (1980) 3011–3017.
[9] J. Wade-Smith, B.J. Verts, Mephitis mephitis. Mammalian species, Am. Soc. Mammal 173 (1982) 1–3.
[10] G.B. Schaller, The Last Panda, The University of Chicago Press, Chicago, USA, 1993.
[11] M.S. Roberts, D.S. Kessler, Reproduction in red pandas, Ailurus fulgens (Carnivora: Ailuropodidae), J. Zool. 188 (1979) 233–249.
[12] A. Glatston, Rotterdam Zoo, Rotterdam, 2007.
[13] J.L. Gittleman, Are the pandas successful specialists or evolutionary failures, BioScience 44 (1994) 456–464.
[14] D. Crews, Gamete production, sex hormone secretion, and mating behavior uncoupled, Horm. Behav. 18 (1984) 22–28.
[15] W. Heape, The "sexual season" of mammals and the relation of the "pro-oestrum" to menstruation, Q. J. Micro. Sci. 44 (1900) 1–70.
[16] A. Spanner, G.M. Stone, D. Schultz, Excretion profiles of some reproductive steroids in the faeces of captive Nepalese red panda (Ailures fulgens fulgens), Reprod. Fertil. Devel. 9 (1997) 565–570.
[17] H.G. Verhage, N.B. Beamer, R.M. Brenner, Plasma levels of estradiol and progesterone in the cat during polyestrus, pregnancy and pseudopregnancy, Biol. Reprod. 14 (1976) 579–585.
[18] F. Wei, X. Lu, L. Chun, L. Ming, B. Ren, J. Hu, Influences of mating groups on the reproductive success of the southern red panda (Ailures fulgens styani), Zoo Biol. 24 (2005) 169–176.
[19] S. Pradhan, in Cherubs of the Mist. Bedi Films, India, 2006.
[20] S.R. Milligan, Induced ovulation in mammals, Oxford Rev. Reprod. Biol. 4 (1982) 1–46.
[21] J. Bakker, M.J. Baum, Neuroendocrine regulation of GnRH release in induced ovulators, Frontiers Neuroendocrinol. 21 (2000) 220–262.
[22] E.A. MacDonald, L.E. Northrop, N.M. Czekala, Pregnancy detection from fecal progestin concentrations in the red panda (Ailurus fulgens fulgens), Zoo Biol. 24 (2005) 419–429.
[23] E.F. Rissman, X. Li, Olfactory bulbectomy blocks mating-induced ovulation in musk shrews (Suncus murinus), Biol. Reprod. 62 (2000) 1052–1058.
[24] L. More, Mouse major urinary proteins trigger ovulation via the vomeronasal organ, Chem. Senses 31 (2006) 393–401.
[25] J.C. Hu, Reproductive biology of the red panda, J. Sich. Norm. Coll. 12 (1991) 1–5 (in Chinese).
[26] L. Xueqing, Z. Zejun, W. Fuwen, et al., Reproductive behavior variations and reproductive strategy in the captive red panda, Acta Theriol. Sin. 24 (2004) 173–176 (in Chinese).
[27] K. Wallen, Desire and ability: hormones and the regulation of female sexual behavior, Neurosci. Biobehav. Rev. 14 (1990) 233–241.
[28] L. Chun, W. Fuwen, L. Ming, L.I.U. Zueqing, Y. Zhi, H.U. Jinchu, Fecal testosterone levels and reproduction cycle in male red panda, Acta Theriol. Sin. 23 (2003) 116–119 (in Chinese).
[29] S.H. Ferguson, J.A. Virgl, S. Lariviere, Evolution of delayed implantation and associated grade shifts in life history traits of North American carnivores, Ecoscience 3 (1996) 7–17.
[30] M.B. Renfree, G. Shaw, Diapause, Ann. Rev. Physiol. 62 (2000) 353–375.
[31] R. Canivenc, M. Bonnin, Environmental control of delayed implantation in European badger (Meles meles), J. Reprod. Fertil. Suppl. 9 (1981) 229–241.
[32] P. Vogel, Occurrence and interpretation of delayed implantation in insectivores, J. Reprod. Fertil. Suppl. 29 (1981) 51–60.
[33] R.A. Mead, Delayed implantation in mustelids with special emphasis on the spotted skunk, J. Reprod. Fertil. Suppl. 29 (1981) 11–24.
[34] M. Freeman, In: E. Knobil, J. Neill (Eds.), The ovarian cycle of the rat, Raven Press, New York, USA, 1988.
[35] D. Wang, X. Zhu, W. Pan, in: D.G. Lindburg, K. Baragona (Eds.), Life History Traits of Reproduction of Giant Pandas in the Qinling Mountains of China, University of California Press, 2004.
[36] G. Weiss, Relaxin, Ann. Rev. Physiol. 46 (1984) 43–52.
[37] C. Tyndale-Biscoe, L. Hinds, C. Horn, Fetal role in the control of parturition in the tammar, Macropus eugenii, J. Reprod. Fertil. Suppl. 82 (1988) 419–428.

CHAPTER

9

Placentation of the Red Panda

Kurt Benirschke

University of California, San Diego, USA

The placenta of the red panda has never been studied. The present description is apparently the only record from this species; even in the extensive review of comparative placentation by Mossman [1], *Ailurus* does not appear. I have described it earlier, however, on my website (http://medicine.ucsd.edu/CPA) and it is that placenta that finds its description here. Perhaps the placenta of the red panda has not been available because of placentophagy, as many animals consume the placenta after birth and this is especially so in carnivores. On the other hand, since numerous animals have been bred in zoological gardens, it is a little surprising that the afterbirth had not been available. Nor has it been possible for me to refer to a publication of the findings at autopsy of a pregnant animal. The placental findings related here indicate that there are great similarities between the *Ailurus* placenta and those of the *Mustelidae*.

This placenta comes from a twin gestation at the San Diego Zoo in which one fetus, the male, weighed 180 g, but the female twin, weighing only 108 g, was stillborn. This placenta was detached from the surviving neonate very shortly after its birth and it was thus well preserved. It had a discoid shape and was quite unlike the completely zonary placenta of raccoons, dogs, and cats. It weighed 5.9 g, measured 4.5 × 3.5 cm, was 2–3 mm thick and had a 3.5 cm umbilical cord attached in addition to the torn placental membranes. It is not known whether the mother consumed any portion of it, although the organ appeared to be complete (Figures 9.1–9.3). Surprisingly, in this delivered placenta, the surface vessels were very prominent and filled with blood. Some other vessels that attached to the membranes were not completely identified. They may be the remains of vitelline vessels, as these exist in some *Mustelidae* (e.g., the wolverine [2]) but I am not certain of this.

The umbilical cord of this red panda afterbirth measured 3.5 cm in length and was only slightly spiralled. Watson [3] remarked how short the cord is in the raccoon and this thus equally applies to the red panda placenta. The umbilical cord contained two arteries, a vein, and a large allantoic duct (Figures 9.4–9.6). Similar to the cord of raccoon placentas, it contained no omphalomesenteric duct. There were numerous small and capillary blood vessels that were remarkably concentrated around the allantoic duct; some of these smaller vessels had a considerable amount of musculature as is true of many other species, e.g., the

Red Panda. DOI: 10.1016/B978-1-4377-7813-7.00009-4

FIGURE 9.1 This is the appearance of the delivered placenta with the disrupted membranes and cord attached.

FIGURE 9.2 Closer look at the fetal surface and lateral insertion of the umbilical cord.

dolphin. The allantoic duct was also accompanied by many bundles of smooth muscle; its lining was of a urothelial nature. A thin amnion covered the umbilical cord and, in the free amnionic membranes, it was for the most part composed of a thin layer of squamous epithelium on loose connective tissue. Here and there, a few patches of squamous metaplasia were present. The amnion is avascular and is loosely attached to the allantoic sac's membrane (Figure 9.7). The latter contained connective tissue and an ample vasculature; its epithelium is flattened, urothelium-like. No remnants of yolk sac structures were identified. A cross-section of this placenta is shown in Figure 9.8. These twins were of different

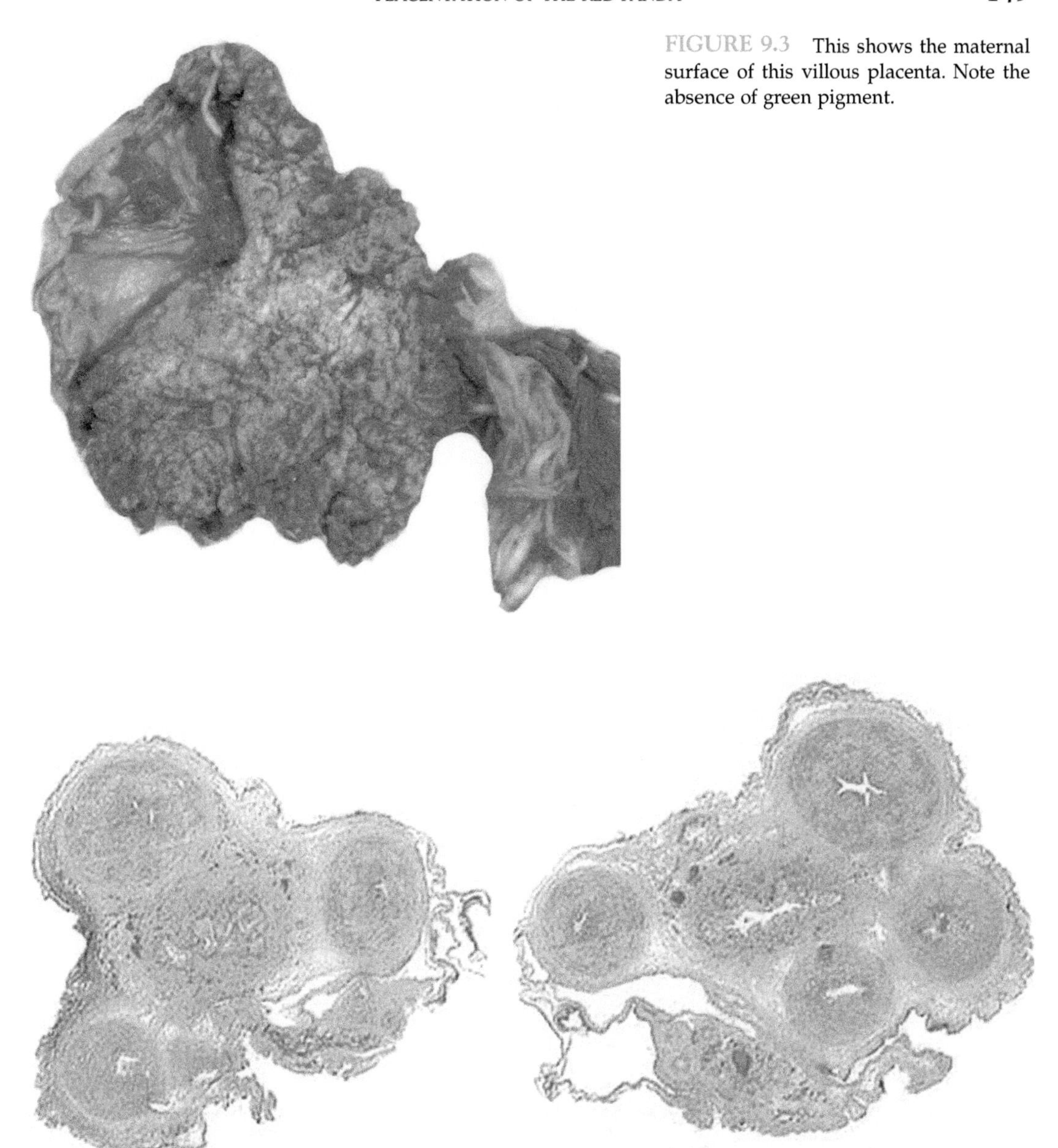

FIGURE 9.3 This shows the maternal surface of this villous placenta. Note the absence of green pigment.

FIGURE 9.4 Two low-power sections through the umbilical cord with adjacent placental membranes.

sex and they were thus surely dizygotic, but the relationship between the two placentas remains unknown. It is not even known whether the placentas were implanted in different uterine horns as implantation of the embryo occurs in a bicornuate uterus as this uterus was described; the ovaries are enclosed in a bursa [4]. The placenta of the stillborn fetus was apparently consumed by the mother and it was thus not available.

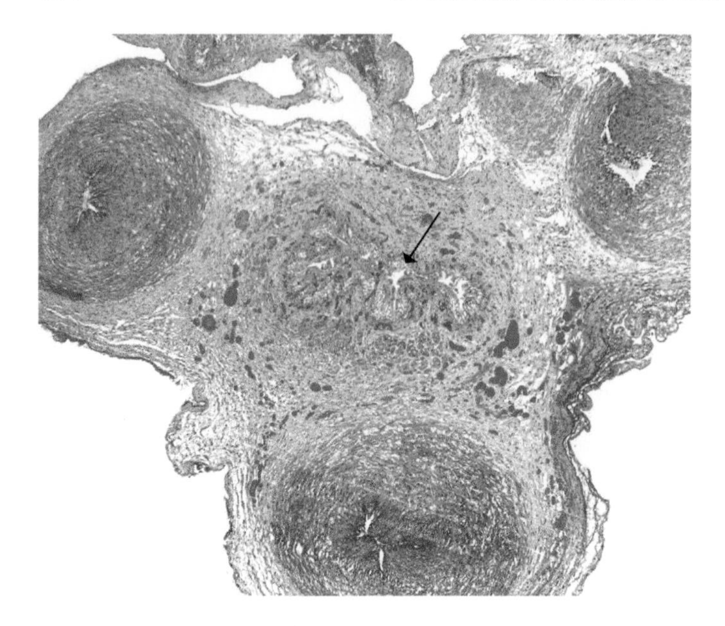

FIGURE 9.5 Higher magnification of the umbilical cord with the arrow pointing at the allantoic duct.

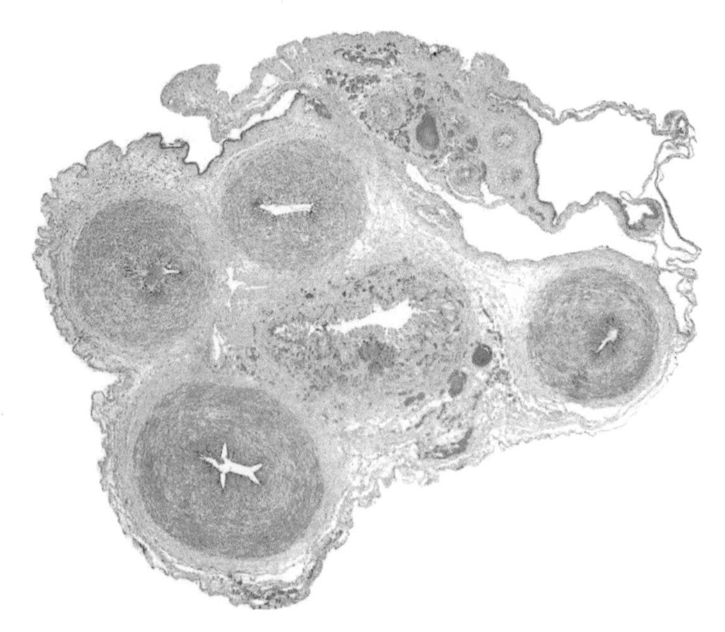

FIGURE 9.6 Allantoic duct in the umbilical cord. Note that the smaller cord (allantoic) vessels surround the allantoic duct while few other blood vessels are seen in the remainder of the cord.

Microscopically, this is a labyrinthine organ with a somewhat confusing histology; it is unlike that found in most other Carnivora. Large maternal vascular channels of nearly uniform diameters traversed to the undersurface of the chorion, even in this delivered organ. Because of the prominence of the basement membrane of the fetal blood vessels, the designation "vaso-chorial" placenta has been used for these types of placenta, rather than referring to them as having an "endothelio-chorial" relationship. Beneath the placental

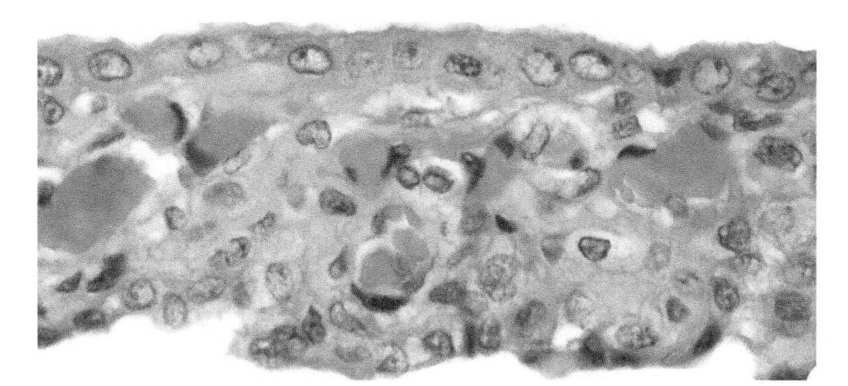

FIGURE 9.7 Histological section of the membranes with a markedly vascularized allantoic sac.

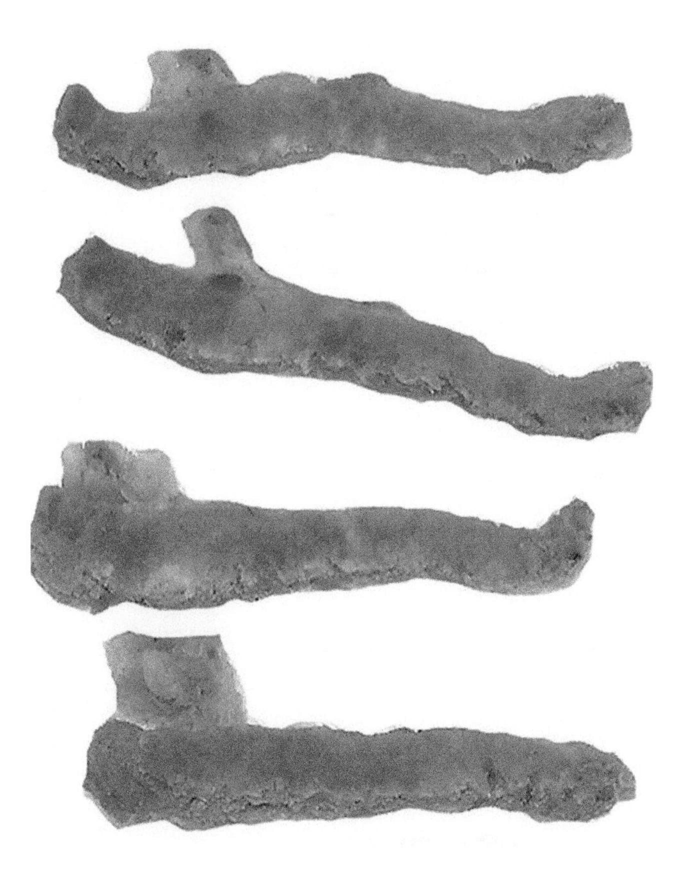

FIGURE 9.8 Cross-sections of the fixed organ.

labyrinth one should expect to find the "transitional or junctional zone", the region in which the placenta normally separates; but only little of this region was present in this detached organ unlike that found in the usual "endothelio-chorial" placentas. This area contained much degenerate material and it is speculated that some of the degenerated material of the junctional zone in carnivores is absorbed for fetal nutrition.

There was no green discoloration at the edge of the disc, as is prominently seen in dog and cat placentas, a feature whose lack had already been remarked upon for the raccoon placenta by Watson [3]. Biggers and Creed [5], however, were critical of some of Watson's descriptions and they obtained subsequently much better material for their own studies [6,7]. They affirmed the zonary nature of the raccoon placenta, being certain that it was not interrupted. In addition though, they disputed the "epitrichium" described by Watson [3] and considered it to be the closely applied amnion. These authors were also the first to introduce the notion of the "haemophagous organ", a prominent structure that is seen especially in dogs. This region also exists prominently in young raccoon gestations but it gradually involutes there towards term. Here, the trophoblast is believed to invade the uterus, thus creating focally a haemochorial organ with the deposition of much green pigment. In later studies [6,7], the authors identified similar structures in several *Mustelidae* and suggested that the presumed iron pigment is phagocytized and is thus made available for red cell production by the fetus. In the red panda placenta, however, only small foci of yellow-staining material were found.

The large vessels are lined by a much hypertrophied, cuboidal endothelium that is placed on a prominent basement membrane. This basement membrane, presumably analogous to that of the raccoon placenta, stains deeply with PAS [8]. Because of the prominence of this basement membrane, the designation "vaso-chorial" placenta has been used. The trophoblast is usually cuboidal and relatively thin and is only focally syncytial. It covers the connective tissue of the very sparse "villous" connective tissue that is filled with fetal red cells. Binucleate trophoblastic cells were not found. There has long been a controversy regarding the nature of the syncytial cells – are they of maternal or fetal origin? The arguments have swung in favour of a trophoblastic derivation [9].

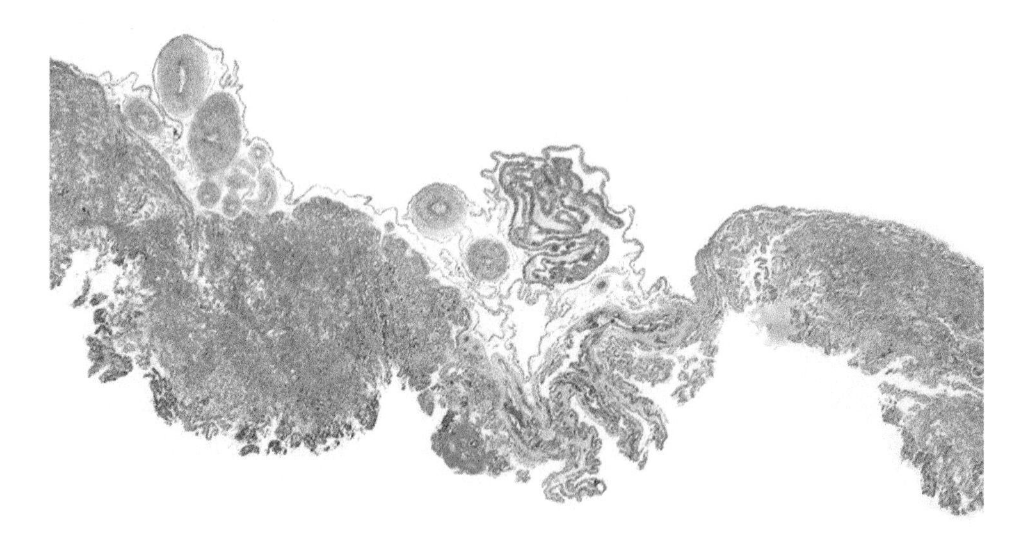

FIGURE 9.9 The disrupted maternal surface of the placenta with the apparent "gap" between villous tissues.

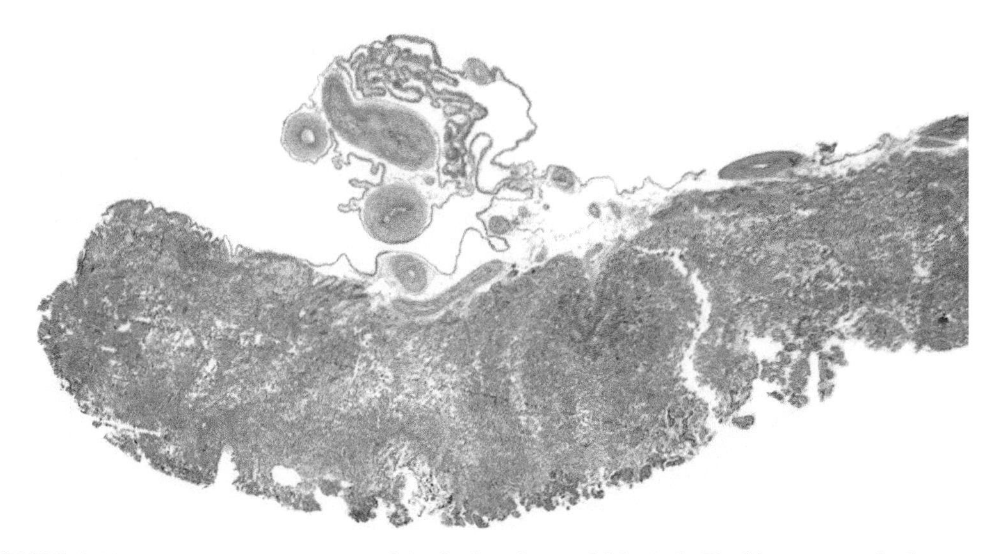

FIGURE 9.10 Microscopic appearance of the fetal surface and labyrinth. The blue areas at the base are portions of the disrupted junctional zone.

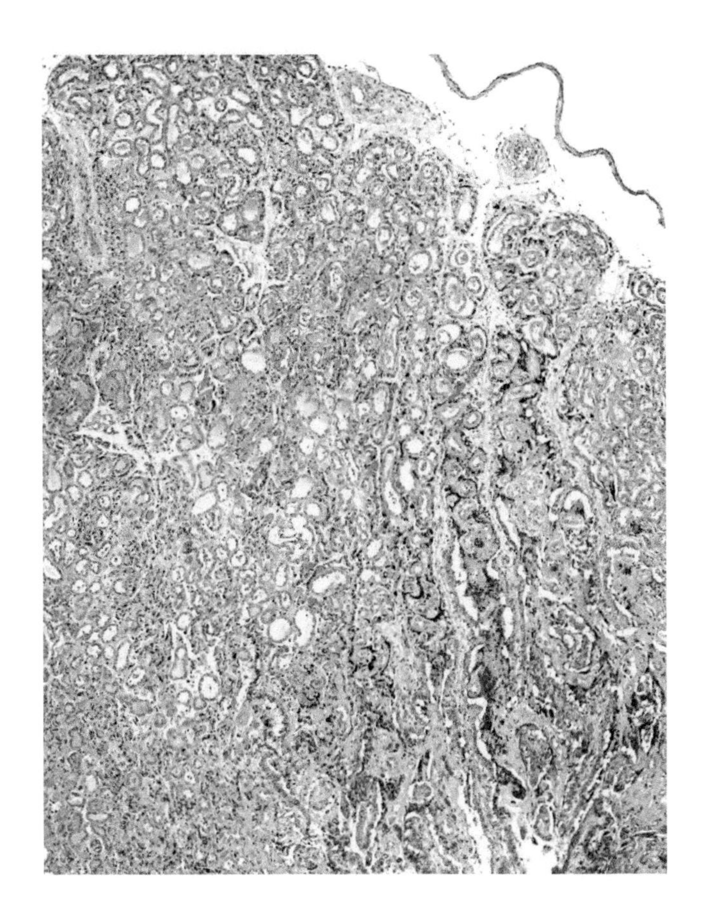

FIGURE 9.11 Histological appearance of another slice of the placental labyrinth with fetal side up.

The placenta of raccoons falls into the general arrangement of other carnivore placentas [9], but the placenta of the red panda is quite different from that of dogs and cats. The filiform arrangement of villi, for instance, is lacking and the placenta does not have the ring-shaped, zonary form as is seen in numerous other carnivore species. It is rather discoid (see Figures 9.1–9.3) with a small central sulcus connecting the villous tissue to a strip of membranes – perhaps this is the remnant of a former zonary type (Figure 9.9). Moreover, in its histology, it also appears to be more similar to the placenta of the primitive carnivore *Zorilla striata* that was first described by Rau [10]. A nearly bidiscoid (rather than zonary) placenta was found in that species, and this is true of the placenta of the lesser panda. In other respects it is also not too much different in its structure (Figures 9.10–9.12) from that of *Zorilla*. The hypertrophy of the maternal capillary endothelium and the gross morphology are nearly identical to many other carnivore species' placentas but, as already stated, it is quite different from those of cats and dogs (Figures 9.13 and 9.14). It must be emphasized, however, that it will be essential to witness an implanted placenta in order to be able to rule out a more ring-shaped organ than that being portrayed here. It is thus also noted that in those other animals possible yolk sac vessels are seen in the cord, but no remnant of yolk sac morphology was identified in the red panda specimen described here. In further consideration of this morphology, it is interesting to note that some genetic studies have also likened the

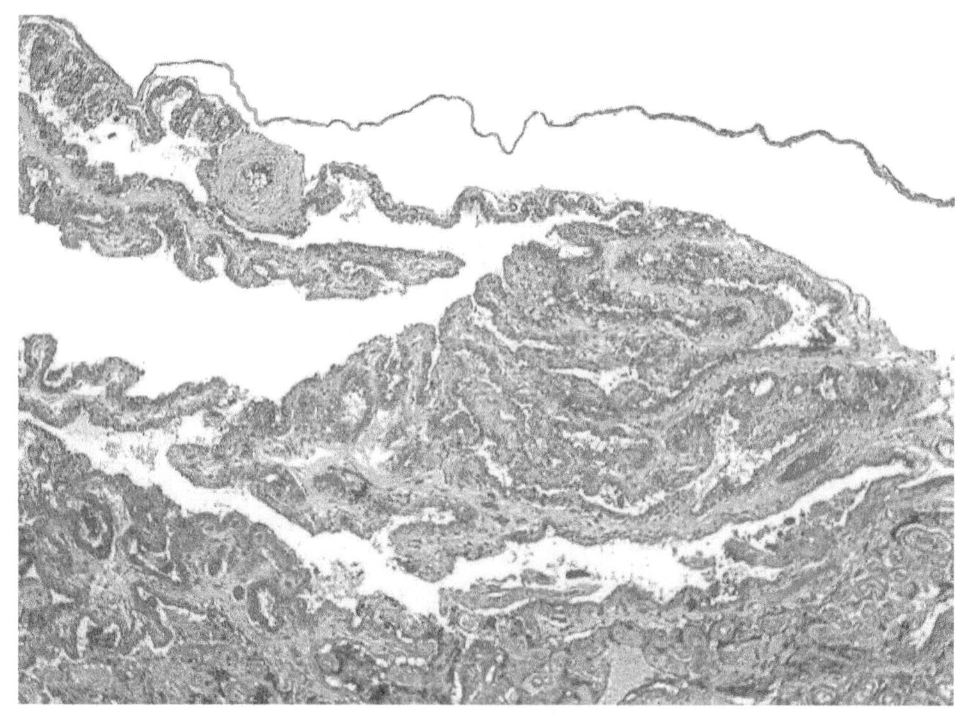

FIGURE 9.12 Microscopic appearance of another region of the labyrinth only with the fetal surface showing the thin amnion.

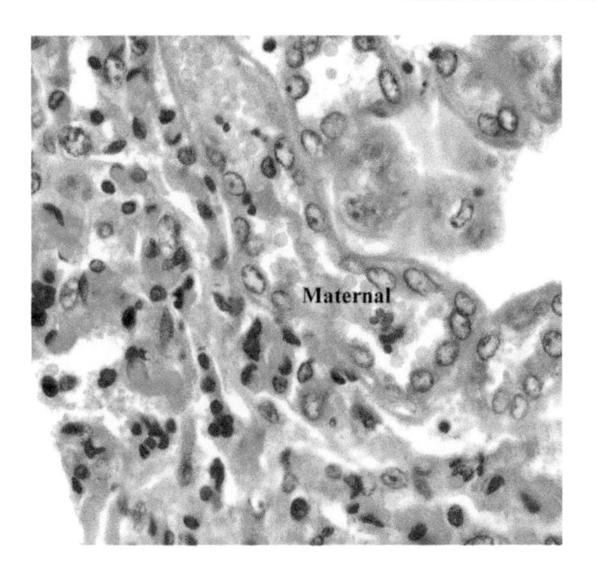

FIGURE 9.13 Microscopic appearance of a maternal vessel (labelled) and adjacent fetal labyrinth.

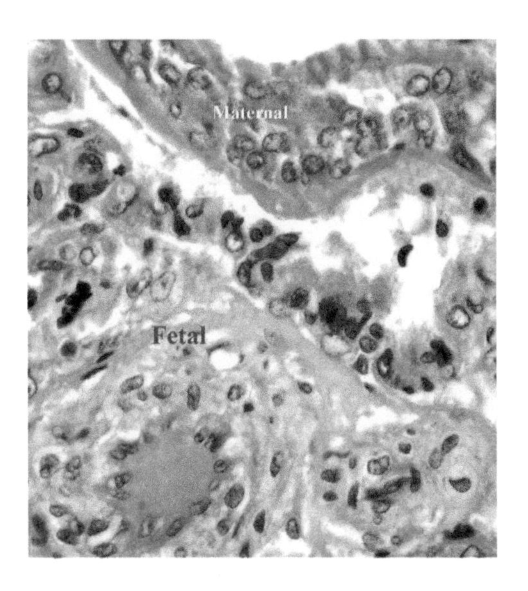

FIGURE 9.14 Higher magnification of maternal vessel (labelled) and the congested fetal (labelled) labyrinth.

lesser panda more closely to the *Mustelidae*. Watson [3], who described the raccoon placenta, stated that a lush decidua was present in the non-pregnant horn of the uterus of the raccoon he examined. Finally, the possibility exists that delayed implantation takes place and/or that matings occur during pregnancy. These two factors have been considered to explain the wide span of gestational dates [11,12].

It is apparent from this brief first description of a red panda placenta that much more reproductive anatomical study is needed. It would be especially helpful if a pregnant uterus with an implanted placenta were described.

References

[1] H.W. Mossman, Vertebrate Fetal Membranes, MacMillan, Houndmills, Basingstoke, UK, 1987.
[2] G.B. Wislocki, E.C. Amoroso, The placenta of the wolverine (Gulo gulo luscus) (Linnaeus), Bull. Museum Comp. Zool. (Harvard) 114 (1956) 93–100.
[3] M. Watson, On the female organs and placentation of the raccoon (Procyon lotor), Proc. Roy. Soc. London 32 (1881) 272–298.
[4] H.W. Mossman, K.L. Duke, Comparative Morphology of the Mammalian Ovary, University of Wisconsin Press, Madison, Wisconsin, USA, 1973.
[5] J.D. Biggers, R.F.S. Creed, Two morphological types of placenta in the raccoon, Nature 194 (1962) 103–105.
[6] R.F.S. Creed, J.D. Biggers, Development of the raccoon placenta, Am. J. Anat. 113 (1963) 417–445.
[7] R.F.S. Creed, J.D. Biggers, Placental haemophagous organs in the procyonidae and mustelidae, J. Reprod. Fertil. 8 (1964) 133–137.
[8] R.F.S. Creed, R.J. Harrison, Preliminary observations on the ultrastructure of the raccoon (Procyon lotor) placenta, J. Anat. 99 (1965) 933 (abstract only).
[9] E.C. Amoroso, Placentation, in: 2nd edn, A.S Parkes (Ed.), Marshall's Physiology of Reproduction, Vol. 2, Little, Brown & Co., Boston, USA, 1961, pp. 127–311.
[10] A.S. Rau, Contributions to our knowledge of the structure of the placenta of mustelidae, ursidae, and sciuridae, Proc. Zool. Soc. London 25 (1925) 1027–1070.
[11] W. Puschmann, Zootierhaltung, Vol. 2, Säugetiere, VEB Deutscher Landwirtschaftsverlag, Berlin, Germany, 1989.
[12] R.M. Nowak, Walker's Mammals of the World, 6th edn., The Johns Hopkins Press, Baltimore, USA, 1999.

CHAPTER

10

The Early Days: Maternal Behaviour and Infant Development

Axel Gebauer

Naturschutz-Tierpark Görlitz, Germany

OUTLINE

INTRODUCTION

To date there are few data available in the literature of the behavioural development of red panda cubs. In the field, the collection of such information is limited by the difficulties in finding nesting places and observing these secretive animals. Apart from some observations of a single panda nest at Singhalila National Park (Darjeeling, India) nothing is known about the early days of infant development in the wild [1]. Zoos, on the other hand, would seem ideal locations for obtaining more detailed information on the denning period,

Red Panda. DOI: 10.1016/B978-1-4377-7813-7.00010-0

however, few data have been published and these are confined to observations made outside of the nest or to weights and measurements of hand-reared animals or young removed briefly from the mother. Roberts has made detailed studies of the reproductive biology of red pandas held at the National Zoological Park, Washington and there are some unpublished student studies of cub behaviour at Knoxville Zoo [2–11]. However, due to the sensitivity of red panda dams during the denning period, studies of the behaviour of mother and cubs in the den have been non-existent. This is the first published investigation of maternal behaviour, cub development and activity budgets of red pandas during the denning period.

METHODS

The observations were made on red pandas which are housed in a large and naturalistic exhibit (Figure 10.1A,B) at Naturschutz-Tierpark Görlitz in Germany [12–15]. The enclosure has an area of 1500 m^2, and is divided into three areas: (1) enclosure for red pandas and Chinese muntjacs (*Muntiacus reevesii*) – 1100 m^2; (2) aviary for white-eared pheasants (*Crosspotilon crossoptilon*) and blue magpies (*Urocissa erythrorhyncha*) – 165 m^2; and (3) a visitor area (China pavilion, playground) – 300 m^2. The exhibit is surrounded by 100-year-old rhododendrons (*Rhododendron* spp.) and contains yews (*Taxus baccata*), a Canadian pine (*Tsuga canadensis*), a dove tree (*Davidia involucrata*), Chinese juniper (Juniperus × media), mountain ash (*Sorbus vilmorinii*) and aralia (*Aralia chinensis*). The enclosure contains a pond, an 80 m long stream, open grassy areas, bushes (*Berberis* spp.) and herbaceous plants to encourage foraging behaviour. There is also a sleeping crate and a heated feeding box for the pandas and a covered feeding station for the muntjacs. Two hollow logs about 3–4 m long and 30–50 cm wide are provided as breeding dens.

In 2007, we installed infrared and daylight security cameras in both nest sites and also digital HDV movie- and photo-cameras. This equipment was carefully prepared in separate boxes (Figure 10.2) behind the breeding dens around 4 weeks prior to the expected birth date. The activities of mother and the single cub in the dens were continuously recorded from 9 June, 2007 (cub 3 days old) to 10 February, 2008 (cub 8 months old) using a hard disc recorder. All recordings were analysed and all behaviours were noted; important recordings were saved. From February to May 2008 (cub age 8–11 months) only recordings of play behaviour were saved. The indirect observations in the den were supplemented with irregular direct observations of behaviour in the exhibit, which was also filmed and photographed as required.

The male of the breeding pair is kept continuously in the enclosure, but encounters with the cubs are rare. To complete the 2007/2008 data, this chapter also refers to incidental observations of four litters during 2001–2005 outside the nestbox.

MATERNAL BEHAVIOURS

Nest Building

Red pandas are born blind and helpless, they develop very slowly and so are dependent, for at least 3 months, on a good den to protect them from predators and bad weather. The selection of a suitable denning site is of prime importance to a

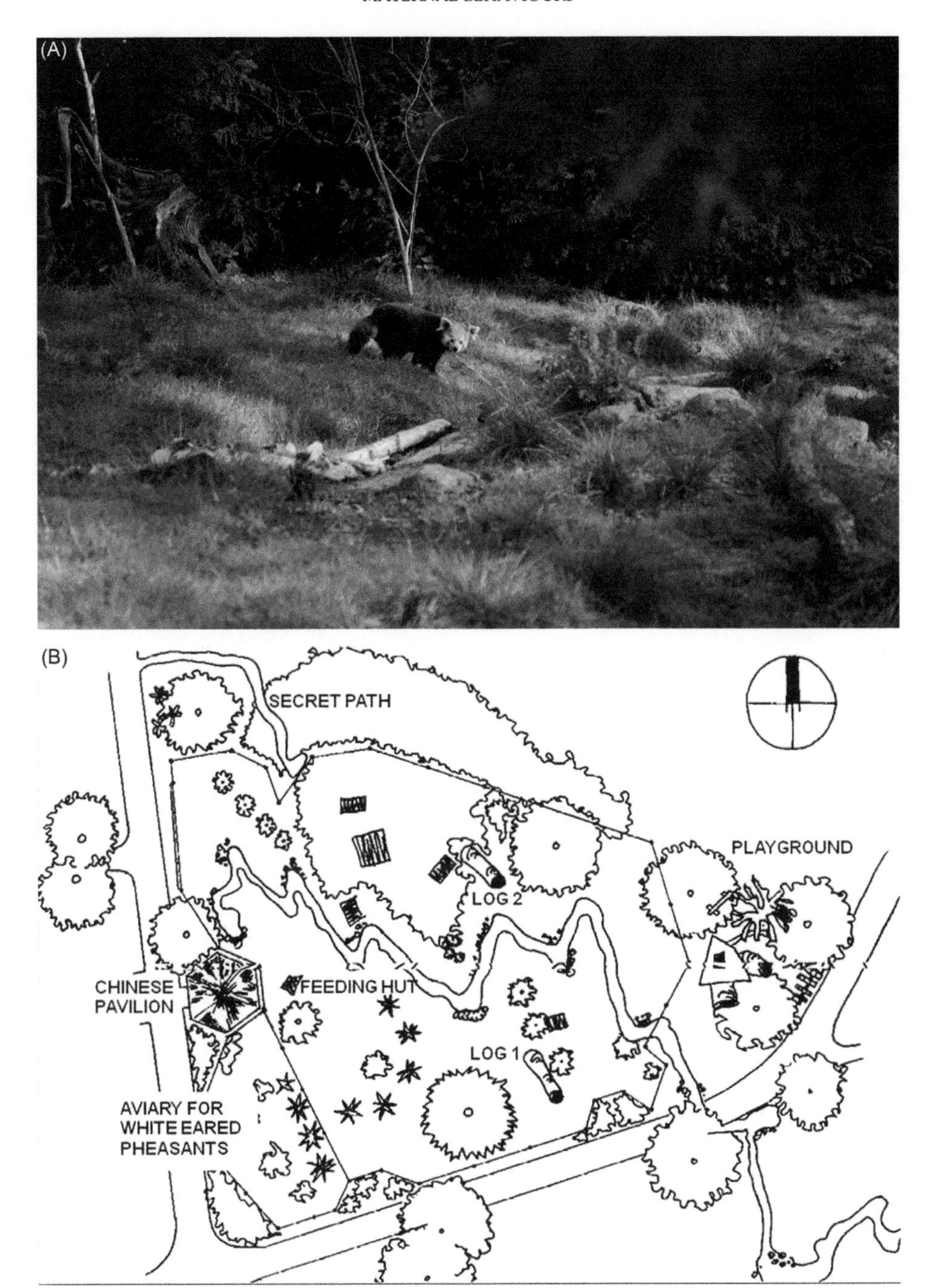

FIGURE 10.1 (A) Impression of the red panda exhibit of Naturschutz-Tierpark Görlitz; (B) sketch of the red panda exhibit of Naturschutz-Tierpark Görlitz.

FIGURE 10.2 Camera box (left) and nesting log (right) which were combined and separated by a Plexiglas window.

female red panda. In the wild, red pandas are reported to nest in tree hollows, hollow logs and stumps, and in rock crevices [1,16–18]. The collection of nest building material precedes the birth by several weeks but becomes more intensive around 2 days before parturition [2,3,19] The behaviour continues after the young are born throughout the denning period. It occurs more frequently when moving to a new nest or around weaning, especially after suckling or play sessions. Even temporary nests are lined with nest material, for example, our female collected fresh yew twigs, placed them in a hollow tree stump (Figure 10.3A) and covered her 64-day-old cub to hide it. Twigs, sticks and branches with or without leaves, up to 40 cm and 5 cm wide, together with bits of bark, leaves, grass and moss are all used as nesting materials (Figure 10.3B–D).

Nesting materials are sought in proximity to the den (see Figure 10.3E) and carried there in the mouth, sometimes assisted by one forepaw (see Figure 10.3D). After depositing the material in the den, the animal scrapes the substrate with its forepaws (similar to the slow scratching movements of a cat before defaecating). This can continue for several minutes. The animal also pushes its snout into the substrate until a depression arises. During both these behaviours the panda adopts a particular arched back posture. The female will also roll and wriggle on her back kicking her legs or bracing them against a vertical object (wall of the den) to form the nest (see Figure 10.3F–H).

FIGURE 10.3 (A) Temporary nest (with a 4-day-old cub) made by the dam in a tree stump which was lined with yew twigs; (B) the female carries a piece of wood; (C) or a large fresh leaf into the nest log; (D) and lifts the nest material with the forepaw; (E) collecting nest material occurs in the den surroundings (*Courtesy of Ann-Kathrin Wirth*). (F–H) the female rolling on the ground as nest building behaviour prior to parturition.

Time in the Den

Red pandas are helpless at birth so all early maternal care takes place in the den. In this study, we found that the time the mother spent in the den with her cub decreased from around 16 hours (67%) per day after parturition to 4 hours (17%) per day when the cub was 70 days old and started to leave the den (Figure 10.4). Over this period, the time in the den can vary greatly; it decreased when external temperatures were high (around day 40), and increased after disturbances by the zoo staff (around days 47 + 55) or the presence of a beech marten (*Martes foina* – day 7).

Rest and Sleep

The immediate postpartum period is characterized by the female resting curled up around her newborn cubs for several hours [1,3]. During the first weeks, the female moved her cubs onto her chest or belly for sleeping (Figure 10.5A). Other forms of contact during sleep such as lying side by side occurred regularly during the first 4 weeks. From 2

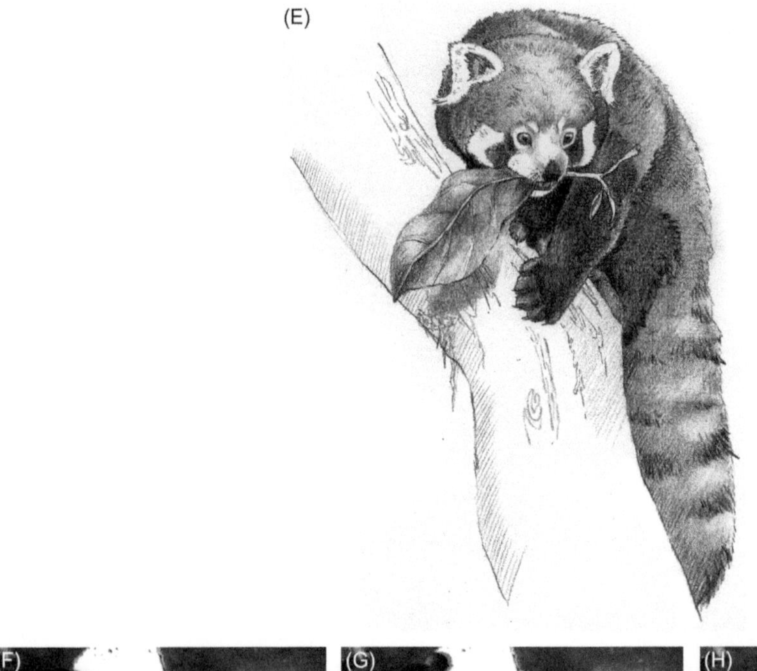

FIGURE 10.3 (Continued).

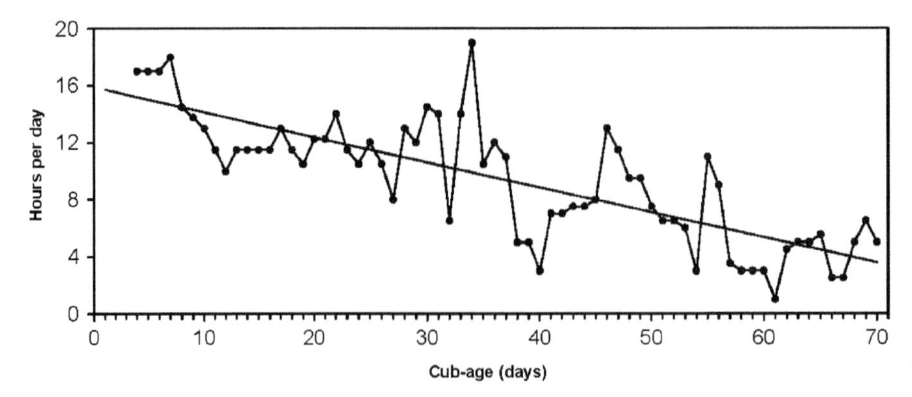

FIGURE 10.4 Denning time of dam (stay of female in the den) between days 4 and 70 of cub age. (Note that the slope indicated in the figure is for descriptive purposes only. It is based on a parametric regression analysis and not a Spearman's rank correlation.)

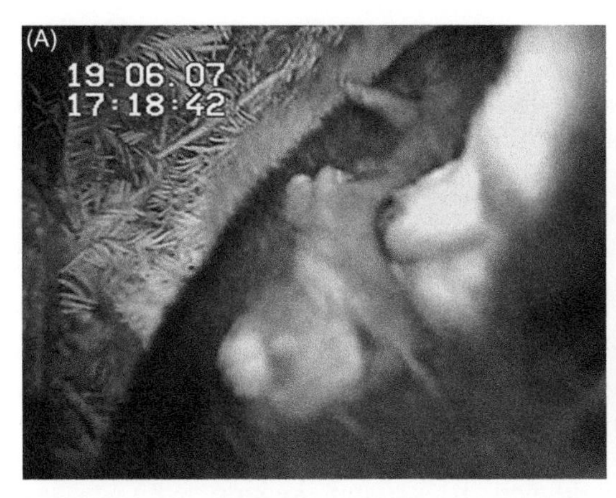

FIGURE 10.5 (A) The female moves the cub during the first weeks on her belly for sleeping, and (B) raises her legs during sleeping in periods with high outside temperatures to maximize thermal discharge.

FIGURE 10.6 During anus licking the female (top) holds her cub (lying, 5 months old) down with a forepaw.

months onwards, the dam can be seen using her cub's back as a pillow during sleep. Even when the cub is 8 months old, the mother will continue to sleep with it regularly in the nest. This continued, but with decreasing frequency, until the cub was 11 months old.

In order to maximize heat loss during hot weather, the mother would rest on her back with her legs raised (Figure 10.5B).

Allogrooming

The female stimulates urination and defecation by licking the ano-genital region of the cub. She consumes the waste products, thereby keeping the nest clean. This behaviour precedes or occurs at the start of suckling. The female rolls the young and holds it with her forepaw to keep it in a suitable position. This behaviour usually lasts between 2 and 4–5 minutes (Figure 10.6).

During nursing, the mother continues to allogroom other parts of the cub's body. Allogrooming is one of the most common behaviours during the denning period. At Knoxville Zoo, grooming sessions of red panda females had an average length of about 8 minutes with a high variability (SD 10, max. 60 minutes) [11]. Allogrooming persisted in the female of Görlitz Zoo until the young were weaned.

Cub Transport

The carrying of cubs in the mouth of the dam is a well-known carnivore behaviour pattern. Red pandas will use the mouth to pick up the cub (Figure 10.7A,B), often assisted by careful use of a forepaw. The female grasps the infant at the back but often on the side of the neck (Figure 10.7C). Normally, an infant will fall into immobility a few seconds after being lifted by the mother. As the cub becomes older, it may try to prevent carrying by becoming rigid (with an arched or straightened back) and kicking with its legs (Figure 10.7D). This behaviour, combined with the increasing weight of the young means that carrying becomes rare from day 70 onwards and stops totally around day 105. Observations at Washington Zoo showed that the mother would retrieve cubs up to the age of 110 days, and return them to the nest if they emitted calls or played [2].

Roberts reported that during the first 2–3 weeks postpartum, captive red panda females moved their cubs frequently from one nesting site to another, sometimes up to six to eight times per day (mean 3.3 per day for the first 35 days) [2]. Frequency drops steadily after the first week and, by the 4th week, she would move them no more than once or twice in a 24-hour period. One should expect that such frequent carrying is an artifact due to captivity, however, but no data are available from natural habitats for reference. Conway reported that frequent carrying behaviour in captive red pandas was due to disturbance [20]. In the wild or under conditions without (human-caused) disturbances, the frequency of cub retrievals is much lower. In Görlitz, we only recorded excessive carrying when the female was disturbed by a beech marten and after the installation of lamps, cameras, etc. in the box behind the den. The frequency of cub manipulation (not only carrying to a different den but also moving the infant within the den) was most frequent in the first month and decreased significantly after this time (Figure 10.8).

The multiparous female in this study carried the juvenile out of the den for nursing around day 70, and carried it back to the den if the cub tried to leave. A cub of this age is already difficult for the mother to lift, and so she would drag it over the ground (see Figure 10.7C).

CUB BEHAVIOURS

During the first 3 months the cubs are confined to the den. During the first 2 weeks, this cub was mostly observed sleeping and suckling and making some attempts at crawling. Locomotion becomes more directed at about 3 weeks but remains slow and uncoordinated for the first 6 weeks. Between days 40 and 50 the cub became more active and began to explore the nest, and climb on different den structures, groom and play. Around day 70, it was carried out into the enclosure by the dam.

(B)

FIGURE 10.7 During the first weeks the cub falls into immobility during carrying (A, B). After 2 month, cubs may try to prevent carrying by going rigid and kicking with legs (C, D). *(B and D Courtesy of Ann-Kathrin Wirth.)*

FIGURE 10.7 (Continued).

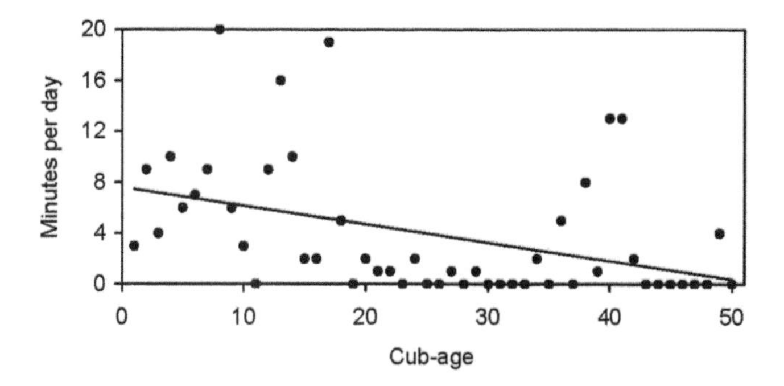

FIGURE 10.8 The frequency of carrying in the den decreases during cub growth. (Note that the slope indicated in the figure is for descriptive purposes only. It is based on a parametric regression analysis and not a Spearman's rank correlation.)

FIGURE 10.9 Cub (66 days old) handling a piece of wood with left forepaw, playful training of nest-building behaviour.

Nest-building

It seems that nest-building movements are innate in red pandas. At about day 60, the cub was first observed manipulating nest material with its mouth and forepaws (Figure 10.9), and showing the nesting behaviours described above (Figure 10.10). Interestingly, these movements coincided with the nest-building behaviour of the dam; she brought nest material into the den and exhibited nesting behaviour, this seemed to act as a stimulus for the infant to do the same. These nest-building activities of the cub seemed to be more frequent between days 60 and 70.

Rest and Sleep

During the first 3–4 weeks, infants can sleep in a stretched, half-curled or curled up position on the female's chest or belly (see Figure 10.5A), next to or even under the mother or alone or next to a sibling when she is out of the den. The previously mentioned positions can also be seen in older infants when they are sleeping on their own (Figure 10.11A–C). Sometimes they use pieces of wood or other objects as pillows (Figure 10.11D) or put their forelegs/paws on the head for protection against mosquitoes.

FIGURE 10.10 Nest building behaviour of a cub (66 days old): arching/humpback posture during mouth pushing in the ground.

From day 100 onwards, cubs regularly sleep outside. After the denning phase, they prefer to sleep in similar places to the adults; horizontal branches or forks of trees but in bad weather (heavy rain, snow, wind...) they still return to their den.

We analysed the pattern of sleeping behaviour during the first 3 weeks (Figure 10.12). There is an uncoordinated rhythm in the first week, but a more stable amount of sleep developed in the following two weeks (60% of the day). One would expect this to decline as the cub becomes older and locomotion and play increase.

Comfort Movements

Stretching and yawning are physiological conditioned reactions with an uncomplicated sequence of movements. They are seen in an early stage, already from day 10 onwards. Stretching of older animals occurs normally after yawning and consists of an initial hump-back-posture and a following long-extended posture.

The more complex patterns of grooming and licking are developed later. First attempts at scratching and self-grooming were seen on days 29 and 39, respectively. Allogrooming of the mother was seen as early as 22 days old. Only after 2 months did the cub exhibit longer grooming bouts (first observation at day 59). By 3 months, we recorded 40 minutes of grooming per day (around 3% of activity budget). Intense grooming is typical of adult animals and may take up to 16% of the daily activity budget [15].

Locomotion

During the first week of life the infant moved very little. There is little strength in the hind limbs in the early days after birth, as it has less than 50% of the muscle of the forelimb [21]. This is stronger due to the need to support the newborn's enormous head. The infant could be seen dragging itself by the forelimbs by day 2 and by day 5 could crawl onto the female's chest to nurse. By day 8, it was able to right itself and, by day 10, we observed active crawling/climbing on the female's belly when searching for the teats. At

FIGURE 10.11 Sleeping positions of red panda cubs at different ages: (A) 4 days; (B) 66 days; (C) 46 days old, 2 cubs of the litter from 2005; (D) 66 days.

FIGURE 10.11 (Continued).

an age of 18 days, the infant crawled a distance of 10 cm after the mother when she was leaving the den. Between days 25 and 40, the cub could be seen moving in a slow and clumsy manner. It would sometimes try to jump but mostly failed and toppled over. The cub was able to walk in a stable fashion by day 43 and between days 50 and 70 gradually became more active. It was interesting to note that the cub seemed to learn its climbing skills by using both existing den structures and pieces of wood or bark brought to the den by the dam. These activities gradually moved from the rear to the entrance of the den in preparation for the first steps outside the den, which begins between days 65 and 92 (Table 10.1). By 3 months, coordination had improved enough for the infant to exhibit locomotor play (see later). At 4 months the cub was able to climb on thin branches and to swing and play in the trees. When the cubs first emerge from the nest they tend to follow the dam closely during walking and climbing (Figure 10.13). This following behaviour may persist up to 10 months.

Individual Play

Physical activity play is the most common form of play observed among mammals [22]. It includes the simplest form of play, the individual locomotor play, where an individual exhibits playful movements without participation of a conspecific or the use of objects. The first true locomotor play was observed when the cub was 40 days old: it was seen lying on its back and kicking with all legs. Such leg kicking and wriggling became more pronounced after day 50 (Figure 10.14). These behaviours were seen mostly after suckling bouts and after awakening or before sleeping. Running, jumping, pouncing, and climbing are other variants of locomotor play that were observed inside the den in the first months. These behaviours continued outside the nest, starting more on the ground or horizontal branches but progressing later to vertical structures; on day 141 we saw the cub climbing alone in the tree, climbing up and down a nearly vertical branch, and jumping from one branch to the next.

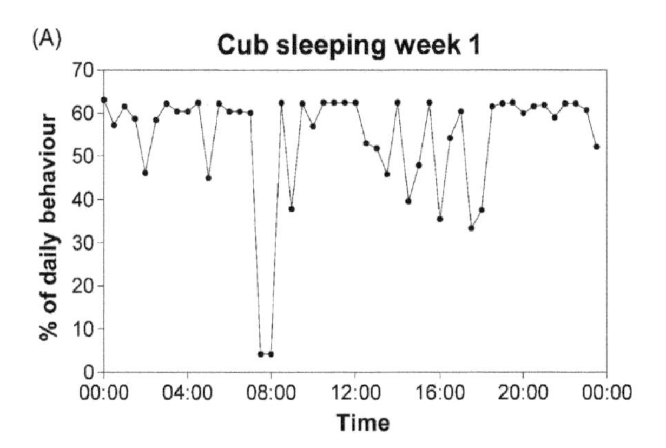

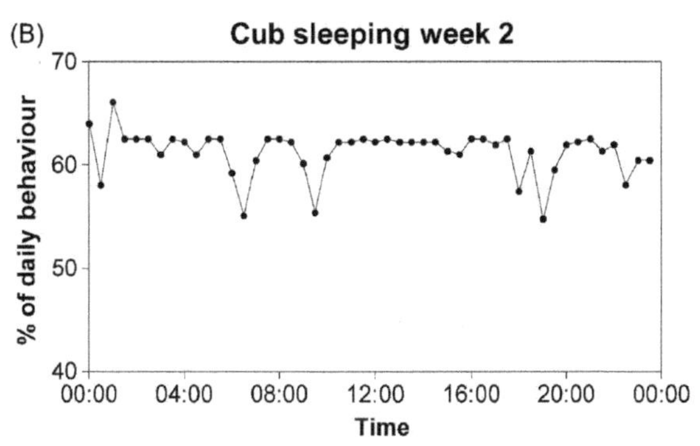

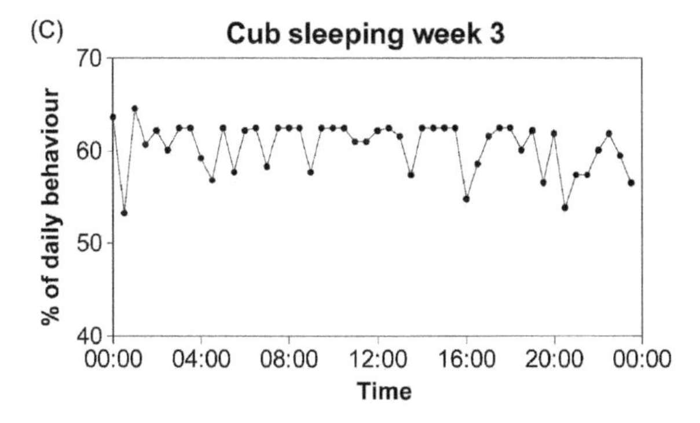

FIGURE 10.12 Daily amount of sleeping of a cub during the first 3 weeks after birth.

Manipulation of inanimate objects (so-called solitary object play) develops in parallel with locomotor play. On day 40, the cub was seen playing for the first time with a twig, this was the opportunistic reaction to the twig falling on its body. After day 51, object play was seen more regularly. The infant manipulated or chewed sticks and leaves, which the mother regularly brought into the den (Figures 10.15 and 10.16). Inanimate objects were observed to be shaken, climbed, or swatted with the paws; it was even seen to use a piece of wood as a sparring partner in a mock fight.

Also after the denning phase, the infant was observed to play with new or interesting things in the enclosure. Even when the cub was 11 months old, object play was still seen:

TABLE 10.1 Age at which a behaviour pattern was observed for the first time (present study and findings of other authors)

	Present study	[2–5]*	[27]*	[19]*	[20]*	[40]*	[41]*	[39]*	[38]*	[23]*
LOCOMOTION										
Crawling	d 5	d 2								
Crawling on/behind dam (directed)	d 18	d 25								
Walking	d 43		d 30						d 52	
Climbing inside den	d 52	d 55								
Climbing outside	d 90	w 13					w 14	w 10–12	w 14	
Investigating nest entrance		d 80								
Leaving den first time	d 71	d 88	d 80	d 90	d 74		w 12	d 60	d 79	d 65–92
Leaving den regularly	d 90	d 118								
Activities away from nest for at least 5 hours	d 100									
Rest away from nest regularly	d 113	d 135	m 3							
Rest away from nest consistently	m 8	m 4								
Single rests in den until	m 12									
NEST BUILDING										
First nest-building movements	d 65??									
First handling nest material	d 69									
NURSING										
First outside nursing	d 66									
Last observed nursing bout	d 240 - in den!	d 163								
WEANING										
Functional weaning	w 20	w 11								
Nutritional weaning	w 30	w 18	m 6–7	m 5					w 22	w 13
Social weaning	m 8–11	m 8		m 6	m 6		m 6	m 6–7		

FEEDING								
First feeding by dam								d 104
First time solid food	d 105 yew-berries	d 90			d 100	d 135	d 116	d 114–131
Independent feeding	m 5 in feeding hut	m 4–5		m 4			m 4	
DEFECATION								
First independent defecation			d 68					
SCENT MARKING								
Scent-secretion		d 25						
Scent-marking		m 6		d 98				
ANTAGONISTIC BEHAVIOUR								
Against dam or enemy (attack)	d 43	d 16						
PLAY								
Locomotory play	d 45		d 45					
Object play	d 40	d 60	d 45					
Social play								
First play with dam	d 48	d 55						
First play fighting (dam repelling cub)	d 69							

d = day, w = week, m = month; * = reference

FIGURE 10.13 A 4.5-month-old cub following the dam close during climbing in a yew tree.

FIGURE 10.14 Solitary locomotory play of a 66-day-old cub: lying on back and kicking with legs.

in one case the cub stood by the enclosure pond and hurled algae out of the water with its forepaws.

MOTHER–INFANT BEHAVIOUR AND COMMUNICATION

Suckling and Weaning

Red pandas may lie, sit, or stand to suckle their young. The suckling posture seems related to age: in the first 4 weeks the female takes the cub in her mouth and then lies or sits in a corner of the den and places it on her belly, hind legs, or the tail to nurse (Figure 10.17A); from then to the end of the denning period the mother will either sit

FIGURE 10.15 (A) and (B) Object play with sticks (cub 4.5 months old).

(67%) (Figure 10.17C,D), lie (13%) (Figure 10.17B), stand (4%) or use a combination of these positions (16%) during suckling. Manipulation and moving of the cub decreases, as it gets bigger. Similarly, the cub also adopts different positions during suckling; it may lie (on back, side, or belly), sit, or stand during suckling. The positions may change several times in a single bout, especially in older cubs. A kneading, the milk tread, is sometimes visible but does not seem to have the intensity seen in kittens. During suckling, the cub would change teats repeatedly. It is not clear if there are teat preferences or if a special suckling order exists in litters with more than one cub.

The female generally uses a favourite place in the den for nursing behaviour. This is mostly one of the farthest corners, where the ground is hollowed by the frequent use. Nursing usually occurs shortly after the dam arrives in the den but also occurs after sleeping (especially in the first weeks after parturition) or a play session. Data on daily distribution, number, and time of nursing sessions within the denning are presented in Figure 10.18. The nursing bouts were distributed irregularly throughout the day over the

FIGURE 10.16 Object play with pieces of wood brought by the dam (top cub 4.5 months old, bottom cub 68 days old).

whole denning phase. The daily number of nursing bouts per day (mean 3.6, range 1–8, n = 232) increases slightly up to day 70, which may be explained by a higher demand of the cub to suckle at the end of the denning phase. By comparison, the daily nursing time (mean 1.03+0.24 hours, n = 232) decreases a little, maybe the result of the female holding milk production at the same level. The mean length of suckling bouts was 17 minutes (n = 232, range 1–51 min) between days 3 and 70.

On day 66 of cub age, we observed the first nursing outside the den. The female carried the young to a yew trunk and suckled and groomed it there. There were about three preferred "nursing-places" in the yew tree where suckling regularly took place. Regular suckling outside the den occurred from the third month onward. Around day 150, the female began to refuse to suckle the cub in the den although she continued to do this outside in a tree. Nutritional weaning has been identified by different authors as between weeks 13 and 22 (see Table 10.1). However, the latest age at which we observed the cub to suckle was 240 days. The cub begins to chew on twigs brought by the mother to the nest between days 50 and 60. At 5 months, the female carries bamboo twigs and presents them to the cub. An incident of bamboo feeding of a 104-day cub in Cologne Zoo was described as

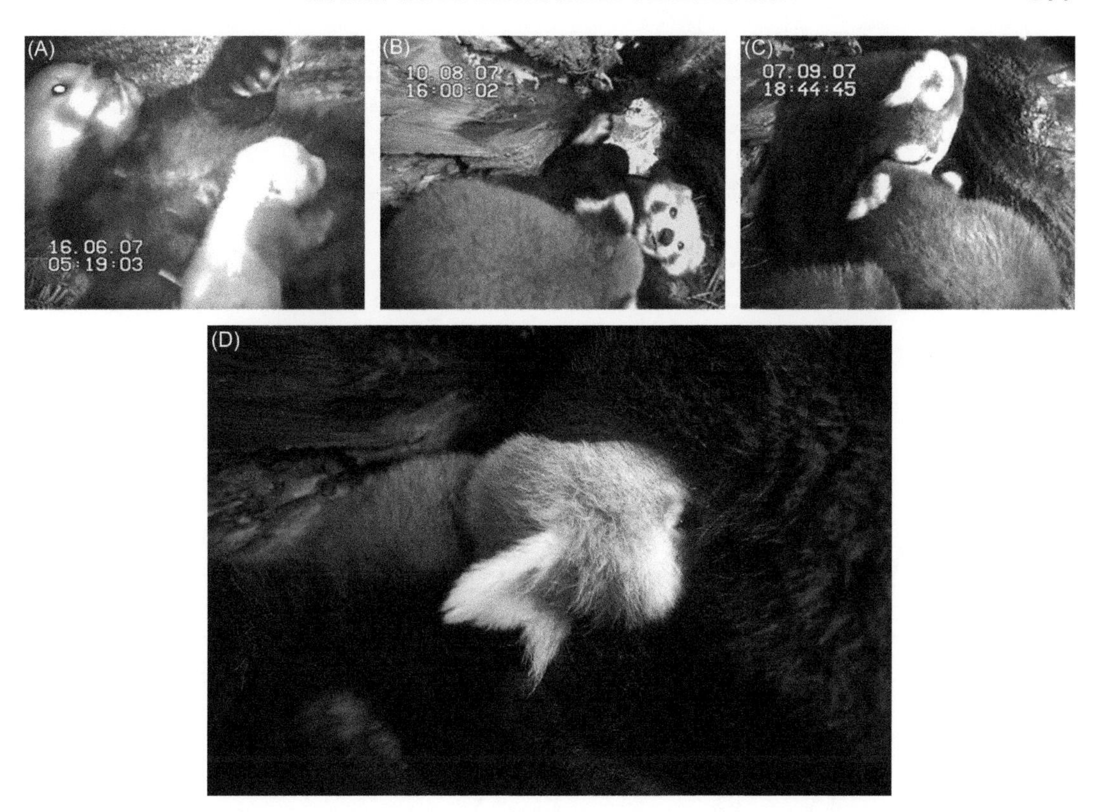

FIGURE 10.17 Different suckling positions of the female: (A+D) lying, (B+C) sitting.

follows: "The female came with bamboo leaves, chewed in her mouth, and the cubs licked repeatedly for 2–3 minutes this food-mash from her mouth" [23].

Although interacting with and chewing on potential food (bamboo twigs, leaves...) occurred as early as day 59, a regular consumption of solid food was not observed until 4 months. At 5 months the cub ate bamboo twigs brought to it in the yew tree by its mother. At the beginning of den leaving (within 10 weeks), the cub independently fed on yew berries. It was first seen using the feeding hut (where fruits, vegetables and other items are offered) at an age of 163 days.

Social Play

Social play is play that is directed toward conspecifics. Play fighting is one form of social play, which resembles behaviours that generally characterize aggressive encounters. Although studies of social play have been published for several carnivore species, they are absent for the red panda with the exception of two unpublished reports from Tennessee University [7,10,24]. Wilson conducted a detailed investigation of play fighting in giant

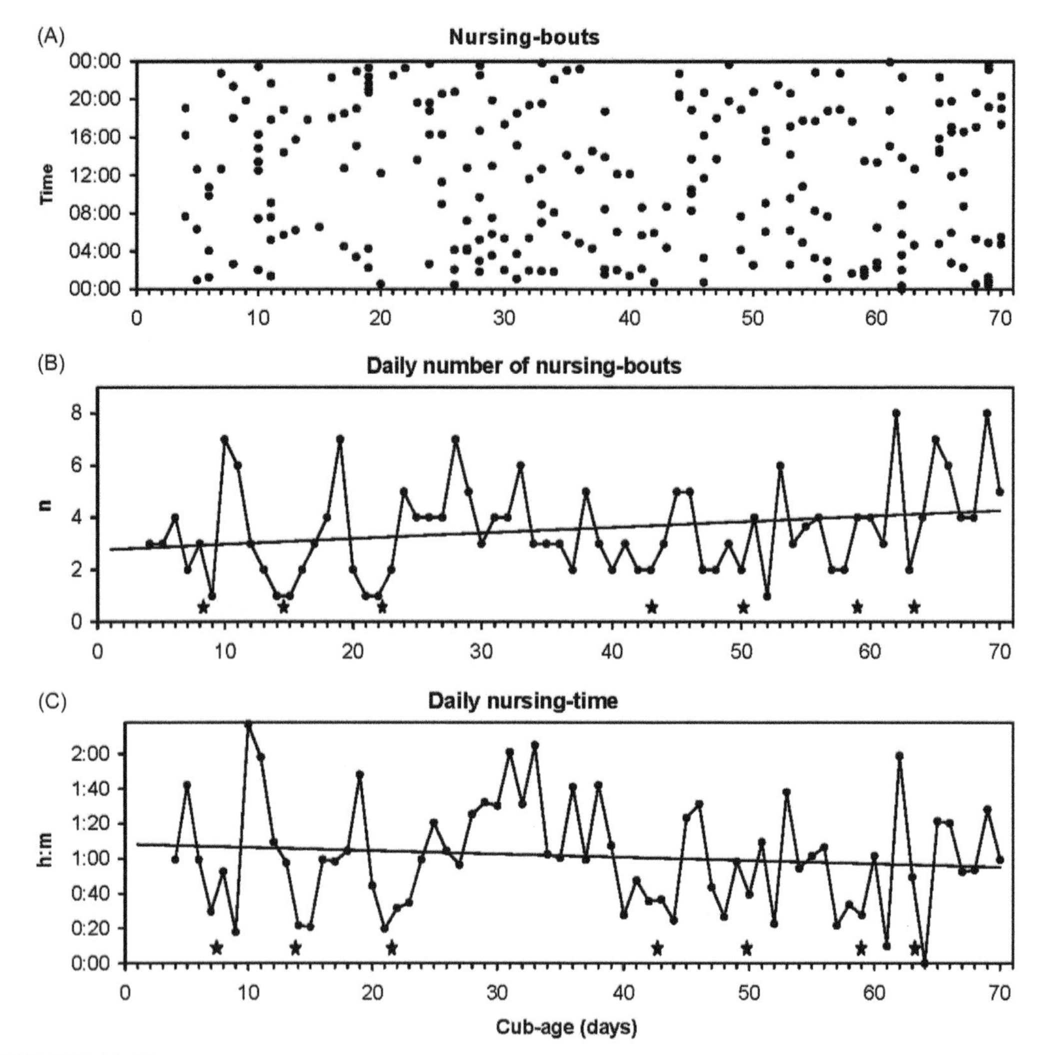

FIGURE 10.18 The nursing bouts are distributed irregularly (top), their daily number increases (centre) and the daily nursing time decreases (bottom) up to day 70 of cub age. (Note that the slope indicated in the figure is for descriptive purposes only. It is based on a parametric regression analysis and not a Spearman's rank correlation.)

panda cubs [25]. We use Wilson's terminology for our description of social fighting in the red panda (Table 10.2 – a summary of all behaviours connected with play fighting).

The first attempt at play fighting was seen on day 48; the cub swatted the dam twice and bit her ear. The movements were slow and awkward and the play bout very short. Over the following days, social play became more frequent but was one sided as the dam did not participate; the mother would ignore these attempts at play and continue what she was doing (Figure 10.19). It was not until day 69 that she began to gently repel the play

TABLE 10.2 Ethogram of social play of red pandas

INITIATION BEHAVIOURS	
Bite move	Incomplete biting with open mouth (does not bite the other animal)
Head shake	Shaking head back and forth/up and down (oriented toward another animal)
Paw laying	Careful and markedly slow paw laying on the back of the other animal
Paw move	Incomplete paw swat: Holding up of paw, but does not swat the opponent
Backwards approach	Approaching to the opponent with (back)side
PLAY BOUT BEHAVIOURS	
Climb	With paw contact to another animal
Stand over	On hind legs and perpendicular to the other animal's body, contact with paws
Bite	Mouth on some part of the other animal: pulling, visibly stretching the opponent's skin or chewing its fur
Claw	Swiping movements at opponent's body with front and/or back paws, from laying position, and thrusting body forward
Paw swat	Beating another animal with paw(s) – making brief physical contact, standing on hind legs
TERMINATION BEHAVIOURS	
Break away	Breaking contact with other animal, remaining without contact or orientation
Push	Lifting of fore or hind paws and placing pressure on the opponent to break a bite-hold
Turn	Twisting and/or rolling to break a bite-hold

This ethogram was adapted from the classification developed by Wilson for Giant Pandas.

attacks of the infant. From day 90 onwards, the mother actively participated in play-fights (Figure 10.20A) although we never observed her initiating any of these interactions (Figure 10.20B). After day 100, the female began to show clearer defensive behaviour towards the cub when it was too rough in its play behaviour; she would bite the cub on the nose, an ear, or a leg. Otherwise, there was no antagonistic behaviour seen between mother and cub. In general, aggression seems to be rare even between red panda adults.

At day 68 we first observed the cub pouncing on its mother's tail as she was leaving the den. The cub hung on to her tail and was dragged along for a short distance. This behaviour continued to be observed until the end of the denning period. No information is available from this study over interactions between siblings. However, Eason reported no quantitative differences of play behaviour between male and female cubs [10]. On the other hand, after the cubs of previous litters had left the den, we regularly saw play behaviour between them and their father (Figure 10.21). Carr also reported that red panda fathers initiated a lot of play bouts [7].

FIGURE 10.19 Two clips of play fighting video scenes: Coming closer (A), fixation (B) and paw swat (C) – cub age 69 and 70 days. Dam is sitting and grooming (left) and does not participate in play.

Antagonistic Behaviour

Antagonistic motivated behaviours can be seen in intra- and in inter-specific situations. We observed the first two attacks at day 43, when the dam entered the log and came to the cub. In both cases the juvenile was startled by the dam, it jumped, leaped on its hind legs, and tried to swat the mother with one or both forepaws. During the short jump the infant uttered special noises (see vocalizations). Once after such a serious attack, the cub exhibited a submissive movement (body and head down, ears back) with a following curling.

We also observed inter-specific aggression when a muntjac entered the den or when we inspected the den or installed cameras, lamps and other equipment in the box behind the log (Figure 10.22). The behaviour is characterized by a short jump, a special alert face with ears back and open mouth and/or tongue testing and followed by puffing and smacking/ chewing as well as tongue testing.

NON-VOCAL COMMUNICATION

Olfaction is an important sense for the red panda, as a solitary species it relies heavily on chemical signals [26,27]. Therefore, it is not surprising that tongue testing was seen commonly, especially when the mother initiated a new contact with her cubs. This behaviour also was regularly seen when the dam entered the den starting with the den environment and ending with the infant. The tongue is protruded slightly, as if licking, and the face appears extra alert (Figure 10.23).

FIGURE 10.20 (A) From day 90 onwards the mother (right) actively participates in play fighting bouts. (B) Play fighting between cub (left) and dam is always initiated by the cub.

Olfactory inspection of the mother becomes more frequent after 3 months; the cub was seen to behave the same way as the dam on entering the den by sniffing and tongue testing the area. Naso-nasal contact was common after the female returned to her cub in the den. Usually, it follows on from the initial sniffing and tongue-testing, although Conover did not mention this behaviour in her report on the behaviour of mother and cubs [8]. In adults naso-nasal contact may be part of the olfactory examination of conspecifics but it also seems to be a form of greeting. We saw naso-nasal contact (without any sniffing) several times in the den, for instance at day 67 (Figure 10.24A) and outside the den immediately after the infant arrived by its mother or vice versa (Figure 10.24B). From an age of 140 days onwards, infants may perform a different greeting ceremony when the female arrives. This "paw-up gesture" consists of a slow lifting of a foreleg and carefully moving it towards the female's head (Figure 10.25). A similar movement of the hind legs can follow this when the cub is lying on the ground. The dam closes her eyes and exhibits tongue testing and/or head shaking and starts to groom the cub immediately. A quite similar

FIGURE 10.21 Play fighting scenes between siblings of the litter from 2001 (10 months old), observed by the father.

behaviour sequence was seen during an encounter of a pair immediately after copulation; the female lifted her forepaw to the male's head [1].

A mutual mouth-licking behaviour ("kissing"), which has only been reported for the red panda, was seen twice during the study, the first occasion was on day 65 and the second time on day 138. On both occasions it was associated with play behaviour.

VOCALIZATIONS

Vocal communication in red pandas is poorly known. Descriptive reports indicate that the species has a repertoire of seven vocal signals although to date no sonograms have been published [4,28–30]. There seem to be similarities in the structural patterns of the high-pitched bleats of giant and red pandas, as well as those of the procyonids [30,31]. Red panda females and males are known to call during mating (a twitter) and in aggressive encounters (huffs and snorts). In this study, we found that vocalizations of the dam during mother–infant encounters were rare and we have not described them here, although when the cub was 101 days old we heard her produce clear, loud whistles as an answer to similar calls uttered by the cub during climbing in a tree.

Although adult red pandas rarely use vocal communication, infants regularly utter calls. Squeaking and whistling sounds have been reported immediately after birth [1]. We

FIGURE 10.22 (A) Attack face of a cub (ears back and open mouth) – a reaction to installation noises in a camera box behind the log. (B) Attack of one of the two siblings of the 2005 litter (left) with snorting and paw swat against a person who is inspecting the den.

recorded them both prior to and during suckling bouts up to an age of about 3 months (Figure 10.26). Around day 150, when the mother tries to refuse cubs more often from nursing, we registered a higher calling activity again. The infants ask the dam to continue nursing by squealing and bleating. A special nursing sound, which is known as a common series of noisy (normally low-pitched) calls in giant pandas (Bruckner, personal communication, [28]), does not exist in *Ailurus*. Occasionally a quiet, snoring sound is audible from red panda infants (Figure 10.27). Red panda cubs whistle when they are abandoned or stressed (Figure 10.28). Roberts heard this call, which he termed "Wheet", after day 30 when the cubs were handled by the keeper [2]. This call type could also be heard frequently between days 70 and 100 when the cub was leaving the den.

FIGURE 10.23 When the female enters the den she exhibits tongue testing as olfactory inspection combined with a special alert face and erected and slightly forward directed ears.

When close to the dam or following her, infants may chirp or thrill for contact (Figure 10.29). This call sounds like the precursor of the adult mating twitter.

Finally, the older cub would also sniff, exhale, or chomp when excited and huff or snort in antagonistic situations (Figure 10.30).

Discussion

The red panda is an altricial species as is the giant panda. Both are similar in their degree of helplessness. The giant panda has been referred to as the most altricial of all eutherian mammals; more than 99% of body growth and 99.9% of brain growth are postnatal in Ailuropoda [32]. In the red panda, 98% of growth is postnatal, the maturation of the cub is attenuated [33]. The first 3–4 months of life are spent in the den, and red panda mothers face substantial challenges in rearing their offspring. The data provided in this chapter refer solely to zoo animals; there is virtually no information available on maternal behaviour in the wild. From this chapter, it can be said that the ontogeny of behaviour of red pandas can be divided in two main phases which have several distinct stages:

1. The denning phase (first 3 months) which is characterized by physical maturation and the development of locomotion, grooming, nest building, vocal, social, and play behaviours. Around days 60–70, when red panda cubs begin to leave the den for the first time, these activities gradually increase in frequency. The social play bouts with both siblings and parents become very boisterous. This continues after the denning phase.
2. The juvenile phase or the time between leaving the den and social weaning (3–approx. 8 months) is characterized by high levels of exploratory behaviour, perfection of locomotor abilities (especially climbing), learning to forage and to eat solid foods and by training social behaviours during play or encounters with unknown adult conspecifics. In the zoo, the bond between mother and young may continue beyond one year of cub age. It is completely unknown how long this phase lasts in the wild (see Table 10.1).

FIGURE 10.24 Nose to nose contact as a greeting ceremony between dam and cub: (A) at day 67 in the den (cub left), (B) at day 117 outside in a tree (cub right).

FIGURE 10.25 "Paw-up-gesture" – a greeting ceremony performed by cubs when the dam enters the den: slow lifting of a foreleg and careful moving to the female's head.

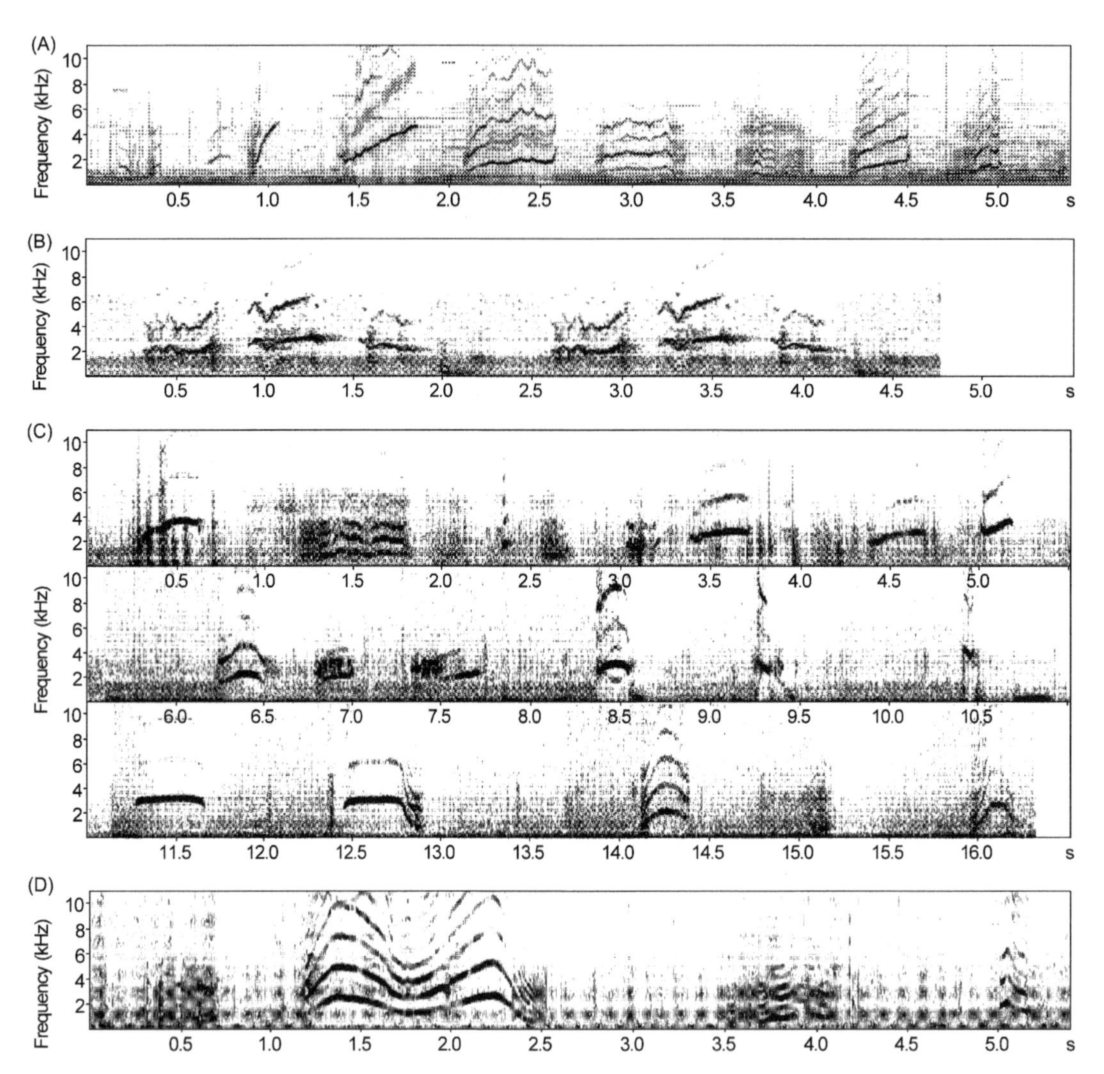

FIGURE 10.26 Squeaking sounds of red panda cubs at different ages and situations: (A) Squeak-twitters of a newborn (?) cub in the den, Darjeeling, Singhalila NP, cut from video "Cherub of the mist" [1]; (B) high-pitched squeaky notes of a cat-reared cub (age 3 weeks) scrawling between kittens, Artis Zoo Amsterdam; (C) high-pitched loud squeaking calls of a cub (age 7 weeks) immediately before suckling begins; (D) low- and high-pitched squeaking of a cub (age 8 weeks) during suckling.

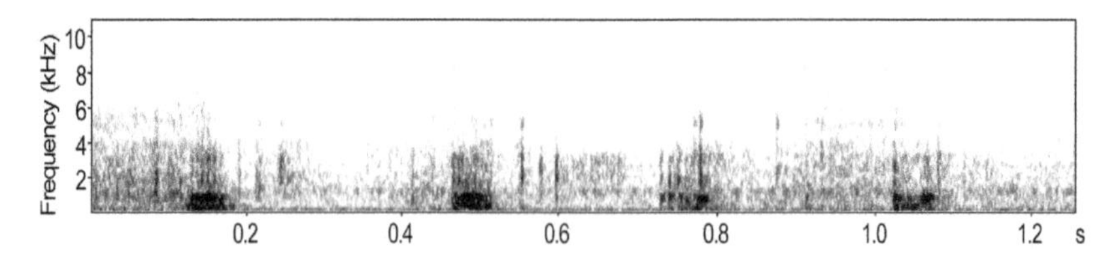

FIGURE 10.27 Snoring sound of a cub (7 weeks old) during suckling.

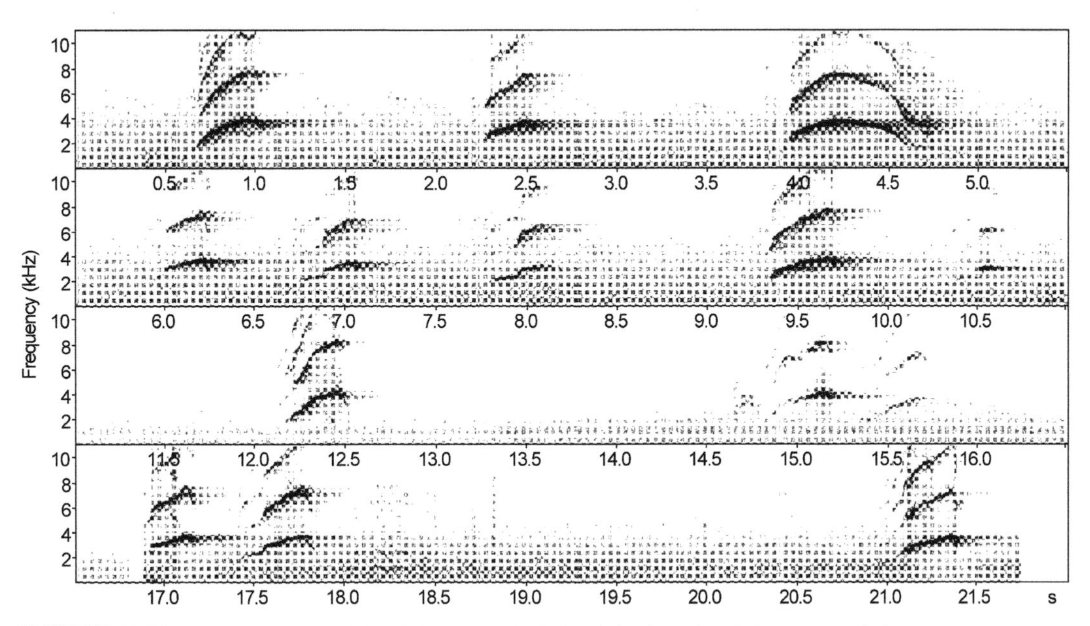

FIGURE 10.28 High-pitched and loud distress calls (whistles) of a sick cub (age 3 months).

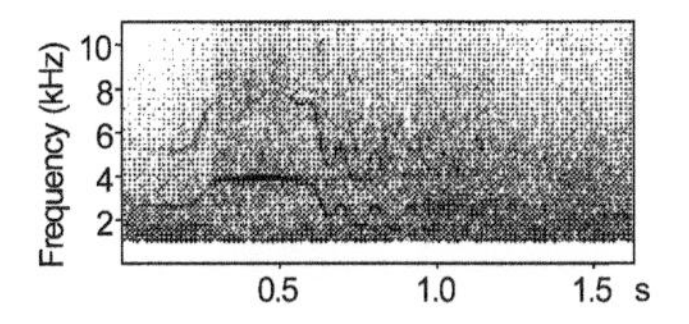

FIGURE 10.29 High-pitched contact thrill of a probably 3-month-old cub following the dam. Darjeeling, Singhalila NP, cut from video "Cherub of the mist" [1].

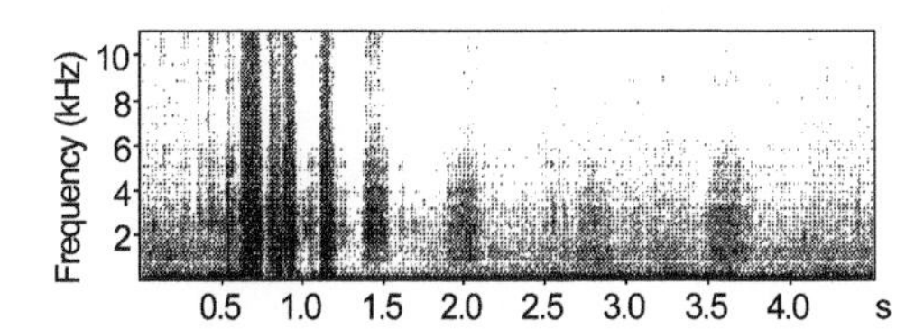

FIGURE 10.30 Puffing/attack noises of a cub (age 7 weeks) against person who is managing the camera.

Numerous descriptive data from zoo studies can be found about the denning time of red panda females. During the first few days after birth mothers remain with the cubs 60–90% of the day [2,4,5,23,35–38]. After the cubs are about 1 week old, the female gradually spends more time away from the offspring, returning every few hours to nurse and groom them, and the number of nest checks and moves declined significantly after day 35 [2,4]. Until now, only one paper has been published with quantitative data about daily distribution of denning [42]. The dam in Madrid Zoo, where summer temperatures are fairly high, mainly spent the nights (22:00–06:00) with her cub for the first 50 days after parturition. At Edinburgh Zoo, a female spent 30% of daytime in the nest box in weeks 3 and 4,

43% in weeks 5–8 and less than 20% until week 16 [43]. Our analysis shows a clear decrease of duration of stay in the den (see Figure 10.4).

Investigations at National Zoological Park Washington showed that nursing visits from the mother became more brief and nocturnal as lactation proceeded [44], and that even after the onset of feeding solid food the female continued to nurse the cubs, although at reduced frequencies [2]. A slightly different picture is produced by our detailed studies. The nursing bouts are distributed more or less irregularly in daytime and increase during the entire denning phase (see Figure 10.7).

There are some investigations into circadian rhythms of adult red pandas [15,36,45,46]. The bimodal daily activity pattern (crepuscular with substantial nocturnal activity in summertime) was described for wild animals in Wolong and for zoo-kept animals [15,41]. This pattern may be different in remote areas with less disturbance as is reported by Wei (see Chapter 11). For this chapter, we were only able to analyse cub sleeping behaviour during the first 3 weeks of life (see Figure 10.12). There is an uncoordinated rhythm in the first week, but a more or less stable quantum of sleeping time per day, which is not yet similar to that of adult animals.

Detailed descriptions showed that captive red panda females react with excessive cub carrying on external disturbances like visitors, traffic, etc. [2,3,20]. In Görlitz Zoo, after the installation of lamps, cameras etc. in the box behind the den, the female carried the cub sometimes to the second log and/or out of the den to a tree, a tree trunk or other parts of the enclosure. One time, the female brought the cub to a hollow yew trunk and covered the juvenile with a fresh yew twig to hide it. The negative consequences of such disturbances can be seen in changed activity budgets for the duration of 2–6 days (see Figure 10.8). Cavities in mature conifers provide sheltered dens inaccessible to large felid or canid predators, but smaller ones like yellow-throated marten (*Martes flavigula*) could enter such dens [18]. A stone marten (*Martes foina*) climbing on the roof of the den in Naturschutz-Tierpark Görlitz caused a great concern in the female and, finally, a retrieval of the cub to the second log. Two and a half hours later the marten came into the den for sniffing/controlling. Also environmental stress and abnormal offspring behaviours can induce cub carrying. With the rise of external temperatures, the female spends significantly less time in the nest [47]. High air temperatures or a wet nest hole due to heavy rain can also cause cub carrying. In September 2002, we observed an excessive carrying of one of the two cubs (age 3 months). This juvenile suffered from a liver disease and died a short time later. Mid-July 2007, the young did not suckle enough because of high temperatures in the den. Obviously, this changed behaviour led to an increased carrying by the mother. Young retrieval was also observed in wild European otter (*Lutra lutra*). We found a 2-week-old cub with a fungal disease on a field far from the nest site. Such behaviour makes sense in order to prevent a contamination of siblings. Cub carrying can also serve as a training of the cubs during den leaving time. From a cub age of about 65 days, the female carries the young out of the den, possibly to teach/show the offspring the first steps in the nest surroundings.

Zoo-kept males, which are typically left in the same enclosure, exhibit increased avoidance behaviour towards the females after parturition [3]. In rare cases, they sleep in the same nest boxes as the cubs. Usually interactions between offspring and its male parents begin to occur after the denning phase, at about 12 weeks of cub age. In encounters at the beginning, males exhibit avoidance behaviour and then gradually interact with the infants and may also participate in play sessions [3,8]. Participation of males in red panda play fighting is well known but not yet studied in detail [2,3,5,39]. Keller observed adult social and locomotor play behaviour especially during the courtship period [19]. The male displays running, hanging upside down, and somersaulting as "show-playing" and holding and wrestling sometimes end with copulation.

Finally, we want to mention some inter-specific behaviours. In a few cases, other animal species used the dens in Naturschutz-Tierpark Görlitz for resting, feeding, or only visiting. At the beginning of the denning phase, a common toad (*Bufo bufo*) slept over daytime in the den, once even on the back of the red panda mother! She never showed any aggressive reaction but sniffed the toad intensively. Only once she touched it carefully with a forepaw. After the denning phase (but den still used as sleeping place), we saw a shrew (*Crocidura* spp.) and a striped field mouse (*Apodemus agrarius*) in the den. Regularly adult or juvenile Reeve's muntjacs (*Muntiacus reevesi*), which inhabit the same enclosure, walked by the den entrance or visited the den. This did not elicit special reactions of the female but did in the cub. When the cub slept deeply, no reaction could be seen, but normally, the incoming muntjac elicited alert or aggressive reactions from the cub. The stranger was attacked with a short jump and puffing/snorting sounds.

The infant exhibited the same type of antagonistic behaviour against humans, for example, when installations were done in the box behind the den or when the keeper had a look into the log from the entrance. The first time we saw this was at day 43 of cub age, but Roberts noted the first "Quack-snort" of two siblings already at days 16 and 18, simultaneously with partial opening of the eyes [2].

Although our female and her daughter allowed us some new insights into the life in a red panda den, we are far from a complete understanding of maternal care and infant development. For example, we know nothing about individual differences or the influence of the dam's experiences or the number of cubs on behaviour. Our understanding of the complexity of social behaviours and vocal communication is limited; never mind about the happenings in a nest hole of wild red pandas. For the moment, we should try to find out more in zoo-kept animals to minimize disturbances at the dens of wild-living conspecifics. And we should use all our insights (and the pictures) of one of the most handsome animals on earth for its conservation.

References

[1] N. Bedi, R. Bedi, Cherub of the Mist. Movie, Bedi Films/Visuals, India, 2006.

[2] M.S. Roberts, Growth and development of mother-reared red pandas, Internat. Zoo Yearbook 15 (1975) 57–65.

[3] M.S. Roberts, Breeding the red panda (*Ailurus fulgens*) at the National Zoological Park, Zool. Gart. Neue Folge 50 (1980) 253–263.

[4] M.S. Roberts, The reproductive biology of the red panda (*Ailurus fulgens*) in captivity. MS Thesis, University of Maryland, 1981.

[5] M.S. Roberts, J.L. Gittleman, Ailurus fulgens, Mammal Species 222 (1984) 1–8.

[6] M.S. Roberts, D.S. Kessler, Reproduction in red pandas, *Ailurus fulgens* (Carnivora: Ailuropodidae), J. Zool. London 188 (1979) 235–249.

[7] A. Carr, Panda play behavior. Unpublished archive report. Comp. Anim. Behav. Lab., University of Tennessee, Knoxville, 1997, 1–16.

[8] G. Conover, The maternal role in the care of juvenile red panda. Unpublished archive report. Comp. Anim. Behav. Lab., University of Tennessee, Knoxville, 1985, 1–13.
[9] N. Dietz, The red panda (*Ailurus fulgens*), Environmental influences on behavior. Unpublished archive report. Comp. Anim. Behav. Lab., University of Tennessee, Knoxville, 1982, 1–16.
[10] C. Eason, Sex differences in the growth and play behavior of captive red panda cubs growth. Unpublished archive report. Comp. Anim. Behav. Lab., University of Tennessee, Knoxville, 1987, 1–11.
[11] C. Okhuysen, Maternal grooming in captive red panda *Ailurus fulgens fulgens*. Unpublished archive report. Comp. Anim. Behav. Lab., University of Tennessee, Knoxville, 1994, 1–15.
[12] A. Gebauer, Anlage für Kleine Pandas, Weiße Ohrfasane und Moschustiere, Ciconia, Jahresbericht Tierpark Görlitz 11 (1998) 24–27.
[13] A. Gebauer, An exhibit for red pandas at Naturschutz-Tierpark Görlitz, EAZA News 38 (2002) 26–27.
[14] A. Gebauer, An exhibit for red pandas at Naturschutz-Tierpark Görlitz. <http://www.zoolex.org/zoolexcgi/view.py?id=311>, 2008.
[15] A. Gebauer, S. Engler, Anmerkungen zum Nahrungsverhalten und zur Aktivität Roter Pandas (*Ailurus fulgens fulgens*) in einem großen Freigehege, Zool. Gart. Neue Folge 71 (2001) 209–220.
[16] G.B. Schaller, H. Jinchu, P. Wenshi, Z. Jing, The Giant Pandas of Wolong, University of Chicago Press, Chicago, USA, 1985.
[17] G.B. Schaller, The Last Panda, University of Chicago Press, Chicago, USA, 1994.
[18] D.G. Reid, J. Hu, Y. Huang, Ecology of the red panda, Ailurus fulgens, in the Wolong Reserve, China J. Zool. London 225 (1991) 347–364.
[19] R. Keller, The social behaviour of captive lesser pandas (*Ailurus fulgens*) with some management suggestions, in: A.R. Glatston (Ed.), The Red or Lesser Panda Studbook No. 1, The Royal Rotterdam and Botanical Gardens, Rotterdam, The Netherlands, 1980, pp. 39–55.
[20] K. Conway, Supplemental feeding of maternally reared red pandas, Internat. Zoo Yearbook 21 (1981) 236–240.
[21] T.I. Grand, Altricial and precocial mammals: A model of neural and muscular development, Zoo Biol. 11 (1992) 3–15.
[22] R. Fagen, Animal Play Behavior, Oxford University Press, New York, USA, 1981.
[23] T. Pagel, Der Kleine Panda Ailurus fulgens – Haltung und Zucht im Zoologischen Garten Köln, Zeits. Kölner Zoo 39 (1996) 139–155.
[24] K.R. Greenwald, L. Dabek, Behavioral development of a polar bear cub (*Ursus maritimus*) in captivity, Zoo Biol. 22 (2003) 507–514.
[25] M.L. Wilson, An investigation into the factors that affect play fighting behavior in giant pandas. PhD thesis. Georgia Institute of Technology, 2005.
[26] C. Li, X.M. Wang, Behavioral responses of the red panda *Ailurus fulgens* to opposite- and same-sex scents, Acta Zool. Sin. 52 (2006) 794–799.
[27] H. Bartmann, W. Bartmann, Gelungene künstliche Aufzucht eines Kleinen Pandas (*Ailurus fulgens*) im Zoologischen Garten Dortmund, Zeits. Kölner Zoo 20 (1977) 107–112.
[28] G. Peters, A comparative survey of vocalization in the giant panda (*Ailuropoda melanoleuca*, David, 1869), Bongo 10 (1985) 197–208.
[29] G. Peters, A note on the vocal behaviour of the giant panda, Ailuropoda melanoleuca (David, 1869), Zeits. Säugetierkunde 47 (1982) 236–246.
[30] D. Kleiman, G. Peters, Auditory communication in the giant panda: motivation and function, in: S. Asakura, S Nakagawa (Eds.), Giant Panda. Proceedings of the Second International Symposium on Giant Panda, Tokyo Zoological Park Society, Tokyo, Japan, 1990, pp. 107–122.
[31] O. Sieber, Vocal communication in racoons (*Procyon lotor*), Behaviour 90 (1984) 80–113.
[32] J.L. Gittleman, Are the pandas successful specialists or evolutionary failures? BioScience 44 (1994) 456–464.
[33] T.I. Grand, Altricial and precocial mammals: A model of neural and muscular development, Zoo Biol. 11 (1992) 3–15.
[34] G.S. Mottershead, Interesting experiments at the Chester Zoo, Zool. Gart. Neue Folge 24 (1958) 70–73.
[35] A.H.M. Erken, E.F. Jacobi, Successful breeding of lesser panda (*Ailurus fulgens*) and loss through inoculation, Bijdragen Dierk 42 (1972) 92–95.
[36] R. Keller, Beitrag zur Ethologie des kleinen Pandas (*Ailurus fulgens*, Cuvier,1825). PhD Thesis, University of Zurich, Switzerland, 1977.

[37] F. Wall, Himalaya catbears (*Ailurus fulgens*), Bombay J. Nat. Hist. Soc. 18 (1908) 903–904.
[38] Y. Iwaki, The growth and development of KOROKORO, an A. f. styani cub, in: A.R. Glatston (Ed.), The Red or Lesser Panda Studbook No. 9, The Royal Rotterdam and Botanical Gardens, Rotterdam, The Netherlands, 1995, pp. 13–18.
[39] M. Lopez, C. Talavera, D.C. Taylor, The breeding and rearing of Red pandas in Madrid Zoo, in: A.R. Glatston (Ed.), The Red or Lesser Panda Studbook No. 8, The Royal Rotterdam and Botanical Gardens, Rotterdam, The Netherlands, 1994, pp. 7–17.
[40] P. Vogt, C. Schneidermann, B. Schneidermann, Hand-rearing a Red panda *Ailurus fulgens*, Internatl. Zoo Yearbook 20 (1980) 280–281.
[41] P. Müller, Nachzucht bei den Kleinen Pandas. Panthera, Mitteilungen aus dem Zoologischen Garten Leipzig, Germany, 1987, pp. 10–12.
[42] C.M.J. Fernandez Gonzalez, Maternal behavior of a red panda (*Ailurus fulgens*) at Madrid zoo: the first three months, in: A.R. Glatston (Ed.), The Red or Lesser Panda Studbook No. 4, The Royal Rotterdam and Botanical Gardens, Rotterdam, The Netherlands, 1986, pp. 13–18.
[43] M. Stevenson, L. Annes, J. Hanning, N. Smith, Red pandas at Edinburgh Zoo, in: A.R. Glatston (Ed.), Red Panda Biology, SPB Academic Publishing, The Hague, The Netherlands, 1989, pp. 103–114.
[44] J.L. Gittleman, Behavioral energetics of lactation in a herbivorous carnivore, the red panda (*Ailurus fulgens*), Ethology 79 (1988) 13–24.
[45] K.G. Johnson, G.B. Schaller, J. Hu, Comparative behavior of red and giant pandas in the Wolong Reserve, China J. Mammal. 69 (1988) 552–564.
[46] B. Holst, The activity patterns of the red pandas at Copenhagen Zoo, in: A.R. Glatston (Ed.), Red Panda Biology, SPB Academic Publishing, The Hague, The Netherlands, 1989, pp. 115–122.
[47] A.R. Glatston, Designing for red pandas, in: P.M.C. Stevens (Ed.), Fourth Symposium of Paignton Zoo and Botanical Gardens – Zoo Design and Construction, Whitley Wildlife Trust, Paignton, UK, 1992, pp. 204–211.

CHAPTER

11

Red Panda Ecology

Fuwen Wei[1] and Zejun Zhang[2]

[1]Key Laboratory of Animal Ecology and Conservation Biology, Institute of Zoology, the Chinese Academy of Sciences, Chaoyang, Beijing, People's Republic of China

[2]Institute of Rare Animals and Plants, China West Normal University, Nanchong, Sichuan, People's Republic of China

OUTLINE

INTRODUCTION

Although Fredric Cuvier was the first person who proposed *Ailurus fulgens* as the zoological name for the red panda in 1825 [1], Thomas Hardwicke was perhaps the first to describe the species and its ecological traits and, in 1821, gave a verbal report at a meeting of the Linnaean Society of London [2]. About two decades later, Hodgson [3,4] contributed much to the knowledge of the red panda by studying both live and killed specimens

Red Panda. DOI: 10.1016/B978-1-4377-7813-7.00011-2

which were bought from native hunters in Nepal. For example, he claimed that red pandas were active at dusk, dawn, and at night. After his work, studies on wild red pandas almost entirely ceased for over 100 years until late in the 20th century. Such a long-term cessation aroused from, to a great extent, the difficulty in studying the elusive and solitary animal, which inhabits faraway mountainous ranges with dense forest cover, with isolated and sparse populations and low accessibility for manpower. Of course, political and social conditions contributed, too. By now, relatively systematic and in-depth studies on wild red pandas have been conducted in their inhabited ranges respectively, such as in Nepal [5–8], China [9–20], Myanmar [21], and India [22,23].

Perhaps the most attractive biological feature the red panda offers science is that, although belonging to the order Carnivora, and thus possessing a simple and short digestive tract typical of a carnivore, it feeds on a bamboo diet. How can it subsist on bamboos, which are low in protein and high in cellulose, and what has it developed in ecological traits to meet its daily nutrient and energy demands? In this chapter, we begin with its habitat and food habits, and then introduce its ecological traits including feeding, digestion, habitat selection, movement, home range, time budget, and activity rhythm as well, based on the information that is available so far. Since the red panda is widely sympatric with the giant panda (*Ailuropoda melanoleuca*), another famous bamboo-feeding carnivore in western Sichuan mountainous ranges, China [11,18,19], we finally discuss how they can coexist in sympatry.

POPULATION ECOLOGY

Habitat

Except in Meghalaya, where it is found in tropical forests, the red panda is basically an animal of subtropical and temperate forests [22]. For the subspecies, *A. f. styani*, its distribution in elevation ranges from 1400 to 3400 m [24], but for the nominate subspecies, *A. f. fulgens*, the distribution in elevation extends much wider (in Chapter 7 "The taxonomy and Phylogeny of *Ailurus*", Colin Groves ranked both *styani* and *fulgens* as independent species based on the Phylogenetic Species Concept, namely *A. styani* and *A. fulgens*). Roberts and Gittleman [25] recorded the animal's altitudinal range to 2200–4800 m, Choudhury [22] claimed it can be found at 1500–4800 m, and almost up to the summer snowline at 5000 m. In Meghalaya, the red panda is found at elevations of 700–1400 m [26]. It was said that somebody even observed it at the elevation of 200 m in Siju Sanctuary adjacent to Balpakram National Park [22].

The red panda can be found in multiple vegetations, including evergreen forests, evergreen and deciduous mixed broad-leaf forests, deciduous forests, deciduous and coniferous mixed forests, and coniferous forests as well, which are associated with bamboo-thicket understories [12,13,22,24,25,27–29]. In China, bamboos in its habitats include not less than 40 species in total, primarily belonging to the genus *Fargesia*, *Yushania*, *Bashania*, *Chimonobambusa*, *Qiongzhuea*, *Indocalamus*, and *Phyllostachys* [30,31]. Even so, the red panda usually feeds on one or two bamboo species in its habitat. For example, in the Qionglai Mountains, China, although there are *Bashania faberi*, *Fargesia robusta*, *Yushania brevipaniculata*, and *Phyllostachys nidularia* available, more than 90% of the red panda's annual diet was composed of *B. faberi* leaves [10,17]. In Mabian Dafengding Nature Reserve (hereafter

Mabian Reserve) of the Liangshan Mountains, China, there are *Qiongzhuea macrophylla*, *Yushania glauca*, and *Chimonobambus pachystachys*. However, the red panda there primarily foraged on *Q. macrophylla*, occasionally on *Y. glauca*, and never on *C. pachystachys* [14]. Similar phenomena were found in Yele Nature Reserve (hereafter Yele Reserve) of the Xiaoxiangling Mountains, China, and in Langtang National Park (hereafter Langtang Park), Nepal [8,12,13].

FORAGING ECOLOGY

Diet

It is difficult to track and observe red pandas directly in the wild. As such, most knowledge on their diet was gathered through indirect analysis of components in their fresh droppings [9,10,12–14,17]. The commonly adopted method is monthly to collect fresh droppings left in the environment, and then oven-dry them. The items in these droppings, such as leaves, shoots, fruits, are separated and weighed respectively. Food items in droppings are roughly digested, and therefore can be easily separated [12,13]. The proportion of each food item in their diet is usually estimated by weight of dry matter [9,12,13,17], or by occurrence frequency [10].

In Wolong Nature Reserve (hereafter Wolong Reserve), Sichuan Province, China, Johnson et al. [9] found that their diet consisted primarily of *B. faberi* leaves (99.1%), as shown by a sample of 332 droppings inspected and only 3.9% of droppings collected contained *B. faberi* shoot remnants. At low elevations in the reserve, it was said that a number of droppings collected in spring were wholly composed of *F. robusta* [32].

Three years later, Reid et al. [10] made a more detailed analysis of their diet in the same reserve, showing that leaves of *B. faberi* were the most important food item with remains found in 93.7% of 791 droppings collected. In spring and autumn, besides leaves, *B. faberi* shoots and fruits of some deciduous shrubs or creeping vines frequently occurred in their diet. Occasionally, *F. robusta* shoots emerging in April and May were eaten. Red panda hair and murid rodent hair were occasionally found in their droppings, too.

Wei et al. [14] surveyed their diet from April to October in Mabian Reserve. In spring (April–June), red pandas preferred new shoots of *Q. macrophylla* (about 87% of dry matter by weight in droppings). In summer–autumn (July–October), besides new shoots of *Q. macrophylla*, they ingested large quantities of leaves, usually from *Q. macrophylla*, and occasionally from *Y. glauca*.

In Yele Reserve, *Bashania spanostachya* leaves are the main food resource for red pandas, which accounted for 89.9% of their annual diet, and especially from November to April, leaves were the only food in their diet [12,13] (Figure 11.1). In addition, they also preferred new shoots during the germinating period, with shoot proportion starting at 17.8% in May, increasing to 42.5% in June and then dropping to 9.4% in August. From late June to October, fruit remains were sometimes found in their droppings, which were judged to be from *Prunus pilosiuscula*, *Prunus* spp., *Rubus foliolosus*, *R. innominatus*, *R. flosculosus*, *Rosa sericea*, *Sorbus koehneana*, *Ribes meyerri*, *R. tenue*, and *R. longiracemosum*. Surprisingly, fir fruits of *Sabina squamata* were found in their droppings occasionally, although they are hard for the red panda to digest.

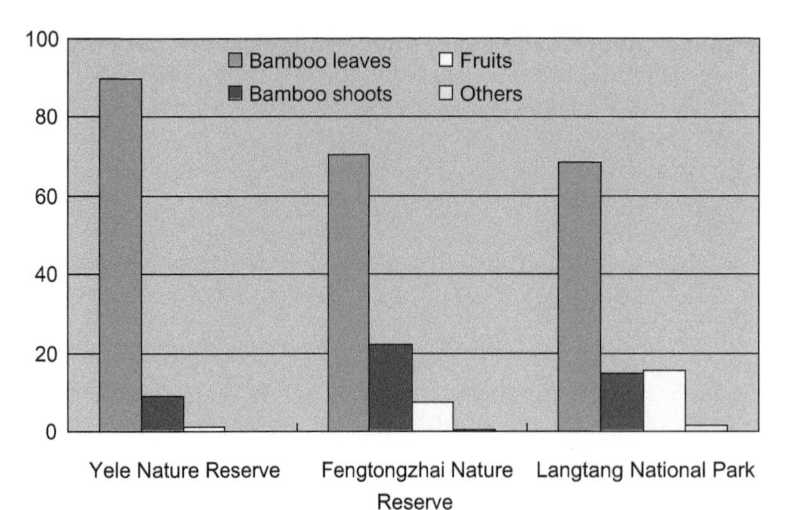

FIGURE 11.1 Annual diets of red pandas in Yele, Fengtongzhai Nature Reserves and Langtang National Park.

Yonzon and Hunter [7] surveyed red pandas' diet in Langtang Park, indicating red pandas there mainly fed on bamboo leaves (68.4%), shoots (14.6%), and arboreal fruits, most commonly *Sorbus* spp. berries (15.4%) (see Figure 11.1). In Fengtongzhai Nature Reserve (hereafter Fengtongzhai Reserve), which borders the Wolong Nature Reserve on the north, red pandas' diet consisted of bamboo leaves (70.5%), shoots (22.1%), fruits (7.2%), and hair (0.2%) during May–November [17] (see Figure 11.1).

In summary, red pandas in different ranges exhibit a similar diet as follows:

- Bamboo leaves are the most important food item in their diet, being almost the only food in winter and usually the most common across all seasons
- Bamboo shoots and some fruits, especially those of *Sorbus* spp., are two important food sources. The former primarily occurs in their diet in spring, and the latter primarily in late summer and autumn
- Maybe depending on availability, they can occasionally ingest some small mammals, birds, eggs, blossoms and acorns, and other parts of their range [3,9]
- Some items in their diet perhaps are ingested by accident, for example, moss and hair.

Dependence on Water

Water plays an essential role in animals' daily life. As such, how closely do red pandas depend on water in their environment? In Wolong Reserve, the water content in their droppings was 72.4±5.5%, and that in *B. faberi* leaves 59.7±8.1% or 12.7% less that in the droppings [9], implying red pandas may have to drink at intervals to compensate for this discrepancy [9]. In the wild, their traces, including faeces, foraging stations, were frequently found at sites close to a water body. For example, in Mabian Reserve, red pandas were frequently found to forage at sites less than 200 m away from a water source [14]. In Langtang Park, observations indicated that proximity to water may be an important habitat requirement because 90% of droppings were found within 100 m of the nearest source of water [8]. Similarly, Pradhan et al. [23] reported that the presence of water bodies at a range of 0–100 m from the point of an animal centred plot was found in 79% cases, indicating the presence of water in the preferred sites is an important habitat requisite.

Feeding

To access bamboo leaves easily, red pandas usually use some elevated objects, such as shrub branches, fallen logs, or tree stumps, to lift their body. In Yele Reserve, 107 of 185 dropping sites (57.8%) were found on shrub branches, 49 (26.5%) on fallen logs, and only 29 (15.7%) on the forest floor [12] (Figure 11.2). Droppings were always found on elevated sites of 1–3 m above the forest floor, occasionally on trees over 12 m [12]. In addition, microhabitats selected by them were also characterized by abundant fallen logs and tree stumps [11,18,19].

When ingesting bamboo leaves, they meticulously nip leaves off branches, not like the giant pandas [9,12,14], which bite off mouthfuls of leaves and branch tips at each foraging bout [33]. However, whether to nip off leaves from the top of branches seems locality dependent (Figure 11.3). In Wolong Reserve, 91.6% of leaves were nipped off intact by red pandas from the joints between the leaves and branches [24], and this is the case in Fengtongzhai Reserve, too [17]. By contrast, in Mabian Reserve, Wei et al. [14] found for 895 leaves surveyed, red pandas usually ingested only half of a lamina (constituting 85.9%). At 15 feeding sites in Yele Reserve, the proportion of half-eaten laminae was also very high, making up 67.4% of the total leaves eaten (n=7316) [12].

Red pandas exhibit a significant preference for bamboo ages. In Fengtongzhai Reserve, they preferred leaves on one- or two-year-old stems during April–June and leaves on old shoots during September–November [17]. Of 226 bamboo stems foraged by them in Mabian Reserve, 80.1% were one or two years old, and the rest were more than three years old [14].

Red pandas prefer newly formed bamboo leaves. In Fengtongzhai Reserve, 89.7% of 981 leaves foraged were newly formed, including expanded leaves (53.4%) and curled ones

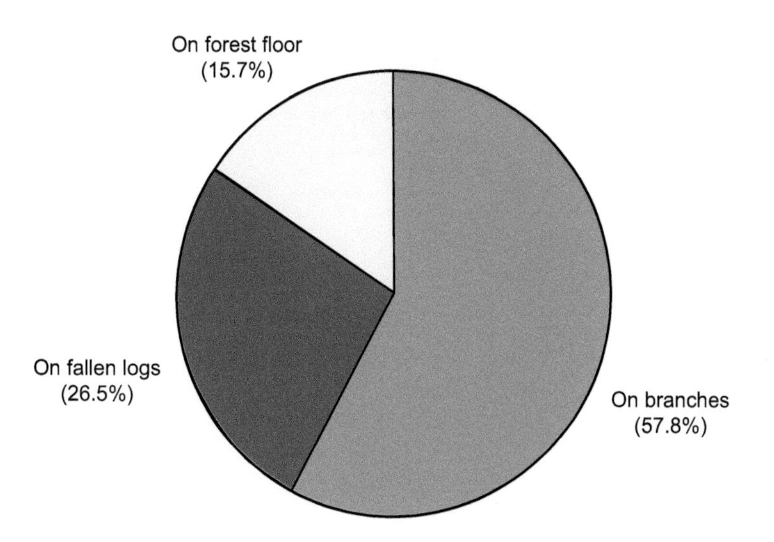

FIGURE 11.2 Occurrence sites of faeces left by red pandas in Yele Nature Reserve.

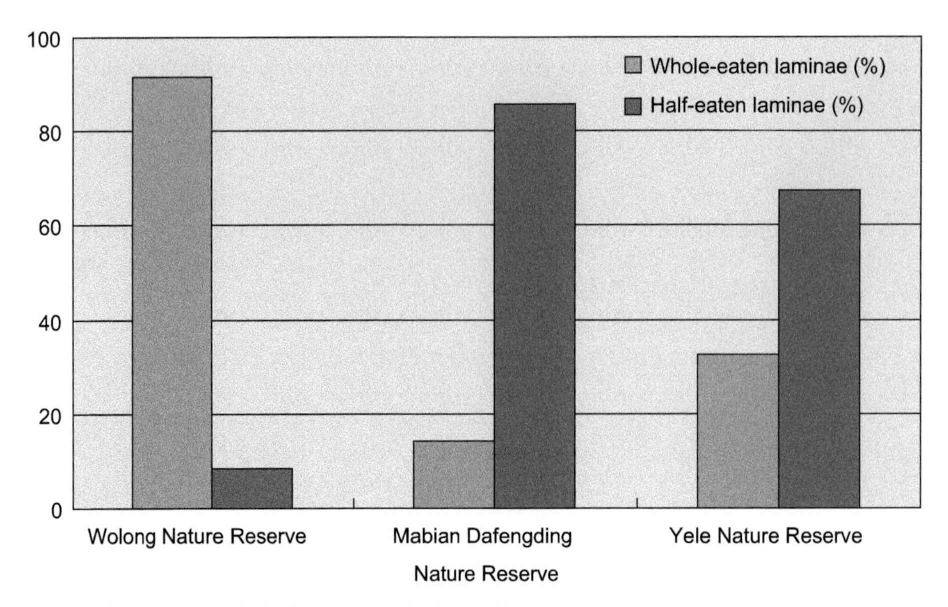

FIGURE 11.3 Percent (%) of whole-eaten and half-eaten laminae by red pandas.

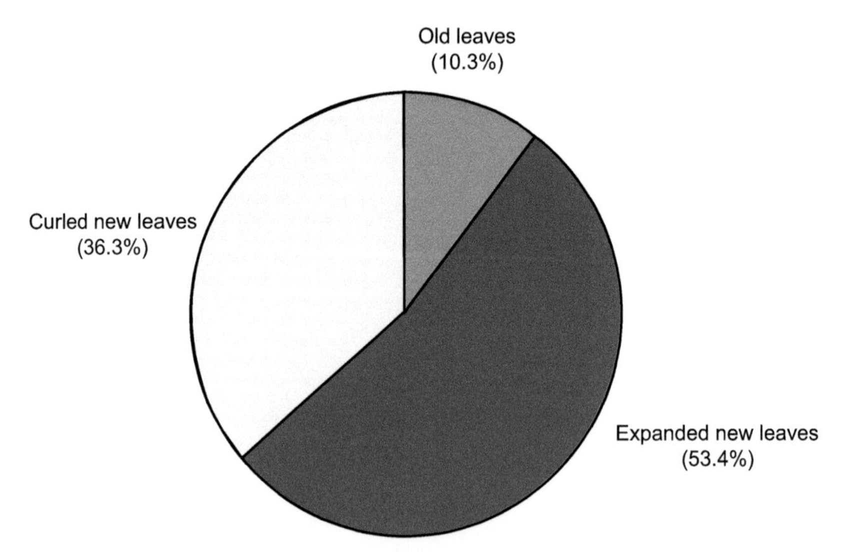

FIGURE 11.4 Selection of bamboo leaves by red pandas in Fengtongzhai Nature Reserve.

(36.3%) [17] (Figure 11.4). At five feeding sites in Mabian Reserve, Wei et al. [14] found they often foraged on tender leaves (78.09%, n = 1018) [17], including unexpanded ones (12.38%). Newly formed leaves are more nutritious than old ones [9,12–14,20,34]. For example, new leaves of *B. faberi* contain 17.5% of crude protein, withering leaves, however, contain only 9% [9,12]. In the Qinling Mountains, new leaves of *B. fargesii* also contain more protein (15.2%) than old ones (10.2%) [34].

Whether red pandas prefer to forage leaves on the upper or lower part of bamboo stems seems controversial. Yang et al. [17] reported that they preferred leaves on the upper part of *B. faberi* stems. However, Johnson et al. [9] found an average of 85.7% of *B. faberi* leaves were eaten at the bottom two nodes as compared to 47.3% at the top two nodes. As there was neither much nutritional difference between leaves near the bottom and top of stems nor a marked difference in growth pattern, Johnson et al. [9] further argued that red panda's leaf selectivity appeared to be related to access, a small animal feeding on a tall food source.

The selectivity when ingesting new shoots is significant, too. In Yele Reserve, red pandas prefer new shoots of *B. spanostachya* with a basal diameter above 10 mm, especially those above 16 mm. In Fengtongzhai reserve, they prefer new shoots of *B. faberi* with a basal diameter above 3.0 mm, and with the height ranging from 10 to 70 mm [17]. Compared with tender ones, robust shoots contain more shoot cores, and can provide more nutrition [13,32].

Mastication

Compared with the giant panda, red pandas chew bamboo leaves more deliberately, with 69.7±5.6% of droppings passing through a 0.84 mm sieve, in contrast with only 6.8±0.6% of giant panda droppings passing through [9]. In Mabian Reserve, 74.68±5.6% of droppings passed through a 1.0 mm sieve (n = 153) [14]. In Wolong Reserve, when red pandas ingested fruits of *Maddenia hypoleuca* and *Cotoneaster moupinensis*, the skin was little chewed, and the seed remained intact [10].

Ingestion and Digestion

During the different seasons, food ingested daily by red pandas varies significantly in Yele Reserve. On average, each panda can consume 1622.4±199.4 g fresh leaves and excrete 1808.8±276.0 g of fresh droppings each day, with dry matter ingested and digested 611.6±67.3 g and 183.8±203 g, respectively [13] (Figure 11.5). In spring, an animal can eat an average of 4242.2±106.7 g and excrete 2598.2±71.5 g each day, with dry matter ingested and digested 402.3 g and 185.9 g, respectively [13] (Figure 11.5). In terms of energy intake and digestion, the daily metabolic energy requirement for a red panda varied from 2603.3 kJ in spring to 3139.8 kJ in summer–autumn and to 2740.8 kJ in winter. To maximize its daily energy intake, a red panda can ingest 10145.8 kJ in spring, 12045.1 kJ in summer–autumn, and 12276.9 kJ in winter [16] (Figure 11.6).

For different food items, the rate of digestibility varies. For *B. spanostachya*, the digestibility rate of leaves is 29.67±0.79%, lower than that of new shoots (45.98±2.77%) [13]. As for nutrient digestibility, crude protein and crude fat possessed the highest digestibility rate, ranging from 70 to 85% (% of dry matter) (Figure 11.7). The digestibility rate of cellulose and lignin is extremely low, indicating that microbial digestion only played a very minor role during digestion (Figure 11.7).

The transit time of new shoots is faster than that of leaves (147.5±33.38 vs 208.0±35.03 min) [13]. Compared with red pandas, the transit time for giant pandas is

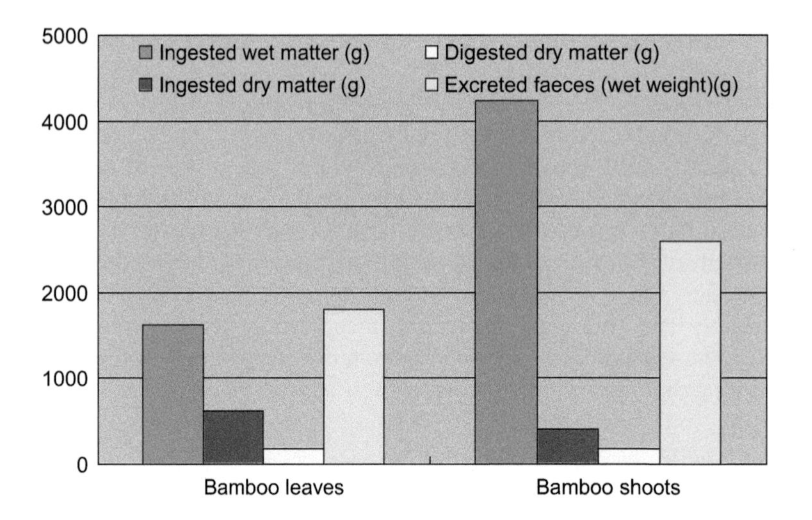

FIGURE 11.5 Ingestion and digestion of bamboo leaves and shoots by red pandas per day.

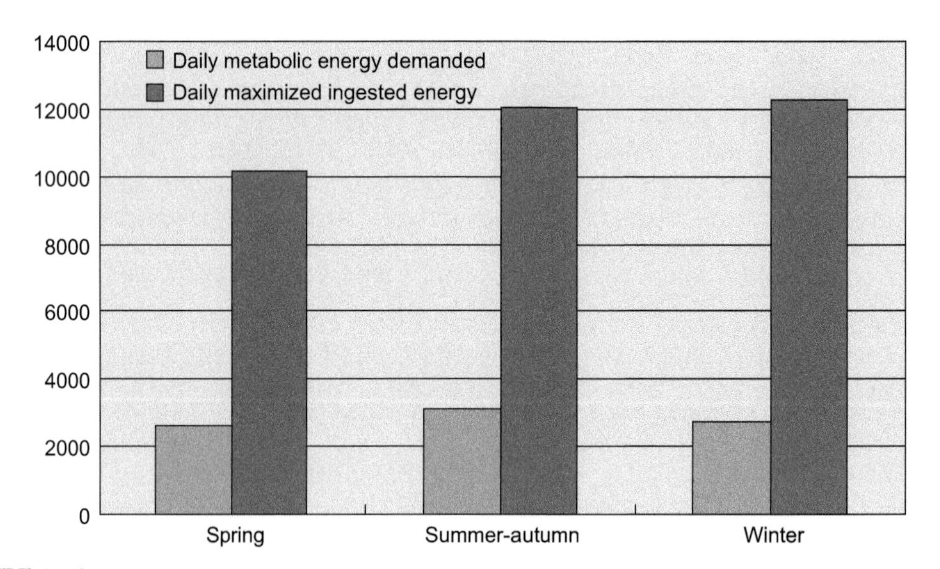

FIGURE 11.6 Daily energy demands and ingestion (kJ) during different seasons.

much longer. For example, in Wolong Nature Reserve, Schaller et al. [32] noticed that the passage time for giant pandas averaged 7.9 h for shoots, 13.8 h for leaves and 10 h for stems (Figure 11.8). However, transit time for the two herbivorous carnivores was much faster than that for the true herbivores, for example, the wapiti *Cervus elephus canadensisi* passed its food in about 51 h [35].

Cherries of *Prunus vaniotii* and *P. brachypoda* were well digested with pits often cracked. The fleshy fruit of *Clematoclethra tiliaceae, Rubus mesogaeus, R. pileatus, Ribes moupinense,* and *R. longiracemosum* were almost completely digested, leaving only the seeds in droppings [10].

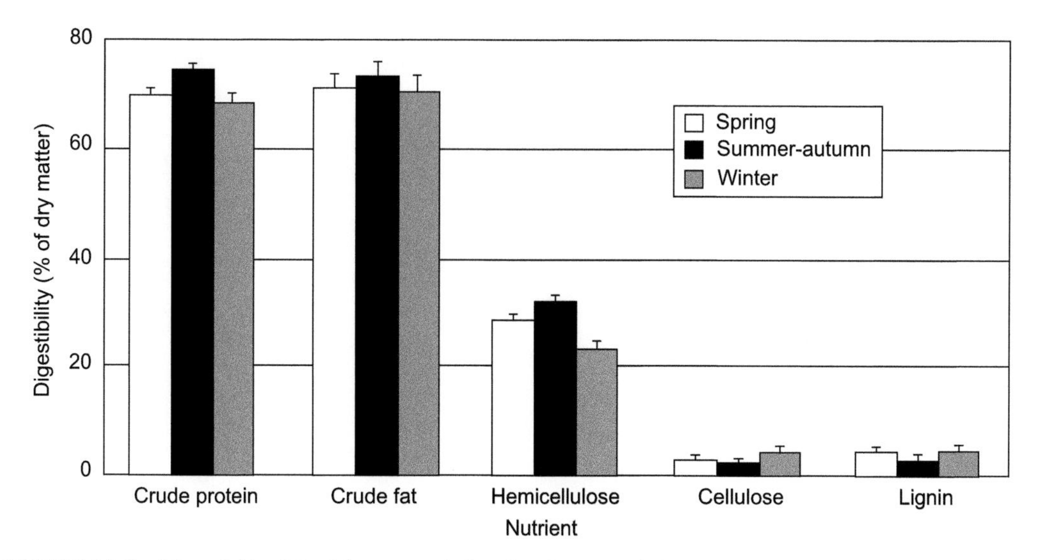

FIGURE 11.7 Digestibility (%) of dry matter in bamboo leaves and shoots.

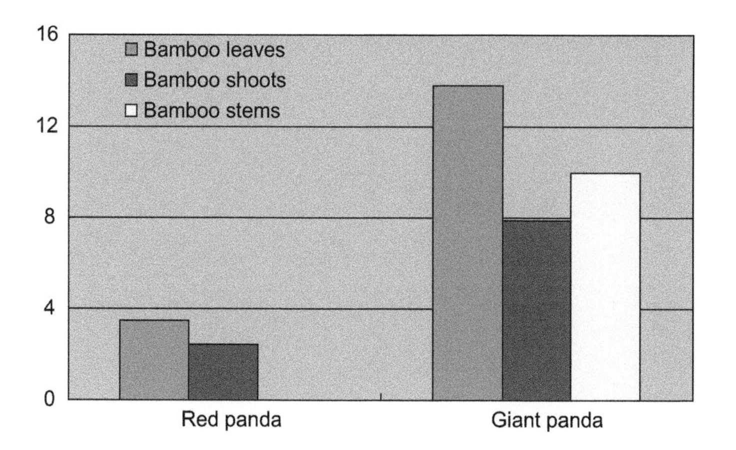

FIGURE 11.8 Comparison of transit time of food items between red and giant pandas.

Foraging Strategies and Nutritional Component Analysis

More than one bamboo species is distributed in the habitats inhabited by red pandas (Table 11.1), but usually only one of them is the primary food source. In Wolong Reserve, the red panda mainly feeds on *B. faberi*, and in Mabian Reserve, Yele Reserve, and Langtang Park, mainly on *Q. macrophylla*, *B. spanostachya*, and *Himalayacalamus falconeri*, respectively (Table 11.1). The only exception is in Singhalila National Park, Darjeeling, India, where two bamboo species, namely *Arundinaria maling* and *A. aristata*, are distributed, and both are the primary food sources for red pandas [23].

The bamboos foraged by red pandas usually have a higher nutritive quality than the rest in sympatry (see Table 11.1). For example, in Wolong Reserve, crude protein in leaves of

TABLE 11.1 Stem Height and Crude Protein Content of Sympatric Bamboo Species in Red Pandas' Habitats

Locality	Staple-food bamboos			Other bamboos in sympatry					
	Species	**Height (m)**	**Crude protein in leaves (%)**	**Species**	**Height (m)**	**Crude protein in leaves (%)**	**Species**	**Height (m)**	**Crude protein in leaves (%)**
Wolong NR	*B. fabeir*	1.0–2.5	14.5	*F. robusta*	2.0–5.0	13.2			
Fengtongzhai NR	*B. fabeir*	1.0–2.5	14.5	*Y.brevipaniculata*	3.0–6.0	13.25			
Mabian Dafengding NR	*Q. macrophylla*	1.5–3.0	15.45	*Y. glauca*	3.0–6.0	13.85	*Chimonobambusa pachystachys*	3.0–6.0	
Yele NR	*B. spanostachya*	1.0–3.5	14.65–16.57	*F. dulcicula*	3.0–4.0		*F. exposita*	3.0–4.5	
Langtang National Park	*H. falconeri*	2.95	15.9	*Thamnocalamus aristatus*	3.7	13.5			

Data on bamboo height from [7,30,31]; date on crude protein in leaves from [7,14, 17,33]

B. faberi is higher than that in *F. robusta* (14.5% vs 13.2%). In Mabian Reserve, they prefer leaves of *Q. macrophylla*, with 15.45% crude protein, higher than that of *Y. glauca* (13.6%) [14]. Similar results were found in Langtang Park and Yele Reserve, too (see Table 11.1).

Even so, there seems another explanation on bamboo selection by red pandas. As mentioned above, this animal usually uses fallen logs, tree stumps, and shrub branches to lift its body, for easier access to bamboo leaves, which implies that the height of the bamboo stems constitutes a limiting factor for them to determine what species to feed on. The higher the bamboo stems are, the lower the availability of bamboo leaves for the animal. So it is not surprising that the red panda would be forced to forage the relatively shorter bamboos in its habitats (see Table 11.1). Coincidently, all these shorter bamboos, being the staple food source for the red panda, are richer in nutrients than their sympatric higher ones (see Table 11.1). In fact, it is not difficult to understand the coincidence, since there is a wide negative relationship between digestibility, and crude protein with twig diameter [36–38].

Among the different parts of bamboos, leaves are the most nutritious, with the highest proportion of crude protein and the lowest of cellulose (Table 11.2). New shoots are the second most nutritious, with the proportion of crude protein higher than branches and stems (Table 11.2). Now that the red panda feed on bamboo leaves all year round and primarily on new shoots in spring, it seems that the animal has developed a set of optimal foraging strategies to maximize its nutrient and energy intakes.

Little is known about the demand of red pandas for trace elements in their food. However, as their year-round food, bamboo leaves usually contain the most abundant trace elements than the other parts. For example, in Yele Reserve, contents of Zn, Cu, Mn,

TABLE 11.2 Nutritional Components in Staple-Food Bamboos of Red Pandas (% of Dry Matter)

Staple-food bamboos	Locality	Items	Composition (% of dry matter)				
			Crude protein	Semicellulose	Cellulose	Lignin	Ash
B. faberi	Wolong NR	Leaves	15.21–15.74	34.93–35.96	27.45–28.07	8.37–8.73	7.89–8.75
		New shoot	14.82	33.90	35.95	5.95	6.41
		Branch	6.15–7.06	28.47–31.59	34.66–35.80	12.07–13.84	5.97–7.06
		Stem	2.08–3.62	22.00–25.02	43.36–47.12	13.58–16.73	1.84–3.10
Q. macrophylla	Mabian Dafengding NR	Leaves	15.45	36.56	28.02	8.80	7.59
		New shoot	13.85	33.30	38.02	5.46	6.35
B. spanostachya	Yele NR	Leaves	15.68	34.24	27.43	8.10	6.61
		New shoot	13.86	30.27	34.54	5.06	5.93
		Stem	3.23	25.03	47.78	15.88	2.34

Data from [33], and [13,14].

Ca, Fe, K, and Mg are all higher in leaves than in shoots, branches, and stems of *B. spanostachya* [39]. Similar results are discovered in *Q. macrophylla* and *B. faberi*, too [20,40]. Red pandas appear to meet their needs for trace elements through feeding on bamboo leaves.

No information on nutritional composition is available on fruits ingested by the animal, although they are assumed to have a higher caloric density than bamboos, which was assumed a necessary supplemental food in summer–autumn, especially for lactating females [10].

Habitat Selection

The basic goal of studies on habitat selection by red pandas is to understand the relationship between their occurrence and environmental characteristics. Johnson et al. [9] surveyed resting sites utilized by red pandas in Wolong Reserve, which were determined by presence of droppings, compacted ground, and other spoors, and found that seven (38.9%, n=18) were on top of or adjacent to fallen logs, three (16.7%, n=18) on or inside stumps, three (16.7%, n=18) at the sheltered base of a large leaning tree, two (11.1%, n=18) among the root systems of standing trees, and one (5.6%, n=18) each in an open bamboo patch, on top of a large tuft of dead grass, and in a fir (*Abies*). Ten (55.6%) of these rest sites were elevated; nine averaged 92±45 cm above ground and one was an estimated 13 m up a tree. In Mabian Reserve, Wei et al. [14] surveyed 52 feeding sites determined by foraging traces and droppings, indicating that these sites were characteristic of high slope (34.6°), moderate tree canopy (47.8%) and bamboo density (19.5 culms/m^2), and close to a water source (<200 m).

In addition, Wei et al. [11] developed systematic sampling techniques through plot establishment to study habitat selection by the animal. In Yele Reserve, Wei et al. [11] compared habitats selected by both the pandas, as primarily determined by faeces they left, through construction of $20 \times 20\ m^2$ plots, and found red pandas occurred at sites on steeper slopes, with higher density of fallen logs, shrubs, and bamboo culms. These sites were close to fallen logs, shrubs, and tree stumps, too. In Fengtongzhai Reserve, Zhang et al. [18,19] surveyed habitats selected by the animal through the same sampling methods, indicating the animal preferred habitats with higher densities of fallen logs and tree stumps in *B. faberi* bamboo forest.

In summary, the red panda exhibits a distinct preference for its habitats, which usually were characterized by abundant fallen logs and tree stumps. Concerning its small body size and the need to reach leaves on bamboo stems for foraging, these environmental characteristics can help them to easily access leaves [11,18,19].

Movement and Home Range

Concerning the difficulty in directly tracking the animal in the wild, many researchers tried to understand its movement and space use through radio-collar techniques. A female red panda, weighing 3.5 kg (subadult), was captured, immobilized and radio collared in Wolong Reserve on 20 February, 1984 [9]. During the subsequent 9-month period, its daily movement distance averaged 481±312 m, ranging from 393 to 635 m (Table 11.3). About 3 years later, an adult female and male were radio collared in the same reserve, and their

TABLE 11.3 Daily Movement Distances (m) and Home Ranges (km^2) of the Red Panda

Localities	Daily movement distance (m)		Home ranges (km^2)	
	Female	Male	Female	Male
Langtang National Park			1.0–1.5	1.7–9.6
Wolong Nature Reserve	481±312*		3.4*	
	235±169	325±210	0.94	1.11
Fengtongzhai Nature Reserve	460.9±329.5	431.5±307.8	2.02	2.66

**Values for a subadult red panda in Wolong Nature Reserve; others are all adults.*
Data from [8,9,10] and our unpublished results in Fengtongzhai Nature Reserve.

daily movement distances averaged 235±169 m (for the female) and 325±210 m (for the male) respectively [10]. Based on sparse observations on the spot, Reid et al. [10] noticed both pandas moved slowly, and appeared carefully deliberate. In Fengtongzhai Reserve, six red pandas (three adult females and three adult males) were captured and radio collared in April 2002. The daily movement distances for the males and females averaged 460.9±329.5 m and 431.5±307.8 m respectively (unpublished data). Males seemed to move more distance per day than females.

As far as we know, there have been four projects aimed at understanding space use by wild red pandas through radio-collar techniques (see Table 11.3). However, the results differed to some extent. Size of home ranges seemed to be influenced by age. For example, in Wolong Reserve, a subadult female (8 months old) occupied a home range of 3.4 km [32], much larger than the adult female (0.91 km^2) and male (1.11 km^2) [10], and it was assumed that the subadult may not have established a stable home range. In addition, Reid et al. [10] argued that red panda home range most closely approximate predicted use of space for omnivores, although it is primarily herbivorous, which perhaps reflects its inefficiency at converting the energy of primary production into metabolizable energy when limited by a simple digestive tract.

In Wolong Reserve, the male and the female shared much of their home ranges, with 63.8% of the male's overlapping that of the female, and 75.5% of the female's that of the male [10]. In Fengtongzhai Reserve, home ranges of six red pandas overlapped extensively, too (10.7–70.6%) (unpublished data). Indeed, as bamboo resources seem extremely abundant in their habitats, it is not necessary for them to develop some behavioural strategies to protect their home ranges from intruders, just like the giant panda [33]. On the other hand, territory behaviours are usually costly, low energy intake and assimilation from bamboos is another possible constraint to the development of such kind of behaviours for them.

Time Budget and Activity Pattern

More than 160 years ago, the animal was said to be active at dusk, dawn, and at night [3]. In Wolong Reserve, a subadult female was found to be primarily crepuscular and nocturnal, too [9]. Captive red pandas were reported to exhibit a similar activity pattern [41,42].

Two other studies, on the contrary, have produced different results. In Wolong Reserve, the female and the male showed similar daily activity patterns, with levels of activity consistently higher in daylight than at night, and crepuscular levels generally being intermediate [10]. In addition, Reid et al. [10] further argued that the difference between diurnal and nocturnal activity levels was related to their diets, for example in the late summer, when their diets were mainly composed of arboreal fruits and bamboos, the difference of activity level was largest, and in the late winter and spring when of bamboo leaves and shoots, the difference least. In Fengtongzhai Reserve, the six red pandas were more active in the daytime than at night, with crepuscular activity levels being intermediate, too; 44.8% of activity period was distributed in the daytime, 30.2% at dawn and dusk, only about 25.0% at night (unpublished data).

Inference on activity pattern of wild red pandas from captive ones is perhaps biased, concerning the common human interference under captive conditions, including food supply, cleaning, animal viewing. In addition, the focal animal in Johnson et al.'s study [9] was assumed not to have established a stable home range [43]. Its crepuscular and nocturnal activity pattern may reflect the need to avoid being predated by its sympatric enemies when moving widely, such as *Panthera pardus, Neofelis nebulosa, Cuon alpinus, Mustela flavigula, Catopuma temmincki,* and *Aquila chrysaeos* sympatrically in the habitats. In one word, wild red pandas seemed to exhibit a more diurnal lifestyle.

Difference of daily activity rate between adult males and females was least. However, the subadult was significantly less active than adults (Table 11.4). All studies showed that

TABLE 11.4 Time budget and activity pattern of wild Red Pandas

Items	Johnson et al. (1988) [9]	Reid et al. (1991) [10]	Our study (2002–2003)
Focal animal	1 subadult female	1 adult male, and 1 adult female	3 adult males, and 3 adult females
Locality	Wolong Nature Reserve	Wolong Nature Reserve	Fengtongzhai Nature Reserve
Duration of field monitoring	9 months	9 months	1 year
Activity rate (%)	36.5	Male: 45; Female: 49	Males: 49.5; Females: 47.9
Monthly variation in activity rate	Not significant	Significant	Significant
Activity rate in different parts of the day	Crepuscular>night> daytime	Daytime>crepuscular>night	Daytime>crepuscular>night
Frequency of long rest per day	2.1±0.9	Male: 1.7±0.7; Female: 1.5±0.7	2.23±0.69
Duration of long rest per day	4.2±2.3	Male: 4.6±2.3; Female: 4.4±2.0	4.02±1.24
Seasonal variation in frequency of long rest	Not significant	Not significant	Significant

red pandas rest for most of the day, and numerous periods of rest are interspersed with frequent activity periods. Such kind of time budget and activity pattern probably promotes the conservation of energy while keeping the digestive tract filled with bamboo [12].

At low environmental temperatures, red pandas can reduce their metabolism without an appreciable reduction in core body temperatures, like other small or medium-sized mammals, such as Matschie's tree kangaroo (*Dendrolagus matschiei*) and the binturong (*Arctictis binturong*) [43–45]. If individuals reduce peripheral circulation to conserve energy in winter, this presumably occurs only at rest and should be short-lived to avoid necrosis [43]. Both Johnson et al. [9] and Reid et al. [10] claimed that the average number and duration of long rests did not vary markedly between seasons, although data in winter were not included in their analyses. However, the study in Fengtongzhai Nature Reserve uncovered a different pattern, showing that, in winter, red pandas had a lower circadian activity rate and a higher frequency of long rests, which was assumed to reflect an adaptation to cold environmental conditions by reducing heat loss (unpublished data).

COMMUNITY ECOLOGY: HOW CAN RED AND GIANT PANDAS COEXIST IN SYMPATRY?

Being specialized on a bamboo diet, both pandas are peculiar in the order Carnivora [11,18,19]. In the western Sichuan mountains, they are sympatric and usually feed on the same bamboo species. How they can coexist in sympatry has attracted many research efforts [11,18,19,46]. Indeed, both pandas do not exclude each other from their habitats. In Wolong Reserve, core area overlap between one red panda and one giant panda was up to 30.8%, and on 7 days they were radio-located within the same 1-ha quadrant [9]. In addition, they were found within 100 m of each other on 59 days [9].

Difference in Diet

Resource partitioning between sympatric species may reduce interspecific competition and is often cited as being responsible for the coexistence of multiple species [9,12,47–49]. Although resource niche overlapped partially between the two pandas (with the overlap index 0.347), Wei et al. [12] found they have significantly different patterns in resource utilization.

First, both pandas exhibit some differences in their food habits although they feed mainly on the bamboo species. Giant pandas might sample some animals and plants; however, their annual diets consisted of over 99% of bamboo species in the Xiangling [12], Liangshan [46], Qionglai [32], and Minhan Mountains [50]. By contrast, the annual diet of red pandas consisted of less than 98% of bamboo species in the sympatric areas [9,10,12,14,17]. In addition, red pandas preferred different kinds of fruits in summer and autumn in the Xiangling [12], Qionglai [10,17], and Liangshan Mountains [14].

Secondly, giant pandas utilized almost every part of bamboo species such as leaves, shoots, branches, and culms. Red pandas, however, only used leaves and shoots. Bamboo leaves made up 89.9% of the red panda's annual diet in the Xiangling [12], 91.4% in the Liangshan [14], and over 90% in the Qionglai Mountains [10]. By contrast, leaves only made up 34.7% of the annual diet of giant pandas in the Xiangling [12], 38.8% in the Liangshan [46], and 40.7% in the Qionglai Mountains [32].

Thirdly, both pandas selected different kinds of bamboo shoots during the germinating period. Giant pandas preferred tall and robust shoots. They selected shoots of *B. spanostachya* taller than 30 cm and larger than 10 mm in basal diameter in the Xiangling Mountains [12], shoots of *F. robusta* taller than 25 cm and larger than 10 mm in basal diameter in the Qionglai Mountains [32]. Red pandas, on the other hand, preferred short and robust shoots which giant pandas usually did not prefer. For instance, red pandas selected shoots of *B. spanostachya* lower than 50 cm and with a basal diameter larger than 10 mm [12].

Separation in Microhabitats

Both pandas exhibit different patterns in microhabitat utilization. In Yele Reserve, the giant panda occurred at sites on gentle slopes with lower density of fallen logs, shrubs, and bamboo culms; sites were also close to trees and far from fallen logs, shrubs, and tree stumps [11]. By contrast, the red panda occurred at sites on steeper slopes with higher density of fallen logs, shrubs, and bamboo culms; sites were also close to fallen logs, shrubs, and tree stumps [11]. In Fengtongzhai Reserve, there were six variables that differed between microhabitats selected by both pandas [18]. The giant panda preferred microhabitats with gentle slope, taller bamboos, thicker trees, and greater fallen log dispersion; however, microhabitats commonly selected by the red panda were characteristic of higher densities of fallen logs and tree stumps [18]. Intense food competition between the red and giant pandas likely does not occur currently because the microhabitats selected by them were apparently different, allowing the two species to feed at different locations [18].

Partitioning of habitats is one of the most common forms of sympatric-species separation, and habitat-use strategies are often cited as the means by which sympatric species avoid competition [18,51–53]. Although divergent habitat-use strategies may reflect evolutionary mechanisms to reduce interspecific competition [51], they may also merely reflect different physiological or ecological requirements [54]. The preference for microhabitats with a gentle slope by the giant panda was widely considered for saving energy when moving [11,18,19,33], or reflecting the need to sit and free its fore-limbs to grasp bamboo culms when feeding [55], and the red panda usually used fallen logs, and tree stumps to access bamboo leaves easily [10,11,18,19,33], the difference in microhabitats utilized by both pandas was more perhaps related to their own physiological or ecological requirements, not aroused from the adjustment from the intense interspecific competition [18].

Circadian Activity Rhythm

Both pandas exhibited distinct circadian activity rhythms. Johnson et al. reported that the red panda had a bimodal daily pattern essentially crepuscular but with a dawn peak preceded by substantial nocturnal activity [9], whereas giant pandas were intermittently active day and night [32,33]. In Wolong Reserve, two other red pandas were more active in the daytime than at night [10].

The overlap of the most active periods between both pandas is less. For the giant panda, two peaks with the highest activity rates fell between 04:00–06:00 and 16:00–19:00 respectively [33].

In Fengtongzhai Reserve, the two active peaks for the six radio-collared red pandas fell between 7:00–10:00 and 17:00–18:00 (unpublished data). Partial separation in time niche between both pandas also contributes to their coexistence in sympatry.

CONCLUSIONS AND PERSPECTIVES

Although diet composition differed a little among different ranges inhabited by red pandas, bamboos are undoubtedly the most important food that they subsist on in their life. They also ingest some other kinds of plants, such as fruits or berries, as their seasonal supplement. A simple tract typical of carnivores implies, on one hand, they almost only can utilize cell solubles (protein, fat, lipid, etc.) in their food, and on the other hand, microbial digestion plays only a very minor role in this animal's energy metabolism, which was also confirmed by field studies.

Like the giant panda, the red panda is a "false herbivore" concerning its physiological traits and phylogenetic position. However, in view of its food habits, it is not a "true carnivore" either. The conflict in its physiology and ecology (especially the diet) of course raises the question of how it can subsist on bamboos, which are low in protein and high in cellulose. So far, field studies have uncovered the animal has developed some optimal foraging strategies to meet its daily need in nutrient and energy as follows:

1. feeding on bamboo species with higher nutrition (usually with easier access, too)
2. intake of the most nutritious and digestible parts, such as shoots and leaves
3. meticulous selection of individual leaves and extensive mastication
4. ingestion of large quantities of food, and rapidly passing through the digestive tract.

Bamboo resource is abundant in the environment, keeping an almost constant nutrition level throughout the year. Each panda occupies an area ranging about 1.0–10 km^2, in which it can find food, water, shelter and its mates as well. Perhaps due to food abundance and energy constraint, home ranges among adjacent individuals overlap extensively. Red pandas rest for more than half a day, especially in winter, when long rest (>2 h in duration) occurs more frequently than in the other seasons. Such kind of time budget can help to save energy consumption, a necessary supplement to its low-quality food source.

Although both pandas are sympatric extensively in the western Sichuan mountains, and usually feed on the same bamboo species, no intense competition between them was observed. Some factors, for example, diet difference, microhabitat separation and activity asynchrony as well, contributed to their coexistence. The separation in microhabitats between them seemed to arise from their own physiological or ecological differences, not from the adjustment due to interspecific competition between them.

Generally, efforts made to understand wild red pandas' ecology are very biased so far. As for research sites involved, most studies were carried out in the western Sichuan mountains, China. In addition, some studies were carried out in Langtang Park in Nepal about two decades ago. By contrast, only sparse information, including status, conservation, and habitat is available for wild red pandas in India and Myanmar at present. So far as we know, no studies have been carried out in Bhutan, another distribution range of the red panda.

Efforts made on the two subspecies are biased, too. Indeed, the most comprehensive and in-depth information on the animal mainly came from *A. f. styani*, the subspecies endemic to China. Greater efforts should be given to *A. f. fulgens* by local scientists and conservationists. In addition, red pandas in the tropical forests, Meghalaya, deserve special attention, too.

As for research contents, previous efforts were biased on their habitat, diet, foraging behaviour, nutrition analysis, energy demand, space use and activity pattern. Knowledge on their social communication, population structure, dispersal, and breeding biology is extremely rare. From a historical viewpoint, some technical deficiency contributed to such a bias. Fortunately, with the support of some modern biological techniques, for example, molecular markers, infrared cameras, and GIS as well, future studies undoubtedly can help to make up our knowledge gaps on these aspects.

References

[1] D. MacClintock, Red Pandas: A Natural History, Charles Scribner's Sons, New York, USA, 1988.
[2] T. Hardwicke, Description of new genus of the class Mammalia: from the Himalaya chain of hills between Nepal and the snowy mountains, Transact. Linnean Soc. 15 (1827) 161.
[3] B.H. Hodgson, On the cat-foot subplantigrades of the sub-Himalayas, J. Asiatic Soc. 16 (1847) 1113.
[4] B.H. Hodgson, Addendum on the anatomy of *Ailurus*, J. Asiatic Soc. 17 (1848) 573.
[5] J. Fox, P. Yonzon, Mapping conflicts between biodiversity and human needs in Langtang National Park, Conservat. Biol. 10 (1996) 562.
[6] P.B. Yonzon, J.M.L. Hunter, Conservation of the red panda *Ailurus fulgens*, Conservat. Biol. 5 (1991) 196.
[7] P.B. Yonzon, Human disturbance and activity pattern of the red panda in Lang Tang National Park, Nepal-Himalayas, Presented at the Symposium of Asian-Pacific Mammalogy, Beijing, China, 1988 (unpublished).
[8] P.B. Yonzon, J.M.L. Hunter, Cheese, tourists, and red pandas in the Nepal Himalayas, Biol. Conservat. 57 (1991) 1.
[9] K.G. Johnson, G.B. Schaller, J. Hu, Comparative behavior of red and giant pandas in the Wolong Reserve, China, J. Mammal. 69 (1988) 552.
[10] D.G. Reid, J. Hu, Y. Huang, Ecology of the red panda in the Wolong Reserve, China, J. Zool. 225 (1991) 347.
[11] F. Wei, Z. Feng, Z. Wang, et al., Habitat use and separation between the giant panda and the red panda, J. Mammal. 81 (2000) 448.
[12] F. Wei, Z. Feng, Z. Wang, et al., Feeding strategy and resource partitioning between giant and red pandas, Mammalia 63 (1999) 417.
[13] F. Wei, Z. Feng, Z. Wang, et al., Use of the nutrients in bamboo by red panda (*Ailurus fulgens*), J. Zool. 248 (1999) 535.
[14] F. Wei, W. Wang, A. Zhou, et al., Food selection and foraging strategy of red pandas, Acta Theriol. Sin. 15 (1995) 259.
[15] F. Wei, Z. Wang, Z. Feng, Comparison of habitats selected by giant and red pandas in Xiangling Mountains, China. Acta Zool. Sin. 46 (2000) 287.
[16] F. Wei, Z. Wang, Z. Feng, et al., Seasonal energy requirements of red panda (Ailurus fulgens), Zoo Biol. 19 (2000) 27.
[17] J. Yang, Z. Zhang, J. Hu, et al., Feeding behavior and nutrition strategy of red pandas in Fengtongzhai Nature Reserve, Sichuan, China. Acta Theriol. Sin. 27 (2007) 249.
[18] Z. Zhang, F. Wei, M. Li, et al., Winter microhabiat separation between giant and red pandas in *Bashania faberi* bamboo forest in Fengtongzhai Nature Reserve, J. Wildlife Manage 70 (2006) 231.
[19] Z. Zhang, F. Wei, M. Li, et al., Microhabitat separation during winter among sympatric giant pandas, red pandas, and tufted deer: the effects of diet, body size, and energy metabolism, Can. J. Zool. 82 (2004) 1451.
[20] A. Zhou, F. Wei, P. Tang, et al., A preliminary study on nutrients in food of red pandas, Acta Theriol. Sin. 17 (1997) 266.

[21] A. Rabinowitz, S.T. Khaing, Status of selected mammal species in northern Myanmar, Oryx 32 (1998) 201.
[22] A. Choudhury, An overview of the status and conservation of the red panda *Ailurus fulgens* in India, with reference to its global status, Oryx 35 (2001) 250.
[23] S. Pradhan, G.K. Saha, J.A. Khan, Ecology of the red panda (*Ailurus fulgens*) in the Singhalila National Park, Darjeeling, India. Biol. Conservat. 98 (2001) 11.
[24] J. Hu, F. Wei, Foraging behavior of red pandas, Sichuan Teachers' Coll. 13 (1992) 83.
[25] M.S. Roberts, J.L. Gittleman, *Ailurus fulgens*, Mammalian Species 222 (1984) 1.
[26] A. Choudhury, Red pandas *Ailurus fulgens* F. Cuvier in the north-east with an important record from Garo hills, J. Bombay Nat. Hist. Soc. 94 (1997) 145.
[27] G.M. Allen, The mammals of China and Mongolia, American Museum of Natural History, New York, USA, 1938.
[28] Z. Feng, G. Cai, C. Zheng, The mammals of Tibet, Science Press, Beijing, 1978.
[29] D. Mierow, T.B. Shrestha, Himalayan flowers and trees, Sahayogi Prakashan Co., Kathmandu, Nepal, 1986.
[30] T. Yi, Taxonomy and distribution of staple-food bamboos of giant pandas (I), J. Bamboo Res. 4 (1985) 11.
[31] T. Yi, Taxonomy and distribution of staple-food bamboos of giant pandas (II), J. Bamboo Res. 4 (1985) 20.
[32] G.B. Schaller, J. Hu, W. Pan, et al., The Giant Panda of Wolong, University of Chicago Press, Chicago, USA, 1985.
[33] J. Hu, G.B. Schaller, W. Pan, et al., The giant panda of Wolong, Sichuan Science and Technology Press, Chengdu, China, 1985.
[34] W. Pan, Z. Gao, Z. Lu, The refugia of giant pandas in the Qinling Mountains, Peking University Press, Beijing, China, 1988.
[35] Z. Jiang, R.J. Hudson, Digestive responses of Wapiti *Cervus elaphus canadensis* to seasonal forages, Acta Theriol. 41 (1996) 415.
[36] K.R. Searle, N.T. Hobbs, L.A. Shipley, Should I go and should I stay? Patch departure decision by herbivores at multiple scales, Oikos 111 (2005) 417.
[37] L.A. Shipley, S. Blomquist, K. Danell, Diet choices made by free ranging moose in northern Sweden in relation to plant distribution, chemistry, and morphology, Can. J. Zool. 76 (1998) 1722.
[38] L.A. Shipley, A.W. Illius, K. Danell, Predicting bite size selection of mammalian herbivores: a test of a general model of diet optimization, Oikos 84 (1999) 55.
[39] A. Zhou, F. Wei, P. Tang, A preliminary study on trace elements in the staple-food bamboo of giant pandas in Yele Nature Reserve, China. J. Sichuan Teachers' Coll. 17 (1996) 1.
[40] Q. Fu, A study on trace elements in staple-food bamboos of giant pandas, J. Nanchong Teachers Coll. 21 (1988) 169.
[41] B. Holst (Ed.), The activity patterns of the red pandas at Copenhagen Zoo: a preliminary report, SPB Academic Publishing, The Hague, The Netherlands, 1989.
[42] M. Stevenson, L. Arness, J. Hanning, et al., Red pandas at the Edinburgh Zoo, SPB Academic Publishing, The Hague, The Netherlands, 1989.
[43] D.G. Reid, J.C. Hu, Y. Huang, Ecology of the red panda in the Wolong Reserve, China, J. Zool. London 225 (1991) 347.
[44] K.G. Johnson, G.B. Schaller, J.C. Hu, Comparative behavior of red and giant pandas in the Wolong Reserve, China, J. Mammal. 69 (1988) 552.
[45] B.K. McNab, Energy conservation in a tree-kangaroo (*Dendrolagus matschiei*) and the red panda (*Ailurus fulgens*), Physiol. Zool. 61 (1988) 280.
[46] F. Wei, C. Zhou, J. Hu, et al., Selection of bamboo resources by giant pandas in Mabian Dafengding Nature Reserve, China, Acta Theriol. Sin. 16 (1996) 171.
[47] P.S. Giller, S. McNeill, Predation strategies, resource partitioning and habitat selection in British Notonecta, J. Anim. Ecol. 50 (1981) 789.
[48] T.W. Schoener, Resource partitioning in ecological communities, Science 185 (1974) 27.
[49] B. Van Horne, Niches of adult and juvenile deer mice (*Peromyscus maniculatus*) in seral stages of coniferous forests, Ecology 63 (1982) 417.
[50] G.B. Schaller, Q. Deng, K. Johnson, et al., Feeding ecology of giant pandas and Asiatic black bears in the Tangjiahe Reserve, China, Cornell University Press, Ithaca, USA, 1989.
[51] R.D. Dueser, H.H. Shugart, Microhabitats in forest-floor small fauna, Ecology 60 (1978) 108.

[52] A.C.W. Marsh, S. Harris, Partitioning of woodland habitat resources by two sympatric species of Apodemus: lessons for the conservation of the yellow-necked mouse (*A. flavicollis*) in Britain, Biol. Conservat. 92 (2000) 275.
[53] S. Sébastien, P. Nicolas, N. Cornelis, Winter habitat selection by two sympatric forest grouse in Western Switzerland: implications for conservation, Biol. Conservat. 112 (2003) 373.
[54] E..M. Wendy, R.D. Chris, Competition and habitat use in native Australian Ruttus: is competition intense, or important? Oecologia 128 (2001) 526.
[55] D.G. Reid, J.C. Hu, Giant panda selection between *Bashania faberi* habitats in Wolong Nature Reserve, Sichuan, China, J. Appl. Ecol. 28 (1991) 228.

CHAPTER

12

A Brief History of the Red Panda in Captivity

Marvin L. Jones

Formerly of San Diego Zoological Gardens, San Diego, CA, USA

OUTLINE

The red panda or *Ailurus fulgens* as it is known to science, was first described by Cuvier in 1825. The first known specimen to have arrived alive at a zoological collection outside its native land, was that acquired by London Zoo on 22 May, 1869. It was the sole survivor of a group of three collected earlier that year near Darjeeling. It died 12 December, 1869 and was followed eight years later by a second individual which arrived on 16 February, 1876. This animal fared better and survived until 17 May, 1881. With the exception of these two animals and one noted as living at the Calcutta Zoo in 1877, there is no record of any other red panda in captivity until Philadelphia Zoo brought the first one to America in 1906.

RED PANDAS IN CAPTIVITY: 1908–1940

In the period between 1908 and 1940 fewer than 50 red pandas were brought into captivity and these were held in a mere handful of the world's zoos. A list of these animals is given in Table 12.1. In addition to these, Crandall [1] notes that a female which had arrived pregnant gave birth to two young at the small zoo in Darjeeling in May, 1908. Further births also occurred in Calcutta in 1919 and in London Zoo in 1919, 1920 and 1921. These appear

Red Panda. DOI: 10.1016/B978-1-4377-7813-7.00012-4

TABLE 12.1 Red panda in captivity 1908–1940.

Sex			Dates of arrival/ departure or note of exhibition	Collection	Comments
Male	Female	Unknown			
1			1 Apr 1908–1 Oct 1908	Artis Zoo, Amsterdam, Netherlands	
	1		1 Apr 1908–11 Oct 1909	Artis Zoo, Amsterdam, Netherlands	
		1	1908	Cologne Zoo, Germany	
		1	1910	Copenhagen Zoo, Denmark	
		2	1 Mar 1910–8 Apr 1910	Rotterdam Zoo, Netherlands	returned to dealer
		1	1911	Melbourne Zoo, Australia	
		1	? Jul 1911	Antwerp Zoo, Belgium	to Brussels Museum
		1	10 Jul 1911–	Bronx Zoo, NY, USA	
		1	1912–	Dublin Zoo, Ireland	
		1	1915–	Melbourne Zoo, Australia	
1	1		1917–	London Zoo, GB	
		1	1917–1918	Melbourne Zoo, Australia	
		1	1918–	Cologne Zoo, Germany	
		1	1919–	Belle Vue Zoo, Manchester, GB	
		1	26 Jun 1919–?	London Zoo, Great Britain	born in London Zoo
		1	1919–	Calcutta Zoo, India	born in Calcutta Zoo
		2	1920–	London Zoo, Great Britain	born in London Zoo
		2	1921–	London Zoo, Great Britain	born in London Zoo
		1	1921–	Pretoria Zoo, South Africa	
1			6 Apr 1922–9 May 1931	Philadelphia Zoo, PA, USA	
		1	1922	National Zoo, DC, USA	
	1		5 May 1913–9 Dec 1924	Milwaukee Zoo, WI, USA	
		1	1926	Edinburgh Zoo, Scotland	
1	1		May 1929	Hagenbeck Zoo, Hamburg, Germany	
		1	1931	La Plata Zoo, Argentina	
		1	27 Mar 1931–19 Apr 1940	London Zoo, Great Britain	
1			5 Apr 1933–2 Jan 1934	Berlin Zoo, Germany	
1			5 Apr 1933–28 Oct 1938	Berlin Zoo, Germany	
	1		5 Apr 1933–24 Apr 1934	Berlin Zoo, Germany	
1			12 Jul 1933–20 Nov 1934	St. Louis Zoo, MO, USA	

(Continued)

TABLE 12.1 *(cont'd)*

Sex			Dates of arrival/ departure or note of exhibition	Collection	Comments
Male	**Female**	**Unknown**			
1			12 Jul 1933–15 Dec 1934	St. Louis Zoo, MO, USA	
1			12 Jul 1933–2 Jan 1937	St. Louis Zoo, MO, USA	
1			Apr 1936–30 Oct 1937	St. Louis Zoo, MO, USA	
	1		Apr 1936–?	St. Louis Zoo, MO, USA	
1			?– Mar 1937	Brookfield Zoo, IL, USA	to Field Museum
		1	NOV 1937–	Brookfield Zoo, IL, USA	
		2	3 Mar 1937–5 Mar 1937	Artis Zoo, Amsterdam, Netherlands	
		1	?– Feb 1939	Schönbrunn Zoo, Vienna, Austria	to Nat. Hist. Museum
		1	12 Jan 1939–30 Sep 1942	Frankfurt Zoo, Germany	
		1	1940–	Chicago Lincoln Park Zoo, IL USA	
		1	? 24 Mar 1940	Schönbrunn Zoo, Vienna, Austria	

to be the only instances of captive births occurring prior to 1940. It will be noted from Table 12.1 that, in the cases of those animals for which we have both arrival and death dates, lifespan in captivity was relatively short. However, the longevity of pandas in many collections is uncertain as they often just made a note that the species was exhibited.

On 21 August 1940, Karl Koch returned to San Diego Zoo with a large shipment of animals from a number of Asian zoos; amongst this transport were four red pandas from Calcutta Zoo. Unfortunately the available records do not indicate how old the animals were, although it is probable that they were at least one year old. Their arrival signaled the onset of a period of improved management of the species which resulted in increased longevity and successful breeding (Figure 12.1). From available curatorial notes and cards the history of these remarkable animals can be partially recreated. We know that the imported male died on 16 December 1953, and the females on 21 April 1950, 1 January 1951 and 30 March 1953, respectively. These animals were all given names by the then keeper Georgie Dittoe, but it is not always possible to match these names to the known death dates. One of the benefits of the studbook and of the International Species Inventory System (ISIS) is that this problem will not arise in the future as each individual animal is separately registered together with information on its parentage, birth and death dates, etc. This should make the job of future panda watchers much easier.

HISTORY OF THE SAN DIEGO PANDAS

The four animals were captured in Nepal and taken via Lucknow to Calcutta where they were handed over to Mr. Koch on 4 May 1940. Thus, May 1940 was the date when

FIGURE 12.1 Keeper Georgie Dittoe and two of the San Diego Zoo pandas.

the animals arrived in Calcutta and not the date on which they supposedly arrived in San Diego Zoo, as has been reported by a number of other authors.

On arrival the animals were all housed together in a single large cage. Breeding commenced within one year. The first litter, which consisted of two young, was born on 29 May 1941. Both these infants died of distemper in July of the same year. In 1942 two further litters were produced, this time of one infant each. The infant born on 1 June died at an unknown date while the male born 12 June survived in the collection until July 1949.

In 1943 all three females gave birth. Two infants were born on 4 June. We do not know exactly what happened to these animals, but it is possible that one of them remained in the collection until 2 July 1949 when an undetermined individual was reported dead. Two further litters of two were born on 10 June. One of these four infants was stillborn, one died on the day of birth, and a third on 11 June. The survivor was a female called Loppy. She later bred at San Diego and lived until June 1955 when she succumbed to pregnancy toxaemia.

The reproductive success of the San Diego animals continued with the birth of a male and a female on 12 June 1944. However, by this time the San Diego panda population was becoming quite large: in December 1944 four adults and five youngsters were recorded in the zoo collection. As a result the zoo finally decided to send some of the young pandas away to other zoos. The 1944 young were the first to go. They were sent to St. Louis Zoo on 7 May 1946 where, unfortunately, they both died in November of the same year.

1945 saw the birth of three litters: one, consisting of a male and a female, was produced by the 1943 zoo-born female, Loppy. These were sent to Brook-field Zoo on 25 May 1949. The male of this pair survived until early 1951 but there is no record of what happened to the female. Two infants of unknown sex were produced by one of the original imported females on 22 June. However, there is no record of what happened to them. Another of the imported females gave birth to a male and female on 3 July. These animals were sold to San Francisco Zoo in May 1949. Unfortunately, the records of this zoo do not indicate their dates of death. I should like to take this opportunity to add that the combination of the Second World War and personnel problems were probably the main reasons for the lack of complete records at many US zoos at this time. In 1946, again, three litters were produced; two of these were known to have been born to imported females. However, the mother of the third is unknown although there is no mention of Loppy having given birth. Two of the young were born on 4 June. One of these animals is known to have died of pneumonia at some unknown date but the fate of the other is unknown. Two further infants were born 23 June; these died on their day of birth as did the single infant born on 26 June.

Two litters, each comprising one infant, were born in 1947: a male on 13 June and a female on 24 June; these both went to San Antonio Zoo on 11 December of the same year. In 1948 three litters were born, one mixed litter on 14 June, another on 17 June and a single male on 29 June. The first litter went to St. Louis Zoo on 28 April 1949, where they did much better than their 1946 predecessors. The male died in St. Louis on 15 August 1962 and the female on 20 August 1955. However, there was no breeding success in St. Louis. The second pair went to Lincoln Park Zoo, Chicago, on 19 April 1950 where there is no record of their fate. There is no record of what happened to the single male. However, San Diego Zoo supposedly had 17 specimens in early 1949. Loppy produced young again on 2 July 1949. The female of this litter went to the Bronx Zoo in February 1950, in exchange for a male which arrived in San Diego on the same date. This animal, which died in July 1959, sired only one young. This was born to Loppy on 6 July 1954 but it only survived for two days.

Unfortunately with the death of the imported quartet in the early 50s, the very successful San Diego panda colony came to an end. San Diego did acquire new animals in later years, but there was no further breeding.

One interesting fact regarding the management of red pandas in San Diego zoo at that time was, as mentioned above, that all the individuals lived together in a single large cage. All the infants which survived were mother-reared and they remained in the family cage as long as they were in San Diego. It was also noted that often a mother, other than the one which gave birth, reared the young.

RED PANDAS IN CAPTIVITY POST-1950

The advances in red panda management which enabled San Diego to keep the four imported animals for a more than average period of time in captivity and which also resulted in the births of many litters, were of use to the many other collections which were beginning to exhibit this species. As of 1950 the number of specimens coming onto the animal market increased notably. Table 12.2 gives an indication of the pandas held in zoos between 1950 and 1969. These data have been gleaned from various zoo archives. As such they do not provide a complete listing of all pandas exhibited, but they do show the majority of specimens in major collections during this period. It can be seen that some 250 pandas were exhibited during the 1950–1969 period. Six of these are also listed in the red

TABLE 12.2 Red panda in captivity 1950–1969.

Sex			Dates of arrival and/or departure/death	Collection	Comments
Male	Female	Unknown			
		1	13 Mar 1950–13 Mar 1950	Antwerp Zoo, Belgium	
		1	13 Mar 1950–4 Jun 1951	Antwerp Zoo, Belgium	
		1	28 Feb 1951–20 Nov 1955	Antwerp Zoo, Belgium	
1			6 May 1962–11 Oct 1965	Basel Zoo, Switzerland	
1			6 May 1952–4 Jun 1957	Basel Zoo, Switzerland	sent to Manchester Zoo
1			11 Mar 1953–24 Apr 1953	Hagenbeck, Hamburg, Germany	sent to St. Louis
			24 Apr 1953–4 Jan 1953	St. Louis Zoo, MO, USA	
	1		11 Mar 1953–24 Apr 1954	Hagenbeck, Hamburg, Germany	sent to St. Louis
			24 Apr 1953–28 Jun 1953	St. Louis Zoo, MO, USA	
			11 Mar 1953–24 Apr 1953	Hagenbeck, Hamburg, Germany	sent to St. Louis
			24 Apr 1953–13 Jul 1953	St. Louis Zoo, MO, USA	
	1		11 Mar 1953–24 Apr 1953	Hagenbeck, Hamburg, Germany	sent to St. Louis
			24 Apr 1953–26 Oct 1955	St. Louis Zoo, MO, USA	
1			20 Mar 1953–29 Jan 1964	Copenhagen Zoo, Denmark	
		1	23 Jun 1953–12 Oct 1957	National Zoo, DC, USA	
	1		23 Jun 1953–?	National Zoo, DC, USA	
		1	22 Nov 1953–21 Jan 1954	Artis Zoo, Amsterdam, Netherlands	
		1	8 Feb 1954–?	Hagenbeck, Hamburg, Germany	
	1		20 Feb 1954–2 Jun 1957	Copenhagen Zoo, Denmark	returned to dealer Krag
1			1 Mar 1954–21 Dec 1962	Rotterdam Zoo, Netherlands	
		1	1 Mar 1954–?	Rotterdam Zoo, Netherlands	
1			6 Mar 1954–?	Hagenbeck, Hamburg, Germany	
	1		6 Mar 1954–27 Jul 1955	Hagenbeck, Hamburg, Germany	
			27 Jul 1955–3 Nov 1955	Rotterdam Zoo, Netherlands	
	1		12 Mar 1954–20 Dec 1966	Zurich Zoo, Switzerland	

TABLE 12.2 (cont'd)

Sex			Dates of arrival and/or departure/ death	Collection	Comments
Male	Female	Unknown			
1			12 Mar 1954–13 May 1955	Artis Zoo, Amsterdam, Netherlands	
1			12 Mar 1954–23 Mar 1961	Artis Zoo, Amsterdam, Netherlands	
	1		12 Mar 1954–14 Mar 1954	Artis Zoo, Amsterdam, Netherlands	
	1		12 Mar 1954–1 Jul 1955	Artis Zoo, Amsterdam, Netherlands	
		2	20 Mar 1954–?	San Francisco Zoo, CA, USA	
		4	24 Mar 1954–?	San Francisco Zoo, CA, USA	
	1		1 Apr 1954–26 Apr 1958	Stuttgart Zoo, Germany	
	1		1 Apr 1954–25 Dec 1958	Stuttgart Zoo, Germany	
1			3 Jun 1954–1 Feb 1961	Philadelphia Zoo, PA, USA	
		1	26 Jun 1954–16 May 1956	St. Louis Zoo, MO, USA	
		1	13 Jun 1954– ?	San Francisco Zoo, CA, USA	
		2	6 Sep 1954– ?	San Francisco Zoo, CA, USA	
		2	1955– ?	San Francisco Zoo, CA, USA	born at San Francisco Zoo
1			10 Feb 1955–10 Nov 1961	Antwerp Zoo, Belgium	
	1		10 Feb 1955–3 Apr 1959	Antwerp Zoo, Belgium	
		1	15 Feb 1955–23 Feb 1955	Artis Zoo, Amsterdam, Netherlands	
		1	25 Mar 1955–21 May 1955	Stuttgart Zoo, Germany	
		1	25 Mar 1955– Jun 1955	Stuttgart Zoo, Germany	
		1	25 Mar 1955–8 Nov 1955	Stuttgart Zoo, Germany	
1			7 Apr 1955 to Amsterdam	Hagenbeck, Hamburg, Germany	
			9 Aug 1955–11 Sep 1955	Artis Zoo, Amsterdam, Netherlands	
	1		7 Apr 1955–to Amsterdam	Artis Zoo, Amsterdam, Netherlands	
			9 Aug 1955–12 Nov 1961	Artis Zoo, Amsterdam, Netherlands	
1			4 May 1955–16 Nov 1965	Hagenbeck, Hamburg, Germany	
	1		4 May 1955–?	Hagenbeck, Hamburg, Germany	

TABLE 12.2 (cont'd)

Sex			Dates of arrival and/or departure/death	Collection	Comments
Male	Female	Unknown			
		2	Jun 1955–Jun 1955	Artis Zoo, Amsterdam, Netherlands	born at Artis Zoo
	1		3 Mar 1956–12 Mar 1956	Hagenbeck, Hamburg, Germany	
1			14 Mar 1956–16 Mar 1958	Cologne Zoo, Germany	
	1		14 Mar 1956–31 May 1958	Cologne Zoo, Germany	
1			17 Apr 1956–3 Jul 1957	Detroit Zoo, MI, USA	
	1		17 Apr 1956–14 Oct 1957	Detroit Zoo, MI, USA	
	1		30 Apr 1956–28 Jul 1956	Hagenbeck, Hamburg, Germany	
		1	30 Apr 1956–21 Mar 1957	Hagenbeck, Hamburg, Germany	sent to Mr. Springer
		1	30 Apr 1956– ?	San Francisco Zoo, CA, USA	
	1		6 Jun 1956–18 Jul 1957	Detroit Zoo, MI, USA	born at Detroit Zoo
		3	1956– ?	San Francisco Zoo, CA, USA	born at San Francisco Zoo
1			24 Apr 1956–27 Jun 1960	Stuttgart Zoo, Germany	
1			22 Dec 1956–9 Apr 1968	Antwerp Zoo, Belgium	
1			1957– ?	Berlin Zoo, Germany	
1			1957–1970	Berlin Zoo, Germany	
	2		1957–1969	Berlin Zoo, Germany	
1			1957–Sep 1958	Cincinnati Zoo, OH, USA	
1	1		? –Jun 1957	Seattle Zoo, WA, USA	from San Francisco where possibly born
		3	9 Jan 1957– ?	San Francisco Zoo, CA, USA	
	1		1957–1961	Rotterdam Zoo, Netherlands	
1			7 Feb 1957–22 Feb 1957	Hagenbeck, Hamburg, Germany	
1			7 Feb 1957–13 Mar 1957	Hagenbeck, Hamburg, Germany	
	1		7 Feb 1957–4 Mar 1963	Hagenbeck, Hamburg, Germany	

TABLE 12.2 *(cont'd)*

Sex			Dates of arrival and/or departure/death	Collection	Comments
Male	Female	Unknown			
	1		7 Feb 1957–7 Sep 1963	Hagenbeck, Hamburg, Germany	
1			7 Feb 1957–26 Apr 1958	Hagenbeck, Hamburg, Germany	sent to Munster Zoo
	1		1 Jun 1957–17 Jun 1957	Copenhagen Zoo, Denmark	
	1		1 Jun 1957–24 Jan 1960	Copenhagen Zoo, Denmark	
	1		1 Jun 1957–3 Oct 1963	Copenhagen Zoo, Denmark	
	1		29 Oct 1957–21 Apr 1958	Cheyenne Mt. Zoo, CO, USA	
1			? –3 Jan 1958	Rome Zoo, Italy	
1			25 Feb 1958–12 Apr 1958	Cheyenne Mt. Zoo, CO, USA	
1			25 Feb 1958–13 Apr 1958	Cheyenne Mt. Zoo, CO, USA	
1			25 Feb 1958–15 Apr 1958	Cheyenne Mt. Zoo, CO, USA	
1			11 Mar 1958– ?	Bronx Zoo, NY, USA	
	1		2 May 1958–16 Sep 1958	Stuttgart Zoo, Germany	
1			14 May 1958–9 Apr 1965	Cologne Zoo, Germany	
1			13 Jul 1958–29 Sep 1958	San Diego Zoo, CA, USA	
	1		13 Jul 1958–?	San Diego Zoo, CA, USA	
1			4 Mar 1959–22 Jan 1971	Frankfurt Zoo, Germany	
	1		4 Mar 1959–24 Mar 1959	Frankfurt Zoo, Germany	
	1		4 Mar 1959–21 Aug 1962	Frankfurt Zoo, Germany	
	1		4 Mar 1959–29 Apr 1969	Frankfurt Zoo, Germany	
	1		17 Mar 1959–4 Apr 1959	Stuttgart Zoo, Germany	
	1		17 Mar 1959–21 Jun 1959	Stuttgart Zoo, Germany	
	1		21 Apr 1959–1 May 1960	Stuttgart Zoo, Germany	
1			?– May 1960	Colombus Zoo, OH, USA	
1	2		4 Jan 1960–17 Jan 1960	Cheyenne Mt. Zoo, CO, USA	

TABLE 12.2 (*cont'd*)

Sex			Dates of arrival and/or departure/ death	Collection	Comments
Male	Female	Unknown			
1			4 Jan 1960–14 Jan 1960	San Diego Zoo, CA, USA	
1			4 Jan 1960–17 Jan 1960	San Diego Zoo, CA, USA	
	1		4 Jan 1960–12 Feb 1962	San Diego Zoo, CA, USA	
	1		30 Jan 1960–20 Jan 1961	Antwerp Zoo, Belgium	
	1		5 Mar 1960–21 Sep 1961	Detroit Zoo, MI, USA	
	1		5 Mar 1960–25 Sep 1961	Detroit Zoo, MI, USA	
1			13 Apr 1960–25 Jun 1960	Cologne Zoo, Germany	
	1		13 Apr 1960–17 Apr 1965	Cologne Zoo, Germany	
1	1		29 Apr 1960–23 Jul 1961	Rome Zoo, Italy	
1	1		19 May 1960–2 Jun 1960	Cheyenne Mt. Zoo, CO, USA	
1			19 May 1960–11 Jun 1960	Cheyenne Mt. Zoo, CO, USA	
	1		19 May 1960–3 Jun 1960	Cheyenne Mt. Zoo, CO, USA	
	1		19 May 1960–8 Jun 1960	Cheyenne Mt, Zoo, CO, USA	
		1	22 Jun 1960–22 Jun 1960	Frankfurt Zoo, Germany	born at Frankfurt Zoo
1			22 Jul 1960–Jan 1974	San Antonio Zoo, TX, USA	from National Zoo
	1		19 Sep 1960–12 Jun 1963	Stuttgart Zoo, Germany	
1			6 Dec 1960– ?	Berlin Zoo, Germany	
	1		6 Dec 1960–21 May 1973	Berlin Zoo, Germany	
	1		7 Mar 1961–8 Mar 1961	Antwerp Zoo, Belgium	
		1	9 Mar 1961–3 Apr 1968	Detroit Zoo, MI, USA	
1			9 Mar 1961–to Madison	Detroit Zoo, MI, USA	
			22 May 1968–28 May 1971	Madison Zoo, WI, USA	
1			24 Mar 1961–18 Dec 1962	Detroit Zoo, MI, USA	
	1		24 Mar 1961–9 Jul 1966	Detroit Zoo, MI, USA	
		1	15 Apr 1961–6 Sep 1968	Stuttgart Zoo, Germany	

TABLE 12.2 (cont'd)

Sex			Dates of arrival and/or departure/death	Collection	Comments
Male	Female	Unknown			
	1		15 Jun 1961– ?	Frankfurt Zoo, Germany	born at Frankfurt Zoo
	1		15 Jun 1961–4 Feb 1962	Frankfurt Zoo, Germany	born at Frankfurt Zoo
1			1962–16 Oct 1963	Copenhagen Zoo, Denmark	
	1		1962–4 Dec 1963	Copenhagen Zoo, Denmark	
1			?	Buffalo Zoo, NY, USA	sent to Cincinnati Zoo
			22 Nov 1962–18 Sep 1964	Cincinnati Zoo, OH, USA	
		2	1962– ?	National Zoo, DC, USA	born at National Zoo
1			5 Apr 1962–29 Jul 1963	Artis Zoo, Amsterdam, Netherlands	
	1		5 Apr 1962–7 Jan 1967	Artis Zoo, Amsterdam, Netherlands	
		1	5 Apr 1962–11 Jul 1962	Artis Zoo, Amsterdam, Netherlands	
		1	5 Apr 1962–12 Jul 1962	Artis Zoo, Amsterdam, Netherlands	
	1		11 Apr 1962–27 Nov 1970	Rotterdam Zoo, Netherlands	
1			12 Apr 1962–1 Dec 1966	Copenhagen Zoo, Denmark	
	1		12 Apr 1962–22 Jul 1964	Copenhagen Zoo, Denmark	
1			18 Apr 1962–11 Jan 1967	Antwerp Zoo, Belgium	
1			18 Apr 1962–9 Sep 1968	Antwerp Zoo, Belgium	
	1		18 Apr 1962–28 Feb 1963	Antwerp Zoo, Belgium	
	1		18 Apr 1962–22 Nov 1963	Antwerp Zoo, Belgium	
1			22 May 1962–23 Sep 1974	Zurich Zoo, Switzerland	
	1		22 May 1962–16 Apr 1973	Zurich Zoo, Switzerland	
	1		9 Jun 1962– ?	Base1 Zoo, Switzerland	
	1		2 Jul 1962– ?	Frankfurt Zoo, Germany	born at Frankfurt Zoo
2			18 Jul 1962– ?	Berlin Zoo, Germany	born at Berlin Zoo
	1		3 Sep 1962–1 May 1967	Columbus Zoo, OH, USA	

TABLE 12.2 (cont'd)

Sex			Dates of arrival and/or departure/death	Collection	Comments
Male	Female	Unknown			
1			12 Sep 1962– ?	Rome Zoo, Italy	
	1		12 Sep 1962–29 Jan 1963	Rome Zoo, Italy	
1			16 Nov 1962–26 Nov 1962	Frankfurt Zoo, Germany	returned to dealer Molinar
1			22 Nov 1962–18 Sep 1964	Columbus Zoo, OH, USA	
	1		30 Nov 1962–15 Aug 1969	Frankfurt Zoo, Germany	
		2	1963 – ?		born National Zoo, DC, USA
1			22 Feb 1963–30 Jan 1966	Wassenaar Zoo, Netherlands	
	1		22 Feb 1963–18 Aug 1966	Wassenaar Zoo, Netherlands	
		1	22 Jun 1963– ?	Frankfurt Zoo, Germany	born at Frankfurt Zoo
1	1		23 Jun 1963– ?	Berlin Zoo, Germany	born at Berlin Zoo
	1		Oct 1963–21 Dec 1963	Stuttgart Zoo, Germany	
	1		26 Mar 1964–18 Jun 1968	Hagenbeck, Hamburg, Germany	
	1		26 Mar 1964–3 Jul 1969	Hagenbeck, Hamburg, Germany	
	1		26 Mar 1964–20 Mar 1970	Hagenbeck, Hamburg, Germany	
	1		26 Mar 1964 to Copenhagen	Hagenbeck, Hamburg, Germany	
			7 Jan 1965–2 Dec 1966	Copenhagen Zoo, Denmark	
		2	30 Jun 1964– ?	Frankfurt Zoo, Germany	born at Frankfurt Zoo
2			8 Jul 1964– ?	Berlin Zoo, Germany	born at Berlin Zoo
1			21 Mar 1965–26 Mar 1965	San Diego Zoo, CA, USA	
1			21 Mar 1965–7 Apr 1965	San Diego, CA, USA	
	1		21 Mar 1965–15 Nov 1969	San Diego, CA, USA	
	1		21 Mar 1965–29 Aug 1971	San Diego, CA, USA	

TABLE 12.2 (cont'd)

Sex			Dates of arrival and/or departure/death	Collection	Comments
Male	Female	Unknown			
1			27 Mar 1965–3 Aug 1969	Winnipeg Zoo, Canada	
	1		21 Mar 1965– ?	Winnipeg Zoo, Canada	
1			15 Apr 1965–17 Oct 1972	Philadelphia Zoo, PA,USA	
1			1 May 1965–9 Jan 1971	Artis Zoo, Amsterdam, Netherlands	born at Berlin Zoo, birth date unknown
1			10 Nov 1965–30 Nov 1965	Milwaukee Zoo, WI, USA	
	1		10 Nov 1965–25 Nov 1965	Milwaukee Zoo, WI, USA	
	1		10 Nov 1965–30 Nov 1965	Milwaukee Zoo, WI, USA	
1	1		16 Dec 1965– ?	Los Angeles Zoo, CA, USA	
1			14 Sep 1966–26 Sep 1966	Los Angeles Zoo, CA, USA	
	1		14 Sep 1966–20 Oct 1966	Los Angeles Zoo, CA, USA	
1			28 Oct 1966–28 Mar 1974	National Zoo, DC, USA	
	1		28 Oct 1966–1969	National Zoo, DC, USA	
	1		1967– ?	Buffalo Zoo, NY, USA	
	1		13 Jan 1967–13 Dec 1968	Rotterdam Zoo, Netherlands	
	1		13 Jan 1967–30 Sep 1970	Rotterdam Zoo, Netherlands	
	1		13 Jan 1967–16 Jan 1972	Rotterdam Zoo, Netherlands	
1			2 Feb 1967– Jun 1968	San Francisco Zoo, CA, USA	
1			2 Feb 1967– ?	San Francisco Zoo, CA, USA	
1			15 Feb 1967–4 Mar 1969	St. Louis Zoo, MO, USA	
1			15 Feb 1967–23 Apr 1971	St. Louis Zoo, MO, USA	
	1		15 Feb 1967–13 Feb 1973	St. Louis Zoo, MO, USA	
	1		15 Feb 1967– ?	St. Louis Zoo, MO, USA	
	1		?–10 Mar 1967	San Francisco Zoo, CA, USA	said to be "old" female

TABLE 12.2 (*cont'd*)

Sex			Dates of arrival and/or departure/death	Collection	Comments
Male	Female	Unknown			
1			19 May 1967–26 Sep 1968	Stuttgart Zoo, Germany	
1			19 May 1967–2 Jul 1970	Stuttgart Zoo, Germany	
	1		19 May 1967–7 Jan 1973	Stuttgart Zoo, Germany	
	1		19 May 1967–29 Jun 1976	Stuttgart Zoo, Germany	
		1	28 Jun 1967– ?	Frankfurt Zoo, Germany	born at Frankfurt Zoo
1			2 Oct 1967–19 May 1969	London Zoo, GB	
	1		2 Oct 1967– ?	London Zoo, GB	
1			7 Oct 1967–5 May 1980	Whipsnade Zoo, GB	
		2	1968–1972	Schönbrunn Zoo, Vienna, Austria	
	1		?–10 Jun 1968	San Francisco Zoo, CA, USA	
	1		?–27 Jun 1968	San Francisco Zoo, CA, USA	
1			?–6 Sep 1968	Stuttgart Zoo, Germany	
	1		29 Jan 1968– ?	Cologne Zoo, Germany	
1	1		30 Jan 1968– ?	Krefeld Zoo, Germany	
1			10 Feb 1968–18 May 1968	Wassenaar Zoo, Netherlands	
1			10 Feb 1968–17 Mar 1968	Wassenaar Zoo, Netherlands	
	2		20 Feb 1968–11 Jan 1971	Artis Zoo, Amsterdam, Netherlands	
		1	1 May 1968–16 Jul 1968	Wassenaar Zoo, Netherlands	
	1		11 May 1968–4 Aug 1972	Duisburg Zoo, Germany	
	1		11 May 1968–	Duisburg Zoo, Germany	sent to Artis Zoo
			22 Nov 1971–12 Feb 1972	Artis Zoo, Amsterdam, Netherlands	
	1		22 May 1968–4 Dec 1977	Cheyenne Mt. Zoo, CO, USA	born at Cheyenne Mt. Zoo
1			18 Jun 1968–26 Sep 1968	Stuttgart Zoo, Germany	
1			18 Jun 1968–2 Jul 1970	Stuttgart Zoo, Germany	

TABLE 12.2 *(cont'd)*

Sex			Dates of arrival and/or departure/ death	Collection	Comments
Male	Female	Unknown			
	1		18 Jun 1968–16 Apr 1976	Stuttgart Zoo, Germany	
1			26 Jun 1968–21 Mar 1970	Milwaukee Zoo, WI, USA	
	1		26 Jun 1968–1 Nov 1971	Milwaukee Zoo, WI, USA	
1			Jul 1968– ?	National Zoo, DC, USA	born at National Zoo
1			10 Aug 1968–5 Oct 1972	Antwerp Zoo, Belgium	
	1		10 Aug 1968–8 Oct 1972	Antwerp Zoo, Belgium	
1			7 Sep 1968– ?	Cologne Zoo, Germany	
	1		19 Sep 1968– ?	Buffalo Zoo, NY, USA	
1			19 Oct 1968–18 Aug 1979	Cheyenne Mt. Zoo, CO, USA	
	1		19 Oct 1968–26 Dec 1968	Cheyenne Mt. Zoo, CO, USA	
1			21 Nov 1968– ?	Los Angeles Zoo, CA, USA	
1			27 Nov 1968– ?	Dallas Zoo, TX, USA	
	1		27 Nov 1968–18 Jun 1970	Dallas Zoo, TX, USA	
	1		21 Dec 1968–?	Los Angeles Zoo, CA, USA	
1			1969–1969	National Zoo, DC, USA	born at National Zoo
		1	1969–8 Oct 1971	Memphis Zoo, TN, USA	
		1	?–17 Jul 1969	Buffalo Zoo, NY, USA	
1			6 Jan 1969–16 Feb 1969	Artis Zoo, Amsterdam, Netherlands	
	1		6 Jan 1969–17 Feb 1969	Artis Zoo, Amsterdam, Netherlands	
		1	11 Jan 1969–31 Jan 1969	London Zoo, GB	
		1	11 Jan 1969–2 Feb 1969	London Zoo, GB	
1			18 Jan 1969– ?	Artis Zoo, Amsterdam, Netherlands	
1			18 Jan 1969–8 Feb 1973	Artis Zoo, Amsterdam, Netherlands	
1			20 Feb 1969–7 Jul 1970	Artis Zoo, Amsterdam, Netherlands	

TABLE 12.2 (cont'd)

Sex			Dates of arrival and/or departure/death	Collection	Comments
Male	Female	Unknown			
	1		20 Feb 1969–13 Nov 1969	Artis Zoo, Amsterdam, Netherlands	
	1		20 Feb 1969–21 Dec 1972	Artis Zoo, Amsterdam, Netherlands	
	1		28 Feb 1969–9 Jun 1980	London Zoo, GB	
	1		28 Feb 1969– ?	London Zoo, GB	
	1		3 Mar 1969–19 May 1969	London Zoo, GB	
	1		3 Mar 1969–14 Aug 1980	Whipsnade Zoo, GB	
1			6 Mar 1969–9 Sep 1973	Copenhagen Zoo, Denmark	
	1		6 Mar 1969–24 Feb 1974	Copenhagen Zoo, Denmark	
1			10 Mar 1969–2 Apr 1969	Miami Crandon Park Zoo, FL, USA	
1			10 Mar 1969–3 Apr 1969	Miami Crandon Park Zoo, FL, USA	
	1		23 Apr 1969– ?	Los Angeles Zoo, CA, USA	
1			19 Jun 1969–19 Jun 1969	Artis Zoo, Amsterdam, Netherlands	born at Artis Zoo
1			19 Jun 1969–20 Jun 1971	Artis Zoo, Amsterdam, Netherlands	born at Artis Zoo
1			24 Jun 1969–stillborn	Dallas Zoo, TX, USA	born at Dallas Zoo
1			24 Jun 1969–25 Jun 1969	Dallas Zoo, TX, USA	born at Dallas Zoo
1			24 Jun 1969–28 Jun 1969	Dallas Zoo, TX, USA	born at Dallas Zoo
1			26 Jun 1969–16 Nov 1972	Cheyenne Mt. Zoo, CO, USA	born at Cheyenne Mt. Zoo
1			1 Jul 1969–19 Jun 1971	Dallas Zoo, TX, USA	born at Dallas Zoo
	1		15 Jul 1969–4 Aug 1969	Frankfurt Zoo, Germany	
	1		16 Jul 1969–31 Jan 1978	Cheyenne Mt. Zoo, CO, USA	
1			17 Jul 1969– ?	Buffalo Zoo, NY, USA	
	1		24 Jul 1969– ?	Madison Zoo, WI, USA	
	1		30 Jul 1969–21 Mar 1971	Cheyenne Mt. Zoo, CO, USA	
1			30 Jul 1969–30 Dec 1975	Cheyenne Mt. Zoo, CO, USA	sent to Denver Zoo

panda studbook and these animals have descendants in the current captive population. In the period between 1950 and 1969 some 20 specimens were born in captivity. It must be noted that in general longevity also increased over this time, although far too many animals still did not long survive the rigors of the voyage from their native habitat.

All of the animals listed in the tables and discussed in the preceding sections belong to the nominate race, *A. f. fulgens*. However, there is a second subspecies of red panda originating from China. This subspecies was first brought to the attention of science when Père David sent its skin to the Paris museum. It was first thought to be a separate species and as such was named *Ailurus refulgens* by Milne Edwards in 1874. However, the Chinese red panda was eventually given subspecific status and the subspecific name *styani*, given by Thomas in 1902, is the one that we use today. Specimens of *A. f. styani* have only been exhibited in captivity outside of China in recent years and thus all are listed in the studbook. It is interesting to mention that Floyd Tangier Smith collected a number of *styani*, together with giant panda, *Ailuropoda melanoleuca*, Tibetan takin, *Budorcas taxicolor tibetana*, and tufted deer, *Elaphodus cephalophus cephalophus*, for the Field Museum in 1931 and that all four species are currently living in the San Diego Zoo.

ACKNOWLEDGEMENTS

This chapter originally appeard in 'Red Panda Biology' (1989) Editor A. R. Glatston. Permission to reprint this chapter was kindly granted by Backhuys Publishers, Leiden, The Netherlands.

References

[1] L. Crandall, Management of Wild Animals in Captivity, University of Chicago Press, Chicago, 1964.
[2] J.R. Ellerman, T.C. Morrison-Scott, A Checklist of Palearctic and Indian Mammals 1758 to 1946, London, 1951.

CHAPTER

13

Red Panda Husbandry for Reproductive Management

Kati Loeffler

International Fund for Animal Welfare, Chaoyang District, Beijing, China

OUTLINE

Reproductive success in the *ex situ* red panda population varies greatly worldwide (see Chapter 17). Red pandas are private, sensitive animals whose husbandry has always been challenging. As we have learned about the natural history of the species and have developed our care of individuals in captivity over the years, reproductive rates have risen. We understand now the necessity for privacy of mothers and their sensitivity to any kind of disturbance which easily results in maternal failure. We are beginning to understand the types of environmental and social stresses to which the species is sensitive. We have developed supplemental milk products and improved weaning methods for cubs whose mothers need assistance. Our understanding of the reproductive physiology of red pandas has improved. Nonetheless, we still have great room for improvement in serving this species, and the complexity of its requirements is evident in the difficulty that we still have in enabling this remarkable animal to reproduce in our breeding facilities.

Age-specific fertility data analysed from the international red panda studbook records of 1984 to 2004 [1] (Figures 13.1 and 13.2) suggest that the *ex situ* red panda population is

Red Panda. DOI: 10.1016/B978-1-4377-7813-7.00013-6

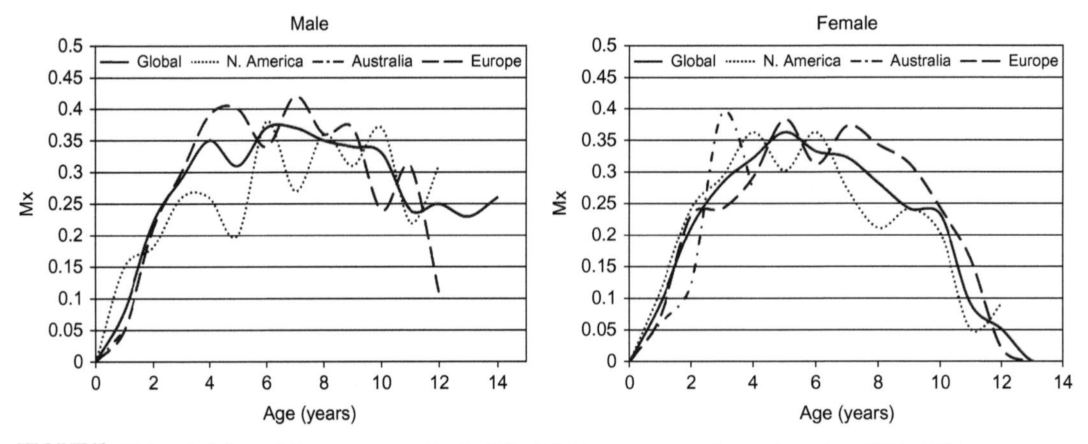

FIGURE 13.1 *A. fulgens fulgens*, age-specific fertility (Mx) by age group for each region, 1984–2004.

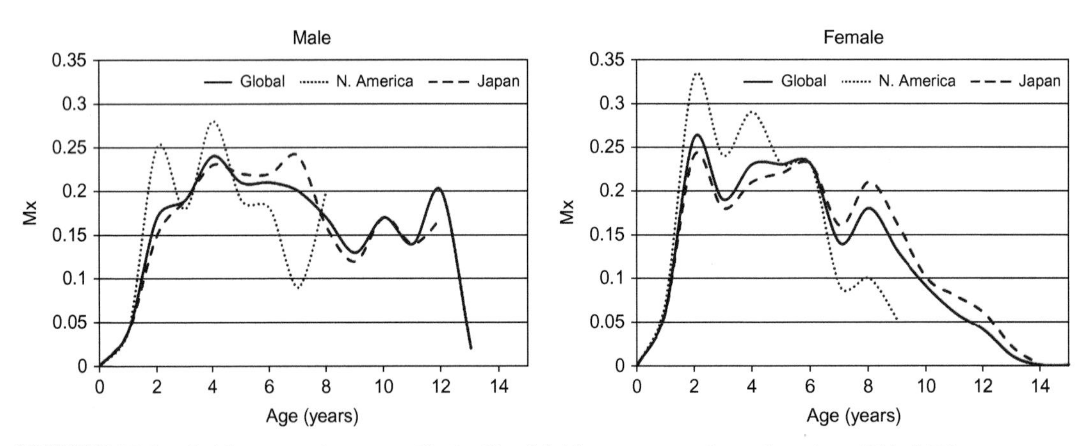

FIGURE 13.2 *A. fulgens styani*, age-specific fertility (Mx) by age group for each region, 1984–2004.

not breeding as happily as it might. Age-specific fertility, or Mx, is calculated as the average number of offspring of the same sex as the parent that are produced by an individual animal in a given age class. Broadly, it may be considered the average number of animals in a given age class that will reproduce [1]. At peak reproductive age, little more than a third of males and females in the global *fulgens* subspecies population are reproducing; for *styani* subspecies, this figure is only 23%. (*Fulgens* populations are calculated from the European, N. American and Australia–New Zealand populations, while *styani* includes the N. American and Japanese groups. China's captive red pandas are primarily *styani* but there are no reproductive data available for this population.) The context for these statistics is not easy to evaluate, as natural reproductive rates are not well understood, much less a realistic expectation under captive conditions. Given that females are capable of reproducing annually, however, with an average of two offspring per litter, it would seem

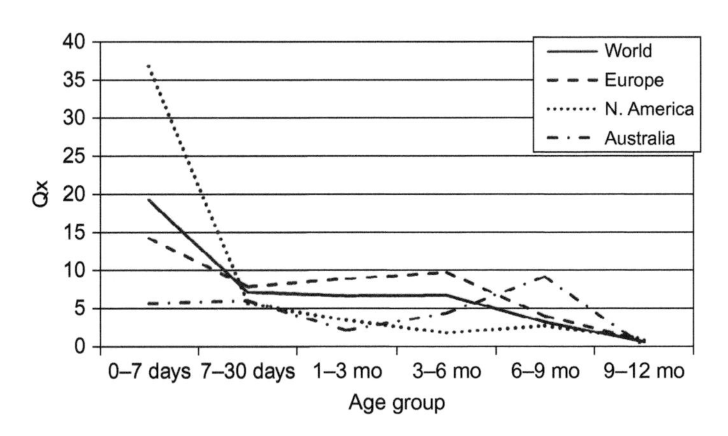

FIGURE 13.3 Probability that a red panda cub (*fulgens* subspecies) in a given age group will die before reaching the next age group, 1997–2007.

that the current rates could be improved. The strong regional differences in reproduction (discussed below) further support the suggestion that husbandry factors are limiting the success of the species *ex situ*.

Figures 13.1 and 13.2 show that male *fulgens* and *styani* appear to begin to reproduce at age 1 or 2 years, peak at 4 years to 7–9, and then drop off gradually until 12 years, when there is a sharp decrease in reproduction. Age-specific fertility is not calculated with consideration for mortality, so the decline in the curves indicates a reproductive senescence that begins at around 8 years of age.

Female *fulgens* also begin to reproduce at age 1 or 2 years, peak from 4–8 years, after which their reproductive output declines. The female *fulgens* curve is more dome-shaped than that of the males, and peaks slightly lower. The higher peak Mx in males may be due to a larger male population or because males breed with multiple females, thereby increasing their chances of producing male offspring. Australia's Mx cannot be assessed relative to that of the other regions because, as for Qx (age-specific mortality or the proportion of individuals entering an age-class which die before they enter the next age-class), the Risk Mx (population size for calculation of Mx) is too low after the first year (males) and fourth year (females) of life. Europe's females appear to have a somewhat longer reproductive life than those of the global and N. American populations.

In *styani*, maximum Mx is similar between males and females, although the shape of the female curve is more peaked than that of the males' (see Figure 13.2). Females in N. America have a higher maximum Mx than those in Japan during peak reproductive years.

The Mx curve for *styani* is shifted to the left relative to that of *fulgens* and is considerably lower in both males and females (see Figures 13.1 and 13.2). Peak fertility is reached earlier in *styani* than in *fulgens* but is similar in duration (2–6 years of age in *styani* vs. 4–8 years in *fulgens*). After the peak, however, the decline in Mx is slower in *fulgens*.

The second factor that appears to contribute most strongly to compromised population growth is infant mortality. As discussed in Chapter 17, this varies greatly among the four regions. Analysis of the global studbook records from 1978–1992 [2] and 1984–2004 indicates that the highest rate of mortality is in the first week of life. Li's [2] analysis indicated a second mortality peak between 30 and 150 days, but this was not seen in the evaluation of a more recent data set (Figure 13.3).

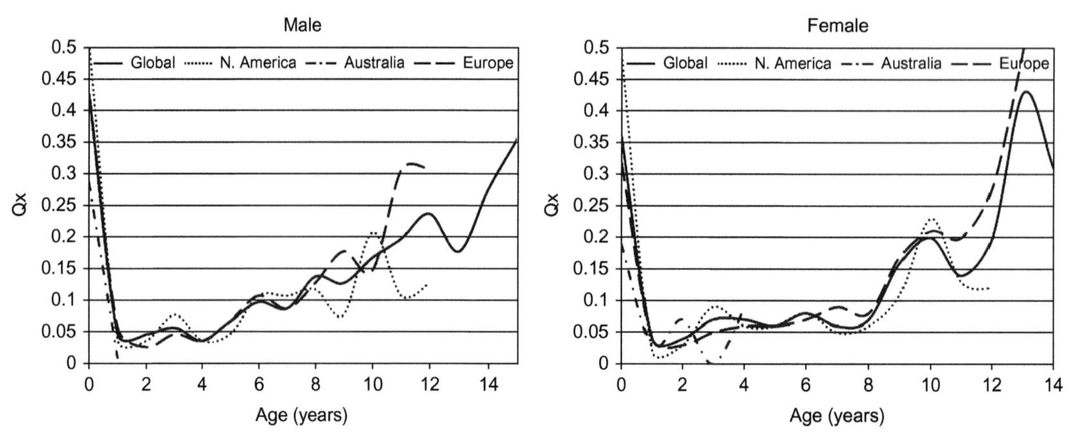

FIGURE 13.4 *A. fulgens fulgens*, age-specific mortality (Qx) by age group for each region, 1984–2004.

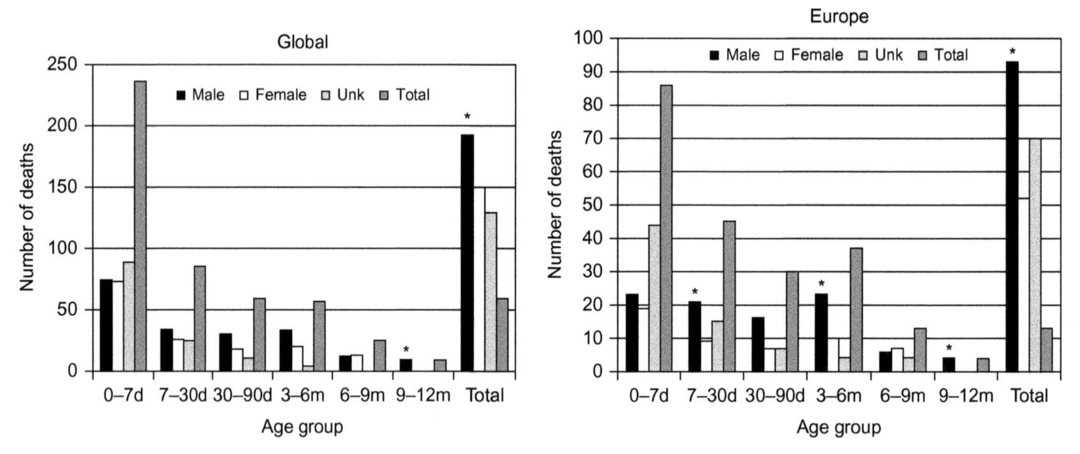

FIGURE 13.5 First year deaths of male and female *A. fulgens fulgens* by age group (d = days; m = months). Asterisk indicates significant difference between males and females in that age group ($\chi^2 \leq 0.05$).

Total first-year mortality for *fulgens* (Figure 13.4) is higher in males than in females globally and in all regions except in N. America, where it is equal between the sexes (Figure 13.5); the difference appears to be significant only in the European population. Plots of age-specific mortality (Figure 13.4) and cumulative first-year mortality (Figure 13.6) indicate that the N. American population has the highest infant mortality rate of all regions: 50% compared with Europe's 42% (male) and 32% (female) and with Australia's 28% (male) and 19% (female).

For *styani*, the first-year mortality difference between the sexes is reversed: 34% for females and 28% for males (Figures 13.7 and 13.8). *Fulgens* male infants suffer higher mortality rates globally than *styani* males (42% vs. 28%), while for female infants Qx is similar

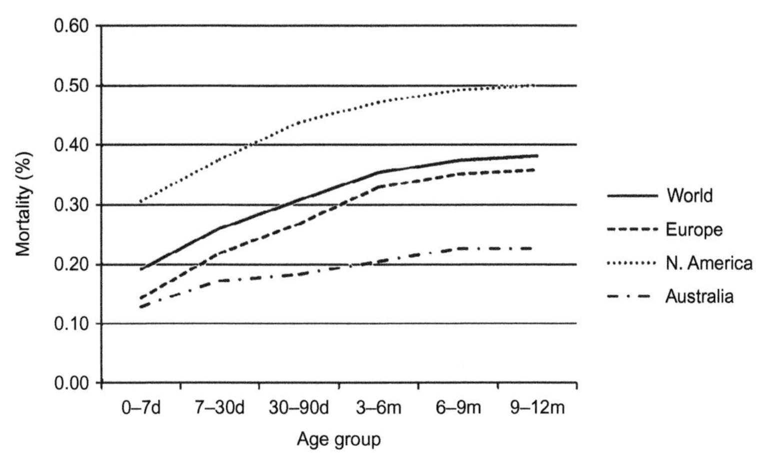

FIGURE 13.6 Cumulative first year mortality (cumulative deaths/total births) for *A. fulgens fulgens*, 1984–2004.

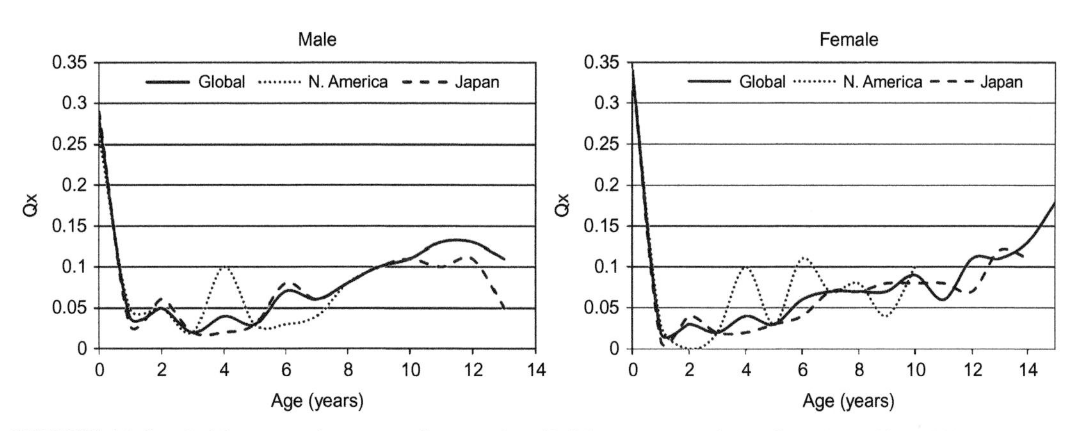

FIGURE 13.7 *A. fulgens styani*, age-specific mortality (Qx) by age group for each region, 1984–2004.

between the subspecies (34% and 36%). Cumulative mortality appears to level off after 3 months of age for *styani*, while for *fulgens* it continues to rise until 9 months (Figures 13.9 and 13.6, respectively). The continued increase for *fulgens* appears to occur principally in the N. American and European populations (see Figure 13.6).

The data plotted for Qx and Mx include only those values for which the Risk Qx and Risk Mx (n value for each group) are >20; with smaller Risk groups, the statistical analysis becomes unreliable. It is interesting, that the Risk Qx for the Australian male population (*fulgens*) begins lower than that of females in the first year of life and falls below 20 after age 1 while, for females, it remains above 20 until age 4 (Figure 13.10). This suggests that either more females than males are born in this population, or that male mortality is higher in the early parts of the first year and that they are therefore removed from the population records sooner, or both. Neither of these explanations is supported by the data, however: sex differences in mortality (Figures 13.10 and 13.5) and total births

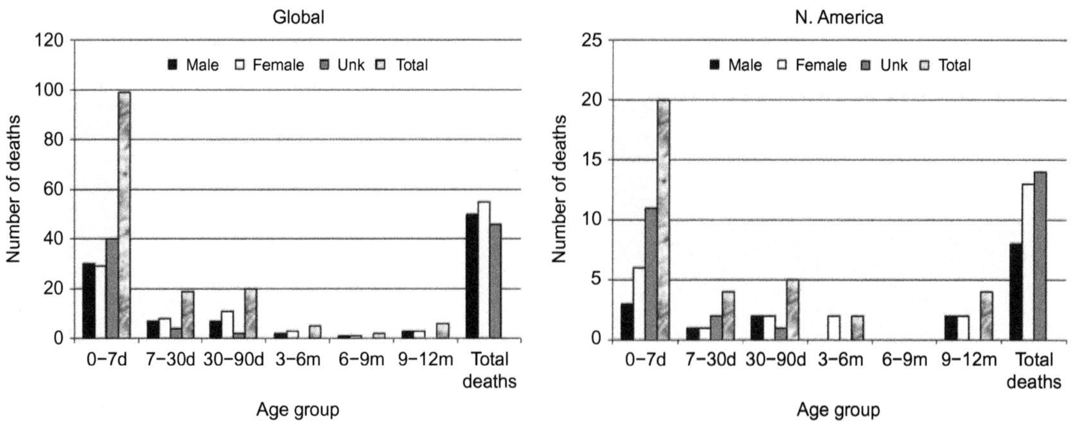

FIGURE 13.8 First year deaths of male and female *A. fulgens styani* by age group, 1984–2004 (d = days; m = months). Differences between males and females were not significant in any age group ($\chi^2 \leq 0.05$).

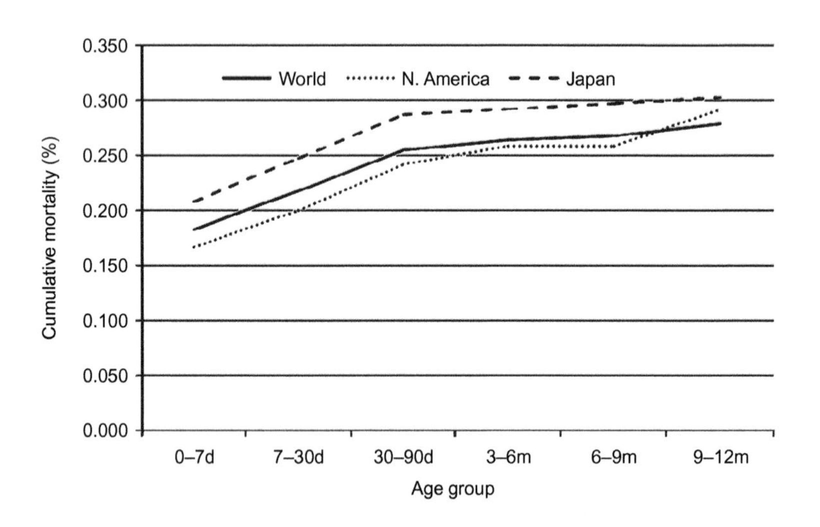

FIGURE 13.9 Cumulative first year mortality (cumulative deaths/total births) for *A. fulgens styani*, 1984–2004.

(Figure 13.11) are not significantly different. Mortality appears to be higher in males than in females in the 7–30 day and 3–6 month age groups (see Figure 13.5), but again, the n values for this observation are too low to determine the significance of this observation. Resolution of the "unknown" individuals into one or other group may alter this information somewhat, but the current data set leaves us with what appears to be a statistical artifact and warrants further investigation once the populations have grown.

The apparent greater number of males born in Europe is also not statistically significant (see Figures 13.10 and 13.11). The difference in male and female population size equals out by the second year. This appears to be due to a greater infant mortality rate in European males, which is statistically significant overall and specifically in the 7–30 day, 3–6 month, 9–12 month age groups (see Figure 13.5). Whether this finding is an artifact of low

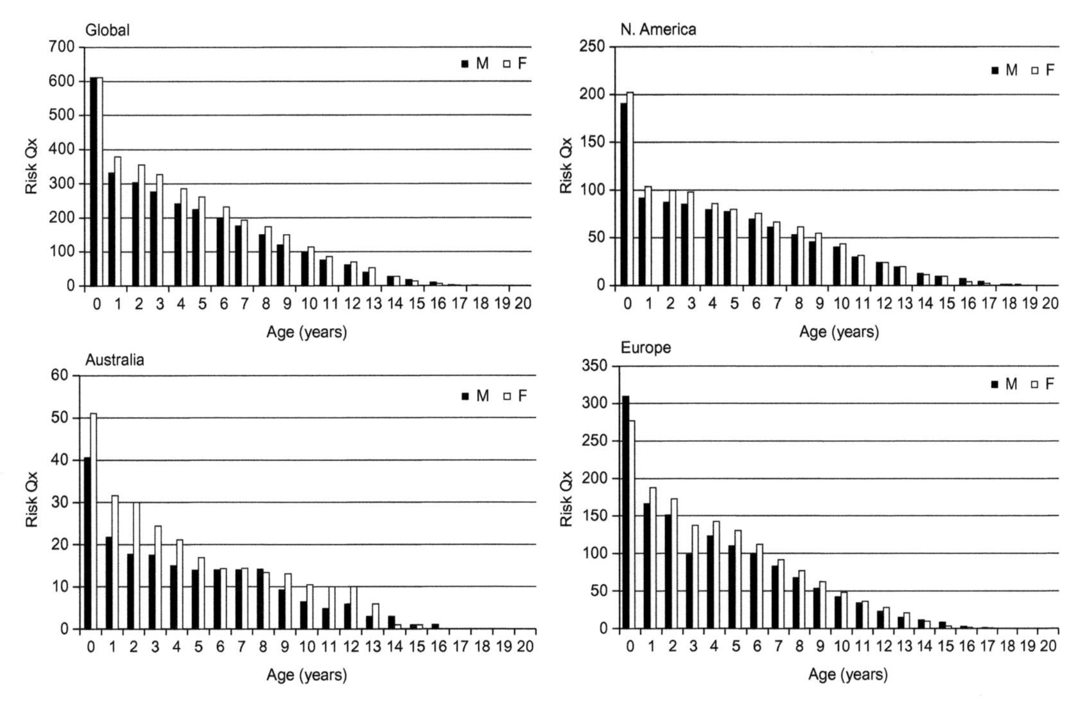

FIGURE 13.10 Comparison of Risk Qx (n from which Qx is calculated) between males and females by age group in global and regional populations of *A. fulgens fulgens*, 1984–2004.

population numbers or whether it is real and warrants investigation into management practices remains to be determined. Japan's *styani* show a similar pattern (Figures 13.12 and 13.8), although none of the differences in this group is statistically significant.

The shape of the Qx curves for *styani* (see Figure 13.7) and *fulgens* (see Figure 13.4) are similar prior to age 8 years; thereafter, mortality increases more quickly in *fulgens* than in *styani*. The age at which populations exceed 10% Qx is 7 and 8 years for *fulgens* males and females, respectively, while for *styani* this is 10 and 12 years. Combined with the Mx data discussed above, it would appear that *styani* enjoy greater longevity than *fulgens* but that this does not prolong their reproductive duration.

All told, these numbers add up to an annual *fulgens* population growth rate (λ) of only 1.01 to 1.11 (female) and 0.99 to 1.07 (male) among the N. American, European and Australian populations. ($\lambda = 1$: no growth; $\lambda < 1$: population growth; $\lambda > 1$: population decline.) Australia's high infant survival is counteracted by a narrow fecundity curve; Europe's broader fecundity curve is compromised by higher infant mortality rates. The balance results in a relatively uniform and, unfortunately, slow annual population growth rate across the regions. For *styani*, the situation is similar, albeit with the male and female lambda values in reverse of those for *fulgens*: a negative growth rate of 0.97 for females, and a barely positive 1.01 for males.

Differences in reproductive and infant mortality rates between the two subspecies may lie in statistical as well as physiological and behavioural factors. Empirically, *styani* females

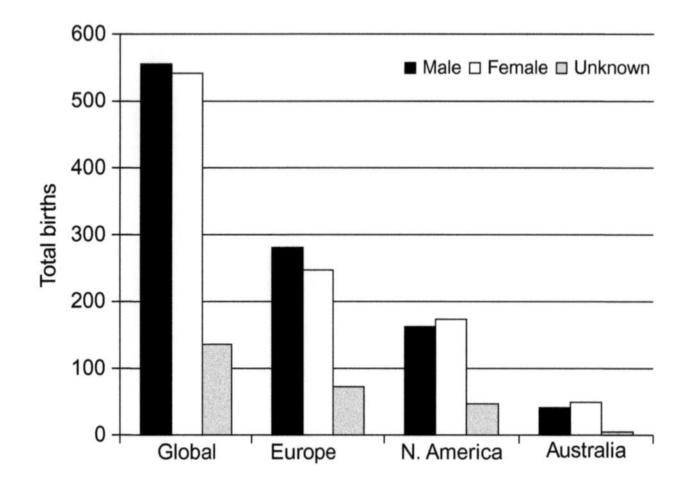

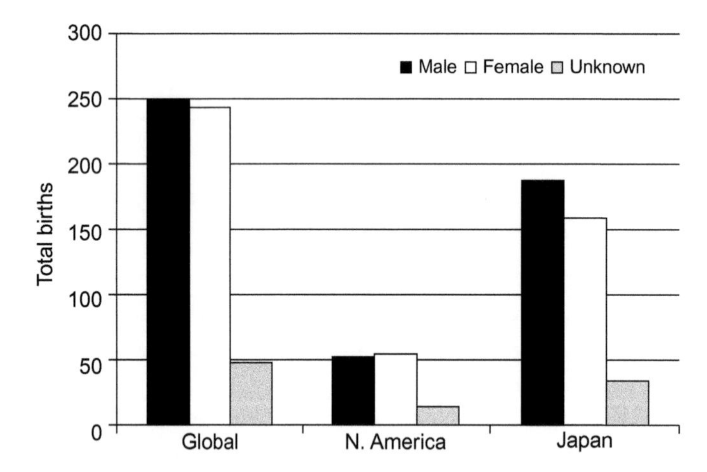

FIGURE 13.11 Birth rates for *A. fulgens fulgens* (top) and *styani* (bottom) globally and regionally, 1984–2004. Differences between males and females are not statistically significant ($\chi^2 \leq 0.05$).

living *ex situ* have a shorter reproductive period than *fulgens* and, as seen from the fecundity data discussed above, their reproductive rate is lower. The infant survival rate reported for *styani* is higher than that for *fulgens*, however. These data are based largely on the relatively small number of *styani* in the few N. American facilities who breed this subspecies. One or two facilities with good infant survival rates will represent a much larger percentage of the whole than for *fulgens*. Similarly, the few facilities which do breed *styani* and do so successfully may not be able to make up for the greater number which does not do so. A renewed effort is now underway to involve the Chinese in the global *ex situ* red panda breeding programme to allow an infusion of new founder members into the population.

The significant question is what husbandry practices contribute to the low reproductive rates and to the high infant mortality. The following is a summary and comparison of the husbandry practices among the N. American, European and Australian red panda facilities. Sources for this comparison are the Red Panda International Husbandry and

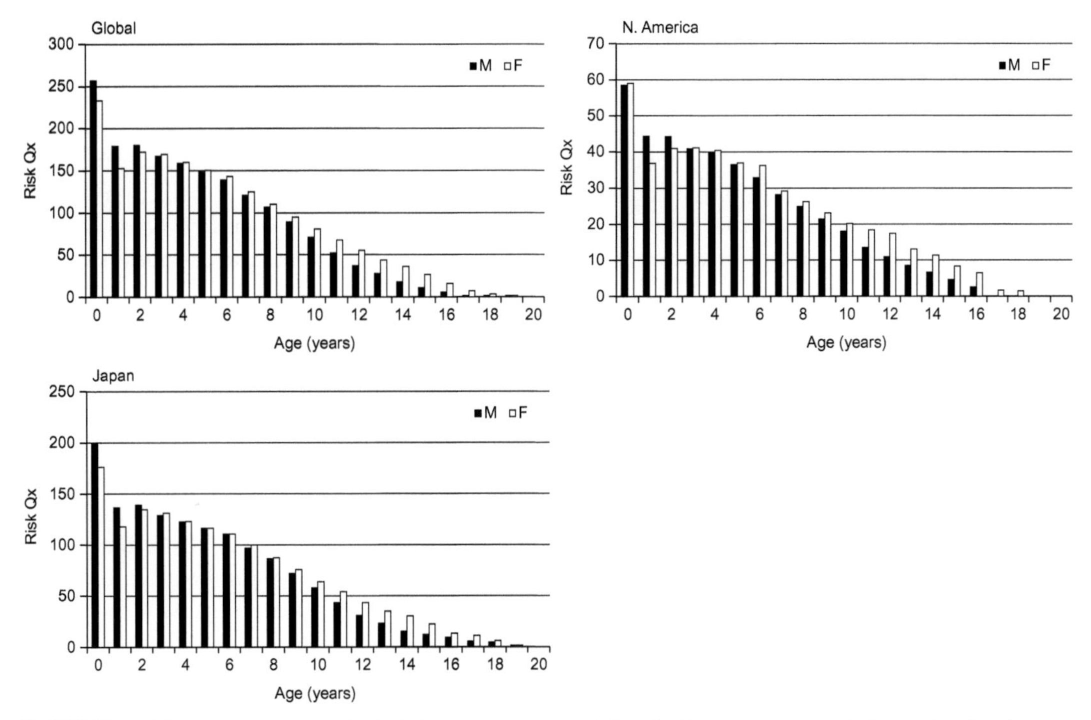

FIGURE 13.12 Comparison of Risk Qx between males and females by age group in global and regional populations of *A. fulgens styani*, 1984–2004.

Management Guidelines [3] (IHMG) and regional protocols, red panda husbandry review articles, and personal interviews with experienced red panda programme managers.

ENCLOSURE SPACE AND DESIGN

Enclosure size and quality, with full time access to the outdoors appear to be critical factors in red panda husbandry. The IHMG recommend a minimum enclosure size of 80 m^2 and 4 m height per single or pair of red pandas, with full-time access to outdoor areas. The N. American Species Survival Plan (SSP) recommends just half of this. Whether small enclosure size is a factor in the high mortality rates in N. American zoos warrants investigation. In a study among European, Australian and N. American zoos [4,5], the average enclosure size was about 300 m^2, although it is unclear from that study how large the enclosures were per pair, per individual red panda or per mother with that year's offspring. Analysis of that study's [4,5] data (assuming an equal number of red pandas in each enclosure) indicates that there are no significant differences among the mean enclosure sizes of the three regions (Student's t-test, df = 44, 9, 14; 1-tailed), nor among their variances (F-test, df = 44, 9, 14). Regardless, the average 300 m^2 horizontal space meets the IHMG for up to three red pandas together, so long as the area has sufficient vertical space, a large enough outdoor area with 24-h access and sufficient diversity of the environment in the enclosure.

The quality of the enclosures is highly important for successful reproduction by all analyses [6] (and personal communications from red panda curators), as it is for most wildlife species maintained *ex situ*. Husbandry protocols and advice from successful red panda breeding facilities state that enclosure space must contain a variety of substrates (e.g., wood, rock, turf, mulch, logs, branches, ropes, vines, pools), must provide opportunity for the animals to use vertical space, and must provide plenty of shade and hiding areas. In the wild, red pandas spend the majority of their time high in the trees and come down primarily to eat and to mate. Providing an environment in which the animals may exercise normal behaviours is paramount to their comfort in captivity. Red pandas also enjoy pools of water, which serve the additional role of providing a cooling medium in warm climates. Interestingly, the aforementioned study [4,5] found that only half of the surveyed zoos provide pools to their red pandas. Shade is also important to breeding success, even in temperate climates [6]. Shade serves not only in the protection from heat, but also in enhancing the animals' sense of obscurity.

A variety of vegetation in the enclosure provides visual barriers, natural structures to climb and to explore, and should be selected to provide foraging opportunities. Red pandas eat a variety of grasses and leaves. Making these available in a naturalistic manner is particularly important if bamboo is not supplied *ad libitum*. The availability of natural forage, moreover, allows the animals to dictate their own activity and feeding patterns rather than relying on the schedules of keepers for feeding. It also provides materials for females to build or to embellish their nests.

A vertical distance and visual barriers from activities and people on the ground are important to the animals' sense of security. The IHMG recommend that exhibits be accessible to public viewing from no more than one or two sides. The study by Eriksson et al. [5] found that 33% of the red panda exhibits in the survey were open to public view on three of four sides, another 33% around half the enclosure, and 16% were accessible all around the exhibit. How many of these open exhibits were breeding enclosures is not clear. Also unclear is how many of these three-quarter or fully exposed exhibits are of sufficient size and provide sufficient privacy within them that the circumferential exposure is offset by the quality of the exhibit. These complex considerations must all be taken into account in a careful analysis that seeks to determine the factors that support high reproductive and survival rates.

Regardless of the availability of data, we know enough about the natural behaviour of red pandas to design enclosures intelligently. Some enclosure designs are therefore rather enigmatic. For example, a new red panda exhibit in a major N. American zoo makes use of the vertical space in the small enclosure with tall trees for the animals to climb (Figure 13.13). However, visitors are able to view the exhibit from ground level and also from a bridge that places them at eye level with red pandas resting in the tree branches. The exhibit is exposed on all sides as well. Red pandas cannot escape the view of the public in this enclosure, particularly in winter when the trees are bare, unless they withdraw indoors. It is not difficult to understand why this exhibit is not conducive to reproductive success.

It is by now well established that red pandas are easily stressed animals with a high requirement for privacy and an undisturbed environment, particularly for females raising cubs. The general rule among red panda managers in N. America is that periparturient

FIGURE 13.13 Red panda exhibit accessible to public view on all sides and from bridge that places visitors on the same level as animals resting in trees. Red pandas normally seek safety and privacy in the height and foliage of trees. In this exhibit, the only option for privacy in this enclosure is to retreat indoors at ground level.

females and females with nursing cubs ideally should not be housed in display exhibits at all. In Europe, the stress of moving a pregnant animal is considered to outweigh the risk of potential exhibit stress, and females are not moved to alternative nursery quarters. Interestingly, one of the zoos with the highest reproductive success in N. America had, until recently, relatively small enclosures (27 m^2 × 2.4 m and 40 m^2 × 4.6 m), and mothers with relatively calm personalities are not removed from the exhibit at all. The enclosures in this zoo are highly enriched, the nest boxes are private, and the keepers have close relationships with the animals (discussed below). That said, this zoo frequently finds it necessary to hand-raise cubs or to supplement feed cubs that remain with mothers. If nothing else, this example illustrates the complexity of the factors that go into husbandry practices conducive to red panda reproduction.

Red pandas are well adapted to coping with low ambient temperatures [7], but have no mechanisms with which to tolerate heat. Protection of red pandas from temperatures above 27°C is critical to the well-being of the animals and for their ability to cope with further stressors. Enclosures should be at least partially in shade that is preferably provided by trees, as already mentioned, and a cool indoor area should be provided for the animals to choose as they wish. Under no circumstances should animals be forced to stay indoors, with the arguable exception of periparturient females who are unlikely to want to leave their nests (see below). Nest boxes in particular should be in shaded areas to guard against excessive heat and to enhance the privacy of the nest areas. Many zoos use misters in hot temperatures to cool the enclosures.

Isolation from species by which red pandas may feel threatened may be as important in the design of red panda enclosures as the opportunity to avoid human visitors. The IHMG suggest a minimum distance of 50 m between red panda enclosures and those of large carnivores or other aggressive species, and that these enclosures should not be located next to one another. Visual barriers and a minimum of 6 m distance between adjacent red panda enclosures are also advised. Eriksson et al.'s study [5] found that heterospecific enclosures were less than 15 m apart in 76% of the surveyed zoos, with nearly 30% of red panda exhibits located adjacent to those of large carnivores of which half contained large cats. One must assume that red pandas in these enclosures, as in enclosures without sufficient opportunity for escape from visitors, must operate under a chronic, at least subliminal, level of stress.

Chronic stress and a learned helplessness in animals who live under inappropriate conditions have been demonstrated [8]. The influence of stress on reproductive function, immune function and behaviour, moreover, is also becoming clear [8–12]. Red panda managers work with the understanding that a minimization of stress is important to the well-being and reproductive success of the animals, but the physiological role that stress plays in reproductive failure in the species is largely unknown. To what extent stress interferes with endocrine regulation of reproduction in red pandas, for example, or in milk production, the quality of passive immunity provided in the mother's milk, or in the complex behaviours of mothers raising young all remain to be understood. Similarly, some of the common medical conditions found in captive red pandas, such as hair loss, may be associated with stress as well. A careful study that examines the relationship between behaviour, physiological indicators of stress, reproductive endocrinology and reproductive success in captive red pandas would be highly valuable. Such a study may help to shed light on some of the husbandry practices that may be improved to raise reproductive rates in the captive red panda population.

None of the husbandry factors identified in the preliminary study by Zidar [4] correlate with infant mortality rates in a statistically significant manner. Enclosure size, proximity to enclosures of predatory species, number of nest boxes, percentage of enclosure accessible to visitor viewing, reported handling of cubs by keepers, ambient temperature or temperature control ... of all the obvious factors that one would consider as making a significant impact on reproductive success in red pandas, none distinguished the institutions with low cub mortality from those with high cub mortality. The clue probably lies in the relationship among multiple factors that will require a careful, detailed study to elucidate. The results of Zidar's and Eriksson's [4,5] studies will help to design the subsequent research strategy and questionnaire design that may bring forward some of the factors that in the pilot study demonstrated a potential suggestion of significance (e.g., enclosure size and proximity of exhibits to predator species). It will be important to encourage the cooperation of more zoos for a subsequent study in order to obtain a greater dataset and thereby enable a more comprehensive and insightful analysis.

NUTRITION

Red pandas are one of the few animals who really will starve to death if not provided with food with which they are familiar and that they find palatable. This is particularly critical at weaning, when red pandas are easily lost to inanition. Their thick fur makes it difficult to visually assess body condition that might alert keepers to weight loss before the condition becomes critical. Excessive weight should obviously be avoided as well, both for health reasons and because overweight females tend to have a harder time becoming pregnant than leaner individuals (empirical observation). Food intake and body weight should be monitored carefully, especially at critical periods when energy demands are changing, i.e., weaning, growth, late pregnancy, lactation and cold weather. For reproductive females and immature animals, this encompasses nearly the entire year. A monthly weighing schedule can easily be developed by training the animals to load into a carrying crate and to be quickly weighed therein or to station on a scale (see section on behavioural management, below). Some keepers work with their red pandas sufficiently to also enable a quick tactile assessment of body condition while the animal is stationed.

Cubs may be weighed more frequently, although this must be carefully managed so as to minimize disturbance of maternal behaviour. Waiting until the mother leaves the nest on her own and then quickly weighing infants, or training mothers to comfortably shift from the nest box for a few minutes are two strategies that work for some zoos. In N. America, even those institutions with hands-off policies for newborns advise weekly weighing of cubs beginning at 8 to 10 weeks of age, particularly if the cubs are part of a litter; in other countries this is not advised. Singletons usually do not have trouble obtaining sufficient milk from the mother, but if there are two or three cubs, they may require supplementation to maintain adequate weight gain. This may be done by providing the cubs with biscuits sweetened with apple juice or a mash of biscuit with banana and other fruit. Alternatively, supplementing cubs by hand feeding may be necessary to ensure that they are receiving enough food. This must be balanced, of course, against the risk of habituating the young animals to sweet, highly palatable food so that one then has the struggle of getting them onto a proper diet once they are at a stable weight.

As mentioned, monitoring weight and supplementation of food becomes particularly critical at weaning. At this age, cubs are still growing, and therefore have high energy requirements, and yet are in a transitional phase of food choices and the maternal relationship that may be very challenging to them. Weaning hand-raised cubs allows a little more control than managing cubs that are with their mother and litter mates who may steal the highly palatable foods before the reluctant cub can get to it. Hand-rearing is discussed below and in Chapter 14 [12].

Zoos often struggle to provide sufficient bamboo for their pandas. In the recent husbandry study [4,5], the provision of bamboo was found to vary from daily to sporadically. In a 1997 survey of N. American zoos, 12% fed no bamboo at all, and 18% of those that did, did so only seasonally [13]. A variety of grasses and browse may be offered to substitute in part for bamboo, and should be planted throughout the enclosures so that the animals can select them at will. In smaller enclosures in which the red pandas destroy the vegetation more quickly than it can re-grow, the use of potted plants facilitates replacement of browse and grazing materials. As mentioned above, providing opportunities for red pandas to forage in their enclosures also helps to satisfy this natural behaviour with which pandas in the wild occupy the majority of their waking hours.

The IHMG recommend at least two feedings per day with multiple feeding stations, or at least separate feeding stations for each red panda in an enclosure. Multiple feeding stations help to encourage a foraging type of behaviour. Moreover, if there are multiple animals in an enclosure, multiple feeding stations are necessary to prevent competition and possible fights over food and the possibility that one or the other individual does not receive enough. Feeding stations should also be placed to make it difficult for vermin to access the food. Feeding huts and the coconut feeders used at Taronga Zoo are clever constructions toward this end. The coconut feeders are halved coconut shells hung from branches or other fixed structures. These also allow placement of food at multiple levels and make food acquisition by the animals a bit more challenging than when they simply feed from a fixed station on the ground. Zidar and Eriksson [4,5] found that 19% of the zoos in the study fed only once per day, and nearly 30% of zoos offered only one feeding station. Providing biscuit only once a day may work if the red pandas are accustomed to eating it free-choice and the food is placed in an area where it does not spoil or attract rodents or insects in the course of the day. Bamboo should be offered at least twice a day, regardless, because it will wilt within a few hours and will then no longer be accepted by the animals. For this reason again, the availability of naturally growing vegetation in the enclosure is a great asset in providing a naturalistic feeding regimen.

Red pandas are almost entirely herbivorous, albeit with a somewhat higher direct protein requirement than the other bamboo-verous carnivore, the giant panda. Even with enough bamboo of the appropriate species available to them, captive red panda diets must be supplemented with other foods to meet caloric and protein requirements [12]. Whether protein requirements increase in late pregnancy and lactation is not clear. Some zoos offer high-protein supplements (e.g., day-old chicks) to females in these physiological states but have observed that the animals usually do not eat it.

NEST BOXES AND KEEPER INTERFERENCE WITH NESTS

Nest box design and placement have been given a great deal of thought and are certainly considered principal factors in reproductive success for red pandas. The IHMG prescribe a minimum of three nest boxes of various sizes per mother. The nest boxes may be placed at various elevations or at least in various locations in the enclosure, always in private, quiet areas. The entrance to the nest box should be concealed with vegetation, as mothers will enter nest boxes less often in the presence of people watching them [6]. The N. American nest box survey [13] suggested that the availability of keeper access doors to the nest boxes that are separate from the doors the animals use is an important factor in minimizing the disturbance of the mother. That said, the subterranean nest areas used by the Australian zoos with high infant survival rates have no separate keeper access doors and none of the nest boxes are elevated as they would be in the wild. Some of the maternal enclosures have only two nest areas rather than the IHMG-recommended three. The nests are, however, private and cool, and the husbandry protocols at those institutions dictate a hands-off management practice for mothers with newborns.

The two factors most consistently demonstrated to influence cub survival in previous studies are temperature in the nest box and disturbance of nest by keepers and visitors [6,13]. The amount of time that mothers spend in the nest is considered a critical indication of the quality of maternal care [6]. High temperatures and disturbance of the nest area increase the amount of time the mother spends away from her cubs. Some zoos provide air conditioning in the nest boxes, and this certainly does appear to help with cub survival [13]. High temperatures (above 27°C or perhaps even less) in the nest box cause mothers to spend more time away from their cubs, stress the mother, and may result in heat stress of the cubs themselves. The subterranean nest boxes used in Australia are not air conditioned, but the earthen or cement casing appears to keep them sufficiently cool.

Many of the zoos with successful reproduction programmes suggest that one of the key factors of their success is that they leave the nest alone and do not handle the cubs at all for the first 8–10 weeks. A healthy mother will be intensely protective of her cubs and, even if she is well conditioned to the keepers, she will become stressed at having someone reach into her nest even for the moment that it takes to weigh the cubs. Once the cubs are a bit older, she will become more relaxed about allowing this. A hands-off approach is facilitated by monitoring the nest box with the aid of a video camera. This allows the evaluation of maternal behaviour and the amount of time that cubs nurse, which can provide a good indication of the health and development of the cubs to experienced managers. A nest box design that allows keepers to visually check the cubs through a Plexiglass peek hole or through a side door with minimal disturbance to the mother serves facilities without the technological resources of a nest box camera.

One study also found a correlation between the number of visitors at the exhibit and the amount of time mothers spent carrying their cubs around from one nest area to another [6]. A restless mother who keeps moving her cubs will spend less time nursing them. Moreover, her hyperactivity indicates her level of stress which, in turn, will influence her milk production and the quality of care that she is able to provide her cubs. That said, there are zoos with temperature-regulated nest boxes, the mothers off exhibit and minimal keeper disturbance policies for mothers with young cubs, and yet still have cub mortality rates above 50%. Again, this illustrates the interrelationship of a variety of factors, some of which may not yet be identified, that are necessary to ensure cub survival.

SOCIAL GROUPING

The IHMG recommend that a breeding pair should be kept together year-around, so long as the male does not interfere with the cubs. This is common practice in European and Australian zoos. Some zoos in N. America and those in Japan close their pregnant females indoors just prior to the onset of the birthing season. The rationale for this is to provide the female a greater degree of privacy and to allow keepers to monitor her more closely. Also, because the facility has to keep some of their animals on display, managers may choose to keep the male on exhibit and move the female off for greater privacy. Some zoos are able to make isolation of the mother indoors work successfully. Others see high infant mortality rates that may be related to the stress of moving the mother to a new space, or to other factors. A pregnant female closed indoors should be supplied with several nest boxes, nesting material and a variety of enrichment items to satisfy and to encourage natural maternal behaviour.

New breeding pairs should be placed together no later than 6 weeks prior to the onset of the breeding season to allow them to become accustomed to one another (SSP guidelines). Keepers agree that the pair should be together full time prior to and during the breeding season to allow them to develop a comfortable relationship and to maximize breeding opportunities. Breeding pairs are generally given two or three seasons to reproduce. If no breeding takes place, alternative partners are found for each animal.

Two mature males should not be kept together in the presence of a female, as they are likely to fight. Mature males housed together are inclined to fight anyway, even without the motivating factor of a female over whom to compete. Young brothers have occasionally been housed together for a year or two following sexual maturity, but the success of this depends on the personalities of the individuals. Some facilities, primarily in N. America, house two females with one male. Trios appear to work quite well for combinations of the right personalities and sufficient space for the three of them, so long as one of the females is moved prior to the birth of cubs. The IHMG emphasize that trios are acceptable as temporary arrangements but should not be left together long term. It is generally recommended that, whenever possible, red pandas should be housed as monogamous pairs.

Cubs of the year may stay with the mother until just before the next litter is born. The IHMG and successful red panda breeders suggest that cubs should stay with the mother for at least 10 months to ensure proper socialization. In one European zoo, a mother raised her litter successfully while sharing the enclosure with the father and her cubs of the previous year (A. Glatston, personal communication). Some zoos in N. America remove the cubs just before the breeding season. This is the common practice in Chinese breeding facilities, where cubs are generally removed from their mothers as early as October (3–5 months of age) because their presence is believed to interfere with breeding. Many cubs fail at such early removal, and weaning mortality is high. Moreover, many cubs end up in isolation, which results in depressed or stereotypic behaviour [14]. In many of these Chinese facilities, it must be added, large groups of red pandas live together in a single enclosure: up to 15 or 20 animals together. Their behaviour in such large groups indicates chronic stress. The animals are in constant movement, pacing, marking and a great deal of fighting, particularly in the hot summer months and of course during the breeding season. In such an arrangement there is also no control over breeding pairs, and inbreeding can become a serious issue [14]. A study on the potential influences of polyandry and polygamy in red panda breeding success suggested that both arrangements were unconducive to reproductive success [15]. With consideration for the limitations of that study's dataset, the suggestion that reproductive endocrinology, and probably other endocrine patterns, are compromised in such housing arrangements is interesting and warrants exploration. Such extremely unnatural social situations may exaggerate endocrine (including stress) and behavioural pathologies that may give us clues toward understanding poor reproductive success even in less outrageous situations, e.g., most zoos under discussion in this chapter.

Once removed from their mothers, subadult red pandas should be housed together in groups in order to provide optimal opportunity for normal social development. Litters may be mixed if necessary, even among institutions, in order to avoid singletons. Once sexually mature, however, they should be paired, either with a mate of the opposite sex, or two females often do well together as well if they are not meant to be in the breeding pool. Many zoos have single adult red pandas, and males may be housed singly. So far as we know, this is normal for adult animals in the wild. However, one must keep in mind that a captive setting rarely substitutes for the rich variety of stimuli and activities available to an animal living *in situ*. Keepers must ensure an adequate enrichment schedule for red pandas housed singly.

BEHAVIOURAL MANAGEMENT AND ENRICHMENT

Environmental and behavioural enrichment and behavioural management of red pandas are not discussed in the IHMG except to state that enclosures should have sufficient vertical space and natural vegetation. The provision of sufficient opportunity for foraging and the related climbing and other physical activity appears to be a common element of facilities with successfully reproducing animals. When provided the opportunity, red pandas, particularly young individuals, may be quite active, inquisitive and exploratory. With creativity and effort, even small enclosures may be outfitted to encourage positive behaviours through the frequent rotation of enclosure furniture and a varied, daily enrichment programme. The Knoxville Zoo keeper training manual [13] provides a variety of excellent ideas for this. Intuitively, it would seem that red pandas that are active, occupied, and relaxed will be more likely to reproduce successfully.

Training provides an excellent opportunity for red pandas to become comfortable with key keepers, enables a variety of husbandry and medical tasks to be performed at minimal stress to panda and staff, and serves as a behavioural enrichment activity for the animals. A strong training programme allows a close relationship to develop between the animal and keeper, which then serves to provide relatively stress-free assistance to a mother, for example, whose cubs may need a little help. A trained female may then be asked to shift from her nest box for a moment to allow a keeper to quickly check the cubs and to supplement feed if necessary. Zoos that have such a training programme and whose keepers consequently develop close relationships with their red pandas consider this a major factor in the success of their red panda breeding programmes.

The Knoxville Zoo keeper training manual [13] outlines the elements of the behavioural management programme utilized to good effect at that facility. Keepers ensure that female red pandas are very comfortable with shifting, habituation to the keeper, weighing, targeting or stationing, crate training, and medicating prior to the arrival of the cubs. This facilitates the keepers' being able to closely monitor the infants. This zoo finds it necessary to hand-rear or to supplement feed cubs with some frequency. Working intensively with the mothers prior to the birth of the cubs is therefore critical.

Other zoos maintain a much more hands-off approach to their red panda nests and do not advise training their animals to shift, load and to station for medical examinations. This tends to be the policy of Australian zoos that have the lowest infant mortality rate of the three regions, and of European zoos. This factor alone, however, again does not seem to explain infant survival. Most likely, other husbandry or environmental factors in these zoos make it unnecessary for keepers to assist mothers, in which case they do well to leave the nests alone. It is also possible that in zoos in which periparturient nests are left alone, cubs may be cannibalized imminently after birth and keepers may not realize that there ever was a birth. This is not highly likely, as the staff at most of these zoos know their animals well and are able to deduce from a female's behaviour and appearance if she has delivered cubs. Figure 13.6 also indicates that, while Europe's first-week mortality rate is much lower than that of N. America, the two rates grow in parallel throughout the year. This suggests that nest boxes may require more frequent monitoring in European zoos after all, in order to catch difficulties that infants and mothers may be experiencing. Again, it will be important to determine the causes of death – or better, the causes of morbidity in time to avoid death – in those cubs that survive the first week and who stand in danger of perishing in the succeeding months.

In any case, no one contends allowing mothers to raise their cubs on their own as much as possible. But if conditions do not favour the success of a hands-off policy, then managers are wise to prepare adaptive measures to assist mothers with as little stress to her as possible until we can figure out the compromising predispositions. As mentioned above, the utilization of video cameras in the nest allows keepers to monitor the mother and cubs without having to bother her. But, if the family should require assistance, interference with the nest may prove to be quite stressful to the mother if she is not already habituated.

There is some debate about natural diurnal activity patterns of wild red pandas. Traditionally, it has been believed that red pandas are most active in the early morning and evening, but this may be an artifact of captive conditions and human disturbance during the day. However, if the animals are found to maintain a crepuscular activity pattern despite efforts to minimize disturbance by visitors, then adaptation to this pattern by husbandry schedules may be reasonable. For example, feeding early in the morning and as late as possible in the afternoon and providing plenty of natural vegetation in the enclosure for the animals to forage as they please will ensure fresh food for the animals when they choose to eat. Disturbance of nest boxes should in any case be restricted to those times when the mother is most likely to venture out on her own.

HAND-REARING RED PANDA CUBS

Red panda breeding facilities do well to have carefully thought-out protocols to help keepers determine if and when to pull cubs for hand-rearing [13]. Sometimes mothers will simply need a bit of assistance, or cubs may need food supplementation and the family unit may remain together. Cubs raised by skilled keepers tend to grow more quickly than mother-raised cubs because of a more intensive nutritional programme, but the care of even the most skilled and dedicated keepers cannot substitute for the psycho-social advantages of a cub raised together with its sibling(s) by its mother.

In any event, all efforts should be made to keep the newborn cub with its mother for at least 24 hours to ensure that it receives sufficient colostrum. Colostrum-deprived infants invariably die of bacterial disease that takes root in the immunologically unprotected neonate. Thereafter, pneumonia due to aspiration of milk is the most common cause of death in hand-raised neonates. The IHMG outline the precautionary steps to avoid this problem, namely to tube-feed until the neonate is stable enough to drink from the bottle, to use sterile dextrose before feeding milk to ensure the safety of the handler's technique, and to use nipples with very small holes in order to regulate the rate at which milk is drawn into the cub's mouth. Small nipple holes and consequent prolonged nursing times also help to satisfy the cub's requirement for suckling. A frequent problem with hand-raised infants is that they will suck on their own or other cubs' fur, ears and paws out of an unsatisfied nursing reflex. Skin trauma and infections such as dermatophytosis may result from the constant moisture and physical trauma.

The milk formula used for hand-rearing red panda cubs is discussed in Chapter 14. Guidelines for frequency, formula concentration and volume of feeding neonates are provided in the IHMG. The Knoxville Zoo Keeper Training Manual [13] provides tabulated guidelines with more detailed information. Normal weekly body weight ranges are provided in the IHMG for *fulgens* subspecies. These differ slightly from those provided by the SSP (*fulgens* and *styani*) and by others [16]. Perhaps a consensus may be reached from recalculation of what is now a larger global database. Body weight and general condition of hand-raised cubs must be monitored daily. Feeding amounts should be recalculated at least every 3 days to adjust to the development of the infant. Care must be taken not to over-feed cubs, as they may quickly develop gastrointestinal and possibly metabolic problems. It is better to err slightly on the side of underfeeding than to overfeed. Diarrhoea, constipation, bloating, respiratory abnormalities, skin irritations and other medical conditions must be noticed and treated promptly before the infant becomes too compromised. It is essential that the principal caretaker be a person with experience at least in hand-raising neonates of other carnivore species. The IHMG are adequate for experienced staff to work with; less experienced keepers are advised to seek assistance, and ideally should spend time training with experienced keepers until they are ready to try the techniques on their own.

As with all infants, care must be taken to provide the appropriate microclimate. Supplemental heat, cooling, and humidity control may be necessary.

The weaning method outlined in the IHMG suggests the introduction of gruel into the milk bottle at about 3 months of age, and then teaching the cub to drink gruel of increasing concentration from a bowl. Alternatively, cubs may be taught to drink from a bowl first, and then the gruel is introduced into the milk. Some experienced practitioners of red panda hand-rearing prefer a more "natural" method of which the principal drawback is that it may take longer to wean the cub. They find this method less stressful to the cubs and worry less about the cubs aspirating gruel from the bottle, since the hole in the nipple will have to be enlarged to accommodate the flow of the thicker liquid. The method preferred by these practitioners is to provide dry biscuit, broken in pieces and moistened with water or apple juice, beginning at about 4 months of age. At 5 months, the keeper will begin to slowly decrease the amount of formula, first by reducing the number of feedings per day, and then by decreasing the volume of formula. In this way, the cub makes the transition more gradually and more on its own choice. Regardless of the weaning method, cubs must be weighed several times a week during the weaning process to ensure that they are receiving enough food. As stated earlier, pandas in this stage of development may easily starve if not monitored closely and the weaning process must be intensively managed to meet the requirements of the individual animal. The IHMG and experienced keepers are in agreement that bamboo and fresh produce should be provided as early as 2 or 3 months of age. The cubs will not eat it yet, but they learn to manipulate it and become familiar with it so that when they are ready to try solid food, they are already comfortable and familiar with these items.

MEDICINE

The principal causes of perinatal mortality are associated with maternal failure (agalactia, neglect, cannibalization and injury to cubs) and respiratory disease [2,17]. These are the same causes that dominated the analysis by Montali et al. [18] a decade earlier. Pneumonia and septicaemia are most often associated with failure of passive transfer. In hand-reared cubs, pneumonia also results from aspiration of food materials. These are husbandry-related issues and must be corrected through reduction of maternal stress and improvement of maternal competence.

Analysis of the past 30 years of neonatal mortality records revealed that 19% of neonatal deaths were stillborn or cubs who died on the day of birth [17]. This finding is worth further exploration to elucidate the possible causes of such a high rate of stillbirths. Infectious disease, congenital malformation, inbreeding, toxicology, and maternal hormonal abnormalities are some of the possibilities. Congenital abnormalities accounted for 11% (3/27) of the neonatal deaths in the study by Montali et al. [18]. This is an interesting observation and would be informative to follow up with a larger sample size and more recent data. Genetic analysis of the global population (excluding China) suggests that inbreeding should not play a significant role in the failure of cubs to thrive. Data on the captive population in China are not available, but a recent study in the best-managed breeding facility in that country indicates that inbreeding is a serious problem [14].

In warm, humid climates, cubs may often suffer from dermatophytosis [18]. Early diagnosis and treatment with itraconazole is effective in curing the condition. Left untreated, the infection may result in necrosis of tails and ears and in the death of the cubs. Concern over this and other medical conditions will require that cubs are checked at least once a week in the nest, and so a strictly hands-off approach to nest management would not be suitable. Interestingly, Australian zoos report little problem with dermatophytosis despite the climate there (personal interviews with zoo veterinarians). Again, however, one finds a circular argument with this condition, as the development of dermatophytosis in cubs may be associated with the mother's over-grooming and carrying her cubs about more than normal. The excessive saliva on the skin of the cubs then sets up an environment for the fungal disease to take hold. So, once again, one is at the question of the husbandry and environmental conditions that potentiate maternal stress and consequent medical problems which then require greater human intervention in the nest.

The reasons for which red pandas living *ex situ* fail to reproduce are not well understood. Again, these are generally believed to be related to stress, or at least that the animals are sufficiently uncomfortable in their captive environment that they either do not breed, become pregnant, or maintain pregnancy to term. As mentioned earlier, a comprehensive study that examines the behaviour and endocrinology of red pandas in a variety of captive environments and management scenarios would be highly beneficial in helping us to understand the factors limiting fertility in the species. An empirical factor that has been observed by many red panda managers and keepers is that females who are overweight will not breed. This may be more of an issue with the *styani* subspecies, who (again, empirically), may have more of a tendency to become overweight than the *fulgens*. An appropriate diet that is not excessively high in starch, and an environment that encourages physical and mental activity are essential in weight management.

Measures necessary to protect red pandas against infectious disease depend to some degree on the area and the diseases endemic therein. As discussed in Chapter 15, canine distemper virus (CDV) is the disease of perhaps greatest concern in red pandas, and zoos in N. America vaccinate against it. The difficulty has been in finding a killed vaccine that is safe to use in highly sensitive species such as red pandas. Merial's PUREVAX® has been available for nearly 10 years now and appears to be serving the need well. The vaccine is not always readily available, however, particularly outside of N. America [19]. Interestingly, the Australian groups do not vaccinate against CDV: rather, they use a three-way vaccine marketed for domestic cats against feline panleucopaenia (parvovirus), feline rhinotracheitis virus (feline herpesvirus type 1) and feline calicivirus. Because of the high endemic incidence of CDV in N. America, the risk of vaccinating red pandas against this virus is generally considered to outweigh the risk of not vaccinating. However, the issue of over-vaccinating animals has recently become a topic of debate (reviewed in [20]) and may be relevant for exotic species as well as domestic animals. The concern is all the more relevant when using vaccines in species other than those for whom the vaccine was developed, which is a situation with which we are frequently confronted in wildlife medicine. Perhaps the question of vaccinating red pandas that are not at significant risk for contracting particular pathogens may need to be revisited in this context.

The relay period from maternal immunity to active immunity in red panda cubs is assumed to be similar to that in puppies and kittens, and so vaccination should begin at 10–12 weeks of age. The IHMG state that this may be delayed until 16 weeks of age if earlier vaccination poses too great a risk of disturbing the maternal nest.

Other infectious diseases that may interfere with reproduction may occasionally become an issue, but little of this is documented or investigated. In a study of captive red pandas in China, for example, animals in several facilities were found to carry alarmingly high antibody titres against *Toxoplasma* [21]. Toxoplasma usually spreads to vegetarian, non-definitive hosts via faeces of cats. Alternatively, red pandas that frequently catch rodents in their enclosures may ingest tachyzoites directly. Exposure of naive females to the organism during pregnancy may result in abortion, stillbirth and congenital malformation of fetuses. The disease does not appear to be an issue in western zoos, but whether this is for lack of testing (particularly for stillborn cubs) or whether it really is not an issue is not clear. Moreover, toxoplasmosis is a disease that becomes a clinical issue in animals who are immunologically compromised. Again, the potential compounding effects of stress and latent disease warrant investigation in this stress-prone species.

Parasite prevention protocols will depend on the climate and environment, but care should be taken that mothers are de-wormed prior to parturition to avoid transfer of larvae to offspring through the milk. Infestations with fleas and ticks may seriously compromise the health of red pandas, particularly of cubs in a nest. Cubs may die of flea infestation of sufficient severity. The IHMG recommend changing nesting material frequently and treating the nest material with an anti-flea product just before parturition. This recommendation may need to be revisited. Frequent changing of nest material risks disturbing the mother, particularly in the first few weeks after parturition. Treatment of bedding may also pose some concern in the potential for cubs to ingest the chemical. Regular topical treatment of the mother so that she is clear of fleas before she enters the nest may be the safest option. Again, this is greatly facilitated by training so that application of the drug is quick and stress-free.

CONCLUSION

In summary, the following principal points may be considered with regard to *ex situ* reproduction for red pandas:

1. Reproduction in the captive red panda population is lower than its apparent physiological potential. This appears to be due to low reproductive rates and high infant mortality. The factors responsible for these are poorly understood, and further research is warranted.
2. Age-specific fertility data of the past 20 years' international studbook records suggest that only a quarter to one-third (*styani* and *fulgens* subspecies, respectively) of the captive red panda population is reproducing at peak reproductive age.
3. Infant mortality varies among regions, with that of Australia lowest and North America's highest at nearly 50%. Female infant mortality is similar between the two subspecies, but for males it is higher in *fulgens*. Cumulative mortality appears to level off after 3 months of age for *styani*, while for *fulgens* it continues to rise until 9 months: this rise is principally due to continued mortality in the N. American and European populations.
4. While *styani* subspecies appears to live a few years longer and reach peak reproductive age a bit earlier, on average, than *fulgens*, reproductive senescence appears to allow both subspecies a similar 4 years' duration of peak reproduction.
5. Enclosure size and quality, with full-time access to the outdoors, sufficient vertical space, a diversity of enrichment, sufficient privacy, opportunity to escape heat, and behaviour-appropriate environmental conditions appear to be critical factors for reproductive success in captive red pandas. Identification of specific husbandry factors that correlate with infant mortality requires further research.
6. Nest box placement, options, environment, privacy and design are critical to successful reproduction.
7. Nutrition of growing red pandas may need to be monitored carefully to assist mothers with agalactia or multiple cubs for which there may be insufficient milk. Weaning is a critical age for young pandas, and care must be taken that animals do not die of inanition during this transitional period.
8. The degree of hands-on monitoring of cubs and mothers varies among regions and institutions. Some advise training mothers prior to parturition so that checking cubs or supplemental feeding is not stressful to the mothers; others advise a completely hands-off approach until the cubs are older and begin to naturally spend time away from the mother.
9. Feeding red pandas is ideally done in a manner that encourages natural foraging behaviour and minimizes competition between adult animals in the enclosure.

10. Recommendations for social grouping of red panda pairs vary somewhat among regions, with some keeping pairs together year-round, while others remove females prior to parturition into private areas. Cubs of the year may stay with the mother until the beginning of the next breeding season. Care must be taken that cubs are not isolated too young, and even when removed from the mother that they are housed with age-compatible companions.
11. While all efforts should be made to keep cubs with their mothers, some cubs may require supplemental feeding or, in the case of complete maternal failure, may need to be hand-raised. Supplemental milk formulas, feeding regimens and weaning methods are outlined in the Knoxville Zoo Keeper Training Manual [13]. Red panda facilities in other regions do not recommend hand-rearing, but perhaps the option and guidelines to do so should be supplied in the IHMG.
12. Principal causes of perinatal mortality are associated with maternal failure and respiratory disease. The latter appears influenced primarily by husbandry conditions. Dermatophytosis may be a significant problem for cubs in warm, humid climates, particularly when compounded with maternal stress.
13. Vaccination of red panda cubs against canine distemper virus with a recombinant vaccine is recommended in areas in which CDV is endemic.
14. Diseases that may affect reproduction (e.g., toxoplasmosis) require investigation.

Many years of experience caring for red pandas in captivity, and a gradual understanding of the natural history of the species have brought significant advances in our ability to properly care for these animals *ex situ*. We are at a difficult juncture now, at which we must figure out the more subtle aspects of red panda husbandry that will enable these animals to reproduce more successfully. The characteristics of their physical environment, the behavioural and environmental enrichment that we provide, the delicate balance between leaving mothers alone and still observing them closely enough to provide assistance when it is needed, perhaps also disease prevention and nutrition, all converge to influence the complex and sensitive requirements of successful reproduction in this unique species. To learn how to manage these factors, a close communication network is required among red panda breeding institutions, together with a series of carefully planned and thoughtfully analysed research investigations.

The creation of *ex situ* environments in which animals feel sufficiently comfortable to thrive is the greatest challenge of captive wildlife managers. Decades of science, experience, improvements in our understanding of animal sentience, and transitions in the philosophy of animal husbandry have brought us a long way toward doing right by the animals whom we make dependent on our care. Species like the red panda remind us that we still have a great deal to learn. The differences in reproductive success for red pandas among institutions around the world hold a wealth of information about the physiological and behavioural characteristics of the species. Research on the interrelationships among stress, behaviour and reproduction will help us to understand how better to care for red pandas, as well as for other endangered animals who now rely so heavily on our own species for survival.

ACKNOWLEDGEMENTS

Special thanks to the many red panda managers and keepers in North America, Europe and Australia for their detailed explanations of red panda husbandry practices at their facilities, and for their insights into the issues of red panda reproduction *ex situ*. Many thanks also to A.R. Glatston for analysis of the international studbook records.

References

[1] A.R. Glatston, K. Leus, Captive Breeding Masterplan for the Red or Lesser Panda *Ailurus fulgens fulgens* and *Ailurus fulgens styani*, Royal Rotterdam Zoological and Botanical Gardens, Rotterdam, The Netherlands, 2005.

[2] Y. Li, Infant mortality in the red panda, in: A.R. Glatston (Ed.), Red or Lesser Panda Studbook, No. 8, Stichting Koninklijke Rotterdamse Diergaarde, Rotterdam, The Netherlands, 1994, pp. 18–35.

[3] A.R. Glatston, Husbandry and management guidelines, in: A.R. Glatston (Ed.), The Red or Lesser Panda Studbook, No. 7, Stichting Koninklijke Rotterdamse Diergaarde, Rotterdam, The Netherlands, 1993, pp. 37–66.

[4] J. Zidar, Department of Animal Environment and Health, Ethology and Animal Welfare Programme 32, Swedish University of Agricultural Sciences, Skara, 2008.

[5] P. Eriksson, J. Zidar, D. White, J. Westander, M. Andersson, Current husbandry of red pandas in zoos, Zoo Biol. 29 (2010) 1–9.

[6] A.R. Glatston, Planning enclosures for red pandas, in: P.M.C. Stevens (Ed.), Fourth International Symposium on Zoo Design and Construction, Whitely Wildlife Conservation Trust, Paignton, UK, 1989, pp. 204–211.

[7] B.K. McNab, Energy expenditure in the red panda, in: A.R. Glatston (Ed.), Red Panda Biology, SPB Academic Publishing bv, The Hague, The Netherlands, 1989, pp. 73–78.

[8] G.J. Mason, J. Rushen, Stereotypic Animal Behaviour: Fundamentals and Applications to Welfare, CABI, Oxfordshire, UK, 2006.

[9] C.L. Coe, G.R. Lubach, J.W. Karaszewski, Prenatal stress and immune recognition of self and nonself in the primate neonate, Biol. Neonate 76 (1999) 301–310.

[10] K.A. Terio, L. Marker, L. Munson, Evidence for chronic stress in captive but not free-ranging cheetahs (*Acinonyx jubatus*) based on adrenal morphology and function, J. Wildlife Dis. 40 (2004) 259–266.

[11] R.M. Sapolsky, The influence of social hierarchy on primate health, Science 308 (2005) 648–652.

[12] S.F. Sorrells, R.M. Sapolsky, An inflammatory review of glucocorticoid actions in the CNS, Brain Behav. Immun. 21 (2007) 259–272.

[13] Red Panda Keeper Training Manual, Red Panda SSP Keeper Training Workshop, Knoxville Zoological Gardens, Knoxville, USA, 2004.

[14] Y.Z. Li, S. Fujun, I.K. Loeffler, et al., Genetic diversity and parentage assessment of the captive red panda (*Ailurus fulgens*) with microsatellite DNA markers (in prep).

[15] F. Wei, L. Xiaoping, L. Chun, L. Ming, R. Baoping, H. Jinchu, Influences of mating groups on the reproductive success of the Southern Sichuan red panda (*Ailurus fulgens styani*), Zoo Biol. 24 (2005) 169–176.

[16] A.R. Glatston, An outline of husbandry and management techniques for the red panda, in: J. Partridge (Ed.), Management Guidelines for Bears and Raccoons, Association of British Wild Animal Keepers, Bristol, UK, 1992, pp. 133–142.

[17] K.S. Kearns, C.G. Pollock, E.C. Ramsay, Dermatophytosis in red pandas (*Ailurus fulgens fulgens*): a review of 14 cases, J. Zoo Wildlife Med. 30 (1999) 561–563.

[18] R. Montali, M. Roberts, R.A. Freeman, M. Bush, in: O.A. Ryder, M.L. Byrd (Eds.), One Medicine. A Tribute to Kurt Benirschke, from His Students and Colleagues, Springer-Verlag, Berlin, Germany, 1984, pp. 128–140.

[19] J. Welter, J. Taylor, J. Tartaglia, E. Paoletti, C.B. Stephensen, Vaccination against canine distemper virus infection in infant ferrets with and without maternal antibody protection, using recombinant attenuated poxvirus vaccines, J. Virol. 74 (2000) 6358–6367.
[20] M.J. Day, M.C. Horzinek, R.D. Schultz, Guidelines for the vaccination of dogs and cats, J. Small Anim. Pract. 48 (2007) 528–541.
[21] Q. Qin, F. Wei, E.J. Dubovi, I.K. Loeffler, Serosurvey of infectious disease agents of carnivores in captive red pandas (*Ailurus fulgens*) in China, J. Zoo Wildlife Med. 38 (2007) 42–50.

CHAPTER

14

Red Panda Nutrition: How to Feed a Vegetarian Carnivore

Joeke Nijboer[1] *and Ellen S. Dierenfeld*[2]

[1]Rotterdam Zoo, Rotterdam, The Netherlands

[2]Novus International, Inc., St. Charles, MO, USA

OUTLINE

INTRODUCTION

Red pandas live in the temperate forest zone of the Himalayan ecosystem between 2200 and 5000 metres (see Chapter 11). Their distribution is associated closely with temperate forests having bamboo thickets, which are the main diet of red pandas (Figure 14.1). However, red pandas have a carnivore-type digestive anatomy, specialized for digesting protein and fats (as is found in meat) and no modification for digesting fibres and carbohydrates – the main nutrients of their natural diet. In captivity, it is often not possible to supply enough palatable bamboo on which red pandas can survive. Historically, palatable captive diets were developed based less on natural feeding ecology than on ingredients used in feeding more omnivorous species (i.e., dogs, primates). Such diets have subsequently been proven inappropriate for optimal health of the red pandas.

Red Panda. DOI: 10.1016/B978-1-4377-7813-7.00014-8

FIGURE 14.1 Red panda feeding on bamboo *(Photo Steven Wallace)*.

This chapter provides an overview of the history of captive red panda diet development by discussing the anatomy and basal metabolic rate of red pandas, feeding ecology, nutritional recommendations for red pandas, and the effect of the diet on nutrition-related problems and faecal quality. Additionally, a section is devoted to hand-rearing red pandas.

ANATOMY AND BASAL METABOLIC RATE IN THE RED PANDA

Red pandas are classified in the sub-family Ailurinae within the order Carnivora. The digestive tract is typical of a carnivore [2], although dentition is atypical. The molars are set low in the jaw and the chewing surfaces are rather flat, more like those in herbivores rather than the pointed cusps on the sides of the molars found in most carnivores. Skull size is large compared with that of carnivores of similar body size, such as other members within the Procyonidae (racoon) family. This greater depth of skull may improve bite pressure at the level of the cheek teeth [2,3]. In 1870, Flower [4] described the gastrointestinal tract of a red panda in detail for the first time; he reported a simple stomach (Figure 14.2) and gut length comparable to that of a domestic cat.

The gut length to body size ratio is comparable to that of small cats [1], and is very short compared to most herbivores. There is no indication that red pandas have any

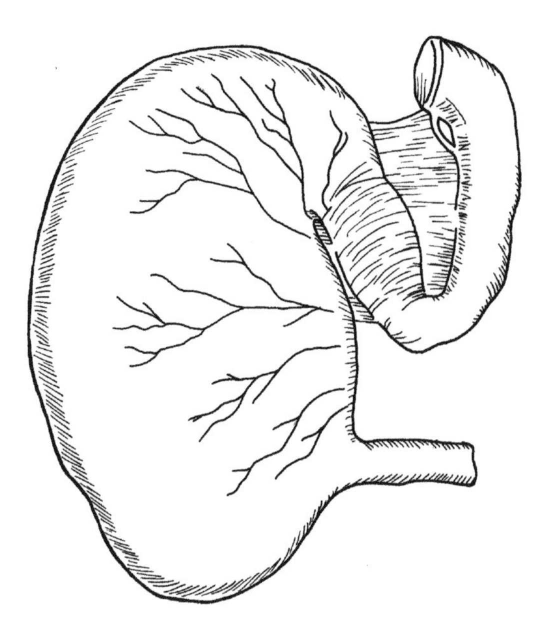

FIGURE 14.2 Stomach of red panda as described by Flower [4].

specialized adaptation(s) to retain microflora to assist in the breakdown of dietary cellulose, suggesting that only simple carbohydrate structures and other nutrients that can be enzymatically degraded can be digested by red pandas [5]. More information on the structure of the gastrointestinal tract can be found in Chapter 6.

The red panda can effectively regulate its body temperature [6], with normal body temperature maintained at $37.6 \pm 0.3°C$ at ambient temperatures between 25 and 36°C. The panda's basal metabolic rate (BMR) is $0.153 \pm 0.006\ cm^3O_2 \cdot g \cdot h$, which is only 39% of the value predicted by the Kleiber [7] equation for a 5.74 kg placental mammal. Thus red pandas have demonstrated low energy needs, which may be successfully met even by poorly digestible diets comprising mainly bamboo. Its rate of metabolism increases when the temperature falls below 25°C, until 19°C, without any decrease in body temperature (intrinsic regulation), whereas below 19°C, the metabolism rate falls without any decrease in body temperature [8]. Thus pandas respond to cold temperatures by decreasing, rather than increasing, energy expenditure. According to McNab [8], the low metabolic rate in pandas probably reflects physiological adaptations for a high-fibre, low-energy diet, low available protein, and a high level of secondary plant compounds.

NATURAL DIETS

The bamboo feeding behaviour of the red panda has been previously described [9] (see Chapter 11). The bamboos which form the main food for red pandas belong to the genera *Phyllostachys, Sinarundinaria, Thamnocalamus, Chimonobambusa* and *Qiongzhuea*. Red pandas also eat small mammals, birds, eggs, blossoms, and berries [9,10]. In the Wolong reserve in China, the pandas eat exclusively *Sinarundinaria fangiana*. The nutrient content of bamboo fluctuates little throughout the year and, overall, bamboo represents a relatively stable nutrient supply. Red pandas eat only the leaves of the mature bamboo plant, and also shoots when available (in spring and autumn), rather than culm fractions [11].

Bamboo leaves contain significantly lower levels of total cell wall constituents, neutral detergent fibre, and dry matter (DM) than culm fractions, and new shoots (when analysed) have been shown to contain similar fibre levels to leaves [12]. Shoots may, however, be more digestible due to a lower degree of cell wall lignification. Table 14.1 presents a summary of the analyses of the protein and carbohydrate fractions of bamboo leaves, including the data of Wei (see Chapter 11).

Vitamins provided by bamboo have not been investigated in much detail as applied to panda nutrition. Beta carotene, identified in bamboo leaves, may provide a ready source of vitamin A activity for these herbivorous carnivores, although enzymatic conversion ability to active vitamin A has not been examined in either panda species. Tocopherols with vitamin E activity have been investigated in bamboo leaves eaten by giant pandas from both Canada and Sichuan Province, China (Table 14.2). Differences were found in seedlings harvested in different months, thus nutrient supplies likely fluctuate seasonally, but vitamin E status or nutritional physiology has not been detailed in either captive or free-ranging red pandas [17].

The mineral content of bamboo leaves has also been evaluated, primarily in reference to giant panda nutrition, rather than specifically targeted for the red panda. Samples were analysed from temperate bamboos in Wolong Reserve (Sichuan province), China, Alberta province, Canada, and Shaanxi province in China. Seasonal variation has been noted in the mineral content of bamboo, with a marked decline in cool months, presumably as grasses were resorbing nutrients for winter dormancy. The data are summarized in Table 14.3. In these studies, leaves contained approximately twice the concentrations of all minerals as culm fractions. Silica is particularly high in bamboo (up to 2.5% of DM in leaves), and varies throughout the year, but its significance to nutrition, selectivity, or digestibility is unknown [12,13,16,17].

Results of field studies of bamboo digestion by red pandas in the Yele Natural Reserve, Sichuan, have shown that digestibilities of dietary DM differ during the seasons, and between components (leaves vs shoots) [18]. In this study, daily metabolic energy requirements varied from 2603 kJ in spring to 3138 kJ in summer/autumn, and 2704 kJ in the winter, whereas corresponding energy intakes were four to five times greater (10 485 kJ in spring, to 12 045 kJ in summer-autumn, and 12 276 kJ in winter) [18]. Digesta passage time was 2 to 4 hours, and resulting DM digestibility of 26.5% for leaves and 44.1% for shoots were similar to values obtained in giant pandas [14,16]. Thus, digestion of bamboo fibre fractions is low, and microbial fermentation appears to play a minor role in digestive physiology of the red panda [5]. Furthermore, the rapid gut transit time and low digestion coefficients suggest that pandas have to consume more than 1.5 kg of fresh bamboo leaves, or 4 kg of fresh shoots daily to meet energy needs, depending on the season. Like their larger cousins, red pandas have been shown able to meet nutritional needs on a diet comprising bamboo, but require vast quantities in order to do so.

TABLE 14.1 Summary of protein and carbohydrate fraction of bamboo

		As % of DM							
Bamboo part	**N (Species)**	**NDF**	**HC**	**ADF**	**Lignin**	**CP**	**Ash**	**Remarks**	**Reference**
Leaves		66				13–22			[12]
Culm		83				Feb-18			[12]
Leaves, China	15	66.5–71.7	36.2–41.2	29.9–40.1	11.0–20.2	12.4–19.4	7.1–10.5		[13]
Leaves N. America	2	59–64	25–35	25–28	4 (July)	13–27	1 (Aug)	Jan–May	[14]
Leaves, China	2					14–16			
Leaves, China	2	72–73	33–36	38–39	9 (Oct)	13–19	8 (Sept)	13 months	[16]
Culms China	2	78–85	22.3–33.0	48.3–64.2	13–16.4	2.5–6	1.2–6.5		[13]
Leaves China		69.8–72.8		35.8–36.8	8.4–8.8	13.2–16.6	6.6–8.8		(see Chapter 11)
New shoots China		69.9–76.8		39.6–43.5	5.5		5.9–6.4		(see Chapter 11)
Branches China		75.2–81.2		46.7–49.6	12.1–13.8		6.0–7.1		(see Chapter 11)
Stems China		79.2–88.9		56.9–63.9	13.6–16.7		1.8–3.1		(see Chapter 11)

NDF=neutral detergent fibre; HC=hemicellulose; ADF=acid detergent fibre; CP=crude protein.

TABLE 14.2 Vitamin composition of bamboo

Bamboo part	N	B-carotene	Vitamin E	Vitamin C	Reference
		mg/kg/DM			
Bamboo		33–66	19.7		[17]
Leaves	4		120.1 ± 21.5		[14]
Shoots	9		38.9 ± 8.8		[14]
Bamboo China				66	[16]

TABLE 14.3 Mineral composition of bamboo leaves

	% of DM
Ca	0.4–0.7
K	0.8–1.7
P	0.1–0.6
Mg	0.1–0.2
Na	0.002–0.1
	mg/kg DM
Cu	15–32
Zn	19–50
Fe	100–312
Mn	50–100

[12,13,16,17]

ZOO DIETS

Historical Diet Information

Potential nutritional problems with red pandas were first described in 1870 by Flower [4], who performed an anatomical investigation of the first red panda that had arrived in Europe at the London Zoological Society on 22 May, 1869. His pathology report suggests that the red panda was inappropriately fed:

> The subcutaneous tissue and the mesentery and sub peritoneal tissue being loaded with fat. The bones generally were soft and spongy in texture, a condition not unusual in animals which die under the abnormal or unhealthy circumstances to which they are subjected in captivity. Numerous haemorrhagic blotches on the mucous membrane of the upper portion of the intestine were the only pathological changes observed in any of the viscera of the animal. Some blood had been extravasated into the intestinal canal.

Flower described obesity and gastrointestinal ulcerations, as well as bone lesions suggestive of rickets. It is likely that the diet contained too much energy in the form of easily digestible carbohydrates and fats. Additionally, inadequate or imbalanced ratios of calcium, phosphorus, and/or vitamin D may have resulted in the described bone pathology. Gastrointestinal lesions may have, additionally, been related to dietary insufficiencies.

A.D. Bartlett [19], Superintendent of the Society's Gardens, described the diet of the panda on which Flower performed anatomical investigations. This article is rarely cited, however, it contains important information on the first artificial diet for red pandas. According to his instructions, the panda should be fed "about a quart of milk per day, with a little boiled rice and grass". But Bartlett had no faith in his basal diet so he started experimenting: "I first tried raw and boiled chicken, rabbit, and other animal substances, all of which he refused to eat. Red pandas do not relish on meat, incidentally they will eat some small birds or some meat". For many years thereafter, and still continuing in many locations, the basic diet of captive pandas consists of milk, boiled rice, sugar, eggs and different kinds of flowers – likely resulting from this initial report. Notwithstanding, Barlett also recorded the following: "He soon began to eat a few leaves and the tender shoots of the roses, and finding some unripe apples that had fallen from the trees, greedily devoured them". However, Barlett did not realize the natural feeding ecology behaviour that this red panda exhibited in wanting to ingest and even possibly requiring dietary plant fibre. Barlett observed very well that the diet of a red panda should contain leaves, however, he did not understand why because at that time there were no suitable dietary studies or analyses for this species.

Until the 1970s and mid eighties, captive red pandas were, in general, fed diets which contained high levels of digestible carbohydrates, supplemented with varying levels of vitamin and mineral mixes, resulting in a high incidence of diet-related mortalities [20].

In 1977, Bush and Roberts [20] published a health survey of red pandas and suggested that pandas require a high-fibre, low-cholesterol diet including bamboo or native/cultivated grasses in order to prevent gastrointestinal disorders including tooth problems and diarrhoea. Similar correlations between dietary constituents and health issues were also described in a European survey on red panda husbandry [1]. Typical red panda diets over those time frames (1970s to 1980s) comprised various mixtures of water, cereals (cooked grains, cornflour, cornflakes, prepared human baby foods), milk products, fruits and vegetables, fibre sources and/or complete feeds (dog or cat pellets), along with a variety of vitamin and mineral premixes, eggs, and miscellaneous other items. In European zoo collections, more than 80 different food components were fed to red pandas [21]. Some zoos even fed red pandas a meat-based diet. Dietary/nutritional reviews published in the red panda studbooks recorded that diets based on porridges/gruels resulted in health problems (dental issues, loose stools), and suggestions were made to exclude or alter the liquid diets, and improve dietary fibre amounts. More nutritional research has been done since that time [6,22–25] and supports the conclusion that red pandas benefit from a high-fibre diet comprising either bamboo, and/or a palatable high-fiber concentrate pellet. Standardized husbandry guidelines were developed for the species through AZA and EAZA that included recommendations for diet changes based on these studies, as well as target dietary nutrient levels [26,27].

Nutrition Guidelines for Red Pandas

Nutritional parameters suggested for diets fed to red pandas can be found in Table 14.4. Commercially prepared high-fibre biscuits containing apple pomace were developed in the USA, and found to be highly palatable for red pandas. Other biscuits containing minimum nutrient levels of (as fed basis): 23% crude protein, 4.5% fat, 13% crude fibre, 1.0% Ca, and 0.6% P have also proven successful in other regions. Advantages of nutritionally complete biscuit-type diets are obvious where year-round supplies of quality bamboo may not be readily available, but also represent a stable food/nutrient supply that can meet flexible management needs, depending on environmental, physiological, or behavioural variables of individual animals.

Minimally, 200–400 g of fresh bamboo should be offered daily to each panda; preferred sorts are *Pseudosasa* and *Phyllostachys* spp. If no bamboo is available, an alternative fibre source (i.e., high-fibre biscuits and/or beet pulp (soaked with biscuits if necessary), other edible grasses/plants) should be fed.

Ideally, bamboo and other fibre products should be available all day for nutritional and for enrichment purposes.

Pandas which are off their food can be offered a slightly sweetened or softened (soaked in water or juice) version of this diet by adding honey or sugar, but sweetening agents should be eliminated when the panda starts eating again to avoid dental health problems [28].

TABLE 14.4 Target nutrient levels for red pandas [27] in the dry matter in % compared to bamboo

Nutrient	Target level	Bamboo	Nutrient	Target level	Bamboo
Crude protein	18.0	13–27	Thiamin	2.5 ppm	
Fat	5.0		Riboflavin	5.0 ppm	
Fibre (ADF)	10.0	25–64.2	Vitamin B6	2.0 ppm	
NDF		64–88.9	Vitamin B12	30.0 ppb	
Hemicellulose		25–41.2	Niacin	30.0 ppm	
Calcium	0.75	0.4–0.7	Folate	600.0 ppb	
Phosphorus	0.6	0.1–0.6	Biotin	100.0 ppb	
Sodium	0.15	0.002–0.1	Choline	1250.0 ppm	
Potassium	0.65	0.8–1.7	Pantothenate	15.0 ppm	
Magnesium	0.1	0.1–0.2	Vitamin A	8000 IU/kg	
Iron	100 ppm	100–312	Vitamin E	220 IU/kg	20–120
Copper	8.0 ppm	15–32	Vitamin D	800 IU/kg	
Manganese	40.0 ppm	50–100	Linoleic acid	15	
Selenium	0.18 ppm				
Zinc	50.0 ppm	19–50			

Missing data indicate no data available

This regimen may also be helpful in weaning young pandas onto solid foods (around 5–7 months of age). During weaning, extra attention should be paid to ensure adequate intakes, but fruits and vegetables should not comprise more than 25% of the diet (weight, as offered) due to poor nutrient composition and possibility of nutrient dilution [27].

More recently, Plump evaluated red panda feeding regimens in 14 British collections. Mean intake was 291 g with a DM percentage of 42.0%; protein level in the DM averaged 15.8%, fat 5.2% and fibre 5.7%. Sixty-one percent of the red pandas were fed twice a day, 22% were fed once a day and 17% were fed three times a day and an average of eight different foods was provided (range 3–13). Most diets contained fruits and vegetables and one-third of the diets included meat. About 90% of the zoos fed bamboo and concentrates, but proportional intakes were not stated. The results of this survey showed that although the nutritional recommendations for red pandas are based on scientific advances, zoos still tend to feed historical diets which are based on traditional or anecdotal nutrition knowledge [29].

Faecal Quality

Johnson described the size and composition of fresh wild red panda faeces, containing approximately 72% water and, in the DM, approximately 99% finely pulverized *S. fangiana* leaves (Figure 14.3). Fresh droppings are well-formed and dark blue/black and shiny on the outside – similar to scat from many herbivores. Bamboo fragments are clearly visible in the droppings [11].

Figure 14.4 shows the droppings of red pandas in Rotterdam Zoo. These stools appear less formed, lighter in colour/lustre, and contain other undigested fibre fractions in addition to bamboo. It is recommended that zoos develop and implement a standardized faecal scoring for evaluating utilization of captive diets by red pandas.

FIGURE 14.3 Droppings of red pandas from the wild *(Photograph: A. Gebauer)*.

FIGURE 14.4 Picture of red panda droppings in Rotterdam Zoo.

Nutritional Disorders

Lubbert et al. and Preece document frequent diet-related problems seen in the first month of life in their pathology reviews (covering time frames 1982–1995), often in combination with other primary lesions [30,31]. Currently, nutritional disorders do not appear to be a main problem among captive young red pandas. Rather, poor suckling reflex of the cub, and insufficient milk production or inadequate maternal care by the dam, play important roles, indirectly related to nutrition. If food intake is not adequate, youngsters must be either hand-reared or supplementally fed (see Chapter 16).

As has been previously mentioned, sweetened gruels can lead to a number of health problems such as caries, periodontitis and circulatory disorders. Diets containing too much fat or soluble carbohydrate, which can result in obesity, should also be avoided [31].

Two-thirds of the zoos reported hair loss in pandas, which has been found in both male and female pandas and is generally restricted to the tail, flanks and hind legs, occurring especially in the summer months and recurring annually. No causes or successful treatments have been reported; nutritional aetiology is unclear. However, it has also been suggested that this is the result of the animal pulling out its own hair. Abnormal behaviour resulting from feeding a too-concentrated diet, with animal feeding time reduced to a few hours per day instead of foraging almost the whole day, may be related to hair-loss incidence [29].

Reddaclif suggests that altering panda diets by including high roughage may have contributed to the improvement in condition and recovery of a red panda with gastric ulcers [32]. Kock also suggests that a high-fibre diet can prevent gastric disorders [22]. Abnormal behaviour resulting from too-concentrated diet because animal feeding time is reduced to a few hours per day instead of foraging almost the whole day can result in pulling out its own hair. Outcomes may be due to direct nutritional influence (improved nutrient balance) and/or in combination with more natural feeding behaviours extended over a longer time period. Lynch reported bone disease in red pandas due to a possible high vitamin A intake [33].

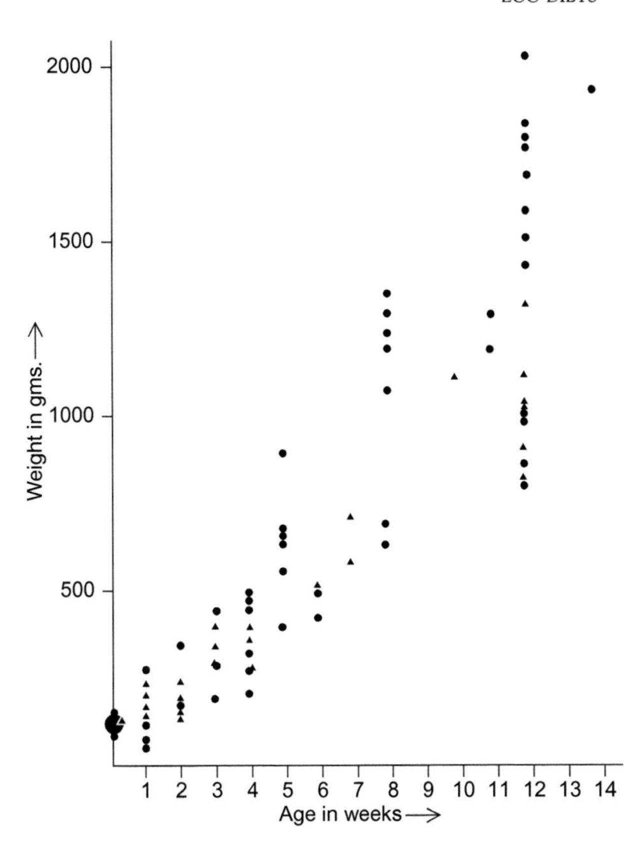

FIGURE 14.5 Weights of mother-reared and hand-reared red panda cubs. Triangles are hand-reared cubs, circles are mother-reared cubs.

Because older pandas can have problems with chewing due to bad teeth, it is recommended to feed them a softer diet like chopped bamboo, soaked pellets or a diet which contains grain mixtures which fulfil the nutrient recommendation for geriatric animals. In any case, maintaining a high-fibre component to the diet is recommended throughout life, beneficial to both oral and gut health. Longitudinal studies should demonstrate whether reports of poor dental health are an inevitable consequence of ageing, or actually result from poorer dietary husbandry in early life. Increasing fibre and decreasing sweet sugar content of panda diets should result in an improved oral environment. On the other hand, the high silica content unique to bamboo can be excessively abrasive and may contribute substantially to dental/tooth wear. An investigation to evaluate possible influence of diet on tooth wear patterns between zoo-reared and free-ranging red pandas using museum specimens may be enlightening in this respect.

Hand-Rearing

Proper hand-rearing protocols address not only appropriate formula selection (types and amounts), but also critical aspects of hygiene, temperature (of the formula and the animal), housing condition(s), animal handling, sucking reflex, feeding frequency, equipment suitability, socialization, and weaning [9,26,34–37], this section focuses only on the milk formula and growth of panda cubs.

According to the authors' best knowledge, no information is available on the milk composition of red pandas.

Giant panda milk composition may be the most appropriate model for red pandas due to similarities in feeding ecologies and digestive physiologies. The milk composition of the giant panda is (on wet basis): 6.7 % protein (5.1–8.3%), lactose 3.2 % (1.1–3.2%), with little change after 23 days (lactose only declines 0.2% and protein about 0.8%) [38]. According to Xuanzhen (referenced in [38]), the nutrient composition of giant panda milk is similar to bear and dog milks. Thus canine milk replacers (with added lactase enzyme) seem to provide an appropriate milk substitute, although constipation has been reported in red panda cubs fed solely a dog milk replacer [38]. Equal parts canine and human milk replacers have also been successfully used to hand-rear red pandas without constipation [38].

Hand-reared cubs with a higher birth weight have a greater chance of survival than smaller cubs. Survival percentage at a birth weight of 100–109 g is 17%, from 110 to 119 g, 67%, and a birth weight >120 g, 75%. Although bacterial overgrowth is a leading cause of death, cubs can also die from an overfilled stomach. Current recommendations suggest feeding 3 ml of formula every 3 hours during the first 5 days, and thereafter on demand [34]. Cross-fostering of infants between females is possible when the infants are young [9,35].

Body weights of hand-reared and mother-reared red pandas are presented in Figure 14.5 and Table 14.5; adult size and mass are attained at around 12 months of age. At about day 30 after birth, both the upper and lower premolars appear. By 6 months they have their complete dentition [9], and can be fully weaned.

TABLE 14.5 Weight gain of hand-reared pandas

Age	Weight (g)
Birth	± 100
2 weeks	160–210
4 weeks	260–360
6 weeks	460–650
8 weeks	650–960
10 weeks	900–1400
3 months	1300–1900
4 months	1800–2600
5 months	2400–3700
6 months	3200–4800

[9]

SUMMARY

Red pandas, although classified within Carnivora, exhibit highly herbivorous dietary habits with a specialization on bamboo leaves and shoots. They possess anatomical, metabolic, and physiological adaptations adapted for this feeding strategy, but no gastrointestinal modifications related to herbivory. Thus, although bamboo is rather poorly digested by red pandas, captive diets containing nutrient profiles that more closely mimic native bamboos, including high-fibre content, result in improved gastrointestinal and oral health in this species, compared with more historical cereal/grain-based diets supplemented with fruits, vegetables, and animal proteins. The use of moderately digestible, nutritionally complete dry high-fibre diets developed through controlled feeding trials, compared to gruels or porridges, contributes to improved faecal quality indices and dental health. Dry diets should be supplemented with 200–400 g bamboo or other natural sources of dietary fibre (when bamboo is not available) to encourage natural feeding behaviours, as well as for nutritional and dietary enrichment. The intake of limited animal proteins and enhanced predatory behaviours in red pandas are often noted more frequently during breeding/reproductive periods. Young red pandas have been successfully hand-reared on various mixtures of commercial canid and/or combinations of domestic carnivore and human milk replacers, with added lactase enzyme and often vitamin supplements. Weaning to the nutritionally balanced biscuit/bamboo diet can commence at 3 months, and be fully accomplished by 5–6 months of age, with adult size attained by 12 months.

References

[1] M.C.K. Bleijenberg, J. Nijboer, Feeding herbivorous carnivores, in: A.R. Glatston (Ed.), Red Panda Biology, SPB Academic Publishing, The Hague, The Netherlands, 1989, pp. 41–50.

[2] B.J. Gregory, On the phylogenetic relationships of the giant panda (Ailopoda) to other actoid carnivore, Amer. Mus. Noitates. 878 (1936) 1–29.

[3] L.B. Radinsky, Evolution and skull shape in carnivores. Additional modern carnivores, Biol. J. Linnean Soc. 16 (1981) 337–355.

[4] W.H. Flower, On the anatomy of *Ailurus fulgens*, Proc. Zool. Soc. London (1870) 752–769.

[5] K.J. Warnell, S.D. Crissey, O.T. Oftedal, Utilization of bamboo and other fibre sources in red panda diets, in: A.R. Glatston (Ed.), Red Panda Biology, SPB Academic Publishing, The Hague, The Netherlands, 1989, pp. 51–56.

[6] B.K. McNab, Energy expenditure in the red panda, in: A.R. Glatston (Ed.), Red Panda Biology, SPB Academic Publishing, The Hague, The Netherlands, 1989, pp. 73–78.

[7] M. Kleiber, Body size and metabolism, Hilgardia 6 (1932) 315–353.

[8] B.K. McNab, Energy conservation in a tree-kangaroo (*Dendrolagus matschiei*) and the red panda (*Ailurus fulgens*), Physiol. Zool. 61 (3) (1988) 280–292.

[9] M.S. Roberts, J.L. Gittleman, *Ailurus fulgens*, Mammalian Species 222 (1984) 1–8.

[10] M.S. Roberts, Red panda website. http://redpanda.cincinattiezoo.org/slmamm.htm, 2007.

[11] K.G. Johnson, G.B. Schaller, J. Hu, Comparative behaviour of red and giant panda in the Wolong Reserve China, J. Mammal. 69 (1988) 552–564.

[12] E.S. Dierenfeld, Chemical composition of bamboo in relation to giant panda nutrition, in: G.P. Chapman (Ed.), The Bamboos, Academic Press, London, UK, 1997, pp. 205–211. Linnean Society Symposium Series.

[13] I.R. Hunter, E. Dierenfeld, J. Fu, The possible nutritional consequences for giant panda of establishing reserve corridors with various bamboo species, J. Bamboo Rattan 2 (2003) 167–178.

[14] E.S. Dierenfeld, H.F. Hintz, J.B. Robertson, P.J. Van Soest, O.T. Oftedal, Utilization of bamboo by the giant panda, J. Nutr. 112 (1982) 636–641.

[15] S.A. Mainka, Z. Guanlu, L. Mao, Utilization of a bamboo, sugar cane and gruel diet by two juvenile giant pandas (*Ailuropoda melanoleucar*), J. Zoo Wildlife Med. 20 (1989) 39–45.
[16] G.B. Schaller, J Hu, W. Pan, J. Zhu, The Giant Pandas of Wolong, University of Chicago Press, Chicago, USA, 1985, pp. 72–81.
[17] S.A. Mainka, L. Mao, Z. Guanlu, Dietary vitamin and mineral concentrations of two juvenile female giant pandas (*Ailuropoda melanoleuca*), J. Wildlife Dis. 27 (1991) 509–512.
[18] F. Wei, Z. Wang, Z. Feng, M. Li, A. Zhou, Seasonal energy utilization in bamboo by the red panda (*Ailurus fulgens*), Zoo Biol. 19 (1) (2000) 27–33.
[19] A.D. Bartlett, Remarks on the habits of the panda (*Aelurus fulgens*) in captivity, Proc. Zool. Soc. London (1870) 769–772.
[20] M. Bush, M. Roberts, Distemper in captive red pandas, Internatl. Zoo Yearbook 17 (1977) 194–196.
[21] J. Nijboer, M.C.K. Bleijenberg, A.R. Glatston, The European red panda diet survey, in: A.R. Glatston (Ed.), Red or Lesser Panda Studbook No. 6, Stichting Koninklijke Rotterdamse Diergaarde, Rotterdam, The Netherlands, 1989, pp. 41–45.
[22] R.A. Kock, P. Spala, J.K. Eva, M. Bircher, M. Ricketts, M. Stevenson, New ideas for red panda diets, in: A.R. Glatston (Ed.), Red Panda Biology, SPB Academic Publishing, The Hague, The Netherlands, 1989, pp. 57–71.
[23] K. Fulton, M. Robert, M. Allen, D. Baer, A.T. Oftedal, S.D. Crissey, The red panda SSP diet evaluation project, Proceedings of the Eighth Dr Scholl Conference on the Nutrition of Captive Wild Animals, Lincoln Park, Chicago, USA, 1989.
[24] L.K. Lambrakis, J.L. Atkinson, E.V. Valdés, Feed intake and diet digestibility of captive red pandas (*Ailurus fulgens*) at Toronto Zoo. Abstract book. First European Zoo Nutrition Conference, Rotterdam, The Netherlands, 1999.
[25] A.C. Plump, Analyses of a recommended captive diet for red pandas (*Ailurus fulgens*). Unpublished paper, 2001.
[26] A.R. Glatston, Husbandry and management guidelines, in: A.R. Glatston (Ed.), Red or Lesser Panda Studbook No. 5, Stichting Koninklijker Rotterdamse Diergaarde, Rotterdam, The Netherlands, 1987.
[27] A.R. Glatston (Ed.), Red or Lesser Panda Studbook No 7, Stichting Koninklijke Rotterdamse Diergaarde, Rotterdam, The Netherlands, 1993.
[28] M.S Robert, A.R. Glatston, International Red Panda Husbandry Manual. International Studbook Library CD rom Version, ISIS & WAZA, 2003.
[29] A.C. Plump, Feeding red pandas, A study of nutrition and diets of red pandas (*Ailurus fulgens*) in captivity (UK and Ireland), Proceedings of the Sixth Annual Symposium of Zoo Research. Edinburgh Zoo, UK, 2004.
[30] J. Lubbert, W. Schaftenaar, A.R. Glatston, A review of the pathology of the red panda, *Ailurus fulgens*, in the period 1982–1991, in: A.R. Glatston (Ed.), Red or Lesser Panda Studbook No. 7, Stichting Koninklijker Rotterdamse Diergaarde, Rotterdam, The Netherlands, 1993.
[31] B.E. Preece, Review of the pathology of the red panda 1994–1995, in: A.R. Glatston (Ed.), Red or Lesser Panda Studbook No. 9, Stichting Koninklijke Rotterdamse Diergaarde, Rotterdam, The Netherlands, 1995, pp. 49–53.
[32] G.L. Reddaclif, Suspected gastric enteritis in a red panda, in: A.R. Glatston (Ed.), Red or Lesser Panda Studbook No. 2, Stichting Koninklijker Rotterdamse Diergaarde, Rotterdam, The Netherlands, 1982, pp. 43–44.
[33] M. Lynch, H. McCracken, R. Slocombe, Hyperostotic bone disease in red pandas (*Ailurus fulgens*), J. Zoo Wildlife Med. 33 (2002) 263–271.
[34] B.J. Gray, Care and development of a hand-reared red panda *Ailurus fulgens*, Internatl. Zoo Yearbook 10 (1970) 134–142.
[35] P. Vogt, C. Schneidermann, B. Schneidermann, Hand-rearing a red panda, Internatl. Zoo Yearbook 20 (1980) 280–281.
[36] J.K. Watson, V.L. Barfield, Supplemental feeding of infant red pandas at the Knoxville Zoological Park, in: A.R. Glatston (Ed.), Red or Lesser Panda Studbook No. 2, Stichting Koninklijke Rotterdamse Diergaarde, Rotterdam, The Netherlands, 1982, pp. 29–35.
[37] A.R. Glatston (Ed.), Red or Lesser Panda Studbook No. 3, Stichting Koninklijke Rotterdamse Diergaarde, Rotterdam, The Netherlands, 1984.
[38] M.S. Edwards, K.J. Lisi, K. Lang, L. Ware, Evaluation of a formula for hand-rearing red pandas (*Ailurus fulgens*). Proceedings of the Nutrition Advisory Group 7th Conference on Zoo and Wildlife Nutrition, Knoxville, USA, 2007, pp. 46–47.

CHAPTER

15

Captive Red Panda Medicine

Joost Philippa and Ed Ramsay

Wildlife Conservation Society, Field Veterinary Program, Viet Nam and Indonesia, Bogor, Indonesia; Department of Small Animal Clinical Sciences, College of Veterinary Medicine, University of Tennessee, Knoxville, Tennessee; Knoxville Zoological Gardens, Knoxville, Tennessee USA

OUTLINE

Red pandas are popular exhibit animals and their veterinary care has attracted a fair amount of study and attention. While the information we have on the health of red pandas relies on observations of a relatively small captive population, there have been regular reviews of the pathology of captive red pandas [1–5] (see Chapter 16). Additionally, reports on the management of clinical diseases have slowly increased.

As with many captive animals, red panda health problems can be categorized into two age-based groups: paediatric and adult. This chapter addresses the broad health concerns of those two groups, followed by sections on parasites, infectious diseases, and vaccination, and concludes with suggestions for chemical restraint and anaesthesia of red pandas.

Red Panda. DOI: 10.1016/B978-1-4377-7813-7.00015-X

PAEDIATRIC HEALTH ISSUES

It has been said that if a red panda does not die within the first year, it is likely to survive and become aged. Certainly, neonates experience a large number of severe health problems and still have a disappointingly high mortality rate. Different authors define the neonatal period, and therefore what constitutes a neonatal death, differently, so comparisons among reviews are difficult. At the Knoxville Zoo, 25% of the 91 red pandas born alive at the Knoxville Zoo died within the first 30 days of life. This neonatal mortality rate is very close to the average 26% seen in the red panda studbook population since 1994 (see Chapter 16), but is lower than the N. American average of 42%.

Many neonatal red pandas suffer from emaciation, inanition, or trauma, or a combination of these [1,2,4,6,] (see Chapter 16). Both emaciation and inanition are assumed to be a result of maternal neglect or insufficient lactation. Poor or insufficient lactation may have several causes. Subclinical mastitis has been described as a cause of decreased milk production in one dam [7]. Malnourished dams may have insufficient milk and stressed mothers also are at risk for poor milk production. Better observation of maternal behaviour and routine weighing of mother-reared infants may help identify cubs at risk before severe problems arise. Supplemental (tube) feeding of infants left with mothers has decreased the number of infants dying of emaciation and inanition, and is performed at many US zoos for cubs with poor weight gains. This allows a cub to stay with its mother and siblings for social contact while continuing to grow at a normal rate. Supplemental feeding is not routinely performed in Europe. Despite aggressive surveillance and interventions, however, 'poor mothering' continues to be a problem for captive red pandas.

The most common traumatic injuries of neonates are superficial bite wounds to the neck, which occur when mothers carry the young. Moist dermatitis at the thoracic inlet, presumably secondary to the mother salivating and/or biting the infant while carrying it, is also common. Severe bite wounds and cannibalism also occur with some frequency. Stress is assumed to the one of the most common factors leading to infant trauma, cannibalism, and decreased nursing [2]. There also seems to be a relationship between temperature and the time the mother spends with infants. If nest boxes are not adequately insulated the dam may not nurse her young adequately in hot weather.

Septicaemia and pneumonia are major causes of captive neonatal deaths. A variety of bacterial pathogens have been cultured from these cases, suggesting that the infections are opportunistic or secondary to other problems. Poor nutrition due to poor mothering may predispose an infant to developing sepsis or pneumonia, particularly if the newborn does not receive sufficient colostrum. Many neonatal and juvenile pneumonias have been caused by, or are suspected to be caused by, aspiration of milk or food. Neonatal red pandas are very aggressive feeders. Better hand-rearing techniques, including careful monitoring of bottle and feeding, and oro-gastric tube feeding of young neonates, has decreased the number of pneumonia cases seen in the Knoxville Zoo collection.

Congenital defects appear to be rare in captive red pandas. An intersex individual [8], a cub with ancephaly, hypoplastic limbs, and truncus arteriosus [1], and cubs with hydrocephalus [9] have been reported. Avascular necrosis of the femoral heads, resembling Legg-Calve-Perthes disease, was diagnosed in one 17-month-old red panda [10].

Juvenile red pandas, those between one month and one year of age, have fewer problems but cases of pneumonia and septicaemia still occur. Weaning can be very stressful for both mother-reared and hand-reared cubs and is a time of potential health problems. Weights of weanlings should be carefully monitored to assure that an individual does not lose too much weight and become compromised. Heat stress has also been identified as a cause of death in neonates and juveniles [4], and the cubbing dens should be air conditioned in hot and humid climates.

Dermatophytosis

One of the most important infectious diseases of captive neonate and juvenile red pandas is dermatophytosis (also called dermatomycosis, or ringworm). A review of 14 cases at Knoxville Zoo revealed that all but one affected animal were less than 4 months old [11] and the senior author is aware of only one clinical case in a red panda over one year of age. *Microsporum gypseum* has been cultured from all cases of red panda dermatophytosis, regardless of institution.

Clinical signs of mild dermatophytosis include diffuse crusting, hair loss, and thickened or flaky skin. Severe lesions may ulcerate or become purulent (Figure 15.1) and covered with a thick crust – a lesion termed a kerion. The latter appears most commonly on the tail and at the thoracic inlet. This disease does not appear to be pruritic in red pandas.

Red pandas with mild lesions of the appendages or face typically respond to clipping the affected area, cleaning the lesion with tamed iodine soaps, and topical therapy. A variety of topical antifungal agents commonly used in domestic carnivores have been used in red pandas with positive results.

Severe lesions may result in severe scarring or, in the case of the tail, the need for amputation if early and aggressive treatment is not instigated. Any suspected dermatophytosis lesions on the tail or at the thoracic inlet should be immediately clipped, cleaned with tamed iodine soap, and treated with topical agents. Animals with these lesions should also be put on systemic antifungal therapy. Itraconazole (5–10 mg/kg p.o. q.12–24 h) has been used for as long as 3 months in a neonatal red panda without adverse effects.

ADULT HEALTH ISSUES

It has been observed that red pandas have "few species specific [disease] peculiarities" [3]. Adult red pandas at Knoxville Zoo (all fulgens subspecies) have lived an average of 8.7 years (range = 1.5–18.1 years; n = 24) although we expect animals to live and breed into their teens. The captive longevity record for the oldest red panda was for a male at the Rotterdam Zoo who died at 21 years, 7 months of age. Red pandas older than 12 years should be considered geriatric. One review indicated that adult deaths occur most frequently in the winter [8]. The reasons for this may be increased weather-related caloric needs, or stresses of being kept indoors for greater periods of time.

Reviews of red panda pathology suggested that stomatitis, periodontitis, gastric ulcers, fatty livers, and enteritis are the most common health problems of the adult captive population [1,2]. The aetiology of gastric ulcers observed in adult animals has not been

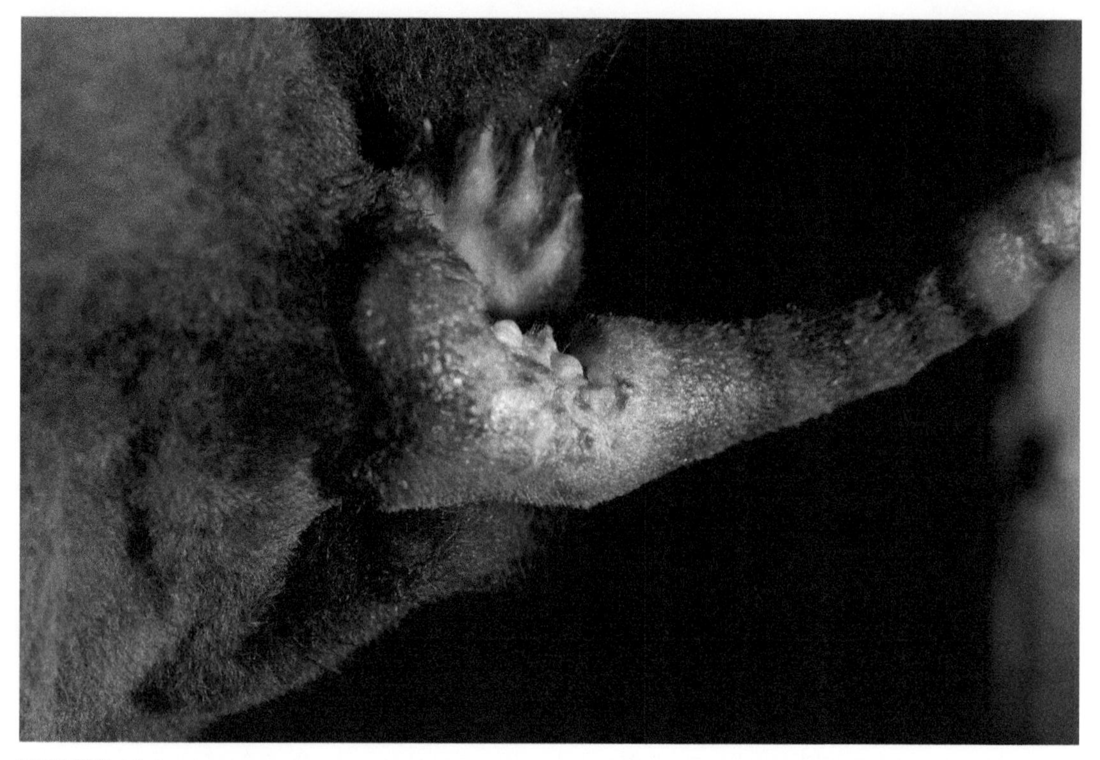

FIGURE 15.1 A severe, ulcerative dermatophytosis of the tail of a red panda cub. *Microsporum gypseum* was cultured from this lesion *(Photo courtesy of Houston Dale, University of Tennessee, Knoxville, TN, USA).*

identified. Inappropriate diet and stress are hypothesized to play roles. It seems that gastrointestinal problems have decreased in number in recent years (see Chapter 16).

Studies of red panda nutritional requirements have resulted in improvements in captive red panda feeding husbandry. The decreased feeding of gruel diets that had large amounts of soluble carbohydrates and fats, and the increased provision of bamboo, regardless of geographical location, has improved the overall nutrition of captive red pandas. The prevalence of dental disease, especially periodontitis, seen in middle-aged pandas also seems related to the feeding of gruels or porridge diets. Periodontitis has become much less common in red pandas in the last decade (see Chapter 16).

A phenomenon of red and giant pandas is a periodic lethargy and passing of mucoid stools [12]. In red pandas, these bouts usually last only one day and can occur at various intervals. Individual red pandas may have mucoid stools as often as once every two weeks but most animals have episodes at monthly or greater intervals. Repeated efforts to identify an aetiology for this syndrome have been made in red pandas and giant pandas, but the cause remains unclear [12].

Other important diseases of adult red pandas include heart disease and osteoarthritis. Congestive heart failure, hypertrophic cardiomyopathy [13], and left ventricular myohypertrophy have been reported in red pandas [2], and cardiomyopathies were

FIGURE 15.2 A captive adult red panda with hair loss. The cause for this condition remains undetermined *(Photo courtesy of the Mill Mountain Zoo, Roanoke, VA, USA).*

reported in over 9% of adult red pandas in a recent pathology review (see Chapter 16). Mild to moderate ankylosing spondylitis and degenerative arthritis of joints have also been seen [1]. Palliative treatment of the latter typically includes oral chondroitin sulphate/glucosamine compounds and non-steroidal antiinflammatory drugs.

Hyperostotic bone disease is an unusual pathology observed in several collections. The Knoxville Zoo and one other collection each had a single animal affected [14], while another zoo had three middle-aged red pandas with lesions [15]. Primary lesions in one of the single animals [14] and the group of three involved the elbow joints and legs. Pathology in these animals resembled hypertrophic pulmonary osteopathy seen in dogs. The other single animal had multiple exotoses of the ribs, especially dorsally and near the spine. Several of the affected animals had either concomitant renal disease or eventually succumbed to renal disease. While kidney disease can cause changes in calcium metabolism, the lesions in the red pandas did not resemble renal secondary hyperparathyroidism. An aetiology for these conditions has not been identified but hypervitaminosis A was postulated as a contributing factor in the group of three cases [15].

Hair loss over the caudal body and tail, or general patchy hair coats, is common in captive red pandas (Figure 15.2). Endocrinopathies can cause symmetrical hair loss and hypothyroidism has been reported occasionally in red pandas [1]. Thyroid function is

TABLE 15.1 Baseline haematological values for healthy male and female red pandas of all age groups

Haematological parameter	No. of samples	Range	Mean ± SEM
Haematocrit (%)	219	29.0–54.0	41.5 ± 0.34
Haemoglobin (g/dl)	214	9.6–17.4	13.5 ± 0.10
RBC ($\times10^6$/μl)	213	4.8–12.8	8.8 ± 0.07
MCV (fl)	210	39.0–55.0	47.0 ± 0.17
MCH (pg)	210	13.0–18.0	15.0 ± 0.05
MCHC (%)	210	23.0–52.0	33.0 ± 0.16
Total protein (g/dl)	226	5.8–9.5	7.2 ± 0.05
WBC ($\times10^3$/μl)	218	3.1–14.2	6.75 ± 0.15
Neutrophils ($\times10^3$/μl)	216	0.8–11.6	3.1 ± 0.13
Relative (%)	216	13.0–87.0	44.3 ± 1.10
Bands ($\times10^3$/μl)	216	0.0–0.2	0.003 ± 0.001
Relative (%)	216	0.0–2.0	0.03 ± 0.015
Lymphocytes ($\times10^3$/μl)	216	0.6–11.1	3.3 ± 0.10
Relative (%)	216	9.0–82.0	49.6 ± 1.10
Monocytes (/μl)	216	0.0–792.0	203.0 ± 10.7
Relative (%)	216	0.0–13.0	3.2 ± 0.16
Eosinophils (/μl)	216	0.0–675.0	78.0 ± 6.19
Relative (%)	235	0.0–8.0	1.3 ± 0.10
Basophils (/μl)	235	0.0–700.0	97.0 ± 8.49
Relative (%)	235	0.0–11.0	1.5 ± 0.12

[17] Table reprinted with permission of the editor of the *Journal of Zoo and Wildlife Medicine.*

usually normal in most animals showing hair loss or patchy coats, and the cause of these cases remains unclear. Many have observed a seasonal hair loss, especially at the base of the tail, occurring when the temperature rises in spring and loss of winter coats begins. A perineal dermatitis has also been observed in yearling female red pandas during the spring. Increased amounts of oestrogen have been proposed as a cause of this syndrome (D. Rost, personal communication).

Chronic renal disease occurs with some frequency in older red pandas. Affected individuals have poor body condition and rough hair coats. Clinical pathology shows azotaemia and urinalyses indicate proteinuria. Animals may survive for years with renal compromise and treatment is usually restricted to limiting the amount of concentrates in the diet and allowing feeding of bamboo *ad lib.* Pathology shows end-stage kidneys with both glomerular and tubular lesions. No aetiology has been identified and this is presumed to be a degenerative disease of old age. Oxalate nephrosis has also been described in a red

TABLE 15.2 Baseline serum chemistry values for healthy male and female red pandas of all age groups

Serum chemistry	No. of samples	Range	Mean ± SEM
Albumin (g/dl)	103	2.0–6.6	4.23 ± 0.07
Alkaline phosphatase (IU/L)	108	1.0–102.0	26.6 ± 2.1
BUN (mg/dl)	109	12.0–55.0	25.0 ± 0.63
Calcium (mg/dl)	109	6.9–11.8	9.2 ± 0.09
Cholesterol (mg/dl)	54	169.0–472.0	281.0 ± 11.4
Creatinine (mg/dl)	108	0.4–1.8	1.1 ± 0.02
Fibrinogen (mg/dl)	210	100.0–500.0	200.0 ± 6.2
Glucose (mg/dl)	108	51.0–281.0	115.9 ± 3.34
Phosphorus (mg/dl)	109	2.0–8.9	4.9 ± 0.13
Potassium (mEq/l)	85	4.1–6.9	5.1 ± 0.05
AST (SGOT) (IU/L)	109	34.0–183.0	61.5 ± 2.07
ALT (SGPT) (IU/L)	108	9.0–165.0	54.1 ± 2.99
Sodium (mEq/L)	86	126.0–150.0	138.2 ± 0.48
Total bilirubin (mg/dl)	109	0.1–1.0	0.24 ± 0.01

[17] Table reprinted with permission of the editor of the *Journal of Zoo and Wildlife Medicine.*

panda [1]. This animal had no known exposure to agents or plants that traditionally cause oxalate toxicity but had been treated with an aminoglycoside antibiotic which is known to cause nephrosis in other species.

Neoplasms are relatively rare in red pandas. A granulosa cell (ovarian) tumour [16], an anaplastic hepatic neoplasm which resembled a histiocytic sarcoma, a hepatocarcinoma (see Chapter 16), a squamous cell carcinoma (see Chapter 16), a thyroid carcinoma (see Chapter 16), and a "lung sarcoma" [2] have been reported. Three haemopoietic system neoplasias have been described, which seems a large number considering the few neoplasias observed in red pandas, overall. Lymphoma [17], a myelogenous leukaemia [18], and a chronic lymphoid leukaemia have each been reported [1].

The clinical pathology of red pandas is generally similar to that of domestic carnivores (Tables 15.1 and 15.2) [17]. A few biochemical parameters differ from domestic dogs. Normal red panda serum sodium concentrations may range slightly lower than those seen in dogs and normal serum urea nitrogen concentrations typically range a bit greater [17].

PARASITES

Several zoos have reported dirofilariasis in red pandas, either definitively or presumably, caused by the canine heartworm *Dirofilaria immitis* [19]. Red pandas in regions of the USA where this parasite is endemic are routinely given prophylactic anti-microfilaria treatment.

Ivermectin, at a dosage of 0.05 mg/kg p.o. q.30 days, therapy is started at 6 months of age and continued throughout the animal's life. In regions with severe winters, treatment may be suspended during the winter months. In the southeastern USA, heartworm prophylaxis is given year-round. In most European countries where red pandas are kept in zoos, *Dirofilaria immitis* is not endemic, although dirofilariasis has been reported in a red panda in southern France (D. White, personal communication).

An occult canine heartworm serological test can be used to identify infection and many zoos do this test as part of routine health exams. Successful treatment of heartworm infection has not been described, to the authors' knowledge. Treatment of dirofilariasis with melarsomine, at the approved canine dosage, should *not* be attempted as at least one red panda has died following melarsomine therapy. If a red panda is not symptomatic but occult heartworm test-positive, long-term, prophylactic dosage ivermectin therapy may be a more prudent therapeutic approach.

Few enteric parasites have been identified in red pandas. The fox whipworm, *Trichurus vulpis,* has been found at necropsy in a Knoxville Zoo juvenile. Trematodes in the small intestine and nematodes in the lung have been described in an animal imported from India [16]. Dermatitis caused by *Filaria taxidaea,* has been described in red pandas in California [20]. It was presumed the panda acquired this parasite from local, wild mustelids, the known hosts for this nematode. Lungworms have been associated with pathology in red pandas and were identified in one survey as a major cause of morbidity in northwestern European countries (D. White, personal communication). Clinical signs of lungworm infections may be subtle and the infection may only be recognized at necropsy. Infections have been variously attributed to *Troglostrongulus* spp., *Angiostrongylus* spp., *Crenosoma* spp., and *Metastrongyloides* spp. [21].

Fleas, presumably domestic carnivores' fleas, are also occasionally found on captive red pandas in North America and Europe. This usually occurs during warm weather, when other animals in the collection are infested. Red panda infestations may be quite severe and deaths due to anaemia secondary to flea infestations have been observed, particularly in mother-raised cubs still in the nest boxes. The pattern of hair loss on the lower back and tail resemble those seen in flea-hypersensitivity in domestic animals, although fleas or flea faeces are not always found. It is recommended that keepers caring for red pandas are not in contact with animals which are commonly infected with fleas. Where this is unavoidable, measures should be taken to avoid contamination of red panda nest boxes. It is also advisable to change red panda bedding regularly. When birth is imminent, treatment of the nest boxes with some anti-flea preparation helps to avoid infant infestations. A number of anti-flea treatments have been used and, generally, those products which are safe and approved for use in domestic cats are effective. Domestic feline dosage regimens for imidacloprid, selamectin, nitenpyram, and fipronil have all been used in red pandas at the Knoxville Zoo without adverse side effects. Caution must be used as at least one zoo killed a panda cub by putting it in a flea bath.

BACTERIAL AND FUNGAL DISEASES

There are early reports of tuberculosis in red pandas [3,16,22]. At least one of these animals was a recent importation from Asia [16]. One was identified as having bovine tuberculosis but it was unclear if cultures were performed [3]. These cases occurred when tuberculosis in zoos and importations of red pandas from the wild were much more common.

There are two reports of Tyzzer's disease, caused by *Clostridium piliformis*, in captive red pandas [23,24]. Both these animals had short clinical presentations with clinical pathology suggestive of hepatic disease. Both infected animals died and had necrotic lesions of the liver and other organs. One of these red pandas had a concurrent *Trypanasoma cruzi* infection, which appeared to have been the primary cause of death [23].

Other bacterial infectious conditions which have been reported in red pandas include haemorrhagic septicaemia, due to *Pasteurella multocida* [25], salmonellosis, and septicaemia and pneumonia caused by *Klebsiella pneumonia* [3]. Pneumonia in red pandas has also been attributed to *Pneumocystis carinii*, now considered a fungus [26]. Leptospirosis was diagnosed in a red panda which presented with an acute haemolytic crisis [27].

VIRAL DISEASES AND VACCINATION

Canine Distemper Virus

Although little published information exists on the prevalence and significance of infectious diseases on the red panda, the most significant infectious agent appears to be canine distemper virus (CDV). The red panda has high susceptibility to natural infection with CDV [22,28–31] and serological surveys have shown natural exposure to this virus [32,33]. Interestingly, since the 1970s, a similar number of outbreaks have been attributed to prophylactic vaccination with modified-live virus (MLV) vaccines [34,35].

Canine distemper virus is a morbillivirus (closely related to measles virus), which has a world-wide distribution, and a very broad host range: species in all families in the order Carnivora (Canidae, Mustelidae, Procyonidae, Hyaenidae, Ursidae, Viverridae, and Felidae) are susceptible to CDV infection, and it is one of the most significant diseases in many of these species.

Clinical signs of distemper vary depending on species, viral strain, environmental conditions, and the age and immune status of the host. In red pandas, the clinical presentation includes depression, anorexia, (oculo-) nasal discharge (serous to mucopurulent), tachypnoea, central nervous signs (convulsions/seizures, paresis/paralysis, incoordination, myoclonus), hyper- or hypothermia [28]. Affected individuals may also have skin lesions similar to those present in seals with morbillivirus-associated dermatitis [36].

The most significant lesions – catarrhal pneumonia and acute necrotizing inflammation with inclusion bodies in several visceral organs and lymph nodes – are very similar in both naturally infected animals and those with vaccine-induced distemper [28,34]. On gross pathology, the lesions have included consolidation and oedema of the lungs with frothy fluid and mucopurulent exudate in trachea and bronchi, splenomegaly, and congestion of the gastrointestinal tract and liver [35]. Histopathological lesions include catarrhal bronchopneumonia, degeneration of epithelial cells of the bronchi, and passive congestion and fatty degeneration of the liver. Eosinophilic cytoplasmic and intranuclear inclusion bodies have been detected in epithelial cells of bronchi, bronchial glands, bronchioles, alveoli, oesophagus, stomach, small intestine, bile duct, pancreatic duct, urinary bladder, epididymis, and uterus [28]. Lesions of the central nervous system (CNS) (cerebral lesions with intranuclear inclusion bodies in nerve cells) have only been found in naturally infected animals [31].

A problem faced in the prophylaxis against distemper in exotic carnivores is the variation between and within species in their reaction to MLV vaccines, with possible lethal consequences. MLV vaccines have been designed to be minimally virulent, while retaining maximal immunogenicity in their domestic counterparts. When used in highly sensitive species or delivered by another route, the residual virulence may cause disease [37]. There are two widely used attenuated CDV vaccine strains: the Onderstepoort strain, attenuated in avian, and now more commonly co-cultivated in Vero cells; and the Rockport strain, attenuated in canine cells. Although the avian-attenuated CDV vaccines are generally safer when used in mammals, neither is safe for use in non-domestic species, as both types have caused fatal disease in a variety of species [38]. Published vaccine-induced distemper in red pandas has been caused by the Rockport strain [34,35]. While MLV vaccines are not recommended for use in non-domestic species, currently MLV canine distemper vaccines continue to be used in Chinese facilities, despite questionable efficacy and safety [32], and reports indicate that vaccine-induced disease and mortality is common [33].

A safer alternative is an inactivated vaccine, which cannot cause an infection, but the efficacy of inactivated vaccines has long been questioned [39,40]. An experimental, adjuvanted, inactivated, CDV vaccine has been used in red pandas and giant pandas in several zoos, and appeared to be safe, but produced low titres with inadequate durability, requiring booster vaccinations two to three times annually [41]. This vaccine is no longer produced. At present, there are no inactivated CDV vaccines commercially available, due to their low immunogenicity – and therefore low demand – in domestic dogs, and the market for non-domestic animals being too small [42]. In Germany, a small amount of inactivated vaccine is produced for use in zoos (Geyer and Matern, personal communication, 2002).

An experimental subunit vaccine incorporating the CDV fusion (F) and haemagglutinin (H) surface proteins into immunostimulating complexes (ISCOM) has been developed, and proven to be capable of producing humoral and cellular immunity in dogs and seals [43,44]. Vaccination protected seals against a lethal challenge infection, although the immunity achieved was not sterile: upper respiratory tract infection occurred in vaccinated, experimentally infected animals [44]. The CDV-ISCOM vaccine has since been shown to be safe in several species in European zoos (W. Schaftenaar personal communication) [45,46]. Red pandas reacted to vaccination with higher antibody titres when the total antigen concentration of the CDV-ISCOM was increased from 5 μg/ml to 10 μg/ml, suggesting a positive dose-response [45], although the concentration should not exceed 10 μg/ml to minimize local reactions to the vaccine. Three vaccinations with a 10 μg/ml vaccine are recommended with a 3-week interval, e.g. young animals at 8, 11, and 14 weeks of age, with one annual revaccination [45].

Recently, a monovalent canarypox-vectored vaccine (Purevax™, Merial, Duluth, USA) expressing the H and F surface antigens of CDV has become commercially available in the USA. Its safety and immunogenicity (in terms of antibody response) in giant pandas [47], European mink (*Mustela lutreola*) [46], and black-footed ferrets (*Mustela nigripes*) × Siberian polecat (*Mustela eversmanni*) hybrids has been documented [48,49] and vaccination of Siberian polecats has protected them from experimental challenge infection [50]. This vaccine is registered for use in domestic ferrets in the USA, but its off-label use in susceptible species in zoos is recommended by the American Association for Zoo Veterinarians [51] and the Veterinary Specialist Group of the IUCN [52]. However, in the European Union, its use is not permitted as it uses a non-registered genetically modified organism [45,53]. Three vaccinations are recommended with 3–4-week interval, e.g. young animals at 8, 11, and 14; or 8, 12, and 16 weeks of age, with annual revaccinations.

The main advantage of Avipox-vectored vaccines like Purevax™ is their safety. The Avipox-vector is an avian virus with a host-restriction to avian species. Virus replication in mammalian cells is blocked at a late stage, importantly leaving the synthesis of viral proteins unimpaired [54]. Protective cellular and humoral immunity is induced in the absence of the complete virus, therefore eliminating the possibility of infection with CDV. Due to the host restriction, there is no dissemination of the vector virus within the vaccinated mammal and therefore no excretion of the vector virus to non-vaccinated contacts or the environment [55].

Rabies Virus

Rabies virus belongs to the genus Lyssavirus in the family Rhabdovirus and causes an acute fatal encephalomyelitis in an extremely broad host range of mammals. Red pandas have been diagnosed with rabies virus infection, and rabies virus has been isolated from red pandas [56]. Clinical signs are not definitive or species-specific beyond acute behavioural alterations. Wildlife may lose their apparent wariness and caution around humans and domestic species, alter their activity cycles, seek solitude, or become more gregarious. Head tilt, head pressing or butting, "stargazing", and altered phonation may be observed [57].

Vaccination recommendations depend on location, risk of exposure, or possible outbreak [51,58,59]: in areas where the incidence of rabies in local wildlife (skunks, raccoons, foxes) is high, vaccination is recommended. In certain countries, local veterinary authorities should be contacted regarding the legal aspects of extra-label vaccination, as some areas may have restrictions [53]. Red pandas can be vaccinated with a commercially available, inactivated rabies virus vaccine approved for ferrets (Imrab™, Merial, Inc.) [60], at 3–4 months of age, with a booster-vaccination after one year, then triannually [61]. The use of modified-live rabies virus vaccines is contraindicated, due to documented vaccine-induced rabies infections in several species.

Parvovirus

Recently, a novel parvovirus has been isolated from red pandas in China [62]. No clinical signs were seen in the infected red pandas, nor is the pathophysiology of the disease and epidemiology of the infection known for this species. There is also a necropsy report of a red panda having feline panleukopaenia in the 1960s, but this infection has not been reported in recent times (see Chapter 16). Vaccination is not recommended in the USA and Europe at this time, as infection with this novel strain occurred without any clinical signs.

Other Viruses

Published reports of other virus infections in red pandas are rare, with low prevalences. A Reo-like virus was seen in one red panda with enteritis [1]. During an outbreak of West Nile Virus (WNV) disease in a zoo, one red panda had serum antibodies to WNV [63]. Serological surveys have shown low prevalence and low or "suspect" antibody titres to canine adenovirus, canine coronavirus, and influenza A virus in red pandas in China [33]. The detection of virus-specific antibodies indicates possible susceptibility to these viruses, although no clinical disease was observed in these pandas.

IMMOBILIZATION

Several drugs and drug combinations have been used to chemically restrain red pandas for short procedures or for the induction of general anaesthesia. Neonates are usually "masked down," by placing a face mask over their face, and having them breathe 5% isoflurane in oxygen. Alternatively, an induction chamber (an opaque plastic box with a tight top) can be used in all age groups. Anaesthesia is induced by flowing 5% isoflurane in oxygen into the box. The animal is removed from the induction chamber when recumbent, and a face mask or endotracheal tube can be applied to maintain anaesthesia. With either induction method, when the animal is recumbent the percentage of isoflurane is reduced to 1.5–3% isoflurane in oxygen for maintenance of anaesthesia. The chamber inductions are relatively slow but recoveries are rapid and animals are quickly able to be returned to their exhibit. Other new inhalant agents, such as sevoflurane, should work comparatively well.

A combination of ketamine HCl, a dissociative anaesthetic (6.6 mg/kg i.m.), and medetomidine, an alpha-2 adrenergic agonist (0.080 mg/kg i.m.), works well in red pandas for short, non-noxious procedures, such as blood collection, teeth cleaning, or radiology [64]. Typically, the animal is transferred from a crate to a small squeeze cage and the drugs are given intramuscularly, via hand injection. The effects of the medetomidine can be reversed with atipamezole (0.4 mg/kg i.m.), and recoveries are usually rapid and smooth.

Other drug combinations that have been used in juvenile and adult red pandas are: ketamine (6–9 mg/kg i.m.) and xylazine (0.2–0.4 mg/kg i.m); ketamine (10–15 mg/kg i.m.) with xylazine i.m. and followed by i.v. diazepam (0.2–0.5 mg/kg); and tiletamine/zolazepam (4.5–6.0 mg/kg i.m.) [17]. They are sufficient for non-invasive procedures (i.e., obtaining blood samples or cultures) of 10 to 25 minutes' duration. The effects of xylazine can be reversed with yohimbine (0.125 mg/kg s.c., i.m., or i.v.).

If surgery or long procedures are to be performed, an endotracheal tube should be placed and the animal maintained on an inhalant agent. Isoflurane in oxygen is the most commonly used agent at this time. Anaesthetic monitoring is similar to that used in anaesthetized domestic carnivores and red pandas do not have novel problems during anaesthesia.

CONCLUDING COMMENTS

Captive red pandas' health has benefited greatly from improved husbandry and nutrition. More routine veterinary care and better preventive medicine programmes, such as vaccination and parasitology surveillance, have also contributed to red pandas living longer. One can anticipate that more attention will be paid in the future to the treatment of geriatric diseases of red pandas. Poor captive reproduction and infant survival are challenges we still face with this species and must do more to address. Better techniques to evaluate reproductive soundness are something that veterinarians and reproductive physiologists need to develop for this species. Continued efforts are also needed, by managers and health care staff, to improve mothering of cubs, with the aim of reducing the number of cubs that need to be hand-reared.

References

[1] R.J. Montali, M. Roberts, R.A. Freeman, M. Bush, Pathology survey of the red panda (*Ailurus fulgens*), in: O.A. Ryder, M.L. Byrd (Eds.), One Medicine, Springer-Verlag, New York, USA, 1984, pp. 128–140.

[2] N. Lateur, Pathology survey of captive red pandas (*Ailurus fulgens*), in: A.R Glatston (Ed.), Red Panda Studbook No. 4, Rotterdam Zoo, The Netherlands, 1987, pp. 19–26.

[3] P. Zwart, Contribution to the pathology of red panda (*Ailurus fulgens*), in: A.R Glatston (Ed.), Red Panda Biology, SPB Publishing, The Hague, The Netherlands, 1989, pp. 25–29.

[4] K.L. Machin, Red panda pathology, in: A.R Glaston (Ed.), The Red or Lesser Panda Studbook No. 6, Rotterdam Zoo, The Netherlands, 1991, pp. 7–24.

[5] J. Lubbert, W. Scaftenaar, A.R. Glatston, A review of the pathology of the red panda, *Ailurus f. fulgens*, in the period 1982–1991, in: A.R. Glatston (Ed.), The Red or Lesser Panda Studbook No. 7, Rotterdam Zoo, The Netherlands, 1993, pp. 10–18.

[6] L. Yinhong, Infant mortality in the red panda (*Ailurus fulgens*), in: A.R. Glatston (Ed.), The Red Panda Studbook No. 8, Rotterdam Zoo, Netherlands, 1994, pp. 18–35.

[7] P.J. Morris, Subclinical mastitis and apparent agalactia in a lesser panda (*Ailurus fulgens*). Proceedings of the 1st International Conference on Zoology and Avian Medicine, Turtle Bay, Oahu, Hawaii, 1987, pp. 459–461.

[8] G.L. Reddacliff, C.R.E. Halnan, I.C. Martin, Mosaic 35,x/36,XY karyotype and intersex in red panda (*Ailurus f. fulgens*), J. Wildlife Dis. 29 (1993) 169–173.

[9] M.T. Frankenhuis, A.R. Glatson, A review of the pathology of the red panda in captivity, in: A.R. Glatston (Ed.), The Red Panda Studbook No. 3, Rotterdam Zoo, The Netherlands, 1984, pp. 45–51.

[10] M. Delclaux, C. Talavera, M. López, J.M. Sánchez, M.I. García, Avascular necrosis of the femoral heads in a red panda (*Ailurus fulgens fulgens*): Possible Legg-Calve-Perthes disease, J. Zoo Wildlife Med. 33 (2002) 283–285.

[11] K.S. Kearns, C.G. Pollock, E.C. Ramsay, Dermatophytosis in red pandas (*Ailurus f. fulgens*): A review of 14 cases, J. Zoo Wildlife Med. 30 (1999) 561–563.

[12] H. Bissell, B. Rude, M. Carr, J. Ouellette, Understanding the etiology of mucoid feces in giant pandas. www.aza.org/AZAPublications/2005ProceedingsReg/, 2004.

[13] H.M. Gardner, Case: Hypertropic cardiomyopathy in a lesser panda, Proc. Am. Assoc. Zoo. Vet. (1985) 76.

[14] C. Wahlberg, M. Kärkkäinen, Development of multiple exostosis in lesser panda (*Ailurus fulgens*). Proc. 25th Int Symp Erkrank Zootiere, Vienna, Austria, 11–15 May, 1983, pp. 411–415.

[15] M. Lynch, H. McCracken, R.F. Slocumbe, Hyperostotic bone disease pathology in red pandas (*Ailurus fulgens*), J. Zoo Wildlife Med. 33 (2002) 263–271.

[16] L. Griner, Pathology of Zoo Animals. Zoological Society of San Diego, San Diego, California, 1983, pp. 419–422.

[17] M.J. Wolff, A. Bratthauer, D. Fischer, R.J. Montali, M. Bush, Hematologic and serum chemistry values for the red panda (*Ailurus fulgens*): Variation with sex, age, health status, and restraint, J. Zoo Wildlife Med. 21 (1990) 326–333.

[18] J. Sleeman, W.S. Sprague, T.J. Painter, T. Campbell, Chronic myelogenous leukemia in a red panda (*Ailurus fulgens*), Proc. Am. Assoc. Zoo Vet. (1999) 131–133.

[19] G. Harwell, T.M. Craig, Dirofilariasis in a red panda, J. Am. Vet. Med. Assoc. 179 (1981) 1258.

[20] C.H. Gardiner, M.R. Loomis, J.O. Britt, R.J. Montali, Dermatitis caused by *Filaria taxideae* in a lesser panda, J. Am. Vet. Med. Assoc. 183 (1983) 1285–1287.

[21] M. Bush, Veterinary Medicine, Veterinary. http://www.si.edu/natzoo/redpanda/dlvet.htm, (accessed 5.10.98).

[22] R.N.T.-W. Fiennes, Report of the pathologist for the year 1960, Proc. Zoo. Soc. London 137 (1961) 12–196.

[23] C.H. Booney, R.E. Schmidt, A mixed infection: Chagas' and Tyzzer's disease in a lesser panda, J. Zoo Anim. Med. 6 (1975) 4–7.

[24] J. Langan, D. Bemis, S. Harbo, C. Pollock, J. Schumacher, Tyzzer's disease in a red panda (*Ailurus fulgens fulgens*), J. Zoo Wildlife Med. 31 (2000) 558–562.

[25] N.C. Nayal, B. Samaddar, M.K. Bhowmik, Haemorrhagic septicemia in a red panda (*Ailurpoda melanoleuca*) associated with *Pasteurella multocida*, Ind. Vet. J. 65 (1988) 543–544.

[26] F.G. Poelma, *Pnuemocystis carinii* in zoo animals, Z. Parasitenk. 46 (1975) 61–68.

[27] T. McNamara, M. Linn, P. Calle, R. Cook, W. Karesh, B. Raphael, Leptospirosis: an under-reported disease in zoo animals, Proc. Am. Assoc. Zoo Vet. (1997) 248–251.

[28] T. Kotani, M. Jyo, Y. Odagiri, Y. Sakakibara, T. Horiuchi, Canine distemper virus infection in lesser pandas (*Ailurus fulgens*), Nippon Juigaku. Zasshi 51 (1989) 1263–1266.

[29] N.S. Parihar, L.B. Chakarvarty, Canine distemper-like disease in lesser panda (*Ailurus f. fulgens*), Ind. Vet. J. 57 (1980) 198–199.

[30] G. Yun, The canine distemper of mink and lesser panda. Proc. 4th Int. Conf. Wildl. Dis. Assoc., Sydney, Australia, 25–28 August, 1981, pp. 75–76.

[31] Z.X. Zhang, S.L. Gao, F.N. Xu, A.X. Chin, Y.Y. Xu, Z.Q. Chin, Survey and control of panda distemper, Anim. Husband. Vet. Med. 15 (1983) 3–7.

[32] I.K. Loeffler, J. Howard, R.J. Montali, et al., Serosurvey of *ex situ* giant pandas (*Ailuropoda melanoleuca*) and red pandas (*Ailurus fulgens*) in China with implications for species conservation, J. Zoo Wildlife Med. 38 (2007) 559–566.

[33] Q. Qin, F. Wei, M. Li, E.J. Dubovi, I.K. Loeffler, Serosurvey of infectious disease agents of carnivores in captive red pandas (*Ailurus fulgens*) in China, J. Zoo Wildlife Med. 38 (2007) 42–50.

[34] M. Bush, R.J. Montali, D. Brownstein, A.E. James Jr., M.J. Appel, Vaccine-induced canine distemper in a lesser panda, J. Am. Vet. Med. Assoc. 169 (1976) 959–960.

[35] C. Itakura, K. Nakamura, J. Nakatsuka, M. Goto, Distemper infection in lesser panda due to administration of a canine distemper live vaccine, Jap. J. Vet. Sci. 41 (1979) 561–566.

[36] T.P. Lipscomb, M.G. Mense, P.L. Habecker, J.K. Taubenberger, R. Schoelkopf, Morbilliviral dermatitis in seals, Vet. Pathol. 38 (2001) 724–726.

[37] I. Tizard, Risks associated with use of live vaccines, J. Am. Vet. Med. Assoc. 196 (1990) 1851–1858.

[38] S.L. Deem, L.H. Spelman, R.A. Yates, R.J. Montali, Canine distemper in terrestrial carnivores: a review, J. Zoo Wildlife Med. 31 (2000) 441–451.

[39] M.J. Appel, W.R. Shek, H. Shesberadaran, E. Norrby, Measles virus and inactivated canine distemper virus induce incomplete immunity to canine distemper, Arch.Virol. 82 (1984) 73–82.

[40] J.G. Sikarski, C. Lowrie, F. Kennedy, G. Brady, Canine distemper in a vaccinated red panda (*Ailurus fulgens*). Proceedings of the American Association of Zoo Veterinarians' Annual Conference, Calgary, Alberta, Canada, 28 September–3 October, 1991, pp. 292–293.

[41] R.J. Montali, L. Tell, M. Bush et al., Vaccination against canine distemper in exotic carnivores: successes and failures. Proceedings of the American Association of Zoo Veterinarians' Annual Conference, Pittsburgh, USA, 22–27 October, 1994, pp. 340–344.

[42] M.J. Appel, R.J. Montali, Canine distemper and emerging morbillivirus disease in exotic species. Proceedings of the American Association of Zoo Veterinarian's Annual Conference, Pittsburgh, USA, 22–27 October, 1994, pp. 336–339.

[43] P. de Vries, F.G. UytdeHaag, A.D. Osterhaus, Canine distemper virus (CDV) immune-stimulating complexes (Iscoms), but not measles virus iscoms, protect dogs against CDV infection, J. Gen.Virol. 69 (1988) 2071–2083.

[44] I.K. Visser, E.J. Vedder, M.W. Van de Bildt, C. Orvell, T. Barrett, A.D. Osterhaus, Canine distemper virus ISCOMs induce protection in harbour seals (*Phoca vitulina*) against phocid distemper but still allow subsequent infection with phocid distemper virus-1, Vaccine 10 (1992) 435–438.

[45] J.D.W. Philippa, Vaccination of non-domestic animals against emerging virus infections, Erasmus MC, 2007.

[46] J.D.W. Philippa, T. Maran, T. Kuiken, W. Schaftenaar, A.D.M.E. Osterhaus, ISCOM vaccine against canine distemper induces stronger humoral immune response in European mink (*Mustela lutreola*) than a canarypox-vectored recombinant vaccine. Vet. Microbiol. in press.

[47] E. Bronson, S.L. Deem, C. Sanchez, S. Murray, Serologic response to a canarypox-vectored canine distemper virus vaccine in the giant panda (*Ailuropoda melanoleuca*), J. Zoo Wildlife Med. 38 (2007) 363–366.

[48] E.S. Williams, S.L. Anderson, J. Cavender, et al., Vaccination of black-footed ferret (*Mustela nigripes*)× Siberian polecat (*M. eversmanni*) hybrids and domestic ferrets (*M. putorius furo*) against canine distemper, J. Wildlife Dis. 32 (1996) 417–423.

[49] E.S. Williams, R.J. Montali, Vaccination of black-footed ferret×Siberian polecat hybrids against canine distemper with recombinant and modified live virus vaccines. Proceedings of the Wildlife Disease Association Annual Conference, 1998, 107.

[50] J. Wimsatt, D. Biggins, K. Innes, B. Taylor, D. Garell, Evaluation of oral and subcutaneous delivery of an experimental canarypox recombinant canine distemper vaccine in the Siberian polecat (*Mustela eversmanni*), J. Zoo Wildlife Med. 34 (2003) 25–35.

[51] R.E. Junge, Preventive medicine recommendations. American Association of Zoo Veterinarians Infectious Diseases Committee, 1995.

[52] M.H. Woodford, Quarantine and health screening protocols for wildlife prior to translocation and release into the wild, IUCN Species Survival Commission's Specialist Group, OIE, Care for the Wild, European Association of Zoo and Wildlife Veterinarians, Gland, Switzerland, 2001.

[53] J.D.W. Philippa, Vaccination of non-domestic carnivores: a review, in: N. Schoemaker, J. Kaandorp, H. Fernandez (Eds.), European Association of Zoo and Wildlife Veterinarians (EAZWV) Transmissible Diseases Handbook, 2005.

[54] G. Sutter, B. Moss, Nonreplicating vaccinia vector efficiently expresses recombinant genes, Proc. Natl. Acad. Sci. USA 89 (1992) 10847–10851.

[55] E. Paoletti, Applications of pox virus vectors to vaccination: an update, Proc. Natl. Acad. Sci. USA 93 (1996) 11349–11353.

[56] J.S. Smith, F.L. Reid-Sanden, L.F. Roumillat, et al., Demonstration of antigenic variation among rabies virus isolates by using monoclonal antibodies to nucleocapsid proteins, J. Clin. Microbiol. 24 (1986) 573–580.

[57] C.E. Rupprecht, K. Stohr, C. Meredith, Rabies, in: E.S. Williams, I.K. Barker (Eds.), Infectious Diseases of Wild Mammals, Iowa State University Press, Ames, Iowa, 2001, pp. 3–36.

[58] S.E. Aiello (Ed.), Merck Veterinary Manual, Merck and Co, Rahway, New Jersey, 1998.

[59] C.M. Fraser, Vaccination of exotic carnivores, in: C.M. Fraser, J.A. Bergeron, A.M. Mays, S.E. Aiello (Eds.), Merck Veterinary Manual, Merck and Co, Rahway, New Jersey, 1991, pp. 1083–1087.

[60] C.E. Rupprecht, J. Gilbert, R. Pitts, K.R. Marshall, H. Koprowski, Evaluation of an inactivated rabies virus vaccine in domestic ferrets, J. Am. Vet. Med. Assoc. 196 (1990) 1614–1616.

[61] National Association of State Public Health Veterinarians (2006). I. N. Compendium of Animal Rabies Prevention and Control, 2006.

[62] Q. Qin, I.K. Loeffler, M. Li, K. Tian, F. Wei, Sequence analysis of a canine parvovirus isolated from a red panda (*Ailurus fulgens*) in China, Virus Genes 34 (2007) 299–302.

[63] G.V. Ludwig, P.P. Calle, J.A. Mangiafico, et al., An outbreak of West Nile virus in a New York City captive wildlife population, Am. J. Trop. Med. Hyg. 67 (2002) 67–75.

[64] W. Schaftenaar, Short note about the immobilization of the red panda (*Ailurus f. fulgens*), in: A.R. Glaston (Ed.), The Red or Lesser Panda Studbook No. 7, Rotterdam Zoo, The Netherlands, 1993, p. 36.

CHAPTER

16

Red Panda Pathology

Brian Preece

Formerly of Veterinary Surveillance Department, Veterinary Laboratories, UK

OUTLINE

Regular reviews of the pathological findings recorded as the result of post-mortem examination of red pandas that died in captivity have been included in "The Red or Lesser Panda Studbook" since it was introduced in 1978 [1–9]. The studbook is issued every other year and a pathology report on the losses over the previous 2 years has been a standard inclusion with the exception of 1993 when a wider review was included [2].

Over the past 30 years, there have been considerable changes in the gross pathological observations and the pathological conditions recorded in the various post-mortem reports available to the studbook keeper. In the 1970s and the early 1980s, many of the animals held in captivity had been caught in the wild and, in 1985, there were only 160 specimens held around the world, 150 of these were *Ailurus fulgens fulgens* (Aff) and only 10 belonged to the subspecies *Ailurus fulgens styani* (Afs) [10]. Reports from that time indicated that wild-caught animals suffered high mortality in the first 4 months of their arrival in various zoological collections. In more recent times, there has been a thriving population of captive-bred animals living under modern husbandry conditions with 730 individuals registered in the studbook at the end of 2006 in 248 different premises. Of these 453 (213 males, 227 females, 13 undetermined) are *Ailurus fulgens fulgens* and 287 (162 males, 123 females and 2 undetermined) of the subspecies *styani*.

The red panda is listed as an endangered species in the red data book as its survival in the wild is threatened by deforestation, loss of habitat and fragmentation of existing wild populations. However, it has bred relatively well in captivity and successful cooperative

Red Panda. DOI: 10.1016/B978-1-4377-7813-7.00016-1

captive breeding programmes have been established in Europe, North America, Australasia, South Africa and Japan. However, despite this success, the management of the captive population is not optimal and, as an example, 124 of the 486 (26%) post-mortem reports received from the beginning of 1994 concerned red panda cubs that had died within their first month of life.

Analysis of post-mortem reports is an important tool in increasing our understanding of the red panda in captivity and improving our husbandry and management procedures for this species and, in order to facilitate this, those zoological collections holding red pandas are requested to report the death of any animal to the studbook keeper for studbook purposes. At the same time, they are asked to submit a copy of the post-mortem examination report.

Since 1994, the information provided in post-mortem reports has been collated and maintained in a single database and it now contains a total of 495 records which predominantly concern Aff. There are 124 (25%) neonates, of which 26 (21%) are Afs; 79 (16%) juveniles, of which six (7.5%) are Afs, and 283 (57%) adults, of which 47 (16%) are Afs, up to the end of 2006. There is insufficient information to categorize a further nine animals (1.8%) and these have been omitted from further consideration for the purposes of this review.

For the purposes of this database, animals are defined as neonates between day 0, i.e., the day that they are born and less than one month of age; juveniles are those animals aged between more than one month and less than one year and adults are those animals aged over one year. The allocation of animals aged greater than one year to the adult category is somewhat arbitrary given that females begin breeding in the region of 18 months of age; however, this is a suitable cut-off point for descriptive purposes.

This chapter reviews the pathological findings over the 30-year period of the studbook with particular reference to the pathological findings recorded in a database for the years 1994–2006 inclusive. Table 16.1 compares the number of deaths recorded within the studbook for each of those 13 years with the number of post-mortem reports made available to the studbook keeper.

In recent years, there has been an improvement in the amount of information given in post-mortem reports, although standard data, such as sex, age and weight, continue to be missing from a small number of reports. Table 16.2 shows the number and percentage of reports missing standard data during the 13-year period in which these parameters have been recorded within the database.

PATHOLOGY FINDINGS

The records contained within the database were very dependent upon the quality and detail provided to the holder of the studbook. The post-mortem examinations were carried out in a wide range of institutions from around the world, each having their own protocols. The information received ranged from reports consisting of a one word diagnosis, e.g., enteritis, without any information on the clinical signs prior to death or any details of laboratory tests that were carried out or evidence of a causative agent, to reports which include full details of gross pathological observations and all relevant laboratory tests.

TABLE 16.1 Number of red panda deaths recorded in the studbook for the period 1994–2006 inclusive and the number and percentage of post-mortem reports received for each subspecies respectively

Year	*Ailurus fulgens fulgens*			*Ailurus fulgens styani*			Total		
	Deaths	PM reports	%	Deaths	PM reports	%	Deaths	PM reports	%
1994	40	29	72.5	28	16	57.0	68	45	66.0
1995	55	37	67.0	26	2	7.7	81	39	48.0
1996	56	26	46.5	26	8	31.0	82	34	41.0
1997	74	37	50.0	24	0	0	98	37	37.8
1998	60	38	63.0	26	13	50.0	86	51	59.0
1999	55	38	69.0	40	10	25.0	95	48	50.5
2000	55	41	75.0	28	11	39.0	83	52	62.7
2001	48	31	64.5	23	4	17.4	71	35	49.0
2002	69	35	51.0	24	3	12.5	93	36	38.5
2003	72	43	60.0	30	6	20.0	102	49	48.0
2004	56	22	39.0	21	2	9.5	77	24	31.0
2005	54	23	43.0	29	2	6.9	83	25	30.0
2006	59	10	17.0	19	1	5.3	78	11	15.4
Total	753	409	54.0	344	78	22.60	1097	486	44.0

TABLE 16.2 Number and percentage of post-mortem examination reports missing standard data during period 1994–2006

Age group	No. in data set	Weight		Age		Sex	
		n	%	n	%	n	%
Neonates	124	77	62	5	4	50	40
Juveniles	79	37	51	2	3	11	9
Adults	283	144	50	32	11	29	10

Therefore, in many cases, especially where there has been a lack of laboratory tests other than just the gross post-mortem examination, it has only been possible to record data according to a broad definition within one of the body systems and it has not been possible to determine whether the gross and histopathological observations have defined the cause of death. This was particularly so where histopathology was carried out and multi-systemic lesions were reported.

Despite attempts in Studbook 7 [2] and Studbook 9 [4], it has not been possible to achieve a standard post-mortem report form that would allow data to be presented in a consistent manner. In the absence of both a standard post-mortem procedure and consistent case definitions for each of the pathological conditions described, it is not possible to directly compare the post-mortem reports from the various institutions carrying out the necropsy procedure. Neither, because of the small populations held in a diversity of locations, is it possible to discuss group and population dynamics, however, despite these limitations, the information provided does make a useful contribution to our pathological knowledge of the red pandas held in captivity.

The database records the subspecies identification, however, no discernible differences have been noted in the gross pathology and other laboratory test results between the two subspecies. Consequently, no differentiation between them has been considered for the purposes of this review when describing case histories.

In considering general trends, the pathological findings have been grouped together within body systems, with a series of individual cases of interest presented at the end of the chapter.

NEONATES

The first few days of life for red panda cubs appear to be a critical period. Figure 16.1 shows the age at death for 118 of the 124 neonatal records held for the last 12 years. For six of the animals described as neonates the actual age at death was not given. Although there is considerable variation in the figures given in the literature, the evidence from the period 1994–2006 indicates that 118/201 (59%) of those cubs that die within the first year (124 neonates plus 79 juveniles) do so within the first month of life and, of those, 70 (35%) within the first three days. This is consistent with an earlier report [8]. Thereafter losses were fewer and where a cub reached one month of age then its chances of survival were considerably enhanced, barring traumatic injury.

Records show that 24 (19%) of the 124 animals were either stillborn or died on the same day they were born. Classification of the 16 cubs recorded as stillborn was based on the lack of evidence of any area of lung inflation at post-mortem examination. Two females underwent caesarean section; in one case both cubs died of respiratory failure and only one of twin cubs survived in the other case.

There is a paucity of information about the birth weights of red panda cubs due to the difficulties associated with disturbing the dam and risk to the cubs. In the wild, female red pandas repeatedly move cubs if they are at risk of predation from elsewhere and, in the captive population, some females have been observed continually carrying cubs, presumably in search of a safe nest. There is always the risk that any intervention at the nest may result in the female either abandoning the cubs or injuring them by carrying them in search of an alternative nest site, especially in the early days after birth.

Body weights were recorded for only eight of the 24 cubs that were either stillborn or failed to survive their first day, however, another seven had been partially cannibalized and therefore were unsuitable to be weighed. No reason was given for the absence of body weight data for the other nine cubs. Similarly, there was an unexplained absence of body weight data for 13 of the cubs that died within the next 5 days and where there was no evidence of partial cannibalism. The available body weights are given in Table 16.3.

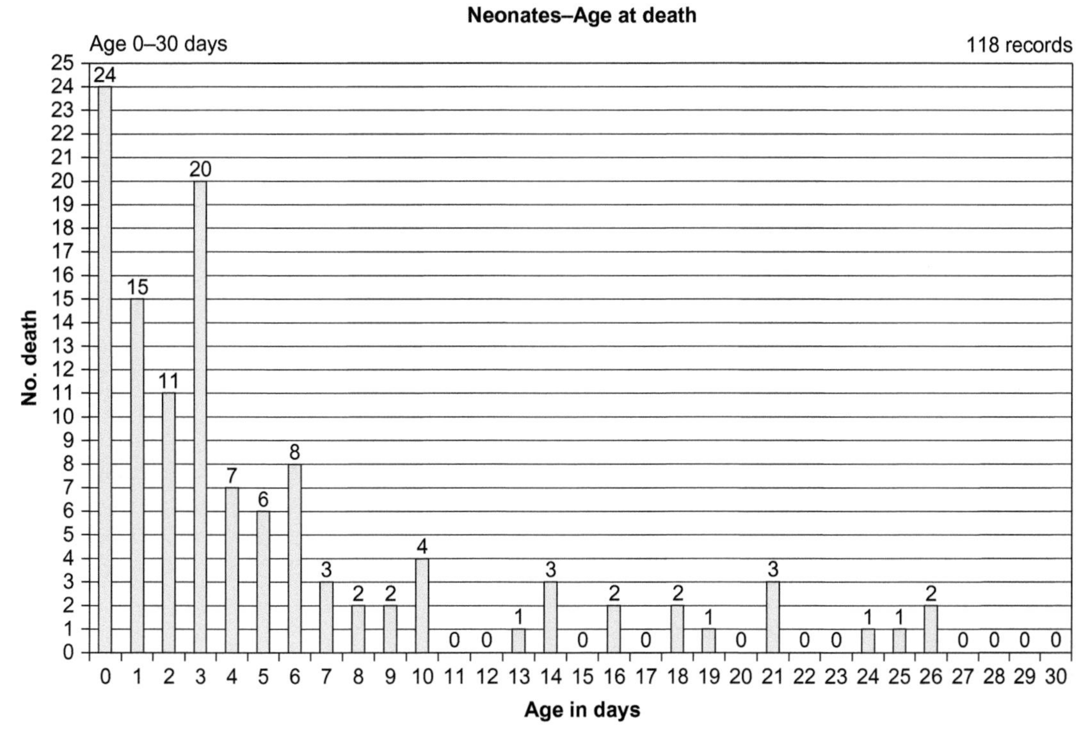

FIGURE 16.1 Age at death for red panda cubs (n = 1 18) who died before 30 days of age, i.e., neonates, during period 1994–2006.

Additional data (A.R. Glatson, personal communication) collected elsewhere over a period of time showed birth weights to fall in the range of 58–150 g (n = 21).

A total of 26 cubs examined out of the 59 (44%) that died between days 1 and 5 had been partially cannibalized and, of these, 13 had evidence of traumatic injuries consistent with bite wounds. These were mainly restricted to the head and neck and had resulted in fractures and attendant haemorrhage. Three cubs, which had survived for a longer period, namely 10–30 days respectively, had injuries which had become infected and led to large abscess formation. In one other case, fungal peritonitis appeared to be associated with a penetrative wound to the abdomen. Although the body weights, recorded at the time of the post-mortem examination, were given for several of these 26 cubs these were not considered suitable for inclusion in the data for Table 16.3.

Within the first 10 days of life the three main causes of death are:

- infanticide often resulting in some degree of cannibalism
- mismothering that leads to starvation and hypothermia
- traumatic injuries consistent with bite wounds.

Early mother–cub bonding is essential for the cub to survive. All those cubs that did not appear to have fed were dead by 5 days after birth and post-mortem examinations

TABLE 16.3 Birth weights for red panda cubs that died within 5 days of birth 1994–2002

Age	No. deaths	Body weight available		
		No (n)	Range (g)	Mean (g)
Day 0 (stillborn)	24	8	52–175	154
Days 1–5	59	20	50–225	101

revealed an absence of milk from the alimentary tract and the complete atrophy of neonatal fat deposits. They can be presumed to have lost weight during that period. Other factors, such as congenital abnormalities, poor sucking reflex of the cub and insufficient milk production by the dam, may also contribute to the high rate of loss of neonates.

There is a single record of a cub with a cyclops congenital abnormality, which survived for 3 days.

In the past there were a number of reports of toxoplasmosis in neonates, however, there have been none in this group since 1994 [7]. However, it is not possible to make any conclusions about this as there is almost no evidence in the post-mortem reports concerning neonates that screening of neonatal brains for evidence of protozoal infection has been carried out during the past 13 years. Although the primary host for toxoplasma is a felid infection, in secondary hosts it can produce abortion or the birth of underweight and weak offspring. In the latter, the suckle-reflex may be poorly developed and this can contribute to malnutrition with consequent hypothermia and starvation.

There were 14 cases of pneumonia, of which five were cases of aspiration pneumonia with evidence of milk globules within the respiratory tract. Four of these five cubs were being artificially reared. Deciding when it is necessary to intervene and attempt hand-rearing is a major dilemma in the management of red pandas. In two of these, nephritis was also present, suggestive of haematogenous spread. Again, prior to 1994, aspergillus pneumonia was evidentially not uncommon and was probably a reflection of the quality of the bedding, but there have been no further reports of this problem since then.

Earlier pathology reviews [1,9] suggested that red pandas are particularly susceptible to infectious diseases, however, that appeared to be based upon a wide range of bacterial species that had been isolated at post-mortem examination from red panda cubs; all of which can be regarded as secondary opportunists. In the past 13 years, there was no evidence of specific bacterial pathogens of particular significance to red pandas.

Somewhat surprisingly, when the problems encountered with other mammalian neonates are considered, there were only two recorded cases of enteritis, cause undetermined, in this age group.

JUVENILES

Figure 16.2 shows the age at death for 72 of the 79 juvenile records held from the last 13 years. No age at death was given for seven cubs described as juveniles.

The literature pre-1994 recorded that traumatic injuries were a significant cause of loss in this age group with a year on year range between 18 and 61%. This no longer appears

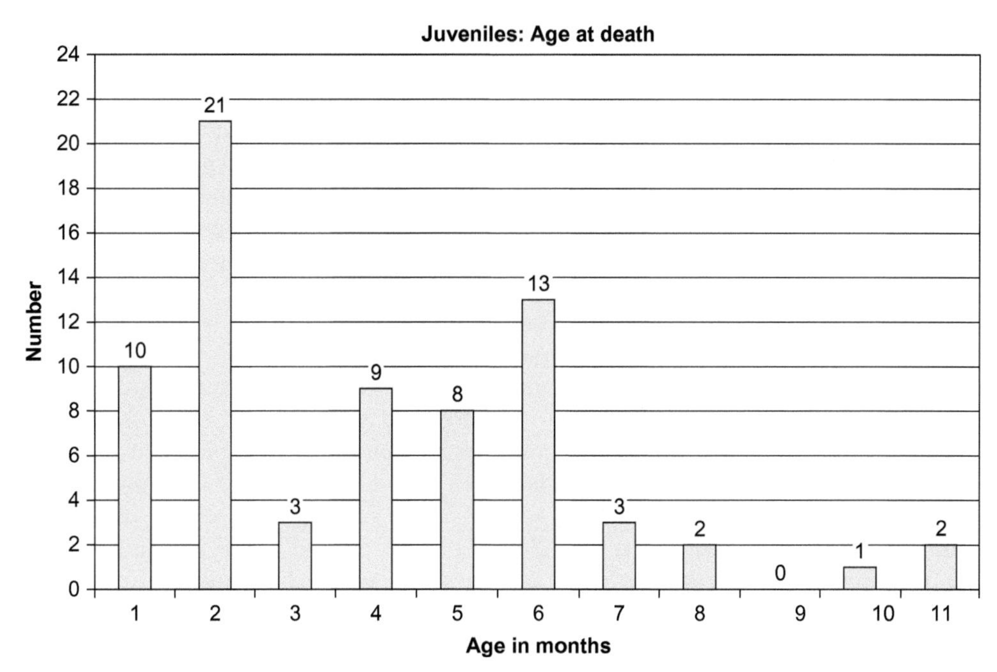

FIGURE 16.2 Age at death for red panda cubs (n = 72) who died between 30 days and one year of age, i.e. juveniles, during period 1994–2006.

to be the case as there were only 6/79 (7.5%) losses due to such injuries in the 13-year period although cubs within this age group may still be subjected to excessive carrying which occasionally results in cervical puncture wounds, abscessation and subsequent septicaemia. Other causes of trauma have been the result of inter- and intraspecies aggression and accidental injuries, such as those that have occurred during attempted capture.

There were regular reports within the early pathology reviews within the studbook that red pandas are prone to gastrointestinal disturbances [3]. Primary causes of death have been attributed to gastritis, erosion and ulceration of the stomach and to enteritis and ulceration of the intestines. It is generally difficult to demonstrate the role of bacterial infection in these cases due to the chronic nature of the lesions, however, *Campylobacter jejuni* biotype 1 was isolated from one 2-month-old cub with enteritis. There was one report of clostridial enteritis in a 3-month-old cub but that was based on the isolation of *Clostridial perfringens* without toxin determination. In other mammalian species, gastric ulceration has been associated with the change from a diet with a low dry matter content such as milk to that with a high dry matter content when animals begin to increase their intake of herbage. Other contributory factors suggested are stress and deficiencies in various vitamins.

Viral enteritis was suspected in a 6-month-old female cub when corona-like viral particles were observed during examination of intestinal content by electron microscopy. The mucosal surface of the jejunum was inflamed and frank blood was present in the intestinal contents.

Although there were sporadic reports of deaths due to rabies and feline infectious peritonitis, in the latter case, contact with infected felids was suspected none have been recorded since the pathology review for 1986/87 [6].

The red panda is susceptible to canine distemper virus and, although there are modified live vaccines that have been developed for use in domestic pets, these are not suitable for use in this species. It has been recognized for almost 30 years that the use of modified live canine distemper vaccines is contraindicated in red pandas [11,12]. Sporadic deaths due to distemper have been recorded throughout the period of the studbook with the latest record being in two 11-month-old cubs in 2000 at the same location. In that case, the cubs had not been vaccinated and infection was considered to be due to contact with infected wild racoons which have previously been found within the premises.

One case noted the difficulty of dealing effectively with flea infestation. Both the parents and a 3-month-old cub were found to be infested and were treated topically. Fleas were still present on the cub one month later and again when it was hospitalized at the age of 5 months. It was found to be severely anaemic and died despite fluid therapy and a blood transfusion. Death due to weight loss and anaemia of unknown cause has also been reported several times elsewhere in juvenile red pandas at the time of weaning.

Two other unusual cases were where a 6-month-old cub died of asphyxiation due to tracheal obstruction by a piece of fruit and elsewhere when a 3-month-old cub died suddenly and was thought to have eaten the berries of yew (*Taxus baccata*).

ADULTS

Figure 16.3 shows the age at death for 256 of the 283 adult records held for the 13-year period between 1994 and 2006. The remaining 27 animals were only described as adults of an unspecified age.

Over the past 30 years, there has been a changing pattern in disease problems seen at post-mortem examination of adult red pandas. In the late 1970s and early 1980s, the majority of captive-held red pandas had been recently caught from the wild and dietary changes were frequently implicated in the hepatic and renal problems that were widely reported as the cause of early death at that time. The most commonly reported problems described were fatty degeneration and hepatitis of an unspecified nature. Whereas wild red pandas which feed mainly, although not exclusively, on a diet of bamboo would have a relatively low-fat diet, captive red pandas in the past received an artificial diet which was highly digestible, low in fibre and high in fat [13].

Since those days much work has been carried out with regard to improving diets and, in the years 1994–2006, only seven out of the 283 (2%) adult records have specifically reported hepatic lipidosis as a suspected contributor to the death of the animal. Of those seven, both age and body weight were given for only three animals and they were a 12-year-old that weighed 6.2 kg, and two 13-year-old adults at different locations that weighed 3.4 kg and 4.0 kg, respectively. When compared to the mean body weight of 4.8 kg, which is derived from the records for 144 adult animals where the body weight was given, the 12-year-old animal would not be considered excessively obese and the other two were below both the mean body weight of animals in captivity and comparable

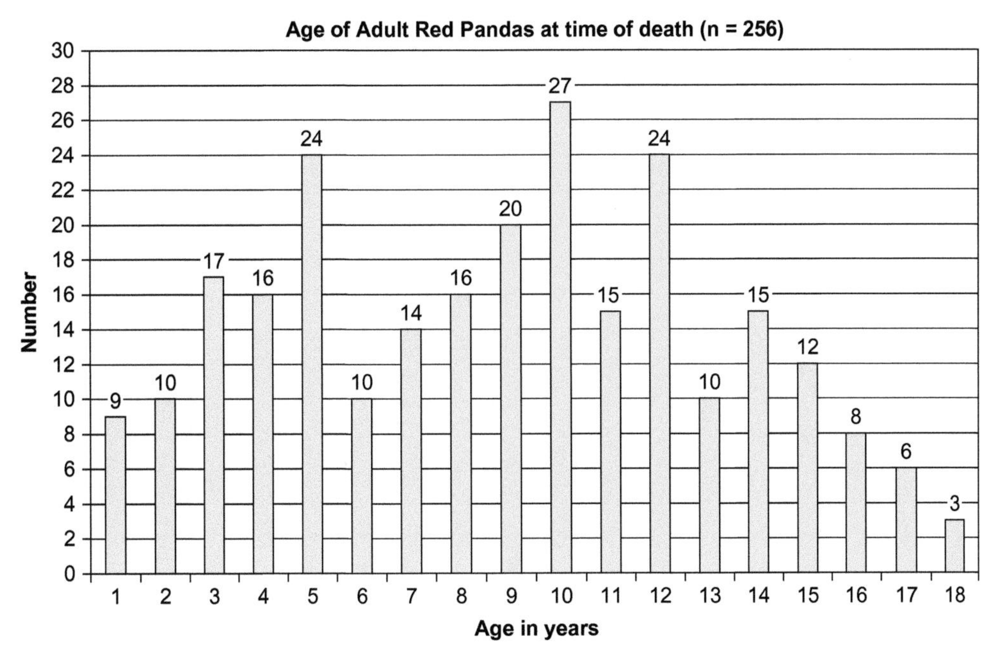

FIGURE 16.3 Age at death for adult red pandas (n = 256) who died during the period 1994–2006.

with those body weights observed in the wild population. With regard to size differential between Aff and Afs, there are insufficient data to demonstrate that adequately as only 21/47 (45%) adult Afs pathology reports provided the weight at death. The weights recorded for 123 Aff and 21 Afs adults are given in Figure 16.4.

In reviewing the pathology findings of any species the influence of diet must be seriously considered and this is of special interest where the red panda is concerned because of the nature of its diet in the wild compared to that available in captivity. While officially classified under the order Carnivora, observations in the wild show that it does in fact eat large quantities of bamboo (possibly 65% of total diet) and that it is only an opportunistic feeder in respect of taking animal protein. These observations are all the more interesting given that the red pandas possess a simple short 'cat-like' intestinal tract which lacks any of the cellulose-splitting adaptations seen in other plant-eating mammals. The lack of such adaptations, together with a short gut length, results in a relatively rapid passage time, poor digestibility and hence a reduced energy intake. Some studies may suggest that red pandas may compensate by eating bamboo at a rate almost equivalent to their own body weight (3.5–4 kg) daily. While increasing food throughput may be possible with a short gut passage time, the red panda's main defence against the extremes of climate would seem to be the efficient control of energy expenditure. While the red panda possesses a dense pelt, it also possesses the ability to reduce its metabolic rate without decreasing its core body temperature. Skin temperature, however, does decrease with the change in metabolism and temperatures of 15°C have been recorded. Other attributes, like low

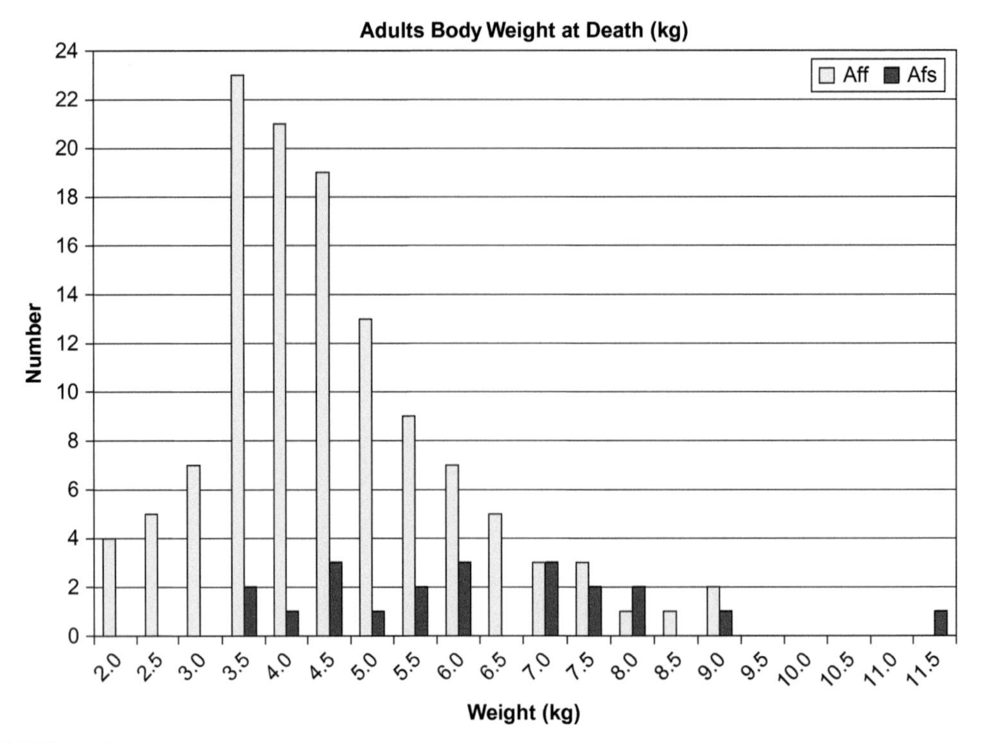

FIGURE 16.4 Weight range of adult red pandas (n = 144) at the time of death (in kg to nearest 0.5 kg), 1994–2006 (mean = 4.8 kg).

fecundity, prolonged gestation and lengthy post-natal development, are all symptomatic of an animal on a nutritional knife edge [14].

In captivity, it is difficult to completely replicate the diet available in the wild. While some zoos do provide their animals with bamboo, most use food items that are low in fibre and high in energy. The long-term use of these products, especially those that have been artificially sweetened to improve acceptance, have been implicated in a high incidence of obesity, and in gastric and hepatic problems. However, the true impact of such diets on the health of red pandas still needs to be fully evaluated, especially from a clinical standpoint. While post-mortem results have been regularly reviewed in the International Studbook for many years, a clinical review has never been accomplished. On a more positive note, the better understanding of red panda biology has resulted in several diets being specially formulated to improve fibre intake and reduce the fat/carbohydrate levels.

In the period prior to 1994, several pathology reviews within the studbook highlighted periodontal disease leading to gingivitis, abscessation and resultant loss of body condition due to feeding difficulties as a significant cause of the loss of adult red pandas [1,3]. It would appear that as a result of significant dietary improvements there has been a much reduced incidence in recent years in that since 1994 there are only seven records citing dental and jaw problems and all were in animals aged 9 years or greater. Two of these, aged 13 and 15 respectively, had unilateral mandibular osteomyelitis.

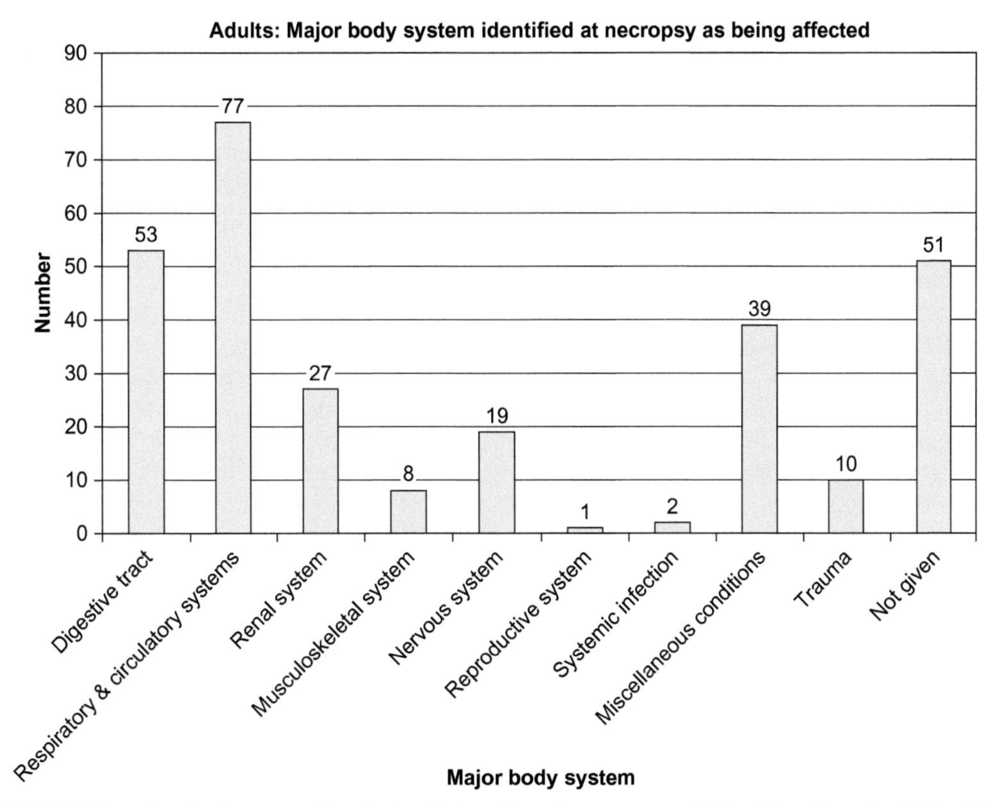

FIGURE 16.5 Major body system identified as being affected at necropsy of 236 adult red pandas (1994–2006).

In one interesting case, where bacterial cultures were carried out on material from a molar abscess, *Streptococcus milleri* was isolated. In other species this organism has been found in the mouth of healthy individuals, however, it may also be isolated from abscesses in various parts of the body and some strains have been associated with dental caries.

In the last 13 years, 93/486 (19%) of the records referred to animals that had survived beyond 10 years of age and it was evident that more chronic conditions had become apparent than were reported previously. Although there was considerable variation in the protocol and the level of further laboratory tests employed at each necropsy, it was possible to identify the major organ system affected in 236 of the 283 (83%) reports for adults, with the caveat that it was not possible to comment as to whether these findings were the specific cause of death due to the reasons described earlier. The primary body system identified in the reports is given in Figure 16.5.

In particular, there has been an increase in the number of necropsy reports noting chronic disease conditions associated with cardiovascular and renal systems, especially in animals considered obese. Cardiomyopathy was noted in 27/283 (9.5%) adults, of which nine of these animals weighed between 5.5 and 9.1 kg, considerably in excess of the mean weight (4.8 kg) of the 144 adults for which records are held. Another 20 adults (7%) exhibited multi-organ failure associated with congestive heart failure.

Renal disease was recorded in a different 27 (9.4%) adults and several of those, where histopathlogy had been carried out, recorded "end-stage" renal lesions similar to those seen in felids. Reference to the kidneys of red pandas appearing to resemble those of cats was made in an earlier contribution to the pathology of the red panda [9].

Over the past 13 years, there have been several reports from premises in warm-temperate and tropical climates of infection with the heartworm *Dirofilaria immitis*. A 6-year-old female, with alopecia of legs and trunk, died suddenly as the result of gastric torsion. The post-mortem examination also revealed the presence of pulmonary obstruction due to the presence of significant numbers of *D. immitis*. Elsewhere, a 10-year-old female died 4 days after surgery to remove an intestinal obstruction despite having made a good recovery from anaesthesia. Necropsy revealed a dilated right heart ventricle containing a large heartworm, a fungal tracheitis and pneumonia. Adult heartworms were also found in a 3-year-old male that became increasingly anorexic over a 6-week period. This animal had a history of intermittent diarrhoea which resolved without treatment. Post-mortem examination revealed a streptococcal pneumonia and heartworms obstructing blood flow. These were present despite treatment with avermectin one month prior to death. Although heartworm is principally an infection of dogs and cats, other canids, felids and some mustelids have also been identified as true hosts. In the USA, bears, raccoons, beavers and humans have been identified as aberrant hosts in which the parasite does not complete its life cycle. Although avermectins are used to prevent infection providing they are given at an appropriate dose on a regular basis they are not adulticidal. In the USA, organic arsenical compounds are used for the treatment of adult heartworm.

In Western Europe, there were several incidents where the canid lungworm, *Crenosoma* spp., was identified and also one incident involving *Angiostrongylus vasorum*. In all these cases of lungworm infestation, where histopathological examination was carried out, there was right-sided heart failure with both auricular and ventricular dilatation, passive congestion of the liver and a secondary glomerulonephropathy due to the chronic deposition of antibody/antigenic complexes.

In 1998, after a 4-year period when there had been no reports of deaths in adults due to canine distemper, three adult males and two adult females at the same location died over a 3-month period. The clinical symptoms of conjunctivitis, erosive glossitis and palatitis, mild crusting dermatitis of the skin on the muzzle, mild gastroenteritis and mild, but focally extensive, pneumonia were common to several of the affected animals. Since then there have been three more incidents involving a 7-year-old, an 8-year-old and a 13-year-old at different locations. All of these cases were reported from the USA.

During this review of the post-mortem reports of red panda deaths there were several interesting cases that deserve a special mention; these are detailed below.

Respiratory distress, cyanosis and depression were displayed by a 3-year-old male. Despite treatment, including the withdrawal of 300 ml thoracic fluid, the animal died the following day. Post-mortem examination revealed the presence of abscessation in lungs, mediastinal lymph node, liver, pancreas and kidneys. Histopathological examination revealed the presence of multiple fungal pyogranulomas thereby demonstrating a disseminated fungal infection. The morphology of the fungal hyphae appeared typical of *Aspergillus* spp. The precipitating or associated factors in this case are unknown, however, in other species, there is usually some underlying factor such as immunosuppression or other disease process.

An 11-year-old female was found one morning to be lethargic with laboured breathing. Radiography of the chest indicated small lung fields and a chest tap removed 350–400 ml of a reddish-brown fluid. This resulted in an irreducible pneumothorax and the decision was taken to euthanase the animal. Gross examination of the thorax revealed a fibrinous pleurisy with small milliary tan-coloured foci on the caudal pleura. The fibrinosupperative pleuritis was due to a bacterial infection and *Actinomyces* spp. was cultured from the thoracic fluid. This red panda also had a low-grade encephalitis associated with a protozoal infection, presumed, but not confirmed, to be toxoplasmosis.

A 5-year-old animal died suddenly during the night and gross pathological examination revealed multiple ulcerations of the gastric mucosa, severe congestion of the jejunum and congestion of the liver. Bilateral nephritis was confirmed histologically. *Clostridium perfringens* and *Clostridium sordellii* were isolated in cultures from intestinal fluid, liver and kidney. The latter of these has been increasingly reported as being associated with acute clostridiosis and sudden death in a wide range of species in recent years.

An 11-year-old female was one of three adults to develop multiple periarticular exostoses in their elbow joints. Although the dietary concentrations of calcium and phosphorus were considered to be adequate, the dietary levels of vitamins A and D appeared excessive and hypervitaminosis A was suspected. No new cases were seen following dietary adjustment.

A 15-year-old female in poor body condition had a severe, locally extensive, chronic pyogranulomatous and granulating osteomyelitis of the left mandible. There was soft tissue swelling over the ventral aspect of the left mandible with three distinct discharging sinuses. In the absence of any neoplastic process this was presumed to be a sequel to earlier periodontal disease.

There have been few reports of neoplastic conditions. One 14-year-old male panda developed a lesion on the right hind foot which progressed to an extensive area of coalescing ulcers revealing friable, nodular red tissue. The animal was euthanased on welfare grounds. The lesion was confirmed histologically as a squamous cell carcinoma. Elsewhere, hepatocarcinoma was diagnosed in a 14-year-old female that had lost weight and only weighed 3.6 kg at death. The liver had a swollen, irregular surface with multiple firm nodules and there were metastases in the mesenteric lymph nodes. An incidental finding was ankylosis of the lumbar and sacral vertebrae.

Two 10-year-old red pandas, one male and one female, at the same location died within 3 months of each other due to *Haemobartonella* infection. The problem was suspected when the male developed a progressive anaemia and died. Histological examination of kidney tissue revealed iron storage in the renal tubules. Clinical disease was then diagnosed in the female and she died as the result of heart failure. Both heart ventricles were dilated and there was a generalized haemosiderosis. As fleas are the intermediary host for *Haemobartonella*, it is possible that the juvenile animal reported earlier died of a similar infection.

A 12-year-old female developed terminal toxaemia as the result of a pyometra. Although regular reviews of the pathology of red pandas have been included in "The Red or Lesser Panda Studbook" since it was introduced in 1978 this was the first report of this condition [5].

Three toxicological cases were reported in the last 13 years. In the first, a 5-year-old female and her cub gained access to chlorine fluid, presumed to be sodium hypochlorite, and shortly thereafter developed acute pulmonary oedema as the result of chlorine toxicosis. The dam died within 5 hours of exposure, whereas the cub, presumably less exposed, survived for one week but eventually died of a severe suppurative bronchopneumonia. In the second incident, two 7-year-old animals, one male and the other a pregnant female, died suddenly. Both were in good body condition and no pathological lesions were identified at necropsy. The two animals had been kept on an island surrounded by tidal salt/brackish water depending upon the rainfall. The female, at least, had been killing and eating waterfowl and crows. Suggested causes of death were either paralytic shellfish poisoning, there being an outbreak in the nearby harbour about the same time and which received the same water, or botulism. There were three municipal dumps within four miles of the premises and the crows were known to commute between both places.

In the final case, pest control personnel placed a fumigant rodenticide, based on aluminium phosphide, into several rodent holes and other holes created by prairie dogs, which had occupied the exhibit prior to the red pandas. The following morning both adult male pandas were found dead on the ground. Abnormalities noted at gross post-mortem examination included cyanosis of mucous membranes and lungs, generalized dark venous congestion and aspiration of fluid gastric contents into the trachea and bronchi. Histopathological findings were consistent with a system insult by a toxic agent and examination of stomach contents and tissue samples by gas chromatography – mass spectrometric methods – produced results consistent with phosphide toxicosis [15,16]. Although the rodenticide was placed in the holes at a depth greater than that considered retrievable by a red panda, further investigation revealed that the product had been on the shelf for some time and some pellet fragmentation had taken place thereby allowing access to product debris.

SUGGESTIONS FOR THE FUTURE

The preceding pages provide a brief summary of the major trends in the health of red pandas over the past 25 years. Such a summary can be valuable in reporting the changes that have taken place as the red panda population has changed from one of primarily wild-captured animals to that of a zoo-born population and thereby highlighting current health and welfare issues.

One of the major changes is that of the increasing longevity and relative high proportion of obesity, and associated health problems, in the captive population. The limited data that have been collected previously suggest that wild adult red pandas weigh between 4.0 and 5.0 kg with few animals reaching the upper limit. In the last 10 years, 14.6% of those red pandas to die in captivity weighed in excess of 5.5 kg and some of these had various problems which had probably reduced their feed intake prior to death therefore suggesting that they had been even heavier when in good health. This suggests that it would be beneficial to weigh captive red pandas at every opportunity in order to monitor their weight and also to increase their need for exercise by reducing the amount of food immediately available to them and increasing activity behaviour through their need to forage for food using enclosure enrichment strategies.

Although considerable advancement has taken place regarding both the veterinary and management matters during the life of the studbook, there are still several areas worthy of further exploration. It would be helpful if all investigations and projects were notified to the studbook keeper so that all interested parties were kept informed of on-going work. Some veterinary areas suggested as being worthy of consideration on a national or regional basis are:

- Coordinated questionnaires on clinical health parameters as animals age, e.g., body weight, teeth examination, pelt condition, ectoparasites
- Development of non-invasive health-screening protocols, especially endoparasites of respiratory and enteric systems
- Setting "normal" values for each of the haematological and biochemical parameters; possibly by using a common laboratory within specific geographical areas.

Care must always be taken when interpreting the results of haematological and biochemical tests by comparing them against published data. Although a great number of laboratories are able to carry out these examinations, it is important to note under what conditions such examinations were carried out before comparing the results from one laboratory with those from another. It is because the conditions of examination vary so much and there is only a small population of animals to be tested that it is difficult to establish a range of "normal" values for each of the haematological and biochemical parameters.

As indicated at the beginning of this chapter, there has been considerable variation in the quality of the reports associated with each post-mortem examination and subsequent laboratory tests. As a consequence, it has often not been possible to do other than to loosely identify the cause of death from the limited information provided. The ideal situation would be to provide a report with the following information:

- Basic data: location, studbook number, age, sex, weight (in kg)
- Gross pathological findings
- Final results of all laboratory tests: bacteriology, parasitology, haematology, serology and histopathology.

Although there have been two previous attempts, via the studbook (Studbook 7 [2] and Studbook 9 [4]), to introduce and encourage the use of a standard report for reporting pathology findings each of those attempts has failed and experience suggests that due to the wide variety of organizations involved in investigating the deaths of red pandas that further attempts to introduce a standard report are unlikely to be successful.

However, if the compiler of the pathology report is to have access to the best possible information in order to make a meaningful contribution towards explaining both the trends and oddities involved in red panda deaths then it requires the cooperation of those involved in keeping them to provide copies of all the available information as indicated above to the keeper of the studbook. Although the report is only compiled on a biannual basis it would be helpful to send those reports as soon as possible so that the compiler can follow up especially interesting cases while matters are fresh in the mind.

With the increasing use of digital photography it would be fairly straightforward and particularly useful to accompany the reports with appropriate images. These would, of course, remain the intellectual property of the person responsible for initiating them and if permission were given for them to be incorporated into the pathology review then they would be correctly attributed according to standard publishing protocols.

ACKNOWLEDGEMENTS

This report has only been possible due to the many contributors of post-mortem reports over the last 25 years and to those who carried out the pathology reviews published in the Red Panda Studbook prior to 1994.

References

[1] N. Lateur, Pathology survey of captive red pandas, in: A.R. Glatston (Ed.), Red or Lesser Panda Studbook No. 4, Stichting Koninklijke Rotterdamse Diergaarde, Rotterdam, The Netherlands, 1987, pp. 19–27.

[2] J. Lubbert, W. Schaftenaar, A.R. Glatston, A review of the pathology of the red panda, *Ailurus fulgens fulgens*, in the period 1982–1991, in: A.R. Glatston (Ed.), Red or Lesser Panda Studbook No. 7, Stichting Koninklijke Rotterdamse Diergaarde, Rotterdam, The Netherlands, 1993, pp. 10–19.

[3] K.L. Machin, Red panda pathology, in: A.R. Glatston (Ed.), Red or Lesser Panda Studbook No. 6, Stichting Koninklijke Rotterdamse Diergaarde, Rotterdam, The Netherlands, 1991, pp. 2–24.

[4] B.E. Preece, Review of the pathology of the red panda, 1994–95, in: A.R. Glatston (Ed.), Red or Lesser Panda Studbook No. 9, Stichting Koninklijke Rotterdamse Diergaarde, Rotterdam, The Netherlands, 1995, pp. 10–19.

[5] B.E. Preece, Review of the pathology of the red panda, 2000–01, in: A.R. Glatston (Ed.), Red or Lesser Panda Studbook No. 12, Stichting Koninklijke Rotterdamse Diergaarde, Rotterdam, The Netherlands, 2002.

[6] W. Schaftenaar, Review of pathological findings in the red panda, 1986–87, in: A.R. Glatston (Ed.), Red or Lesser Panda Studbook No. 5, Stichting Koninklijke Rotterdamse Diergaarde, Rotterdam, The Netherlands, 1989, pp. 7–13.

[7] W. Schaftenaar, A.R. Glatston, Patterns of pathological findings on the red panda, in: A.R. Glatston (Ed.), Red or Lesser Panda Studbook No. 8, Stichting Koninklijke Rotterdamse Diergaarde, Rotterdam, The Netherlands, 1994, pp. 37–49.

[8] L. Yinghong, Infant mortality in the red panda (*Ailurus fulgens*), in: A.R. Glatston (Ed.), Red or Lesser Panda Studbook No. 8, Stichting Koninklijke Rotterdamse Diergaarde, Rotterdam, The Netherlands, 1994, pp. 18–35.

[9] P. Zwart, Contribution to the pathology of the red panda (*Ailurus fulgens*), in: A.R. Glatston (Ed.), Red Panda Biology, SPB Academic Publishing, The Hague, The Netherlands, 1989, pp. 25–29.

[10] M.F. Mörzer Bruyns, The importance of cooperation between zoos and field ecologists, in: A.R. Glatston (Ed.), Red Panda Biology, SPB Academic Publishing, The Hague, The Netherlands, 1989, pp. 73–78.

[11] Anon, Susceptibility of the lesser panda to canine distemper, Internatl. Zoo Yearbook 2 (1960) 107.

[12] M. Bush, R.J. Montali, D. Brownstein, J. Everette, Vaccine induced canine distemper in a lesser panda, J. Am. Vet. Med. Assoc. 169 (1976) 959.

[13] R.A. Kock, P. Spala, J.K. Eva, P. Bircher, M. Ricketts, M. Stevenson, New ideas for red panda diets, in: A.R. Glatston (Ed.), Red Panda Biology, SPB Academic Publishing, The Hague, The Netherlands, 1989, pp. 57–71.

[14] B.K. McNab, Energy expenditure in the red panda, in: A.R. Glatston (Ed.), Red Panda Biology, SPB Academic Publishing, The Hague, The Netherlands, 1989, pp. 73–78.

[15] S. Gupta, S.K. Ahlawat, Aluminium phosphide poisoning – a review, Clin. Toxicol. 33 (1) (1995) 19–24.

[16] K.H. Plumlee, Toxicant use in the zoo environment, J. Zoo Wildlife Med. 28 (1) (1997) 20–27.

CHAPTER

17

Red Pandas in Zoos Today; The History of the Current Captive Population

Angela R. Glatston
Rotterdam Zoo, The Netherlands

OUTLINE

The zoo populations of the two subspecies of red panda seem to be healthy and their future relatively secure. However, as appearances can be deceptive, it was considered to be important to evaluate the current populations in more detail in order to determine if they are indeed safe as they would seem. In this chapter, we investigate the status of the current zoo populations in detail with a view to establishing their long-term viability. However, in order to appreciate how the current captive populations are likely to behave, it is an advantage to have an understanding of their history and development.

Red pandas of the nominate form, *Ailurus fulgens fulgens*, have been maintained in western zoos since 1869 which was the date when the first sorry-looking specimen arrived at London Zoo from Calcutta. The story of their history in captivity has generally not been one of success. It is safe to say that the majority of red pandas that were exported from the wild and arrived in western zoos prior to 1970 did not survive very long. Furthermore, very few of them managed to breed and even those which did so have left no descendants in today's zoo

Red Panda. DOI: 10.1016/B978-1-4377-7813-7.00017-3

population; the oldest founder of the current *A. f. fulgens* zoo population was imported in 1966. The only exception to this fairly bleak scenario was the group of red pandas which were maintained in San Diego Zoo in the 1940s. These were the first red pandas to breed in western zoos with any degree of regularity. The success of this group is discussed by Marvin Jones [1] (see Chapter 12). However, it is worthwhile noting that even this successful breeding group has no descendants surviving today. Most of the red pandas born in San Diego Zoo during the 1940s were sent to other animal collections in North America. Unfortunately, none of the zoos were as successful with them as was San Diego. Even those red pandas which were allowed to remain in San Diego eventually failed to continue breeding; the line gradually died out and was extinct by the mid-1950s. The breeding of red pandas in other zoos remained sporadic at best until well into the 1970s and it was not until the late 1980s or early 1990s that red panda zoo births can be said to have become reasonably commonplace.

It is possible to follow the development of the captive population in considerable detail because all data on ancestry, location and dates of birth and death have been recorded for all red pandas living in zoos since 1978. These data are held in the International Red Panda Studbook which was initiated in 1978 and is maintained by Rotterdam Zoo. Prior to 1978, data for some zoo red pandas were available via the International Species Inventory System (ISIS). These data, which are now integrated into the studbook, provide an insight into the early development of the zoo population.

The current zoo population of *Ailurus fulgens fulgens* can largely trace its origins back to animals which were imported from the wild in the late 1960s and 1970s. During this period, the studbook lists more than 100 red pandas which were exported from Asia to western zoos. However, this number probably only represents the tip of the iceberg. Many zoos, at that time, did not keep detailed records on their animal stocks and, even those which did so, have not necessarily been able, or willing, to provide complete information to the studbook or to ISIS. In the first years of the studbook, data were often collected several years after the red pandas in question arrived at a particular zoo. It is therefore not surprising that those red pandas which had died, without breeding, in the intervening years were forgotten and therefore often omitted from the zoo's report to the studbook. In this way, it is fairly certain that many red pandas were exported from the wild and sent to live out their lives in zoos where they left no record of their passing. In addition, many of the wild-caught red pandas for which records are available left no descendants in the current population. It is highly probable that only a very small percentage of red pandas that were exported to zoos in the 1960s and 1970s managed to breed in captivity at all.

In the same way, today's zoo population of Styan's pandas (*Ailurus fulgens styani*) largely has its roots in those animals which were exported from China in the late 1980s and early 1990s. This trade route was only eventually closed in 1994 when the red panda was included onto Appendix I of CITES (International Convention on the Trade in Endangered Species). This legal instrument effectively forbids the trade of wild-caught and F1 generation specimens of those species listed in Appendix I. The inclusion of the red panda onto Appendix I effectively put a stop to their export from China at that time. The studbook has records of 124 Styan's pandas which were exported from China in the 1980s and early 1990s. Again this could also just represent the tip of the iceberg of these panda exports; animals may have been sent to zoos not participating in the studbook or ISIS. Many of the Styan's pandas arriving in zoos at this time also failed to breed and so have no descendants in the current zoo population. However, a greater percentage of the wild-caught Styan's pandas managed to breed than did of the original *A. f. fulgens* exports.

The first international red panda studbook [2] published in 1980, reported that there were 128 red pandas (including three *A. f. styani*) living in zoos around the world as of 1 January 1978. This figure was later shown to be an underestimate; we now know of 146 *A. f. fulgens* and five *A. f. styani* kept in zoos at that time. However, even these figures are probably lower than the true number of red pandas in zoos at that time for the reasons outlined earlier. In addition, no records are available about the red pandas held in Chinese zoological collections or wildlife rescue centres, a situation which has not improved with time. There was a brief period in the 1990s when the Chinese zoos did provide information to the international studbook. However, even at the time the Chinese data were far from complete. Kati Loeffler (see Chapter 18) has provided a review of the current situation of captive red pandas in Chinese zoos.

By the end of 2006, the studbook population of red pandas numbered 738 individuals. The *A. f. fulgens* population had increased from 146 in 1978 to 456 at the end of 2006, while the *A. f. styani* population had increased from five to 282 in the same period. Most of the increase in the *A. f. fulgens* population was due to captive breeding; only 14 imports from the wild have been recorded since 1978. However, the converse it true for *A. f. styani*; much of the growth in this population has been due to the continued survival of many of the 124 animals exported from China.

Given their numbers, both of these populations seem to be fairly healthy and not in any imminent risk of dying out. However, the health of a population cannot be ascertained solely in terms of numbers, other factors such as demographic stability, genetic diversity and levels of inbreeding are also of importance when considering the health of a population. Also, if a population is not well managed, its status can change fairly rapidly so there is never a good time to be complacent about the future of a captive population of an endangered species. The following sections review the changes which the zoo populations of red pandas have undergone in the past 30 years and demonstrate how quickly the future of a captive population can change in either a positive or negative direction.

THE DEVELOPMENT OF *AILURUS FULGENS FULGENS* POPULATION

When the first International Red Panda Studbook was published in 1980, the future of the zoo population of red pandas looked bleak; the population consisted largely of wild-caught individuals of unknown age, few animals were breeding, many of the young that were born failed to survive and the age pyramid of the population (Figure 17.1), an indicator of demographic viability, was decidedly unstable. An analysis of the population based on the Leslie Matrix and simulations using Monte Carlo methodology were published in the first studbook [3]. It concluded that the population was destined for extinction. However, it also emphasized that these conclusions were based on the analysis of a very small population and that there were therefore insufficient data for accurate predictions. The zoo population of *A. f. fulgens* has been analysed three times since then: in 1988 (data up to 31 December 1983), 1993 (data up to 31 December 1992) and 2005 (data up to 31 December 2003). An overview of these analyses is presented in Table 17.1.

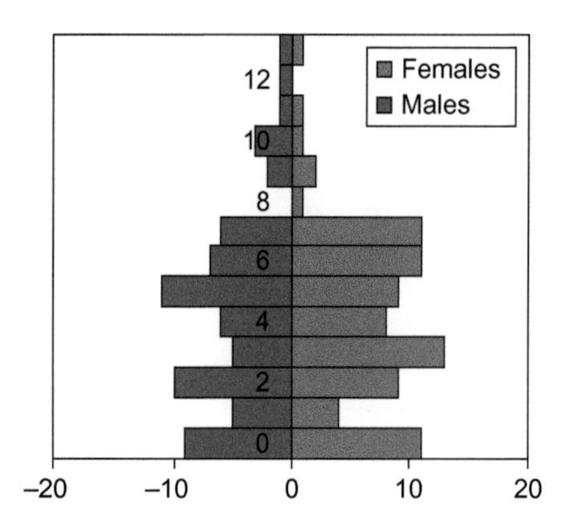

FIGURE 17.1 The age pyramid of the *A. f. fulgens* population in 1978.

TABLE 17.1 Red panda demographics/genetic 1978–2004 from publications

	As of 31-12-1979	As of 31-12-1983	As of 31-12-1992	As of 31-12-2003
Population size	140	144	212	476
% wild caught	78% (01-01-1978)	22%	<1%	<1%
Annual growth rate	0.75	0.93	1.085	1.05
30-day infant mortality				27.2
Number of founders	25*	25	26	28
Genetic diversity			0.954	0.9475
Inbreeding	0.024	0.01	0.026	0.047

*From Zoo Biology *article [4].*

The analysis of the 1983 population was published in an article appearing in *Zoo Biology* [4]. It concluded that the population had not really grown significantly in the 5 years since the first analysis; however, it noted that the percentage of captive-born individuals in the population had increased from 22% to 78%. Furthermore, although λ (the annual rate of change) had improved since 1979, it was still below 1.00, confirming that the population was not growing. The article concluded that the future of the population still seemed uncertain. Indeed, over the next few years it looked as though the red panda zoo population was not going to survive; the number of pandas in captivity continued to hover at about the same level; increasing one year and then decreasing the next (Figure 17.2). In fact, it was not until the late 1980s that a breakthrough seems to have occurred; the number of births and/or the number of infants surviving increased substantially and exceeded the number of adult deaths. This improvement can be correlated with two events: first, the establishment of officially endorsed coordinated breeding programmes in North America (SSP), Europe (EEP) in 1984 and 1986 respectively and shortly thereafter one in Australia (ASMP) and, secondly, the publication of minimum husbandry and management guidelines in 1988 [5] which were later revised in 1993 [6].

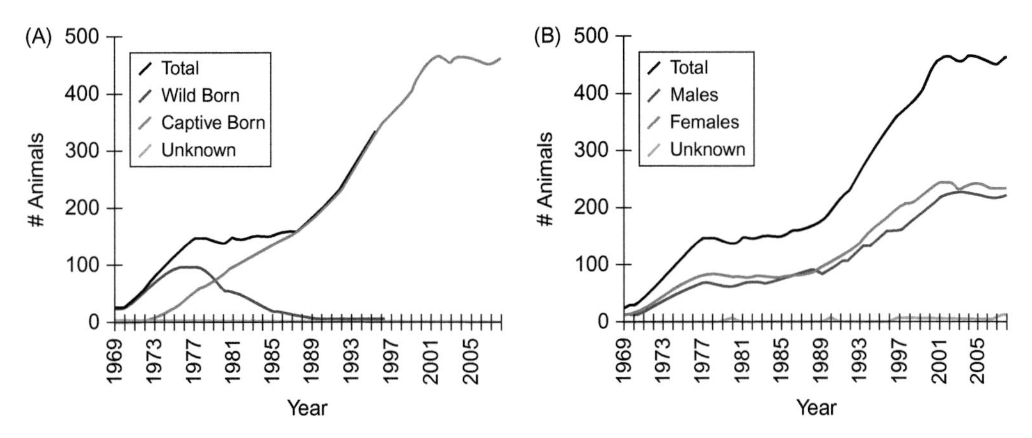

FIGURE 17.2 Changes in the population of *A. f. fulgens* held in zoos around the world over the studbook period: (A) the numbers of wild-caught and captive-born individuals; (B) the numbers of males and females in the population.

By the time that the population was analysed for a third time in 1993 [7], the situation had changed dramatically. At the end of 1991, the *A. f. fulgens* zoo population numbered 212, a substantial increase over the 144 at the end of 1983. In addition, the population was composed almost entirely of captive-born individuals (see Figure 17.2), the age pyramid was more stable and λ was above 1.00 indicating clear growth. This growth started post 1987 (see Figure 17.2). The analyses tended to indicate that the red panda zoo population of that time was demographically secure. This conclusion was confirmed by the results of the fourth analysis in 2004 [8]. At the end of 2003, the captive population of *A. f. fulgens* numbered 476 individuals, a number which exceeded the carrying capacity of 390 planned in the first Global Masterplan [7], and the age pyramid looked healthy (Figure 17.3). However, analysis revealed that λ, the annual rate of change, was lower in 2003 than it had been 1993. It was not clear if this reduction in growth rate was the result of a management decision to control growth after the planned carrying capacity was exceeded, or the inadvertent consequence of changes in husbandry and management or due to complacency on the part of population managers who believed the zoo population was secure. Nevertheless, even with the decline in λ, the future of the population was considered to be very healthy by the report of the second Global Masterplanning session.

Zoo populations of red pandas are managed at a regional rather than at a global level. The majority of the *A. f. fulgens* zoo population is housed in Europe and North America with smaller groups held in Australia, India and elsewhere. An overview of the demographic and genetic analyses of the regional populations is presented in Table 17.2. Regional data were not analysed in the 1979 or 1988 publications. All regional populations increased in size between 1992 and 2003 (Figure 17.4). However, the greatest increase was seen in the European population where the actual population in 2003 far outstripped the planned carrying capacity of 180 places discussed in the 1993 masterplan [7]. The Australian population more or less developed as planned during this period. However, the North American population was substantially lower than its intended capacity of 230

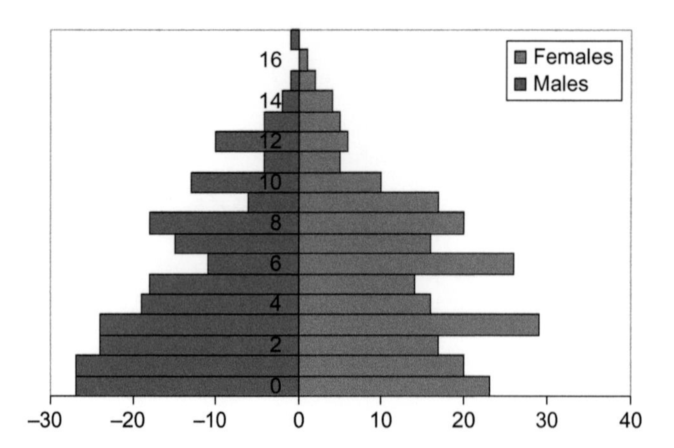

FIGURE 17.3 Age sex pyramid of the *A. f. fulgens* zoo population on 31 December 2003.

TABLE 17.2 Regional red panda population genetics and demographics for the three larger populations from publications

	SSP		EEP		ASMP	
	1992	**2003**	**1992**	**2003**	**1992**	**2003**
Population size	90	105	110	221	30	44
Annual rate of growth	1.064/1.053	1.01/0.99	1.088/1.86	1.05/1.06	1.09/0.85	1.28/1.15
30-day infant mortality		40.2		23.6		10.6
Number of founders	16	23	23	26	18	21
Gene diversity	0.9062	0.908	0.9428	0.9247	0.9053	0.8692
Inbreeding		0.043		0.042		0.070

red pandas of which 160 were to be *A. f. fulgens*. The annual rate of change declined between 1992 and 2003 in Europe and North America but increased in Australia. In Europe, this was due to a concerted effort to reduce population growth in response to the population exceeding the planned carrying capacity. In 2000, the red panda EEP introduced a policy to delay first reproduction until the animals were 5 years old. All young animals were placed in same sex grouping for 2–3 years on reaching sexual maturity to reduce population growth.

The genetics of the *A. f. fulgens* population has also changed over the years. Although no genetic analysis of the population was conducted in 1979, the 1988 publication reviewed the number of founders contributing to the population in 1979 and in 1983. In both cases this was 25, although the percentage contribution of these founders changed quite markedly between the two transects; individual founder contributions had become significantly less equal by 1983. Referring to Table 17.1, it seems as though the population recruited three new founders over the years; by 1992, the number of founders had increased to 26 and, by 2003, to 28. However, this is an oversimplification of the situation, five new founders were recruited into the global population between 1983 and 2003; two

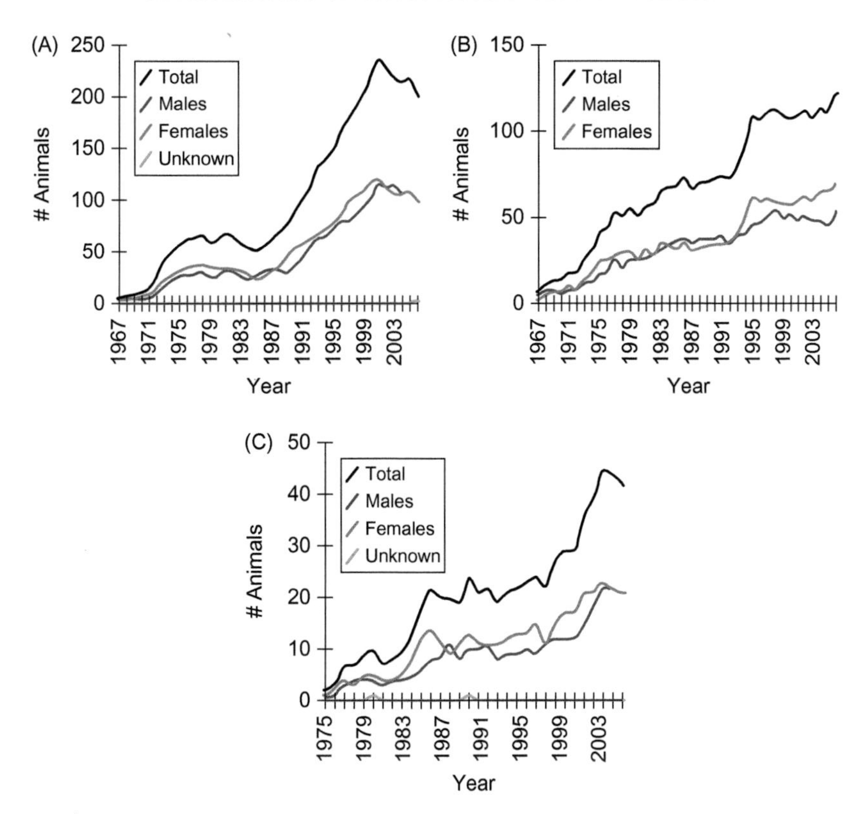

FIGURE 17.4 Number of *A. f. fulgens* in the regional populations: (A) Europe; (B) USA; (C) Australia/New Zealand.

arrived in Europe in the early 1980s as a State gift from Nepal to the people of Spain and the other three were rescued from the wild in India in the late 1990s. This means that the contributions of three of the original founders have been lost to the global population. In addition, the contributions of the most recently added founders have been retained in the Indian regional population and therefore the addition of this new blood has had no impact on the population outside of the Indian borders.

At the regional level, the number of founders contributing to the regional populations has changed markedly over the years as the result of inter-regional exchanges of animals encouraged in the two global masterplans. Table 17.2 shows that between 1993 and 2003 the number of founders contributing to the European population rose from 23 to 26 while those contributing to the Australian population from 18 to 21. This is particularly remarkable when you consider that there were only four founders contributing to this latter population in 1983. In the same period, the number of founders contributing to the North American population rose from 12 in 1983 to 16 in 1993 and 23 in 2003. It is undoubtedly as a result of this marked increase in the number of founders that the genetic diversity of the North American population actually increased between 1993 and 2003 despite the limited growth in population size over the same period.

Another important parameter determined by the genetic analyses of a population is the average inbreeding coefficient. The level of inbreeding increased from 0.01 to 0.05 in the world zoo population of *A. f. fulgens* in the 20 years between 1983 and 2003. Comparative figures for inbreeding levels in the regional populations of *A. f. fulgens* are unavailable for 1993. However, in 2003, inbreeding in the three major regional populations ranged between 0.04 and 0.07. The level of inbreeding is of particular importance to the management of red pandas as it has been shown that infant mortality increases [4, 7] and fertility decreases [7] with increasing levels of inbreeding.

THE DEVELOPMENT OF THE *AILURUS FULGENS STYANI* POPULATION

The studbook history of the zoo population of *A. f. styani* is somewhat shorter than that for the *fulgens* subspecies. Although there have always been a few Styan's pandas registered in the studbook, their numbers were not adequate to form the basis of a viable population in the early years. There were undoubtedly viable numbers of Styan's pandas maintained in Chinese zoos throughout this period, however, these animals were not known to the studbook keeper and so not registered in the studbook.

The captive population of Styan's pandas began to increase in the late 1980s; at that time the numbers of *A. f. styani* registered in zoos in Japan and North America approached 100 for the first time (Figure 17.5). This increase was due to a sudden spate of exports from China to zoos in Japan and North America. These exports occurred just prior to the introduction of CITES restrictions on the trade in wild-caught red pandas. In fact, these imports were the stimulus for providing the red panda with the protection of a CITES Appendix I listing. The zoo population of Styan's pandas has only been analysed twice since that time. The first analysis of the zoo population of *A. f. styani* was published in 1993 [9] and included those animals in captivity in China which had been reported to the

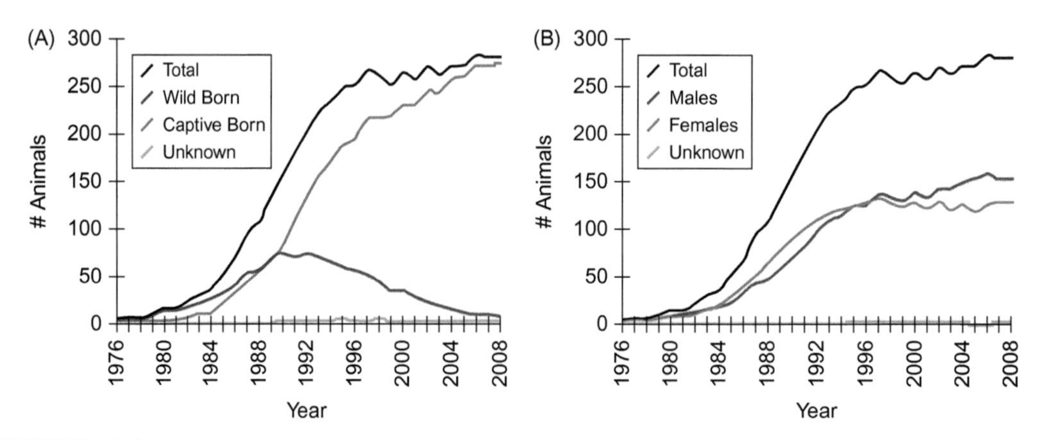

FIGURE 17.5 Changes in the population of *A. f. styani* held in zoos around the world over the studbook period: (A) the numbers of wild-caught and captive-born individuals; (B) the numbers of males and females in the population.

studbook. It is quite clear that the numbers of red pandas held in Chinese collections was substantially under reported at this time, a separate publication [10] indicates that, in 1994, there were 283 red pandas in captivity in China while studbook records indicate only 97 animals in 1991. The second analysis took place in 2003 [8] as part of the global master-planning session for all zoo red pandas; unfortunately, no data from Chinese zoos were available for the second analysis. The results of both these analyses are presented in Table 17.3.

On 31 December 1992, there were 266 Styan's pandas living in captivity of which 212 were in China and Japan, 43 in North America, six in Europe, five in Australia and five in Central and South America. Fifty-two percent of the zoo population at that time was comprised of wild-caught animals. The age pyramid of the population at that time is presented in Figure 17.6. There seems to have been a slight bias towards females in this zoo population. Further analysis of the population yielded a λ below 1.00 indicating the population was not at replacement level.

The second analysis of the Styan population focused on the zoo populations in Japan and North America as no information was available from China and the Styan's pandas which had been living in Europe, Australia and South and Central America in 1992, had since died out. On 31 December 2003, there were 275 Styan's pandas housed in zoos in

TABLE 17.3 Styan's panda population demographics and genetics for publications

	Population 31-12-1992	Population 31-12-2003
Population size	266	275
% wild caught	52	1.5
Annual growth rate	1.034/0.949	1.012/0.968
30-day infant mortality		22.5%
Number of founders	44+	45
Genetic diversity	0.979	0.97
Inbreeding	0	0.01

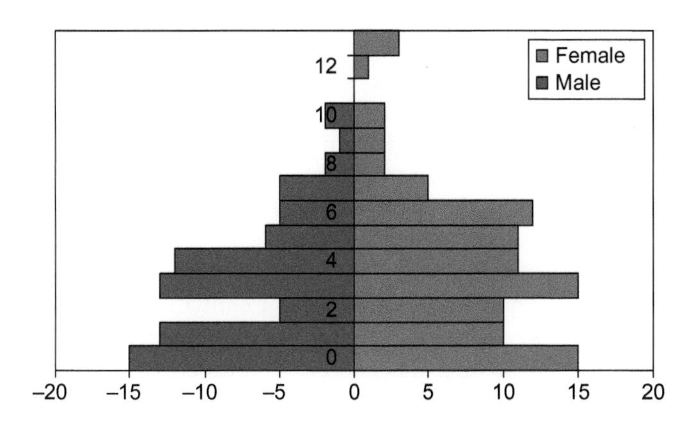

FIGURE 17.6 Age pyramid of *A. f. styani* population in 1991.

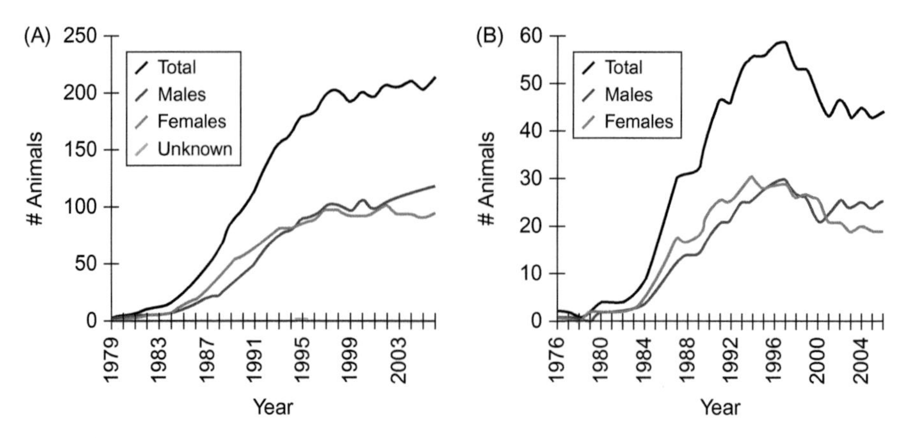

FIGURE 17.7 Numbers of *A. f. styani* in the regional populations: (A) Japan; (B) USA.

TABLE 17.4 Regional populations of Styan's pandas, 31-12-2008

	SSP	**Japan**
Population size	46	229
% wild caught	0	1.5
Annual growth rate	1.003/0.956	1.025/0.989
30-day infant mortality	21.5	25.5
Number of founders	31	38
Genetic diversity	0.9066	0.9697
Inbreeding	0.04	0.0054

North America and Japan. Over the 12-year period between the two analyses the combined Japanese and North American population of *A. f. styani* had increased by 118 individuals (Figure 17.7). The North American population had only increased by three despite the import of eight red pandas from China and Japan in the intervening years while the Japanese population had increased substantially from about 115 to 229. During this period, Japan had only imported 15 animals from China. Furthermore, the proportion of wild-caught animals in the population as a whole had decreased from 52% to nearly zero over the same period. These are positive changes; nevertheless, the annual rate of change, λ, still remained at, or just below replacement level for both regions (Table 17.4). This lack of growth is reflected in the instability of the age pyramids (Figure 17.8).

Despite the lack of growth, the genetic parameters of the *A. f. styani* population were more encouraging than those for *A. f. fulgens*. There were 45 founders contributing to the global captive population of Styan's pandas; 31 and 38 contributing to the North American and Japanese zoo populations respectively. The level of inbreeding at both global and regional population level had remained low.

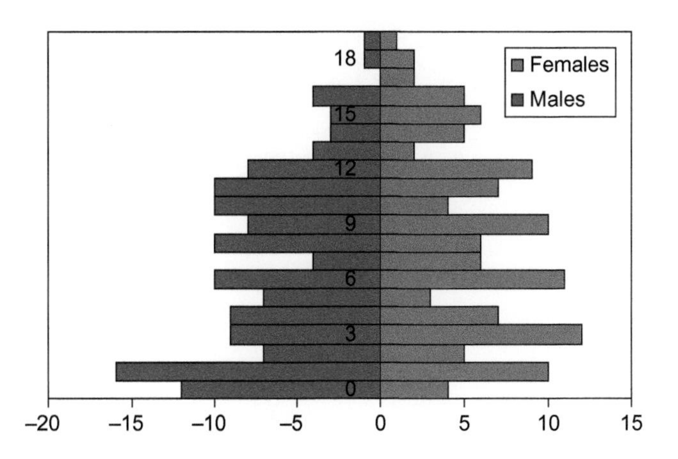

FIGURE 17.8 Age pyramid *A. f. styani* population on 31 December 2003.

THE CURRENT POPULATIONS OF *A. F. FULGENS* AND *A. F. STYANI*

The goal of the masterplanning sessions in 1993 and 2004 was to develop captive red panda populations which would be able to maintain 90% of natural genetic variation for 100 years. The conclusion of the more recent analysis in 2003 was that this was possible in the case of *A. f. fulgens* so long as the population remained adequately managed. However, in the case of the Styan's pandas the prognosis was less optimistic despite the large number of founders; the problem was the lack of growth. In retrospect, this lack of growth seems to be equally problematic for both populations. Growth in the *A. f. fulgens* population seems to have levelled off and even declined since 2000 (see Figure 17.2). In order to investigate the impact of recent changes in population growth on the demographic and genetic parameters of the population an analysis was conducted on the global and regional *A. f. fulgens* and *A. f. styani* populations over the 10-year period, 1 January 1997–31 December 2006 using PM2000[x] and SPARKS[y] software. In order that anomalous results from small age-sex cohorts should not distort the analysis, it was assumed that all red pandas would stop breeding at 15 years and die at 16 years. An overview of results of these analyses is presented in Table 17.5.

At the end of 2006, the global *A. f. fulgens* population numbered 455 individuals. This represents a decrease in population of 21 specimens since the masterplan was revised in 2004. This decrease is of itself not very significant given the size of the population. After the period of rapid growth in the late 1980s and 1990s, the population has stabilized and has remained fairly constant in size since 2000 (see Figure 17.2). Nevertheless, the growth rate of the population has now fallen below 1, which indicates that the population is indeed declining. Furthermore, the 30-day infant mortality rate has increased from 27.2% to 28.2%. These changes are reflected in the age-pyramid which is less stable than that published in 2004 (Figure 17.9). At the same time, there has been little change in genetic diversity despite the fact that two additional founders have been recruited to the population. The level of inbreeding has also risen over the same period from 0.047 to 0.054. Although the global population is not in danger, a change in management may well be needed to ensure the population's future.

TABLE 17.5 An overview of the genetic and demographic parameters of the global and regional populations for the period 01-01-1997 to 31-12-2006

	Fulgens				*Styani*		
	Global	**SSP**	**EEP**	**ASMP**	**Global**	**SSP**	**Japan**
Population size	455	121	203	42	271	44	227
% wild caught	<1%	0	0	0	2%	0%	3%
Annual growth rate	0.992	0.972	0.989	1.006	0.993	0.959	0.995
30-day infant mortality	28.2%	41.7%	25.3%	14%	25.8%	27.8%	25.4%
Number of founders	28	23	24	23	42 + 2	32 + 1	33
Genetic diversity	0.947	0.904	0.926	0.863	0.970	0.926	0.963
Inbreeding	0.054	0.047	0.047	0.080	0.012	0.041	0.007

SSP, N. American population; EEP, European population; ASMP, Australia/New Zealand population.

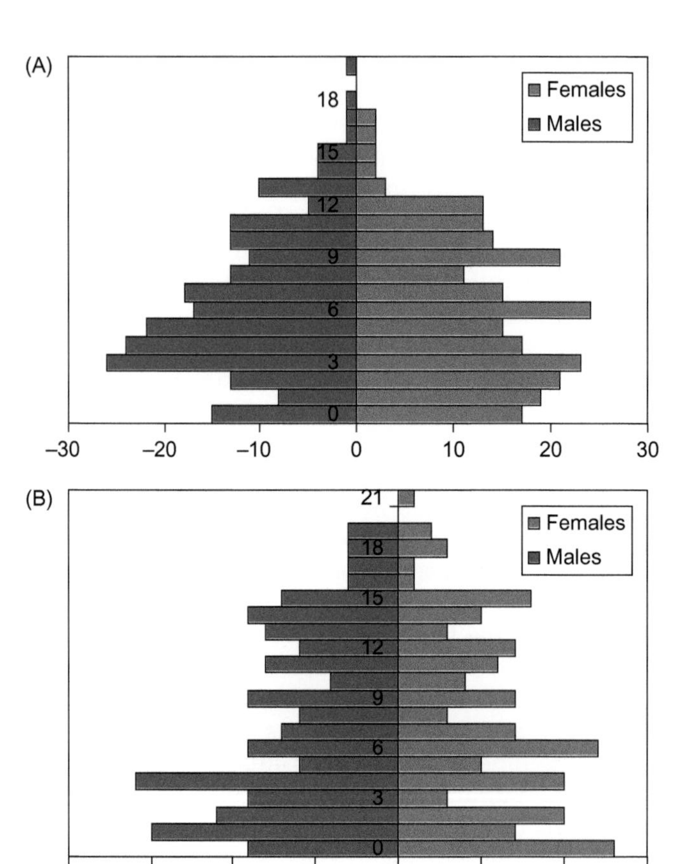

FIGURE 17.9 Age pyramids of (A) *A. f. fulgens* and (B) *A. f. styani* on 31 December 2006.

Looking at the regional populations (see Table 17.5), it would seem that while the population in North America has remained the same size, that in Europe has decreased. In fact, λ is now below 1.00 in both of the larger regional populations and is therefore now below replacement level. However, perhaps the most worrying regional population is that in Europe. This population, which was doing very well at the time of the last masterplan analysis [8], has since declined in numbers and also lost two founders. These changes have led to nearly 2% loss in gene diversity in a comparatively short period of time (4 years). In addition, inbreeding is rising quickly, slightly more so than in the smaller North American population. The other major population of *A. f. fulgens* is housed in the Australian region. This population is markedly smaller that the other two, numbering just over 40 individuals. The Australian population has declined by two individuals since 2004; however, it has recruited two new founders as the result of transfers proposed in the 2003 masterplanning session.

At the end of 2006, the zoo population of Styan's pandas numbered 295 of which 271 were housed in Japan and North America. This means that the managed population in Japan and North America has increased by four individuals since the global masterplanning workshop in 2004. The age pyramid has not changed in shape over this period (see Figure 17.9). The *A. f. styani* population continues to behave much as it has in the past with the annual rate of change remaining below replacement level (see Table 17.5). Nevertheless, genetic analysis of the population remains promising; gene diversity remains more or less the same at 97%. Despite the loss of three founders, gene diversity of the Styan population is significantly better than that of *A. f. fulgens*.

In the period since the masterplanning session, both the North American and the Japanese populations of *A. f. styani* have declined by two individuals. In addition, the annual growth rates of both these regional populations have declined still further; λ was slightly above 1.00 for both populations in 2003, however, this present analysis shows λ is now clearly under 1.00 in both regions.

There have also been significant changes in the number of founders in both populations since 2003. The North American population has recruited one new and one potential founder, as a result the genetic diversity of the population has increased from 90.7% to 92.6%. In the case of Japan, the regional population has lost five founders since 2003 and gene diversity has declined by nearly 0.7% as a result. In both populations inbreeding has remained very low.

THE INDIAN POPULATION OF A. F. FULGENS

The Indian zoo population of *A. f. fulgens* needs separate discussion as it is in the unique position of being a captive population located within a range state. It therefore has a different role to that of any of the other regional red panda populations discussed here. As emphasized in the second global masterplan [8], the Indian zoo population has the potential to provide a link between the captive and wild populations, not only in the facilitation of data and information exchange between zoo and field but also in the movement of animals and/or genetic material between these populations. The Indian zoos have already provided captive-born red pandas for reintroduction and restocking (see

Chapter 25) and undoubtedly will continue to do so in the future. They have also accessed several new founders which are now contributing to the genetic well-being of the captive population. Therefore, this comparatively small regional zoo population held only in two zoos has a very important role in terms of *in situ* and captive conservation.

Red pandas have a fairly long history in Indian zoos. In fact, the first known captive-born red panda was born in Darjeeling Zoo in 1908 to a wild-caught female that was captured pregnant and gave birth in the zoo [11]. Although red pandas have been maintained in zoos in India since that time, breeding success was fairly limited until the initiation of the current breeding programme in 1994. The first international red panda studbook records no *A. f. fulgens* specimens living in Indian zoos in 1978. However, we now know that there were at least six red pandas living in Kanpur Zoo at that time although four of these had died by the end of 1978. In addition, there were also red pandas living in the old Deer Park in Gangtok at that time (van Dam, personal communication) but no detailed records are available. In 1990, Sally Walker, from Zoo Outreach organization, was employed to undertake a survey of red pandas in Indian zoos. She reported that in 1990 there were five Indian zoos housing the species [12]; there were three females in the Padmaja Naidu Himalayan Zoo, Darjeeling (two captive born and one gift), a pair in the old Deer Park in Gangtok and three further zoos each with a single female. The information given for the group of red pandas in Darjeeling later proved to have been incorrect. We now know that a group of one male and three females, all of wild origin, was held there at that time.

The establishment of a viable population of red pandas in Indian zoos was one of the goals of the first captive breeding masterplan as well as of the IUCN-SSC Action Plan for Procyonids and Ailurids [13]. In support of this objective, the first red panda was sent from Europe to India in 1993. This was a young male which unfortunately failed to survive more than a few months. However, a second transport of red pandas from Europe to India was agreed for the following year. This was a group of two males and two females which were donated by Antwerp, Cologne and Rotterdam zoos to the Padmaja Naidu Himalayan Zoo in Darjeeling. These pandas were all somewhat older than the male in the first transfer and they not only survived but thrived in their new home. Furthermore, with the exception of one male they all bred regularly. Their arrival signalled the initiation of the very successful Indian red panda breeding programme.

In 1994, just prior to the establishment of the breeding programme, there were just six red pandas listed in the studbook as living in Indian zoos. At the end of 2006, 12 years after the arrival of the four pandas from Europe, there were 19 red pandas (15 males and four females) in the Indian population. This may not seem to represent a significant increase in numbers over 12 years but, considering that over the same period, the four old, wild-caught individuals died and four zoo-born specimens were reintroduced into the wild, it is quite a respectable increase. Darjeeling and Gangtok, the two zoos in the region, bred a total of 46 baby pandas over this period. In addition, at the peak holding in 2001 there were 30 *A. f. fulgens* (16 males and 14 females) living in these two zoos. It is well known that small populations, such as that of the red pandas in India, are subject to stochastic processes. This is demonstrated by the critical shift in sex ratio from a healthy 16 males to 14 females in 2001 to an almost unsustainable 15:4 by the end of 2006 (Figure 17.10).

Analyses of the Indian population of red pandas have been published twice in the past. The first, in 1997 [14], was a part of a captive breeding masterplan for the Indian red panda population produced for the Central Zoo Authority. The second, in 2003, was part of the second global masterplan. The Indian population has now been analysed for a third time for this publication. This last analysis is based on the 10 years from January 1997 to December 2006 and uses the same analysis methods and criteria as those employed for the other regional red panda populations discussed earlier. Overviews of the results of all the analyses of the Indian zoo population are presented in Table 17.6 and Figure 17.10. The data indicate that, after an initial period of population growth, the Indian population of *A. f. fulgens* has stagnated. As is the case of the other regional populations, λ is currently just below 1.00, indicating non-growth. On the other hand, infant mortality has substantially improved over the years and, with 30-day mortality below 20%, India seems to be nearly as successful as Australia in combating this problem. However, on the genetic side, the Indian population is beginning to encounter difficulties. Despite the fact that a new

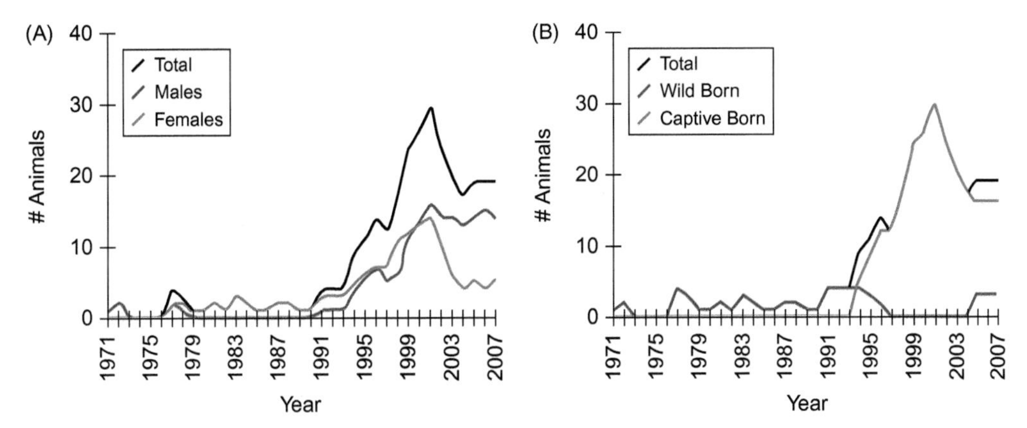

FIGURE 17.10 Numbers of *A. f. fulgens* held in the Indian zoo population: (A) numbers of males and females; (B) number of wild-caught versus captive-born individuals.

TABLE 17.6 Analysis of the Indian regional population of A. *f. fulgens*

	As of 31-12-1996	As of 31-12-2003	As of 31-12-2006
Population size	14	20	19
% wild caught	0	0	15%
Annual growth rate	0.844	1.061	
30-day infant mortality	46%	18.4%	19.3%
Number of founders	22	22	23 + 2 potential
Genetic diversity	0.881	0.850	0.8535
Inbreeding	0.25	0.028	0.0635

founder has recently been recruited into this population, gene diversity has not really improved. In addition, inbreeding continues to rise, a problem which is not helped by the small number of females in the population. These genetic and demographic problems need to be addressed if India is to develop a viable population. The genetic analysis of the population has always indicated that there is more potential in the population than has actually been achieved, for example, in the second global masterplan, the genetic diversity in the population was calculated as 0.850 whereas the potential genetic diversity was estimated as 0.900, 5% higher. Similarly, calculations based on the last 10 years show these figures to be 0.850 and 0.935 respectively making potential gene diversity 8.5% higher than is actually the case. However, given the demography of the current population, only four females, one of which is older than 10 years, improving gene diversity on the basis of the current population would seem to be impossible. Furthermore, if no new females are added to this population within the next few years it is likely that it will die out. However, there is still cause for optimism, the current Indian population includes two recently wild-caught animals (one male and one female) which have yet to breed. Fortunately, Indian zoos occasionally obtain red pandas that have been rescued from the wild. During the last few years three such animals have been received and, undoubtedly, more will arrive in the future. These wild-born animals will improve the viability of the zoo population. Furthermore, two female red pandas will be imported into India from the Australian regional population in exchange for two Indian-born males. Although the Australian "genes" are already well represented in the world population, these exchanges will not only improve the sex ratio in India but will also allow the captive population to "catch" genetic material from important Indian males before they are too old to breed. In this way, genetic diversity will be improved and inbreeding reduced in both regions. However, having said this, it must be noted that this exchange of animals between the Indian and Australian regions has been in the planning for more than 2 years and still has not been realized due to various bureaucratic obstructions. If these problems are not resolved soon, the Indian captive red panda population will be seriously compromised. Also if similar delays occur in the future it will be very difficult to integrate the Indian red population into any attempt at global management.

The Indian government has stated that it is committed to using zoo breeding programmes to complement national efforts to conserve wildlife *in situ*. They have recommended that endangered species, with wild populations numbering less than a few thousand, should be brought into zoos for captive breeding. They want to establish global captive populations for these species which number at least 250 individuals, all of which are to be genetically and behaviourally healthy. Furthermore, 100 of these specimens should be housed in Indian facilities. These captive populations are insurance against extinction in the wild [15]. The global zoo population of *Ailurus fulgens fulgens* easily complies with these criteria. However, whether the ambitious plan to have a zoo population of 100 red pandas in India will prove to be feasible remains to be seen. Nevertheless, a viable global captive population does exist and this does afford the insurance that the Indian government is seeking. The viability of the global population could be improved if some of the new genetic material, which exists in the Indian zoo population as a result of the rescued wild pandas, were distributed to the other regional breeding programmes. More information on the future of the global population is discussed in Chapter 19.

CONCLUSIONS

The future of red pandas in the wild is far from certain. However, we are fortunate in having good-sized populations of both *A. f. fulgens* and *A. f. styani* maintained in the zoos around the world. We need to manage these valuable resources and ensure that we keep them healthy and viable if we are to safeguard this unique species against extinction in the event of a further decline in the wild population. In 2004, the second global masterplan concluded that the zoo population of *A. f. fulgens* was viable and that it had the potential to maintain 90% gene diversity for 100 years. However, to achieve this aim, the population had to be allowed to grow to 585 or better yet 650 individuals. Unfortunately, we do not seem to be able to achieve this; the zoo population of *A. f. fulgens* is not growing. The analyses of the last 10 years show some negative changes which have only become evident in the last few years. The state of the captive population at the end of 2006 was not as good as it had been 10 years previously; numbers had declined, founders been lost and there was accelerated loss of genetic material. Although these negative changes have not proceeded too far as yet, rapid action is necessary if these changes are to be halted. In particular, regional population managers need to take stock of their populations, evaluate the problems and address them at the regional level. Without this commitment the regional populations could decline beyond the point of no return and with them a valuable part of the global zoo population. The same applies to the regional coordinators of the *A. f. styani* populations. Here the problem is even more acute. The registered population of *A. f. styani* is about half the size of that of *A. f. fulgens*. Although this is still a good-sized population with a large number of founders, it is not viable. The second global masterplan concluded that a concerted effort was needed to address the problems of this declining population. Analyses of the Styan population over the last 10 years show that there has been no improvement; in fact, the situation has become worse. Despite the fact that the global and regional populations of *A. f. styani* remain genetically very healthy they are demographically challenged and, if this is not addressed, the population will not survive in the long term. However, there is undoubtedly a sizeable population of Styan's pandas in the zoos and wildlife rescue centres of China. If these animals could be incorporated into the global population we would have an excellent prognosis for the future of the *A. f. styani* population. Integration of these animals into the globally managed population must be a priority for the future.

One of the major factors preventing population growth in the captive populations of both subspecies is infant mortality. This has always been a problem in both red panda populations. The rate of 30-day infant mortality has actually risen in both global and regional populations in the most recent analyses; it currently ranges between 14 and 42%. It is lowest for *A. f. fulgens* in Australia, highest for *A. f. fulgens* pandas in North America and similar for both subspecies in the other regions at about 25–28%. In both global masterplans, regional coordinators were asked to address the problem of infant mortality. We need to discover why the 30-day mortality of *A. f. fulgens* in North America is 42% while that of *A. f. styani* in the same region is only 28% and why Australian zoos have a 30-day mortality of around 14% when the global average is twice that level. These questions would seem to be fairly easy to answer yet we consistently fail to do so. Are there substantial differences in housing, diet or management practices which account for this directly? Or do zoos that operate a non-disturbance policy fail to register births in those cases where the infants die and are consumed by their mothers in the first days thereby under-reporting both births and perinatal deaths? However, the lack of population growth is not solely due to high rates of infant mortality; low fertility also plays an important role particularly in the case of Styan's pandas. We need to know why some zoos fail to breed red pandas while others are very successful. Inbreeding may be a contributory factor in some cases. Raised levels of inbreeding have been shown to be correlated with increases in infant mortality and reduced fertility [4, 5]. This is a matter of concern as there has been a gradual increase in inbreeding levels across the board. Some increase in inbreeding is inevitable in a closed population but it can be minimized through good management which again is the province of the regional population.

A number of recommendations for the management of both subspecies were produced at the end of the 2004 masterplanning session, not all of these have been fully implemented. Some of the inter-regional transfers have taken place but not necessarily with the animals nor between the regions recommended. Another masterplanning session is required within the next few years to re-evaluate the global and regional populations and to determine what steps need to be taken.

Despite the potential to maintain good populations of both subspecies of red pandas, zoos are not yet managing their red pandas optimally. The fact that red pandas can be well managed is clear. There are zoos and regions which are, or have been, very successful. Unfortunately, these successes alone are inadequate to ensure the future of the population. We need to make a greater effort to understand the biology of the red panda and the management conditions required to encourage fertility and reduce infant mortality. In addition, we need the discipline to ensure that those zoos that are unwilling or unable to provide the correct facilities for breeding and rearing red pandas do not get these animals. Population managers are regularly approached for animals by zoos located in regions which are climatically unsuitable for red pandas. They must feel confident enough to refuse these requests and those zoos which are not allocated animals must realize this is not a criticism of their skills or facilities but is related to factors beyond their control.

It will only be when we fully understand the biology of the red panda and are willing and able to include this knowledge in our captive management decisions that we will be able to build and maintain healthy, viable populations of red pandas in captivity.

References

[1] M.L. Jones, A brief history of the red panda in captivity, in: A.R. Glatston (Ed.), Red Panda Biology, SPB Academic Publishing, The Hague, The Netherlands, 1989, pp. 9–24.
[2] A.R. Glatston (Ed.), The Red or Lesser Panda Studbook No. 1, Stichting Koninklijke Rotterdamse Diergaarde, Rotterdam, The Netherlands, 1980.
[3] H. Veeke, A.R. Glatston, Computer simulation of trends in the captive population of red pandas, in: A.R. Glatston (Ed.), The Red or Lesser Panda Studbook No. 1, Stichting Koninklijke Rotterdamse Diergaarde, Rotterdam, The Netherlands, 1980, pp. 11–22.
[4] A.R Glatston, M. Roberts, The current status and future prospects of the red panda (*Ailurus fulgens*) studbook population, Zoo Biol. 7 (1988) 47–59.

[5] A.R. Glatston (Ed.), The Red or Lesser Panda Studbook No. 5, Stichting Koninklijke Rotterdamse Diergaarde, Rotterdam, The Netherlands, 1989.
[6] A.R. Glatston (Ed.), The Red or Lesser Panda Studbook No. 7, Stichting Koninklijke Rotterdamse Diergaarde, Rotterdam, The Netherlands, 1993.
[7] A.R. Glatston, F. Princee, A Global Masterplan for the Captive Breeding of the Red Panda. Part 1: *Ailurus fulgens fulgens*, Stichting Koninklijke Rotterdamse Diergaarde, Rotterdam, The Netherlands, 1993.
[8] A.R. Glatston, K. Leus, Global Captive Breeding Masterplan for the Red or Lesser Panda, *Ailurus fulgens fulgens* and *Ailurus fulgens styani*, Unpublished reference document, 2005.
[9] B. Lu, A.R. Glatston, F.P.G. Princee, Demographic and genetic analyses of the studbook population of *Ailurus fulgens styani*, in: A.R. Glatston (Ed.), The Red or Lesser Panda Studbook No. 7, Stichting Koninklijke Rotterdamse Diergaarde, Rotterdam, The Netherlands, 1993, pp. 30–36.
[10] W. Fuwen, Z. Feng, H. Hu, et al., Brief history of raising and breeding red pandas in China, in: A.R. Glatston (Ed.), The Red or Lesser Panda Studbook No. 10, Stichting Koninklijke Rotterdamse Diergaarde, Rotterdam, The Netherlands, 1998, pp. 22–29.
[11] F. Wall, Birth of Himalayan cat-bears (*Aelurus fulgens*) in captivity, J. Bombay Nat. Hist. Soc. 18 (1908) 903–904.
[12] S. Walker, Red pandas in Indian Zoos: history, status and management, in: A.R. Glatston (Ed.), The Red or Lesser Panda Studbook No. 6, Stichting Koninklijke Rotterdamse Diergaarde, Rotterdam, The Netherlands, 1991, pp. 25–40.
[13] A.R. Glatston, The red panda, olingos, coatis, raccoons and their relatives, Status Survey and Conservation Action Plan for Procyonids and Ailurids, IUCN, Gland, Switzerland, 1994.
[14] A.R. Glatston, A captive breeding masterplan for red pandas in Indian Zoos (revised version). Unpublished paper prepared for the Central Zoo Authority, India, 1997.
[15] B.R. Sharma, N. Akhtar, B.K. Gupta (Eds.), Proceedings of International Conference on India's Conservation Breeding Initiative, Central Zoo Authority, Delhi, India, 2008.

CHAPTER

18

Management, Husbandry and Veterinary Medicine of Red Pandas Living *ex situ* in China

Kati Loeffler

Veterinary and Scientific Consultant for International Animal Welfare and Conservation Programmes

OUTLINE

MANAGEMENT OF THE *EX SITU* RED PANDA POPULATION IN CHINA

Red pandas are commonly found in Chinese zoos and wildlife parks, although their number in captivity is unknown. A partial studbook was kept briefly by Tianjin Zoo, but lack of compliance among institutions to make it comprehensive has left the effort to languish. The difficulty lies in the lack of system for permanent identification of individual animals, group housing, the frequent movement of red pandas among institutions and from the wild into captivity, and the lack of written records. Like most second-class endangered species in China, the red panda does not enjoy the popularity and protection of, say, the giant panda. Resources for, and efforts toward, its protection and captive management are relatively meagre. Much of the Chinese general public are unaware of the existence of red pandas, believing instead that "Xiao Xiongmao" (literally, "Little Bear-Cat") is simply

Red Panda. DOI: 10.1016/B978-1-4377-7813-7.00018-5

a small giant panda ("Da Xiongmao", which means "Large Bear-Cat"). Red pandas are unique to China and the Himalayas and share the giant panda's habitat and food resources (Figure 18.1).

Although classified as CITES-1 and listed on the IUCN Red List, red pandas receive only Class 2 protection status under Chinese law. This allows the capture of individuals from the wild for "farming for utilization" which is part of China's wildlife conservation policy, and "for scientific or conservation purposes" by permit from the provincial governments. It has been standard practice to supplement red panda breeding programmes and zoo exhibits with wild-caught individuals because reproductive success and breeding management in captivity are poor. In 2003, the Chinese State Forestry Administration (SFA; which oversees all issues concerning wildlife living *in situ* in China), suggested to the provincial governments that permits for capture of wild red pandas should be restricted in order to protect the *in situ* population. According to an official in the SFA's wildlife conservation department, the provincial governments complied with this and wild capture has been prohibited since then. That said, the author has learned from certain breeding centres that they have obtained wild-caught red pandas as recently as 2007.

The SFA's prioritization of objectives for the management of endangered species is as follows. The first level of priority is protection of habitat. If one cannot save a species just by habitat protection, then captive breeding is undertaken with the ultimate aim of reintroduction into the wild. This is the current status of the giant panda, for example. The third level of priority is to monitor the use of the species or reasons for its decline, e.g., markets,

FIGURE 18.1 Wild habitat of giant pandas and red pandas, Baodinggou, Sichuan Province © *Sarah Bexell.*

trade, poaching. For some species of economic importance (e.g., Asiatic black bears for the bear bile industry), captive breeding is encouraged to serve the market and, theoretically, to protect the wild population. Poaching of black bears continues, however: cubs to stock bear bile farms, and adults for gall bladders and other parts (www.animalsasia.org). To what extent red pandas are poached in China is unclear. Cubs are sold in pet and wildlife markets as pets, and occasionally one sees a red panda pelt, usually in the shape of a hat (author's observations). How many of the pandas who are caught for breeding actually are obtained by permit is unknown. Reports of red panda poaching are rare, but this is most likely not because poaching is rare, but because even if the poacher is apprehended, the event is not newsworthy. The information that we have comes from personal observations and unofficial reports. Foreigners or young Chinese who are taking an interest in animal issues in their country will occasionally write to someone like the author asking what to do about a red panda that they have just seen in a market.

According to the SFA, a permit for wild-capture of a Class 2 species is issued only after review of a proposal and support from scientists that establish how the project for which the animals are captured will benefit the species. The making of policy that pertains to wildlife issues is becoming more scientific in China. Changing the protection level of a Class 2 species to Class 1 will require extensive scientific documentation of the population status and trend, and then legislative petition for the change. This presents the eternal Catch 22: a species that is already rare, particularly one that is so difficult to study (given the harsh habitat in which it lives and its elusive behaviour), garners little research interest and has a difficult time competing for research funds against other, more popular or more lucrative species. Given the loss of habitat space and quality that is the principal threat to giant pandas, and the empirical observation of nature reserve staff that red pandas are sighted with decreasing frequency, one may assume that wild red panda populations are not stable and that they warrant increased protective measures. Clearly, the SFA recognizes this in its suggestion to the provincial governments to restrict the removal of individuals from the wild. But bringing such suggestions into practice requires additional effort and resources that currently do not appear in the conservation picture for red pandas in China.

Red pandas are bred in zoos, wildlife parks and private breeding centres throughout China. The private breeding centres may also serve as "wildlife rescue centres" to which the provincial government sends red pandas that have been "rescued" from the wild or are otherwise temporarily without a home. The owners of these centres state that they receive no history with the animals; at most they might learn if a particular individual was wild-caught. Zoos and wildlife parks request red pandas from these breeders; breeders sell the animals to them by permit of the provincial government, or loan breeding stock.

The predominant subspecies of red panda in China is *styani*, although *fulgens* are found in southwest China (Yunnan province) as well [1]. There is likely to be capture of both from the wild (considering where wild red pandas are caught) but, according to managers of captive breeding facilities, there is no effort to keep the two subspecies separate *ex situ*. The managers who were interviewed had not been aware that there were different subspecies.

Individual red pandas are not permanently identified in captive facilities (e.g., microchip), and what records exist do not travel with animals when they move from one institution to another. Pedigree information is therefore unknown beyond one or two generations which an astute keeper will remember personally. Facilities with several breeding animals usually house groups of males and females together, or introduce several males to several females during the breeding season. Introduction of new genetic material is achieved by exchanging animals (usually males) with other institutions every few years or purchasing wild-caught individuals. Empirical observations by keepers indicate that a few males will dominate breeding every year in the group mating situations, and there is concern that offspring over several generations may be closely related, if not inbred. A study of microsatellite loci of some of the red pandas at the Chengdu Research Base of Giant Panda Breeding supports this observation through a demonstration of a loss of critical alleles [2], as does a recent pedigree analysis of all the red pandas at this institution [3]. The population at this institution is the largest in China and is one of the better managed ones, but showed that 42% of the 32 tested animals had an inbreeding coefficient of ≥0.5 (e.g., mother/son pairing; scores >0.5 indicate several generations of breeding with closely related individuals). The genetic diversity of the population is relatively high, indicating that the four founding members of the population probably came from the wild, or from widely distributed breeding centres. Breeding management is done by grouping up to 15 red pandas to breed at will, or by placing a pair that has bred well in the past together in a cell for a few days. There are no written records.

RED PANDA HUSBANDRY IN CHINESE FACILITIES

Enclosures for red pandas vary greatly among institutions and with the purpose for which the animals are kept (Figures 18.2 and 18.3). Observations are based on visits by the

FIGURE 18.2 Red panda enclosure in Fuyan, Anhui Province © *Kati Loeffler.*

author to 17 red panda facilities throughout China, including those that are reputed to be the most advanced. Wildlife parks, which are gaining popularity in China and which are generally sufficiently removed from large cities to allow more space, may have outdoor enclosures of 1500 m^2. Most of these contain trees for the animals to climb; some contain free-growing bamboo and other vegetation. Shelters may or may not be available in these enclosures, and are usually in the form of a cement hut on the ground. Exhibit enclosures are usually surrounded by a cement wall with or without a dry moat. Escapes are not uncommon from some of these enclosures. Facilities located near panda habitat may have an exchange of wild red pandas climbing into the enclosure and captive individuals leaving it. One park keeps its red pandas on a small island in a lake.

Zoo exhibits may consist of a small outdoor yard with trees, bamboo or shrubs, or simply grass or dirt, or may be a cement room with a glass front. Enrichment for captive wildlife is still a very new concept in most Chinese zoos, and few facilities have more than a wooden box for the pandas to climb onto if there are no trees. Some of the more advanced institutions have built climbing structures and even small pools and fountains in the enclosures. Visitors enjoy seeing a "harmonious" environment, which encourages zoo officials to outfit the outdoor animal areas accordingly. Off-exhibit enclosures are usually simple cement cells with poor light and ventilation (see Figure 18.3). These rarely contain anything more than a box or crate for the animals to come off of the floor and rarely contain bedding. In some off-exhibit breeding areas, these rooms open into a small yard of dirt, grass or low shrubs.

FIGURE 18.3 Indoor enclosures for red pandas at two institutions © *Kati Loeffler.*

FIGURE 18.4 Large groups of red pandas may share a single enclosure. These are receiving their twice-daily bowl of milk © *Kati Loeffler.*

Many facilities house several red pandas together, mixing males and females throughout the year (Figure 18.4). In the late spring, pregnant females are pulled into nursery houses until the cubs are removed or weaned. Density may be as high as a dozen or more red pandas in an area of 1000 m^2. Fighting and associated injuries are common in these situations (Figure 18.5). Red pandas in these environments also generally move about much more than is normal for the species, pacing and scent marking for many hours of the day. In other facilities where animal density is lower and/or there is sufficient vertical space and foliage, red pandas exhibit less stress and conflict behaviour.

Experienced red panda breeding facilities have recognized the importance of a quiet, private environment for nursing females. Signs placed before nurseries at these zoos politely ask visitors to stay away and explain why. Visitors generally read these notices as an irresistible invitation, however, and it is not uncommon for human fingers and eyeballs and bits of food to poke through a hole in the mesh of a nursery house window, accompanied by loud chatter and abrupt noises meant to get the animals' attention. When possible, private breeding centres in China or nursery outposts belonging to zoos are located away from cities and towns, on a mountain slope or in a farming community. Posting experienced and competent keeper staff here is a major difficulty, as few people with options are willing to stay at these remote locations for the low-paying, demanding work that a nursery facility offers. Caring for animals is generally considered a very low-ranking job in China. Keepers are hired for basic labour and are trained as needed. The rank of the animal in the Chinese popular perception determines the level of a keeper's pay, such that

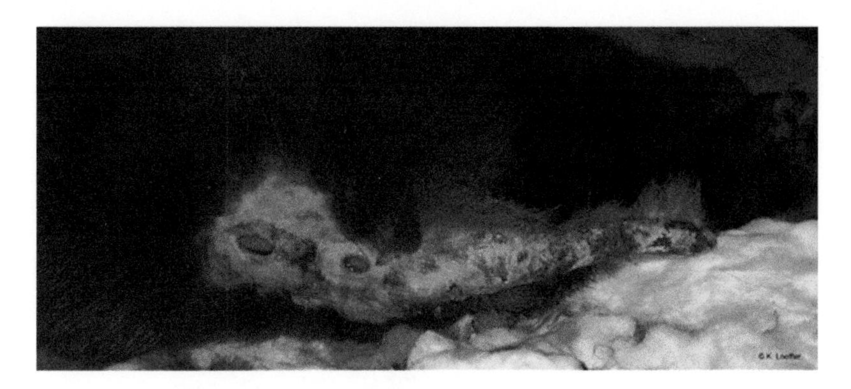

FIGURE 18.5 Bite wounds are common among red pandas living in large groups. This is an older female who has lost half of her tail and the pelage on her hindquarters and tail to bite wounds. She was being treated for severe wounds that had been unnoticed for several days © *Kati Loeffler.*

giant panda keepers, for example, are paid more than keepers of red pandas or antelope or parrots. Delivery of sufficient fresh bamboo to these remote breeding facilities is another complication, as this may require considerable staff time and transportation expense.

Typical nursery rooms consist of a closed cement room with a single box on the floor that the female may use as a nest. To provide privacy, the windows to these rooms are often small or very high, or consist simply of a few small holes in the wall. Poor ventilation and high summer temperatures are a frequent source of stress to nursing mothers. Reproductive success is, expectedly, low, even in the best facilities, with high infant mortality rates being the principal compromise. Mothers stressed by heat, noise, and environments devoid of sufficient privacy and nest choices often kill or neglect their cubs, or overgroom them until the cubs lose ears and tails or develop dermatophytosis (Figure 18.6). Inability of mothers to provide sufficient milk for their cubs, even for a single cub, is also a common reason for a cub to die or be removed for hand-rearing.

At the best red panda breeding facility in China, very young cubs that have been removed from the mother are kept in an incubator that maintains temperature and humidity. By 3 or 4 weeks of age they are moved to a bare wooden box with a mesh lid. Cubs are kept separated so that they do not suck on one another. Raising cubs in isolation of one another is generally considered highly deleterious to the animals by husbandry staff in western zoos who have successful red panda reproduction programmes (see Chapter 13). Supplemental heat may be provided by a 40 W light bulb. At this institution, the diet has recently been changed to a mixture of Esbilac® (Pet-Ag, Houston, Texas, USA) and a lactose-free human infant formula. Survival of cubs on this formula has improved significantly since the change from more traditional diets made of human infant formula alone, or cows' milk or goats' milk. Constipation, bloat, respiratory infections and dermatophytosis are common in hand-reared cubs.

Adult diets vary among institutions. Some facilities continue to feed milk throughout adulthood (see Figure 18.4), with raw eggs, vitamin and calcium tablets, and bread or biscuits made on site. The bread is based on a mixture of corn, soy, rice and wheat with micronutrient supplementation. Others provide a gruel that may be made of any combination of corn, rice, soy, eggs (raw), bamboo powder, silkworm powder, fish meal, meat powder, and ground-up mice caught on site (raw). Some institutions provide bamboo; others do not, or only in limited amounts. Most facilities also feed chopped apples. There is very

FIGURE 18.6 Mothers who feel stressed or uncomfortable with their nursery area often over-groom their cubs or begin to devour them. Cubs may lose ears and tails or develop dermatophytosis from the over-grooming © *Kati Loeffler.*

little variation in the diets, and no effort at behavioural or, with the exception of the naturalistic outdoor enclosures, environmental enrichment.

VETERINARY MEDICINE OF *EX SITU* RED PANDAS IN CHINA

Canine distemper virus (CDV) is, as in most of the world, a serious concern for captive red pandas; the risk for wild red pandas is unknown. The virus is endemic in the domestic dog population, which has been growing exponentially in recent years. Keeping a pet dog or guard dog (in rural areas) is increasingly common, although population control is not. Most dogs, if they are vaccinated at all, are vaccinated only against rabies. Some institutions vaccinate their red pandas against CDV. The common vaccines are either a killed CDV or a multivalent vaccine that includes a modified live CDV. Deaths due to vaccine-induced distemper are common in red pandas vaccinated with the latter, but the vaccine continues to be used. A few institutions have begun to use the Purevax® (Merial, Duluth, Georgia, USA) recombinant CDV vaccine. Its efficacy is unknown in red pandas, although antibody titres and lack of clinical cases of distemper in red pandas vaccinated with this vaccine in North America and Europe suggest that it is protective (R. Montali, unpublished data). A vaccine study is currently underway at the Guanzhou Zoo to monitor CDV antibody titres over 2 or more years of vaccination with Purevax®.

A serosurvey of 10 Chinese red panda institutions in 2004 indicated that CDV titres in unvaccinated red pandas were low (20–100% of each group had titres 1:8 or 1:16, the others were negative). For comparison, the mean antibody titre in vaccinated dogs in the USA is 1:256, with titres greater than 1:32 considered positive. The finding suggests that risk of exposure in the zoos in the 2004 study was low, or that those individuals who do contract the disease do not survive to participate in serosurveys [4]. Titres in vaccinated red pandas were highly variable (negative at 1:8 to 1:512); the same result was found in another study conducted for giant pandas and red pandas who had been vaccinated with the multivalent vaccine (titers up to 1:10 240) [5]. This suggests that the locally produced canine vaccines are not suitable for use in red pandas. A similar pattern was found for canine parvovirus (CPV) and canine adenovirus (CAV). One vaccinated red panda in the study by Qin et al. had a CAV titre of sufficient magnitude (1:1024) to suggest exposure to the wild virus. Canine parvovirus was isolated from faeces of unvaccinated red pandas [6], and antibody titres suggest that natural exposure to this virus does occur [4]. None of the institutions in the study had veterinary records by which to determine if clinical cases with characteristics of active parvovirus infection had occurred in the past.

High serum antibody titres against *Toxoplasma* [4] suggest that this parasite may be an issue of concern, at least in some institutions, and that further research on the source of infection and the clinical effects thereof is warranted. The red pandas in four of the 10 study sites in the study by Qin et al. [4] had positive titres, and these ranged from low positive (1:64–1:128), intermediate (1:256–1:512) and active (greater than 1:1024, with some beyond the limits of the assay at 1:4096). On the basis of clinical experience with domesticated species, low and intermediate titres generally indicate previous exposure to the organism but not active disease. "Active" titres suggest active infection. Feral cats are common, particularly in urban areas, and their excretion of oocysts may be a source of infection for the red pandas. Direct acquisition of trophozoites by ingestion of mice may be another source (rodent infestations are ubiquitous; also see diet, above).

Mucous stools are reported to occur several times a week in some institutions or only once or twice a month at others. As in giant pandas, this is probably associated with the diet. Red pandas receiving milk pass mucous faeces much more often than others. The physiological consequence of these events is not understood.

Mange occurs with high incidence in some facilities, which lose several cubs to the disease each year. The specific mite responsible for the infections has not been identified (skin scrapings for definitive diagnoses are not done) but the condition does appear to be responsive to the standard three-injection course of ivermectin in red pandas that are treated early enough.

Dermatophytosis, or ringworm, is common in young cubs, particularly those living in the very warm areas of China. The cases observed by the author had been treated ineffectively with single injections of antibiotic and had resulted eventually in the loss of a cub's tail or in the death of the cub. Standard treatment with itraconazole rescued cubs uneventfully.

By far the most common veterinary concern in institutions with a large number of red pandas is bite wounds that result from fights in the overpopulated enclosures (see Figure 18.5). Many of these wounds go unnoticed or untreated until severe abscesses have formed and, in several cases, result in the amputation of tails. Some of the older red pandas in these institutions have large hairless areas on their tails and backs and their skin is thick with scar tissue from old wounds.

Heartworms have been found incidentally in red pandas that had died of undiagnosed causes. These were adult roundworms situated in the right ventricle; the taxonomy of the helminth was not determined. Microfilaria were present in high densities in the peripheral circulation of these animals, but the individuals had not shown clinical signs characteristic of heartworm disease. Since these findings, monthly oral dosing with ivermectin has been initiated at this institution. Parasite prevention protocols at other institutions vary greatly, from none at all to administration of ivermectin or an oral anthelminthic one or more times a year.

CONCLUSION

The need for management of the *ex situ* red panda population in China has been recognized in the Global Master Plan that was prepared in 2005 as the result of an international workshop [7]. An important step toward this goal was China's participation in this workshop. Capacity building in animal husbandry, animal management and veterinary medicine are important next steps to improve the quality of the population and the welfare of the individuals living in captivity in China. The implementation of good record-keeping practices will be critical for the development of centralized resources such as a studbook and breeding recommendations.

Research on genetics is necessary for the management of the current and future *ex situ* population. A project to begin this effort has been initiated at the Chengdu Research Base of Giant Panda Breeding in Sichuan Province. Each of the 40 red pandas at this institution was microchipped at the end of 2007, and analysis of genetic relationships within the group is underway. Maintenance of permanent records is in place, and breeding management will follow the genetic analysis. The hope is that management of this population will serve as a model, which will be transferred, together with standards for husbandry and veterinary care of captive red pandas, to other institutions in China.

Diagnosis of diseases of red pandas and definitive determination of the causes of death are important toward the understanding of veterinary medicine of red pandas living *ex situ* in China. A commitment to this effort and capacity building through collaboration with western universities or non-governmental organizations for development of veterinary and laboratory expertise is necessary.

Research to elucidate the ecology, behaviour, population characteristics and veterinary medicine of red pandas living *in situ* in China is urgently needed. This will help us to understand the status and trend of the wild population and threats to the species. Protection of the *in situ* population is the ultimate purpose of the proper management of an *ex situ* population, and both must be addressed in tandem in order to preserve this remarkable animal.

References

[1] F. Wei, Z.J. Feng, Z.W. Wang, J.C. Hu, Current distribution, status and conservation of wild red pandas *Ailurus fulgens* in China, Biol. Conservat. 89 (1999) 285–291.

[2] X. Liang, Z. Zhang, L. Zhang, et al., Isolation and characterization of 16 tetranucleotide microsatellite loci in the red panda (*Ailurus fulgens*), Molec. Ecol. Notes 7 (2007) 1012–1014.

[3] Y.Z. Li, F.J. Shen, L. Zhang, et al., Genetic Diversity and Parentage Assessment of the Captive Red Panda (*Ailurus fulgens*) with Microsatellite DNA Markers, 2010, (in press).
[4] Q. Qin, F. Wei, E.J. Dubovi, I.K. Loeffler, Serosurvey of infectious disease agents of carnivores in captive red pandas (*Ailurus fulgens*) in China, J. Zoo Wildlife Med. 38 (2007) 42–50.
[5] I.K. Loeffler, J.G. Howard, R.J. Montali, et al., Serosurvey of ex situ giant pandas (*Ailuropoda melanoleuca*) and red pandas (*Ailurus fulgens*) in China with implications for species conservation, J. Zoo Wildlife Med. 38 (2007) 559–566.
[6] Q. Qin, I.K. Loeffler, M. Li, K. Tian, F. Wei, Sequence analysis of a canine parvovirus isolated from a red panda (*Ailurus fulgens*) in China, Virus Genes 34 (2007) 299–302.
[7] A.R. Glatston, K. Leus, Global Captive Breeding Masterplan for the Red or Lesser Panda *Ailurus fulgens fulgens* and *Ailurus fulgens styani*, Rotterdam Zoo, Rotterdam, The Netherlands, 2005.

CHAPTER

19

The Global Captive Population of the Red Panda – Possibilities for the Future

Kristin Leus

Copenhagen Zoo and IUCN/SSC Conservation Breeding Specialist Group (European regional office), Merksem, Belgium

OUTLINE

INTRODUCTION

Previous chapters have illustrated the history and current status of the global captive population of the red panda. In this chapter, we will investigate the potential long-term future strategies for the captive populations of the two subspecies.

Most of the red pandas in captivity worldwide are part of regional, cooperative, science-based captive-breeding programmes that function within their respective regional zoo and aquarium associations (Table 19.1). The majority of the *Ailurus fulgens fulgens* captive population taking part in cooperative breeding programmes lives in Europe and

Red Panda. DOI: 10.1016/B978-1-4377-7813-7.00019-7

TABLE 19.1 Names of the regional zoo associations and cooperative breeding programmes in the different world regions with red panda populations

Regional Zoo Association	Name for intensively managed cooperative breeding programme in the region
European Association of Zoos and Aquaria (EAZA)	European Endangered Species Programme (EEP)
Association of Zoos and Aquariums (AZA) (North America)	Species Survival Plan (SSP)
Australasian Regional Association of Zoological Parks and Aquaria (ARAZPA)	Australasian Species Management Programme (ASMP)
African Association of Zoos and Aquaria (PAAZAB)	African Preservation Programme (APP)
Central Zoo Authority (Statutory Body under the Ministry of Environment and Forests, Govt. of India) (CZA)	Conservation Breeding Programme (CPM)
Japanese Association of Zoos and Aquariums (JAZA)	Species Survival Committee of Japan (SSCJ) propagation plan
Chinese Association of Zoological Gardens (CAZG)	

North America, with smaller populations in Australia and New Zealand, South Africa, India and Japan. The managed section of the *Ailurus fulgens styani* global captive population is spread over two subpopulations in North America and Japan. A sizeable number of *A. f. styani* and *A. f. fulgens* live in Chinese zoos, but the individuals in the latter subpopulation are not yet cooperatively managed. The same is true for a handful of red pandas of either subspecies living in other countries around the world.

During the 2004 Global Captive Breeding Masterplan meeting in Rotterdam Zoo, the Netherlands, representatives from the Chinese, Indian, European, North American, Australasian and South African subpopulations of the global captive population together decided that the overall aim of the global captive population for each of the subspecies of the red panda (*fulgens* and *styani*) should be to [1]:

1. Directly contribute to the conservation of the species by providing a genetically and demographically sustainable backup for the wild population and by holding the potential to supply individuals for reintroduction programmes (especially for the captive subpopulations in the range states); and
2. Indirectly contribute to the conservation of the species through education and the raising of public awareness regarding the biology and conservation of the red panda by each of the institutions contributing to the global captive programme.

There are many different ways in which an *ex situ* population can directly contribute to the conservation of the taxon in question, for example, by: providing a genetically and demographically sustainable backup population for the wild population; functioning as a source for individuals for reintroduction or supplementation; being the source for long-term gene and biomaterial banking; and being the subject of research on biological and ecological questions relevant to *in situ* conservation. Depending on the particular situation of the species in question, the relative importance of each of these components will vary.

In this chapter, we will concentrate on how the global captive population of the red panda can directly contribute to the conservation of the species and its subspecies.

A BACKUP FOR THE WILD POPULATION

One of the ways in which an *ex situ* population can directly contribute to the conservation of the species is by providing a genetically and demographically sustainable backup population for the wild population. A sustainable population could be either self-sustainable, or sustainable with the aid of importation of individuals from the wild.

Self-sustainability

A captive population that is demographically self-sustainable is able to maintain at least a stable, and where necessary growing, population size purely on the basis of having a birth rate that is larger than the death rate (in other words without relying on importations). A genetically self-sustainable population is able to maintain a sufficiently high level of gene diversity without further importation of unrelated individuals. Below, we will investigate if it is possible to maintain demographically and genetically self-sustaining global captive populations of both subspecies of the red panda.

Requirements for a Self-sustaining Captive Population of Red Pandas

Geneticists and population biologists use a number of different ways to measure genetic diversity but, for the purposes of this chapter, it is sufficient to know that genetic diversity relates to the number, frequency and distribution of variations in the DNA code within individuals and within the population as a whole. The more genetically diverse a population is, the better. After all, it is only because not all individuals have exactly the same genetic make-up, that when changes occur in the environment of the population (and wild environments are typically quite variable), at least some individuals in the population are likely to have the right genetic characteristics to be able to cope with that change. Therefore, the genetic diversity of a population in a sense represents its long-term evolutionary potential [2]. Genetic diversity makes it possible for the process of natural selection to occur. In the short term, there is also an overall positive correlation between genetic diversity and the fitness of the population [3].

Maintaining genetic diversity is therefore a very important goal in species conservation. In sufficiently large wild populations, the loss of genetic diversity through selection and through chance (referred to as random genetic drift) is balanced by the gain through mutations [2]. To balance the loss of genetic diversity due to drift with the generation of new diversity through mutations, populations are thought to need an effective size (Ne) of at least 500–5000 individuals [2,4–6]. The effective population size is the size of an ideal population that would have the same rate of loss of genetic diversity or inbreeding as is observed in the real population, whereby an ideal population is characterized by the random breeding of all individuals with all individuals (including breeding within sexes and with self), a constant size, a constant number of breeding individuals through the generations, an equal sex ratio, equal family sizes, non-overlapping generations and dispersal into or out of the population [2]. Needless to say, most real-life populations are far from "ideal" and the effective population size is usually much smaller than the true population size (N). The ratio of Ne/N indicates how "effective" the true population size can be in terms of preserving the species. If you find the definition of Ne hard to digest, it is perhaps easier to understand that, for example, an actual population of 500 individuals composed of 480 males and 20 females will have a lower effective population size than one with 250 males and 250 females. Important factors influencing the effective size of a population are the number of breeding animals (the more breeders, the better), the sex ratio (the more even, the better), family sizes or founder representation (the more even, the better) and fluctuations in population size (the fewer fluctuations, the better). There are a number of different formulas to calculate Ne, based on the different factors described above that influence Ne but, in general, one can state that wild populations often have an Ne/N ratio of around about 0.1 [7], meaning that the effective population size is about one-tenth of the true population size. Knowing this, it means that in general, to balance the loss of genetic diversity due to drift with the generation of new diversity through mutations, we would need populations of 5000–50 000 individuals.

What does this all mean in terms of a genetically self-sustaining global captive population of red panda? Would we, in the best case, need 5000 individuals? Luckily, managed captive populations have an important advantage in this regard. The breeding in captive populations can often be proactively managed to benefit Ne (and therefore the retention of genetic diversity). In captivity, it is to a certain extent possible to decide how many and which individuals breed, how much, and with whom. Therefore, captive populations often have an Ne/N ratio that is larger than that of wild populations. The international studbook database for the red panda [8] maintained in the software SPARKS (Single Population Analysis & Record Keeping System) [9] and analysed with the analytical software package Population Management 2000 (PM2000) [10] allows us to calculate the ratio of Ne/N for the global captive population (Table 19.2). For all Ne and genetic calculations for the purposes of this chapter, all living animals of 16 years and older were excluded from the analysis. This is because demographic analysis has shown that relatively few pandas live longer than 15 and that the chances for reproduction at these ages are zero for females and very small for males. Any important genetic material that is present in these animals and that has not yet been passed on to the next generation is therefore unlikely to still be passed on in the future. Leaving them included in the genetic analysis could give a false sense of security.

TABLE 19.2 The effective population size (Ne) and the ratio of the effective over the true population size (Ne/N) for the global captive populations of the two subspecies of the red panda

	Ne/N	Ne	N≤15 years	N TOT
Ailurus fulgens fulgens	0.28	118	423	456
Ailurus fulgens styani	0.29	61	210*	269*

**This is the population in the international studbook for Japan and North America. There are another ≈250* A. f. styani *in Chinese institutions for which there are no detailed data and which therefore could not be included in the SPARKS software dataset.*

All animals of 16 years and older were excluded from the analysis. N ≤ 15 years = Total living population of 15 years and younger. N TOT = Total living population.

An Ne/N ratio of 0.28 and 0.29 for *A. f. fulgens* and *A. f. styani* respectively falls within the range that appears common for captive populations (0.2–0.4) [2,11]. These Ne/N ratios mean that 'only' a minimum of about 1700–1800 red pandas would be necessary to balance the loss of genetic diversity due to drift with the generation of new diversity through mutations. This is still much larger than the current true population sizes (*A. f. fulgens*: 456; *A. f. styani*: 269 in Japan and North America; ≈540 with Chinese animals included), or the estimated world captive carrying capacity (about 580 and 660 for *A. f. fulgens* and *A. f. styani*, respectively) [1]. Because the number of individuals needed to avoid the loss of genetic diversity is almost invariably much larger than there is good-quality captive space available, it was suggested to settle for a compromise in managing captive populations, allowing a small loss of gene diversity and therefore being able to keep smaller populations of more species. Initially, the goal proposed for captive populations was to preserve 90% of the genetic variation in the source population for 200 years, reflecting the estimated time frame needed for wild habitat to recover/become available following conservation actions and a predicted decline in the human population in 100–200 years [12]. The time frame was later shortened to 100 years due to an underestimation of space limitations and a perceived growing potential in the contribution of cryopreservation and assisted reproduction [13]. In the zoo world, a self-sustaining captive population is therefore often defined as one that can maintain a minimum of 90% of the gene diversity of the source population for 100 years, without the further importation of unrelated individuals. Newly imported individuals that are unrelated to the current population are usually referred to as founders. Strictly speaking, they are called "potential founders" for as long as they do not have living descendants and only become true founders once they do have living descendants. Founders are often born in the wild, but could in a sense also come "packaged" in the form of unrelated captive-born individuals from institutions that previously were not part of the breeding programme in question.

How many red pandas of each subspecies are needed to maintain 90% of the source gene diversity for 100 years without the importation of new founders? Apart from the ratio Ne/N, other population parameters that are important to determine these population size targets are:

1. the population growth rate
2. the generation time
3. the current gene diversity [2,13].

This can easily be understood as follows.

1. The only way that gene diversity (variations in the DNA code) can be preserved in the long term is to either literally freeze the genes for later use (which for some species is possible through cryobanking of gametes, embryos, etc.) or to pass them on to future generations through breeding. Since in most higher vertebrates only 50% of the genetic material of the parent is passed on to each offspring, the more offspring an individual has, the more of its genetic material is passed on to the next generation. When many individuals produce many offspring this results in a high population growth rate. Therefore, all things being equal, populations with higher growth rates retain more gene diversity in a fixed time frame than those with lower growth rates.

2. Furthermore, it is very likely that not all individuals breed, and many individuals will have insufficient numbers of offspring to pass on all of their genetic variation to the next generation. In terms of genetic diversity, each generation therefore contains only a sample of the genetic diversity of the previous generation [13]. Each time a population proceeds to the next generation, genetic diversity is lost. Populations or species with long generation times (e.g., elephants, vultures, tortoises) pass through only a few generations in the time span of 100 years. In the same amount of time, species or populations with short generation times pass through many generations, and therefore many steps at which genetic diversity is lost. All things being equal, populations with longer generation times retain more genetic diversity in a fixed amount of time.
3. The current gene diversity is obviously an important determinant of how much gene diversity can be retained after 100 years. This "starting point" is to a large degree determined by the number of unrelated wild individuals that have founded the captive population, as well as by how close to the genetic ideal the population has behaved from establishment to present time. Relatively large future increases in gene diversity are usually only possible by adding new founders, by breeding with underrepresented founders that are still alive or, in young programmes, by breeding with the descendants of underrepresented founder lines. In more established programmes, in which most founders are dead and access to new founders may be limited, only relatively small increases in gene diversity are sometimes possible by breeding preferentially with descendants of underrepresented founder lines (according to the mean kinship principle – see [2,13–15] for more details). More often, however, gene diversity continues to be lost (hopefully gradually) over time and management will be directed towards slowing the rate of loss over time.

A Self-sustaining Global Captive Population of Red Pandas – is it Possible?

The studbook population analysis software Population Management 2000 (PM2000) [10] provides a deterministic calculation of the population size needed to maintain a certain proportion of gene diversity of the source population for a certain amount of time, given the generation time, Ne/N ratio, current population size, current gene diversity and the population growth rate of the captive population in question [16].

For *Ailurus fulgens fulgens*, with the generation time, Ne/N ratio and yearly growth rate projected from the age- and sex-specific birth and mortality data calculated for the period of 1 January 1984 to 31 December 2006, PM2000 tells us that 607 individuals are needed to maintain 90% of gene diversity for 100 years (Table 19.3). This is only 27 animals above the estimated global carrying capacity which did not contain the individuals in China.

The same calculations for *Ailurus fulgens styani* tell us that even with a full capacity of 660 individuals, the gene diversity retained after 100 years will be only 87.9% (see Table 19.3). Even with 1000 individuals, the amount of gene diversity that can be retained does not increase. The main reason for this is the overall projected yearly growth rate, which is practically stable at 0.6% per year. This means that if the current birth and death rates do not change, after 100 years, there will only be about 448 individuals in the population, and a higher carrying capacity therefore does not help in this situation. The low overall growth rate is mainly due to a negative projected female population growth of −2.4% in the North American subpopulation, compared to +1.2% for females in the Japanese

TABLE 19.3 Genetic and demographic population parameters for the actively managed global captive populations of *Ailurus fulgens fulgens* and *Ailurus fulgens styani* as projected from the current population and the age-specific fecundity and mortality rates of the populations during the period 1 January 1984 till 31 December 2006

	Ailurus fulgens fulgens	*Ailurus fulgens styani*
Generation time	6.0	6.4
Lambda (yearly growth rate)	1.033	1.006
Current population size (N) (excluding individuals ≥16 years)	423	210
Effective population size (Ne)	118	61
Ne/N	0.28	0.29
Current % gene diversity retained	94.7	96.9
N needed to maintain 90% gene diversity after 100 years	607	Not possible. Can maintain 87.9% gene diversity with 660 individuals
Carrying capacity	580	660

All animals of 16 years and older were excluded from the genetic analysis. The world population for *A. f. styani* was taken to be the population in the international studbook for Japan and N. America. The individuals in Chinese and other collections are not currently part of the actively managed population.

population. The total estimated global carrying capacity for this subspecies includes the estimated current population size in China, but also 150 spaces in North America. A negative population growth for females in North America would therefore compromise the sustainability of the global population because the SSP population needs to be large enough to achieve this total carrying capacity. The genetic importance of the SSP population for the world population is investigated further in this chapter. Potentially good news is the fairly large (estimated ≈250 individuals) population in Chinese institutions for which precise data are lacking and that could therefore not be included in the analysis. As there are no records of recent importations from China, or of the export of red pandas to China, it is highly unlikely that the individuals in the Chinese population are related to those in the global managed population. This means that the total number of pandas required to achieve the global genetic goal is smaller. The Chinese institutions also have the animal enclosures needed to reach the global carrying capacity. This, however, implies bringing the Chinese pandas under tight population management.

With some husbandry and management improvements to increase the growth rate of – particularly – female *A. f. styani* in North America and detailed management of the Chinese subpopulation of this subspecies, it would appear possible to achieve self-sustaining global captive populations of the two subspecies of red panda. However, is the situation really as "safe" as it appears?

Semi-autonomous Regions – Good or Bad?

The above estimates of genetic retention with various population sizes assume the existence of one large, interbreeding population. In reality, the global captive population of

TABLE 19.4 Total number of red pandas (N) and number of red pandas of 15 years and younger (N ≤ 15 years) for each subspecies and in each subpopulation

N total/N ≤ 15 years	EEP	SSP	ASMP	APP	CPM	SSCJ
A. f. fulgens	203/193	122/106	42/39	27/27	19/17	32/30
A. f. styani		44/39				225/171

EEP: Europe; SSP: North America; ASMP: Australasia; APP: (South) Africa; CPM: India; SSCJ: Japan.

red pandas under active management is distributed over different subpopulations in several geographic regions: Europe, North America, South Africa, Australasia, Japan and India for *Ailurus fulgens fulgens*, and North America and Japan for *Ailurus fulgens styani* (Table 19.4). Each large world population is therefore composed of a series of smaller populations. Is this good or bad? We will first provide some theoretical background to this issue before evaluating the specific situation of the red panda.

In small populations, random genetic drift tends to outweigh genetic adaptation in determining how the genetic variation of the population changes over the generations, meaning that the relative frequencies in which particular variations in the DNA code are present in the population from generation to generation will, to a larger degree, be determined more by pure chance than by natural selection for the most advantageous (i.e., adaptive) variations. Furthermore, inbreeding accumulates faster in smaller than in larger populations, and in the majority of those naturally outbreeding populations studied to date (in captivity or in the wild), inbreeding was found to be associated with inbreeding depression (some form of reduced fitness in inbred individuals, e.g., increased juvenile mortality, smaller litter sizes, reduced sperm count) [17–20]. Inbreeding depression can lead to extinction [21,22]. Large captive populations would therefore appear to fare better than small populations. However, genetic adaptation to captivity tends to happen faster in larger than in smaller populations [23]. Even individuals from captive populations actively managed to minimize kinship [14] (which reduces genetic adaptation to captivity [24]) may still demonstrate reduced fitness when confronted with more stressful conditions, such as upon reintroduction into the wild [23]. Captive and wild environments tend to differ from one another in many aspects, and it is therefore not surprising that genetic adaptation to one of these environments can be detrimental when placed in the other environment. Large continuous captive populations can therefore also have their downsides. In addition, several studies have suggested that, although a series of smaller subpopulations tends to lose genetic diversity quite rapidly in each subpopulation, overall this structure may retain genetic diversity better than one large population with the same total number of individuals. This is because, through random genetic drift, different subpopulations will preserve different variations of the DNA code [25–27].

For the above reasons, it has been suggested that in captive settings it may be best to work with a series of smaller semi-autonomous populations in order to avoid too much genetic adaptation to captivity, with occasional transfers between these populations to avoid too much loss of genetic diversity and inbreeding in each subpopulation [23,26,27]. Smaller populations will accumulate inbreeding faster, but when individuals from different, genetically relatively distinct (although possibly also moderately inbred) subpopulations are bred with each other to produce individuals for reintroduction, this reintroduction stock will be outbred, genetically diverse and have less adaptation to captivity. However, two important conditions need to be met in order for a subpopulation structure to be more beneficial to overall genetic diversity retention than one large population: (1) no subpopulation should go extinct (e.g., through inbreeding depression, demographic instability or catastrophes); and (2) for the effect to be large enough to be significant in practical terms, the subpopulations must remain at fixed effective population size [2] (Lacy, personal communication). Some stochastic computer simulation models have found an increased loss of genetic diversity with semi-autonomous populations compared to one large population, exactly because of increased demographic variation due to the subdivision, leading to decreased Ne of the metapopulation and the subpopulations [28]. Migration between subpopulations did not prevent this effect on loss of genetic diversity. The issue of whether the apparent benefit of population subdivision is really predominantly due to the preservation of different alleles in different populations, or is perhaps predominantly an artefact of the assumptions and conditions used so far in the models to test this (e.g., no extinctions, fixed Ne, constant sex ratio and equal family sizes), is therefore still not clear. This purported genetic benefit of a subpopulation structure still needs further testing with theoretical models, particularly in the framework of real-life, proactively managed, captive breeding programmes. For example, active genetic management may lessen the benefit of differential genetic drift.

Apart from genetic factors, there are other issues to take into consideration when evaluating the advantages and disadvantages of semi-autonomous subpopulations. Working with semi-autonomous regional subpopulations in one global programme significantly increases the availability of cage space for a species. This is important since, for many species, fairly large populations (typically several hundred individuals) are needed to maintain 90% of gene diversity of the source population for 100 years. For practical and resource reasons, it is usually not possible to transfer individuals among regional subpopulations as frequently as within subpopulations. Keeping inbreeding in subpopulations at acceptable levels and avoiding too much loss of genetic diversity may require moving a minimum of one and up to 10 individuals between the populations per generation [29]. This rate of exchange is likely achievable in practice and is significantly less than the number of inter-institutional transfers that would typically happen within one subpopulation in the time span of a generation. Ironically, however, this rate of genetic exchange is so frequent that, in genetic terms, the global population may almost behave as one population, thereby likely compromising the ability of different subpopulations to retain different alleles through genetic drift, because the frequent migrations would prevent subpopulations from becoming/remaining sufficiently genetically distinct (Lacy, personal communication). Another advantage of working with geographically separate semi-autonomous subpopulations is the reduced risk of losing the majority of the captive population through catastrophes such as disease outbreaks, natural catastrophes, or political and socio-economic instability. For example, if the majority of transfers occur within subpopulations, the chances are smaller that a disease outbreak in one subpopulation will be transferred to another subpopulation.

In real life *ex situ* breeding programmes, the combined benefits of a larger global carrying capacity, less widespread effects of catastrophes, and several logistical and resource use advantages may well lead to proportionately bigger advantages for the global population than the possibly small genetic benefit that a series of semi-autonomous subpopulations can bring in terms of differential genetic drift in different subpopulations and reduced and differential adaptation to captivity (Lacy, personal communication). This too needs further testing.

In practical terms, the aim is therefore to find the balance between population size (large enough to limit extinction risk and demographic instability and small enough to limit adaptation to captivity), inbreeding levels (low enough to avoid significant inbreeding depression) and sufficient genetic uniqueness of the subpopulations (subpopulations that are different enough so that the genetic variation in the total metapopulation is significantly higher than that in each subpopulation and so that individuals can migrate among subpopulations to keep inbreeding levels acceptable, but that are also not so different that the extinction of one subpopulation would compromise the genetic integrity of the metapopulation).

Having established that there may still be important advantages to working with semi-autonomous subpopulations in captivity, let us have a look at the subpopulation structure of the two subspecies of red pandas in captivity.

Ailurus fulgens fulgens

Four of the six subpopulations are very small (Australasia 42, Japan 32, South Africa 27, India 19 – including individuals of 16 years and older). Demographic stochasticity by itself significantly increases the probability of extinction in populations with fewer than 100, and certainly fewer than 50 individuals [2]. In such small populations, chance events such as a few consecutive years with more males than females born, or one year with a higher mortality rate or lower birth rate, can have relatively large consequences. Despite the captive setting, where somewhat more control over environmental variability can be expected, these small red panda populations are therefore very insecure, especially if the small population size runs concurrent with accumulating levels of inbreeding (Table 19.5). A census of the world population of *A. f. fulgens* from 1 January 1978 to 1 January 1992 was found to have seemingly lower growth rates projected from life tables for inbred pandas compared to non-inbred pandas, as well as higher first year mortalities in inbred versus non-inbred individuals [30]. More recently, Boakes et al. [20] were able to demonstrate the presence of inbreeding depression with regard to juvenile survival in the North American captive population of red pandas (two subspecies combined).

There tends to be a threshold relationship between incremental extinction risk and inbreeding depression. At low levels of inbreeding, the risk of extinction remains very low,

TABLE 19.5 Mean inbreeding coefficient (F) in the European (EEP), North American (SSP), Australasian (ASMP), South African (APP), Indian (CPM) and Japanese (SSCJ) subpopulations of the two subspecies of the red panda (*Ailurus fulgens fulgens* and *Ailurus fulgens styani*)

Mean F	EEP	SSP	ASMP	APP	CPM	SSCJ
A. f. fulgens	0.0486	0.0467	0.0726	0.0886	0.0536	0.0414
A. f. styani		0.0399				0.0067

but there is a marked and incremental increase in risk of extinction due to inbreeding depression from moderate levels of inbreeding onwards [31]. Hence the general recommendation to exchange animals among subpopulations at least when the average inbreeding coefficient reaches 0.1–0.2 [2]. In the case of the red panda, this is illustrated by the South African and Australasian populations of *A. f. fulgens*, where the average inbreeding coefficient is creeping towards 0.1 (see Table 19.5) and where, in comparison to the larger North American and European populations, more frequent transfers in and out of the population will be necessary to keep the level of inbreeding under control. Although the size of the Indian subpopulation is even smaller, the average level of inbreeding is currently lower due to the relatively high number of founders (22) for the small number of living descendants. However, the living descendants are very related to each other. Unless the three potential founders in this population start breeding soon, and unless further unrelated animals can be added in the future, inbreeding will accumulate very fast in this population. In addition, the Indian subpopulation's sex ratio is currently strongly male biased, with 14 males and three females of 15 years or younger, which introduces another significant factor of demographic instability and extinction risk.

Losing one or two of the smaller subpopulations of *A. f. fulgens* would have a relatively small impact on the demographic health of the metapopulation but might significantly affect the genetic status and potential of the metapopulation if they contained a relatively high amount of genetic diversity that is not present elsewhere in the metapopulation.

The software package PM2000 allows us to investigate the genetic distinctness of the subpopulations of the global captive populations of each of the red panda subspecies.

The percentage of the total genetic diversity of the metapopulation of *A. f. fulgens* that is present as between population divergence rather than within-population diversity is 5.02% (=Fst). In principle, the bigger the Fst the better, with respect to retention of gene diversity across the metapopulation, but the higher the risk if any of the subpopulations goes extinct. Also, if the subpopulations are relatively small and Fst is kept high, this in time implies higher inbreeding levels in the subpopulations, which could lead to reduced fitness through inbreeding depression. If Fst were to be 10%, this would imply that inbreeding in the subpopulations would generally become 10% higher than in the global population if it were to behave as one interbreeding population. Although as indicated above, further theoretical modelling is needed to provide firm guidelines for subpopulation management of captive populations, an Fst of 10% is provisionally considered acceptable. It might therefore be beneficial to still somewhat increase the genetic distinctness among subpopulations of this red panda subspecies. However, let us compare the subpopulations two by two (Table 19.6). The subpopulation pairs with most genetic distinctness between them are India with respectively Australasia, South Africa and Japan, as well as South Africa with Japan, whereby each pair has more than 5% of the total gene diversity of the two populations present as between population divergence rather than within population divergence. These are also the smaller subpopulations (42 or fewer individuals) that are more vulnerable to extinction. It would appear that this would lead to a loss of gene diversity should one or more of these populations go extinct. However, the largest subpopulation (EEP) has low Fst values with all of the subpopulations, including the Indian subpopulation, which means that the EEP population contains the majority of the genetic diversity that is present in the other subpopulations. The EEP population is

TABLE 19.6 Current population gene diversity as a proportion of the gene diversity of the source population (diagonal) and the proportion of the total gene diversity of each pair of subpopulations that is present as between population divergence rather than within-population diversity (Fst) (below the diagonal) for the regional subpopulations of the global captive population of *Ailurus fulgens fulgens*

	EEP	SSP	ASMP	APP	CPM	SSCJ
EEP	**0.9255**					
SSP	0.0232	**0.9053**				
ASMP	0.0189	0.0374	**0.8657**			
APP	0.0150	0.0344	0.0298	**0.8441**		
CPM	0.0121	0.0260	0.0554	0.0679	**0.8306**	
SSCJ	0.0188	0.0346	0.0382	0.0539	0.0705	**0.8351**

EEP: Europe; SSP: North America; ASMP: Australasia; APP: (South) Africa; CPM: India; SSCJ: Japan.

slightly more genetically distinct from the SSP population (Fst = 2.32%), which is in itself also a larger population.

The non-EEP/SSP populations are therefore at the moment only bringing a fraction of additional genetic diversity. In fact, the EEP and SSP population together would by themselves already be able to maintain 88% of gene diversity for 100 years with only 460 individuals, the estimated total carrying capacity of the two populations [1]. This is to a large extent because the EEP and SSP together currently contain almost the same amount of gene diversity as the whole global captive population (0.9401 compared to 0.9469). The largely unmanaged Chinese population, however, is likely to contain a substantial amount of unique genetic material.

Are the larger EEP and SSP populations in danger of accumulating a lot of genetic adaptation to captivity? Both populations are being managed to reduce kinship, but although this has been shown to reduce/slow genetic adaptation, it does not eliminate it [23,24]. The EEP and SSP populations of *A. f. fulgens* have gone through on average about 4.5 generations in captivity, and with a generation time of about 6 years, can be expected to go through about 16.7 generations in the next 100 years. In the study by Woodworth et al. [23], the trends in fitness reduction under wild conditions after captive breeding were already obvious at generation 13 when the first assays were carried out, and became more prominent over time. Although the red panda populations have only gone through a few generations to date, it would appear that over the next 100 years enough generations will pass for some genetic adaptation to start manifesting itself, should the effective population size be large enough for genetic adaptation to overrule inbreeding depression as the dominant cause of reduced fitness under more stressful conditions. In Woodworth et al. [23], after 50 generations of breeding under benign captive conditions, the populations of *Drosophila* that suffered most from predominantly genetic adaptation-caused reduced fitness when placed under wild conditions were those with an Ne of 500. Those with Ne = 250 had very slightly reduced "wild" fitness compared to the populations with Ne = 100, which performed best under the wild conditions. The populations with Ne = 50 or 25 also had reduced fitness compared to those with Ne = 100, but for these small populations this was predominantly caused by inbreeding depression. The EEP and SSP currently have an effective population size of 61.8 and 25.2 respectively. The Ne values for both populations therefore suggest that avoiding inbreeding depression may be of bigger concern in these populations than genetic adaptation to captivity. Further studies are needed to test after how many generations and at which Ne sizes genetic adaptation to captivity starts to be significant in captive populations, including actively managed captive populations in which mating strategies are employed that are designed to slow the loss of gene diversity and to reduce adaptation to captivity.

Purely from a population management point of view it would be wise to:

1. Investigate the possibilities for taking advantage of the fact that the Indian captive subpopulation is located in a range state by building a relatively small but genetically diverse and demographically sufficiently stable subpopulation that makes use of confiscated/rescued wild pandas and that can function as a nucleus for building up stock for reintroduction if and when that becomes a relevant conservation approach (see further in this chapter for details on the potential for the Indian subpopulation). Offspring from these new founder lines can be sent to the other regional subpopulations such that they can become a bit more genetically distinct and can receive additional genetic diversity.
2. Avoid population decline or population fluctuations and gradually introduce some new genetic material in the EEP and SSP populations (from India – see item 1), whereby different founder lines should go to each of the two subpopulations. In this way, the needed global carrying capacity for a self-sustaining population can be reduced.
3. Either:
   - **(a)** Leave the South African, Australasian and Japanese subpopulations as they are, whereby they help to reach the needed global carrying capacity and fulfil an important conservation ambassador and educational role for the species and its habitat. However, because of their small size there is a relatively high risk for catastrophes, demographic instability and inbreeding depression. Regular transfers in and out of these populations will remain necessary to keep inbreeding sufficiently low and to solve possible demographic problems. Should such intercontinental transfers no longer be possible because of disease transfer regulations or shortage of resources, these populations will very quickly run into problems. It would thus be unwise to include much unique genetic material in these subpopulations.

Or,

   - **(b)** Should the carrying capacity be available in these regions, expand the South African, Australasian and Japanese subpopulations to about 80–100 individuals each such that the demographic risk becomes smaller. This would make these subpopulations less vulnerable to problems with intercontinental transfers so that they could also be made more genetically distinct. Ideally, this would include the importation of some new founder lines from India (should they be available), whereby different founder lines should go to each of the subpopulations.

4. Maintain an up-to-date studbook of the Chinese population in SPARKS, and actively manage the population, such that it can be determined how much and how this population can contribute to the global carrying capacity and the global genetic health of the population.

Ailurus fulgens styani

At the end of 2006, the total global captive population of *A. f. styani* for which detailed data could be entered in SPARKS was spread over two subpopulations, one in Japan's SSCJ (N = 225) and one in North America's SSP (N = 44), as well as a handful of individuals spread over other countries that are not part of the formal breeding programme and are therefore not included in the analysis below.

The SSP population of this subspecies is relatively insecure because of its small size, which leaves it vulnerable to demographic stochasticity and inbreeding accumulation, and also because of its overall population decline since 1997, mainly due to a high female first-year mortality resulting in a negative projected growth rate for females and a male-biased sex ratio of the current population (23 males and 16 females of 15 years or younger; four of these females are 12 years or older and therefore have doubtful breeding prospects). The high insecurity of the SSP population could form a significant genetic problem if it were to contain a significant proportion of genetic diversity that is not present in the Japanese subpopulation.

From PM2000 analyses we learned that the percentage of the total gene diversity of the metapopulation of *A. f. styani* (SSP + SSCJ) that is present as between-population divergence rather than within-population diversity is only 1.01% (=Fst). In fact, the Japanese population alone is able to maintain 89% of the gene diversity of the source population for 100 years with 470 red pandas, whereas the two populations together can only retain 87.9% with the same number of pandas (Table 19.7). Therefore, the SSP population contains only a small amount of unique genetic material and is from a demographic point of view currently a drag on the potential of the global population. If the growth rate in the SSP population can be increased, however, so that the growth rate of the combined population reaches that of the Japanese population alone (1.9% per year), then 90% of gene diversity can be maintained for 100 years with a carrying capacity that is not excessive, in

TABLE 19.7 Yearly growth rate (lambda), proportion of effective population size to true populations size (Ne/N), generation time (T), current gene diversity as a proportion of the source gene diversity (GD), percentage of source gene diversity retained after 100 years and the population size (N) needed to maintain this percentage of gene diversity (GD) for 100 years under these conditions of the captive breeding programme for the SSCJ (Japanese) and the SSCJ + SSP (North American) population of *Ailurus fulgens styani*

	Lambda	Ne/N	T	Current GD	%GD after 100 years	N for GD
SSCJ	1.019*	0.29	6.5	0.9640	89.0%	470
SSCJ + SSP	1.006*	0.29	6.4	0.9694	87.9%	470
SSCJ + SSP	1.019**	0.29	6.4	0.9694	90.0%	435

*projected from the data
**superimposed
The total carrying capacity of the SSCJ + SSP population = 470 individuals.

this case 435 pandas (Table 19.7). During the 2004 masterplan, it was estimated that the SSP population has a carrying capacity of about 150 pandas. The carrying capacity of the Japanese population is not known at present. However, if we assume it is at least as much as the present population, then the total carrying capacity for the two regions is about 375. Even if the growth rate of the SSP population can be improved, there is therefore a need for space for an extra 60 pandas. The global carrying capacity for this subspecies mentioned earlier in this chapter (about 660 individuals) included the estimated carrying capacity of the currently not-yet-cooperatively managed Chinese population. Unless both the mortality of SSP females can be addressed (so this population can grow to its carrying capacity) and space for an extra 60 pandas can be found among the SSP and Japanese populations, the Chinese institutions will be needed to achieve the necessary carrying capacity for a globally self-sustaining captive population of this subspecies. If the Chinese population contains unrelated genetic material, this may decrease the required global carrying capacity, but nevertheless also requires cooperative and pro-active management of the Chinese population to make sure this important genetic material is not lost.

The SSCJ population has an effective population size of 49.6 and the SSP of 7, which suggests that avoiding inbreeding depression will be of bigger concern than reducing genetic adaptation to captivity.

Sustainability through Captive Births Combined with Imports of Pandas of Wild Origin

From the above we have learned that self-sustaining captive populations of red pandas typically need carrying capacities of several hundred individuals. However, should it be possible to include a limited number of new founders at regular intervals, the required global carrying capacity would be reduced. In range countries of endangered species, wild individuals can often be obtained through confiscations and rescues, or, only in case of high conservation urgency and importance, actively sought from the wild. This may allow the establishment of an in-country population that represents a reservoir of genetically distinct individuals, the offspring of which can be sent to different subpopulations of the global captive population to help reduce inbreeding and required global carrying capacity and to help maintain the genetic distinctness among the subpopulations at appropriate levels. Such an in-country population could also function as a nucleus to produce and prepare individuals for supplementation or reintroduction if and when this becomes a suitable conservation strategy for the species in question (e.g., [32]). The Indian captive population of *A. f. fulgens* might be a candidate for developing this type of population.

In 2005, the Central Zoo Authority of India set the following targets for India's overall Conservation Breeding Initiative [33]. At the global level, at least 250 properly bred and physically, genetically and behaviourally healthy individuals of each targeted species should be managed in captivity, of which at least 100 should be managed in India, to act as insurance in case of loss of the species in the wild. The targeted species are severely threatened Indian species with few hundreds/thousands (≈fewer than 2500) left in the wild, with a priority allocated to species with localized distributions. The red panda has been selected as one of these targeted species. One major zoo in the habitat range of the species will be the coordinating zoo (Padmaja Naidu Himalayan Zoological Park in Darjeeling will be the coordinating zoo for the red panda) and two to four other zoos in the habitat range of the targeted species will take part in the breeding programme (the Himalayan Zoological Park in Bulbulay, Gangtok, Sikkim and the captive facility at Yachuli in Arunachal Pradesh have been identified for the red panda). Occasionally, animals from the programme will be released in the identified habitats in accordance with the IUCN Guidelines for Reintroduction [34] in order to gain experience and to develop the mechanisms for such operations, so that these can be called upon if and when formal reintroduction or supplementation programmes become appropriate conservation strategies for the species concerned. The development of a satellite facility in India to both manage red pandas and to prepare them for release has been suggested.

Let us investigate what should be the demographic and genetic characteristics of the Indian subpopulation of *A. f. fulgens* in order to achieve the targets described above. The current captive Indian subpopulation contains 15 males and four females, of which one male and one female are 16 years or older and are unlikely to still breed. If these animals are removed from the genetic analysis, the current gene diversity in the living descendant population of the Indian subpopulation is 83.06%. Much of this genetic variation is already represented in the EEP and SSP population. Needless to say, the population urgently needs females to make breeding demographically possible and will need regular infusion of unrelated genetic stock to keep inbreeding levels low. The more this unrelated genetic material can be introduced in the form of new wild-origin pandas, for example, rescued or confiscated animals, the more the Indian subpopulation can function as a help for the global population and as a potential source for animals for reintroduction.

How many new founders would be needed to keep the Indian population genetically healthy? We will first introduce some theoretical considerations and will then present a possible scenario.

The current percentage of gene diversity lost tells us what the mean inbreeding coefficient would be in the next generation if all living descendants (i.e., non-founders) were to randomly breed. In the case of the Indian population of *A. f. fulgens* this would seem to suggest a mean inbreeding level in the next generation of about 17%, almost halfway between the equivalent level of half-sib and full-sib matings, which is starting to be significant. However, in reality, mating would not be random – pairs would, as much as practically possible, be selected according to the mean kinship values of the partners and with avoidance/minimisation of inbreeding. Furthermore, although living founders are by default left out of the gene diversity calculations in PM2000 (Lacy and Ballou, personal communication), one would in reality continue breeding with living founders. For example, among the 17 animals of 15 years and younger are three recently rescued wild pandas of breeding age that have not yet bred. Furthermore, if new founders could be brought in at regular intervals, and one could keep breeding with living founders, one could avoid or minimize inbreeding for longer than would be the case with random mating of living descendants. Therefore, as long as a small number of new founders is being added on a regular basis, inbreeding can be kept low and gene diversity will, after the start-up phase of the population, even continue to increase, but as soon as no more founders are added, gene diversity will decrease rapidly with a concurrent rise in inbreeding, especially if the population size is kept low (and in genetic terms, 100 individuals is still a relatively small population). The Indian subpopulation would therefore need to:

- be large enough to make it demographically stable, which promotes the retention of gene diversity and minimizes the risk of extinction of this genetically important subpopulation through demographic instability or catastrophes. The proposed 100 individuals distributed over at least three institutions should make this possible
- have a growth rate sufficient for it to remain at carrying capacity and, when appropriate, for it to supply individuals that can genetically supplement the other subpopulations and/or supply individuals for reintroduction
- include a small numbers of new founders on a regular basis and base mating recommendations on the principle of minimizing kinship [14,35] combined with limiting inbreeding, in order to keep genetic diversity at an acceptable level.

The following theoretical example created with the aid of PM2000 aims to illustrate this. If we include the three potential founders into the genetic calculations (assuming that they will breed in the near future), PM2000 tells us the current living population has retained 87.99% of the gene diversity of the source population. During the main growth phase of the EEP population, between 1986 and 1999, this population showed a projected annual growth rate of about 9%. Let us therefore assume that when efforts are made to try to make the Indian population grow as soon as possible, a yearly growth rate of 5% should be possible. Analysis of the larger subpopulations also teaches us that a generation time of 6 years and a Ne/N ratio of 0.28 should be feasible. Using these population variables, Table 19.8 presents a few possible theoretical scenarios for the Indian programme projecting 50 years into the future. In this model, each time a founder is added it is assumed it will contribute 40% of a founder genome to the population. If starting 2 years from now and ending in year 35, two founders are added to the population every 3, 4 or 5 years until year 35, then between 79.83 and 84.26% of gene diversity can be retained for 50 years, if the total carrying capacity is 100 individuals. Because this is a subpopulation of a global

TABLE 19.8 Number, frequency and duration of founder addition as well as the percentage of gene diversity of the source population retained after 50 years with a carrying capacity of 100 individuals and different levels of yearly growth for the Indian captive subpopulation of *A. f. fulgens* (modelled using PM2000 [10])

Year to start adding founders	Year to stop adding founders	No. founders per addition event	Years in between addition events	Growth rate (%)	Year in which population reaches 100 individuals	% Gene diversity retained after 50 years
2	35	2	3	5	27	84.26
2	35	2	4	5	30	82.07
2	35	2	5	5	30	79.83
2	35	2	3	2.5	44	82.34
2	2	24	0	5	20	78.38

In this model, each time a founder is added it is assumed it will contribute 40% of a founder genome to the population.

population, and because inbreeding can be kept low for a longer period of time than the retained gene diversity suggests due to the many living founders in this population, this would in that time frame appear to be a sensible level of gene diversity as a goal. The slower the growth rate achieved, the less gene diversity will be able to be retained. A slow growth rate is not only detrimental to the genetic health of the programme, but also means fewer individuals will be available to be sent out of the population, either to genetically supplement the other subpopulations, or for release. Adding a few founders at a time but at regular intervals achieves a better genetic result than adding all of the founders at once near the beginning of the programme. For example, adding 24 founders all at once in year 2 allows the retention of 78.38% of gene diversity, compared to 84.26% if the 24 founders are added two at a time at 3-year intervals. This is logical if one considers that the earlier a founder is added to the population, the more time (and therefore generations) there will be for its genetic contribution to be subjected to genetic drift. Adding many founders (which all have to breed) at one time may also cause logistical problems (space consideration, etc.).

There are of course many different possible scenarios. Nevertheless, the exercise above has taught us some general recommendations for the Indian subpopulation:

- The first priority is to introduce females into the population so that better breeding results become demographically possible. Unrelated wild-origin animals would be the best option genetically. However, should these not be available through rescues or confiscations, individuals from other captive subpopulations are also possible, but whenever possible these should be genetically unrelated or little related to the current stock.
- Additional founders should be added to the programme whenever possible, ideally about two founders every 3–5 years, and these founders must be able to breed (the space and husbandry expertise must be available). Again, these preferentially would be wild-origin individuals obtained through rescues or confiscations. The more new founders that can be incorporated in the Indian subpopulation, without resorting to active capture from the wild for this purpose (which is not necessary and may hold risks to the wild population), the larger its additional benefit to the global captive population. Should such animals not be available, unrelated or little related stock from other regions should be imported to benefit the genetic condition of the Indian population, but this is of smaller benefit to the global population.
- Adding a few founders at a time but at regular intervals gives a better genetic result than adding many founders at one time.
- Aim for as high a growth rate as possible, until the carrying capacity is reached; this is beneficial to the genetic health of the population but also means the population can cope better with functioning as a source for animals for genetic supplementation of subpopulations or for releases.
- Base pairing recommendations first on the mean kinship values of the individuals and second on the relatedness of the mates (to minimize inbreeding).
- Spread the population over the institutions involved to minimize the risk of extinction (or severe reduction) of the subpopulation due to unforeseen catastrophes. Put basic measures in place to limit the risk of disease outbreaks.

REINTRODUCTION CONSIDERATIONS

When reintroductions are appropriate and how they should be planned and executed is described in the IUCN/SSC Guidelines for Reintroductions [34]. Here we will concentrate on a few population biological considerations to be taken into account when identifying individuals for reintroductions.

A group of individuals selected for reintroduction should ideally be genetically very diverse and non-inbred. After all, the higher the genetic diversity of the reintroduced population, the larger its evolutionary potential, which will be especially important when the relatively benign captive environment is traded for the much more challenging wild environment. Furthermore, inbred individuals are likely to experience reduced fitness through inbreeding depression, a process that may show itself much more prominently in more stressful conditions, such as field conditions [36]. On the other hand, the genetically "best" individuals for reintroduction may also be individuals that are still important to maintain the genetic health of the captive population. Since the latter is both the source and the backup for the to-be-created or supplemented wild population, it is vital that its integrity is also maintained. It is therefore essential that the choice of individuals for reintroduction should be based on careful pedigree analysis (e.g., [37]).

The following general guidelines should minimally be followed (for a review of population biological considerations for reintroduction, see [2]):

- During the initial stages of reintroduction when methods are still being developed and experience is low, the risk of mortality for the reintroduced individuals post reintroduction is especially high. During this "experimental phase" it is wise to identify individuals that are overrepresented in the captive population, such that their potential death is not detrimental to the captive source population (e.g., [38]). Even during these experimental phases the IUCN guidelines dealing with the non-population biological aspects of reintroductions must be taken into account.
- Once successful reintroduction methods have been established and larger-scale reintroductions to re-establish locally extinct populations or supplement depleted populations are planned (implying that there is a clear overall *in situ* conservation plan for the species and its habitat, of which the reintroductions are an integral part [34]), the genes of individuals selected for reintroduction should be poorly represented in the wild, and well represented or over-represented in captivity. In addition, individuals chosen for reintroduction should be unrelated to each other. In this way, a genetically diverse reintroduced/supplemented population is created without compromising the genetic health of the source captive population.

In this respect, genetically diverse populations in range states, that have also exported some individuals to different regional populations with different founder lines represented in different regional populations, are helpful instruments. They would represent a genetically diverse source population close to the natural habitat (which reduces logistic costs and possibly offers more similar conditions to the wild environment than captive populations in other regions, thereby reducing deleterious adaption to captivity) that can produce unrelated individuals for reintroduction, while having the safeguard that some of this important genetic material is also present elsewhere in the world population.

GENERAL CONCLUSIONS

Ailurus fulgens fulgens

1. If declines or large population fluctuations in especially the largest subpopulations (EEP and SSP) can be avoided, it appears possible to build a self-sustaining global captive population with just a few individuals above the current estimated global carrying capacity.
2. The following can be recommended for the subpopulations:

- Either: (a) leave the South African, Australasian and/or Japanese subpopulations small, in which case their main role would be to help reach the needed global carrying capacity (not much rare genetic material should then be introduced because of their vulnerability to demographic instability and catastrophes); or, should the carrying capacity be available in these regions, (b) expand the South African, Australasian and/or Japanese subpopulations to about 80–100 individuals each such that the demographic risk becomes smaller. In this case more rare and distinct genetic material could be introduced into these regions, at least initially, with preferably different founder lines going to each of the subpopulations.
- The Indian Central Zoo Authority has set a target of a global population of at least 250 properly bred and physically, genetically and behaviourally healthy individuals, at least 100 of which maintained in India, to act as insurance in case of loss of the species in the wild. Occasionally, individuals will be released in suitable habitats in accordance with the IUCN Guidelines for Reintroduction in order to gain experience and to develop the mechanisms so that these can be called upon if and when formal reintroduction or supplementation programmes become appropriate conservation strategies for the species. In order to achieve this it would be beneficial to:
- Introduce females into the population so that better breeding results become demographically possible.
- For the foreseeable future (e.g., the next 20–30 years), add additional founders whenever possible, preferably small numbers at regular intervals rather than many all at once. The more new wild-origin individuals that can be incorporated in the Indian subpopulation without resorting to active capture from the wild for this purpose (which is currently not necessary and may hold risks to the wild population), the larger its additional genetic benefit will be to the global captive population and its suitability as a diverse in-country population that, when necessary, can produce animals for reintroduction or supplementation. Should such animals not be available, unrelated or little related stock from other regions should be imported in small numbers every few years to benefit the genetic condition of the Indian population, but this is of smaller genetic benefit to the global population.
- Aim for as high a growth rate as possible, until the carrying capacity is reached and in general, base pairing recommendations first on the mean kinship values of the individuals and second on minimizing inbreeding.
- Cooperatively and pro-actively manage the Chinese population so that it can contribute to the global carrying capacity and where possible the genetic health of the global population.

Ailurus fulgens styani

1. Unless the mortality of SSP females can be addressed (so that this population can grow to its estimated carrying capacity of 150) and space for an extra ≈60 pandas can be found among the SSP and Japanese populations, the Chinese institutions will be needed to achieve the necessary carrying capacity for a globally self-sustaining captive population of this subspecies. If the Chinese population contains unrelated genetic material, this may somewhat decrease the required global carrying capacity.
2. Cooperative and pro-active management of the Chinese population is strongly encouraged in order to make sure that the carrying capacity in this region as well as potentially important genetic material is not lost.

ACKNOWLEDGEMENTS

The author would like to thank Robert C. Lacy, Jonathan D. Ballou, Kathy Traylor-Holzer, Laurie Bingaman Lackey and Angela Glatston for their very helpful comments on the manuscript.

References

[1] A. Glatston, K. Leus, Global Captive Breeding Masterplan for the Red or Lesser Panda *Ailurus fulgens fulgens* and *Ailurus fulgens styani*, Royal Rotterdam Zoological and Botanical Gardens, Rotterdam, The Netherlands, 2005.
[2] R. Frankham, J.D. Ballou, D.A. Briscoe, Introduction to Conservation Genetics, Cambridge University Press, Cambridge, UK, 2002.
[3] D.H. Reed, R. Frankham, Correlation between fitness and genetic diversity, Conservat. Biol. 17 (2003) 230–237.
[4] I.R. Franklin, Evolutionary change in small populations, in: M.E. Soulé, B.A. Wilcox (Eds.), Conservation Biology: An Evolutionary-Ecological Perspective, Sinauer, Sunderland, MA, USA, 1980, pp. 135–150.
[5] C.D. Thomas, What do real populations tell us about minimum viable population sizes? Conservat. Biol. 4 (1990) 324–327.
[6] L. Nunney, K.A. Campbell, Assessing minimum viable population size: demography meets population genetics, Trends Ecol. Evol. 8 (1993) 234–239.
[7] R. Frankham, Effective population size/adult population size ratios in wildlife: a review, Genet. Res. 66 (1995) 95–107.
[8] A. Glatston, International Studbook for the Red or Lesser Panda (*Ailurus fulgens*), Royal Rotterdam Zoological and Botanical Gardens, Rotterdam, The Netherlands, 2006.
[9] ISIS [International Species Inventory System], SPARKS (Single Population Animal Record Keeping System Software), Version 1.54. International Species Inventory System, Eagan, MN, USA, 2004.
[10] J.P. Pollak, R.C. Lacy, J.D. Ballou, Population Management 2000, version 1.213, Chicago Zoological Society, Brookfield, IL, USA, 2007.
[11] G.M. Mace, Genetic management of small populations, Internatl. Zoo Yearbook 24/25 (1986) 167–174.
[12] M.E. Soulé, M. Gilpin, W. Conway, T. Foose, The millennium ark: how long a voyage, how many staterooms, how many passengers? Zoo Biol. 5 (1986) 101–113.
[13] R.C. Lacy, Managing genetic diversity in captive populations of animals, in: M.L. Bowles, C.J. Whelan (Eds.), Restoration of Endangered Species, Cambridge University Press, Cambridge, UK, 1994, pp. 63–89.
[14] M.E. Montgomery, J.D. Ballou, R.K. Nurthen, P.R. England, D.A. Briscoe, R. Frankham, Minimising kinship in captive breeding programmes, Zoo Biol. 16 (1997) 377–389.

[15] J. Wilcken, C. Lees (Eds.), Managing Zoo Populations: compiling and analysing studbook data, Australasian Regional Association of Zoological Parks and Aquaria, Sydney, Australia, 1998.
[16] R.C. Lacy, J.D. Ballou, Population Management 2000 User's Manual, Chicago Zoological Society, Brookfield, IL, USA, 2002.
[17] K. Ralls, J.D. Ballou, Extinction: lessons from zoos, in: C.M. Schonewald-Cox, S.M. Chambers, B. MacBryde, L. Thomas (Eds.), Genetics and Conservation: a Reference for Managing Wild Animal and Plant Populations, Benjamin/Cummings, Menlo Park, CA, USA, 1983, pp. 164–184.
[18] P. Crnokrak, D.A. Roff, Inbreeding depression in the wild, Heredity 83 (1999) 260–270.
[19] L.F. Keller, D.M. Waller, Inbreeding effects in wild populations, Trends Ecol. Evol. 17 (2002) 230–241.
[20] E.H. Boakes, J. Wang, W. Amos, An investigation of inbreeding depression and purging in captive pedigreed populations, Heredity 98 (2007) 172–182.
[21] R. Frankham, K. Ralls, Inbreeding leads to extinction, Nature 392 (1998) 441–442.
[22] D.H. Reed, E.H. Lowe, D.A. Briscoe, R. Frankham, Inbreeding and extinction: effects of rate of inbreeding, Conservat. Genet. 4 (2003) 405–410.
[23] L.M. Woodworth, M.E. Montgomery, D.A. Briscoe, R. Frankham, Rapid genetic deterioration in captive populations: causes and conservation implications, Conservat. Genet. 3 (2002) 277–288.
[24] R. Frankham, H. Manning, S.H. Margan, D.A. Briscoe, Does equalization of family sizes reduce genetic adaptation to captivity? Anim. Conservat. 4 (2000) 357–363.
[25] R.C. Lacy, Loss of genetic diversity from managed pops: interacting effects of drift, mutation, immigration, selection, and population subdivision, Conservat. Biol. 1 (1987) 143–158.
[26] R. Lande, Breeding plans for small populations based on the dynamics of quantitative genetic variance, in: J.D. Ballou, M. Gilpin, T.J. Foose (Eds.), Population Management for Survival and Recovery: Analytical Methods and Strategies in Small Population Conservation, Columbia University Press, New York, USA, 1995, pp. 318–340.
[27] S.H. Margan, R.K. Nurthen, M.E. Montgomery, et al., Single large or several small? Population fragmentation in the captive management of endangered species, Zoo Biol. 17 (1998) 467–480.
[28] R.C. Lacy, D.B. Lindenmayer, A simulation study of the impacts of subdivision on the Mountain Brushtail Possum *Trichosurus caninus* Ogilby (Phalangeridae: Marsupialia), in southeastern Australia. II. Loss of genetic variation within and between subpopulations, Biol. Conservat. 73 (1994) 131–142.
[29] L.S. Mills, F.W. Allendorf, The one-migrant-per-generation rule in conservation and management, Conservat. Biol. 10 (1996) 1509–1518.
[30] A. Glatston, F.P.G. Princée, A Global Masterplan for the Captive Breeding of the Red Panda. Part 1: *Ailurus fulgens fulgens*, Royal Rotterdam Zoological and Botanical Gardens, The Netherlands, 1993.
[31] R. Frankham, Inbreeding and extinction: a threshold effect, Conservat. Biol. 9 (1995) 792–799.
[32] R.C. Lacy, A. Vargas, Informe sobre la gestión Genética y Demográfica del Programma de cria para la Conservación del Lince Ibérico: Escenarios, Conclusiones y Recommendaciones. Conservation Breeding Specialist Group, Apple Valley (MN), USA and Ministerio de Medio Ambiente, Madrid, Spain, 2004, http://www.lynxexsitu.es/genetica/documentos_genetica.htm.
[33] B.R. Sharma, Concept paper on in-situ ex-situ linkage – Conservation Breeding of Endangered Wild Animal Species in India, in: B.R. Sharma, N. Akhtar, B.H. Gupta (Eds.), Proceedings of International Conference on "India's Conservation Breeding Initiative", 21–24 February, 2008, New Delhi, India, Central Zoo Authority, New Delhi, India, 2008, pp. 8–14.
[34] IUCN, IUCN/SSC Guidelines for Re-introductions, IUCN, Gland, Switzerland, 1995.
[35] K. Leus, R.C. Lacy, Genetic and demographic management of conservation breeding programs oriented towards reintroduction, in: A. Vargas, C. Breitenmoser, U. Breitenmoser (Eds.), Iberian Lynx Ex-situ Conservation: An Interdisciplinary Approach, Fundación Biodiversidad, Madrid, Spain, 2009, pp. 74–84.
[36] J.A. Jiménez, K.A. Hughes, G. Alaks, L. Graham, R.C. Lacy, An experimental study of inbreeding depression in a natural habitat, Science 266 (1994) 271–273.
[37] S.M. Haig, J.D. Ballou, S.R. Derrickson, Management options for preserving genetic diversity: reintroduction of Guam Rails to the wild, Conservat. Biol. 4 (1990) 290–300.
[38] W. Russell, E.T. Thorne, R. Oakleaf, J.D. Ballou, The genetic basis of black-footed ferret reintroduction, Conservat. Biol. 8 (1994) 263–266.

CHAPTER

20

Status and Distribution of Red Panda *Ailurus fulgens fulgens* in India

Dipankar Ghose[1] *and Pijush Kumar Dutta*[2]

[1]Head, Eastern Himalaya and Terai Program, WWF-India Secretariat, New Delhi, India
[2]Pijush Kumar Dutta, Landscape Coordinator, Western Arunachal Landscape, WWF-India, Tezpur, Assam, India

OUTLINE

INTRODUCTION

The red panda is distributed from Nepal in the west through China, India and Bhutan to Myanmar. In India, only the nominate subspecies, *Ailurus fulgens fulgens*, occurs. It is found in the Eastern Himalayan region, in the states of West Bengal, Sikkim and Arunachal Pradesh with some stray records from Meghalaya (Figure 20.1). It is represented in the following three WWF Global 200 Ecoregions [1] – Eastern Himalayan broadleaf and conifer forests, Naga-Manipuri-Chin hills moist forests and Northern Indo-China subtropical moist forests.

The red panda is designated as vulnerable by the IUCN [2] and is listed as a Schedule I species in the Wildlife (Protection) Act of India, 1972 and an Appendix I species under the Convention on International Trade in Endangered Species of Wild

Red Panda. DOI: 10.1016/B978-1-4377-7813-7.00020-3

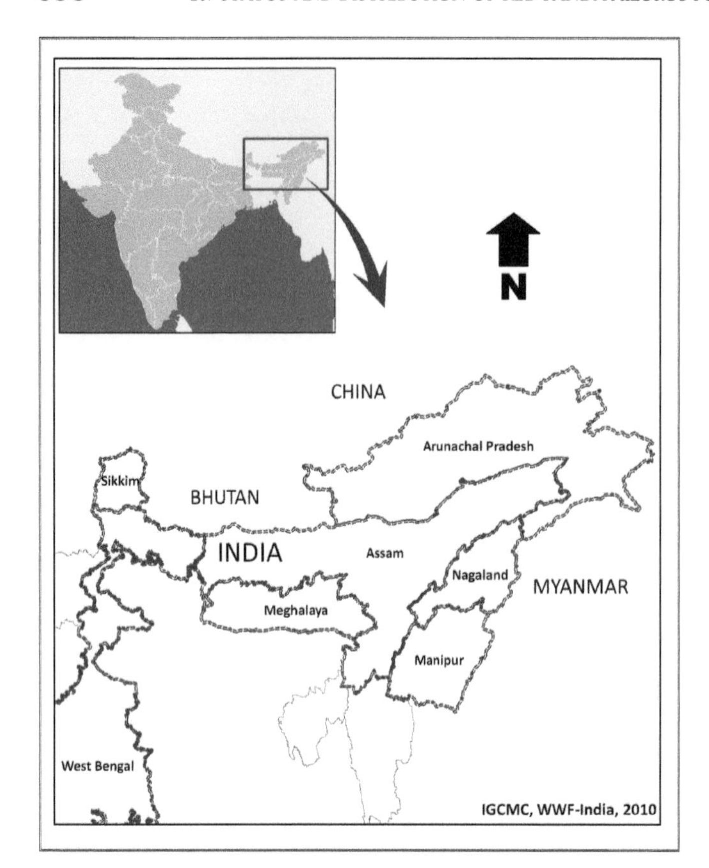

FIGURE 20.1 Indian States in Eastern Himalaya.

Fauna and Flora. Despite being such an important species, information on this animal from India is scanty. Apart from field studies from elsewhere in the range [3–7], there have only been two papers from India dealing with red pandas published in peer-reviewed international scientific journals, both almost a decade old now [8,9]. Since that time, more studies have been conducted on the red panda in India and this chapter reviews the current status and distribution of the red panda in India based on all field information currently available.

METHODS

More than 20 field surveys have been undertaken in northern West Bengal and Sikkim since 2005 to ascertain the presence or absence of the red panda in areas where this species was formerly known to occur and where sightings had not previously been recorded. In addition, extensive surveys were also initiated in Arunachal Pradesh during 2008; the information collected in 2008 has been supplemented by secondary data from this state which were collated during 2006 by the second author. Several methods were used to collect information during these surveys. Secondary data collection was done through

FIGURE 20.2 Capacity building in the field.

FIGURE 20.3 Red panda.

questionnaires and public interviews in the villages around the potential red panda habitats; a survey of the literature and interviews with the Forest Department field personnel about the occurrence of red pandas in the Protected Areas and Reserve Forests. Primary data were collected from potential red panda habitats through transect and trail monitoring [10] from the study sites in Sikkim and Arunachal Pradesh (Figure 20.2 shows training to collect these data). Red pandas are shy animals (Figure 20.3) therefore it was very hard to get records of direct sightings. As a result, these surveys focused on indirect evidence such as droppings, browsing marks, nest sites, pugmarks, skins or pelts, etc. of this species. All these indicators were counted to estimate a crude encounter rate on the abundance of this animal.

RESULTS

Distribution

West Bengal

The red panda occurs in two Protected Areas (PAs) of this state – these are Singalila National Park and Neora Valley National Park, both of which are in the Darjeeling district in the northern part of West Bengal (Figures 20.4 and 20.5 show red panda habitat in the Neora Valley area). Whereas past work in Singalila came up with a density record of one red panda/3.9 km^2, no such density data are available for Neora Valley as this Protected Area has not been properly surveyed for red pandas. At Singalila [8] red pandas were found within the altitude range of 2600–3600 m. Informal records of red panda from Neora Valley indicate that this species is common and the first author has sighted red pandas in three of his six visits to this area. Tourists visiting this Park and the ground staff of the Forest Department also reported sighting red pandas both inside the park and along its periphery which would seem to confirm that the species is common in this area.

Sikkim

Red pandas have been reported in six of the eight PAs in the state; these are Khangchendzogna National Park, which is located in the north and west districts of

FIGURE 20.4 Red panda habitat at Neora Valley.

Sikkim, Singba Rhododendron Sanctuary in the north district, Pangolakha Wildlife and Kyongnosla Alpine sanctuaries in the east district, Maenam Wildlife Sanctuary in the South district and Barsey Rhododendron Sanctuary in the west district. Fambong Lho Wildlife Sanctuary in the east district was reported to have red pandas in the early 1990s [9], however, recent surveys failed to discover any sign of this animal in this PA. Red pandas have also been reported to occur in Rate Chu Reserve Forest, above the Himalayan Zoological Park, Bulbulay (K. Madan Shankar, FEWMD, Govt. of Sikkim, personal communication). Red pandas are occasionally sighted by tourists and trekkers near Tshoka village and along the Yuksum-Dzongri and Yambong trekking trails in the West district of Sikkim which are within the Khangchendzonga Biosphere Reserve. These reports have been confirmed by members of Khangchendzonga Conservation Committee (KCC) and the Sindrabong Khangchendzonga Eco-friendly Society (SKES), both local non-governmental organizations (NGOs). Sightings of red panda from Maenam and Barsey sanctuaries were common until the late 1990s, however, the number of sightings has reduced in recent years. Two red pandas were rescued by Forest Department field staff from forest-fire-affected areas in Barsey some years back (Figures 20.6–20.10 show the Mt Khangchendzonga area and typical red panda habitats in Sikkim).

The red panda is the State Animal of the Indian state of Sikkim and draws a lot of attention in the state. World Wide Fund for Nature–India (WWF–India) initiated a red panda conservation project in collaboration with the Forest, Environment and Wildlife Management Department (FEWMD), Government of Sikkim, during 2005 with support from WWF–Germany and the German Zoo Directors' Association. The aim of this project is to ascertain the current distribution and status of red panda in the state of Sikkim and then work on long-term conservation of this species in the state. Initially, rapid extensive surveys were carried out in Barsey, Maenam, Fambong Lho and Kyongnosla sanctuaries and parts of Khangchendzonga Biosphere Reserve. Red panda evidence was recorded from all but one – the Fambong Lho Wildlife Sanctuary. A questionnaire survey was

FIGURE 20.5 Red panda habitat at Neora Valley.

FIGURE 20.6 Mt Khangchendzonga from Barsey.

FIGURE 20.7 Red panda habitat at Barsey.

conducted in areas around potential or earlier reported red panda habitats to ascertain the present status. Based on these secondary data and earlier field survey results, intensive surveys for red panda were initiated jointly by WWF–India and FEWMD at Pangolakha Wildlife Sanctuary. More than 10 red panda sightings were recorded during a span of a year in this sanctuary. During intensive surveys in Sikkim, evidence of red pandas was found on steeper slopes interspersed with fallen logs, shrubs and bamboo culms, similar findings were also reported by Wei et al. [11] during a study at Yele Nature Reserve, Sichuan Province, China.

FIGURE 20.8 Red panda habitat at Barsey Rhododendron Sanctuary.

FIGURE 20.9 Red panda habitat in East Sikkim.

Arunachal Pradesh

In the Tawang district, red pandas have been sighted on four occasions in areas between Khirmu and Bomdir (Figure 20.11, shows red panda habitat in the Tawang district). A local reportedly killed a red panda in *ringal* bamboo forest at Nuranang valley somewhere between Sela and Jang in 1995, the measurement of the head and body was 58 cm and the tail was 36 cm long. Red panda skins have been reported in local villages and sightings of red pandas in the rhododendron forests were recorded from Thingbu area during the early 1990s. A Forest Department Range Officer sighted red panda on four

FIGURE 20.10 Red panda habitat at Maenam.

FIGURE 20.11 Red panda habitat in Tawang.

occasions near PT Tso in 1997 and once an immature animal was caught from that area and sent to Itanagar Zoo located in the State Capital of Arunachal Pradesh in the same year. A red panda was shot by a local hunter behind the Tawang Monastery in 1997; another was killed and eaten by feral dogs near the army camp at PT Tso in 1998. Similar killings of red panda by feral dogs have been reported from other areas of this district. In fact, red panda skins have been sold in the Tawang market as recently as 1998. Mishra et al. [12] reported red panda sightings from Upper and Lower Nyamjang Chu, PT Tso, Mukto and Mago Chu.

Only one record of red panda exists for the East Kameng district. In 1999, local villagers killed one individual at Chayangtajo. The animal was an adult, the length of the head and body was 79 cm and that of the tail was 43 cm.

In West Kameng district, a red panda was caught in 1984 by villagers near Niyukmadong and another individual was caught in 1985 by the labourers constructing a road near Lamacamp. A third individual was hunted by a villager in Phudung area, near Sangti Valley in 1986, its skin was used to make a cap. A red panda was killed by a domestic dog at Khongbam near Ramalingam camp in 1988. Forest Department staff sighted red pandas five times between 1989 and 1991; on four of these occasions the animals were seen in fruiting oak (*Quercus* spp.) trees, in and around Eagle Nest Pass, which is along the boundary between Eagle Nest Wildlife Sanctuary and Sessa Orchid Sanctuary. Two animals were sighted by the Forest Department Range Officer in 1993 near Bompu at Eagle Nest Wildlife Sanctuary. A red panda was sighted on a rhododendron tree by a Forest Watcher near the fourth milestone to Sundarview, on the border of Eagle Nest and Sessa sanctuaries. More than ten sightings of red panda from Dirang were also reported by the caretaker of the local Circuit House between 1980 and 1996. Observations by an independent researcher in this district came up with 23 droppings from Eagle Nest Pass, 17 droppings from Chakoo and two droppings from Bompu in Eagle Nest Wildlife Sanctuary. Mishra et al. [12] sighted red panda from most of the high-altitude areas of West Kameng district. Sighting and killing of red panda were also reported from Mandla Phudung during 1998. One red panda was killed near Nafra in 1999. At Chander village near Dirang in 1999, a red panda was caught by a local and was released into the forest in the same year. A similar incident was also reported in 2000.

In West Siang district, one red panda was seen by the first author in 1999 near Monigong village. The animal was on an oak (*Quercus* spp.) tree on a steep hillside. A second red panda was seen killed near Pidi by local villagers in 1999.

In the Dibang Valley district, one red panda was killed by a local hunter in 1992. In the lower Dibang Valley, a red panda skin was reportedly collected from a local hunter in 1994 and a live specimen was seen in bamboo forest near Mayodiya Pass in the same year by Forest Department staff.

One red panda was sighted in 1984 at Ditchu Reserve Forest (c. 3000 m) in Lohit and unauthenticated sighting of eight individuals in a single bamboo clump was reported in 1992 from an area along the Wakro-Deban road.

Records of red panda from the other 11 districts of Arunachal Pradesh are scanty.

Meghalaya

Choudhury [9] reported the presence of red panda from Nokrek and Balpakram National Parks in Garo Hills within this state. However, there has been no recent authentic report of sighting or scientific study of red panda in this state.

Population Estimate

There has been no proper population estimate for this species in the recent past for the whole of India except for Singalila National Park in West Bengal where about 64 individuals could be estimated based on the study by Pradhan et al. [8]. Though Choudhury [9]

FIGURE 20.12 Dam construction in N.E. India.

estimated about 5000–6000 red pandas in India based on an encounter rate of one red panda/4.4 km^2, these numbers could not be validated in India currently. There is a considerable difference in red panda encounter rates from adjacent areas. Pradhan et al. [8] found one red panda/1.67 km^2 in Singalila, whereas Yonzon and Hunter [5] calculated an encounter rate of one red panda/2.9 km^2 in Langtang National Park, Nepal. Recent extensive surveys in Sikkim found a mean of 0.2 red panda scat group/hour at Barsey, Pangolakha and Kyongnosla sanctuaries. Based on direct and indirect evidences, the red panda encounter rate in Pangolakha Wildlife Sanctuary was estimated at one red panda/ 2.7 km^2 by a joint team of WWF–India and FEWMD (Partha Ghose and Basant Sharma, WWF–India, personal communication). Given the extent of the potential habitat of red panda available within India in the states of West Bengal, Sikkim and Arunachal Pradesh, we could well say that more than 85% of red pandas in India occur in the state of Arunachal Pradesh.

Threats to Survival

Red pandas have no negative impact on humans [13], however, they are facing threats due to anthropogenic pressures. These threats may be caterogized into two – direct threats, such as habitat destruction and poaching, and indirect threats, such as lack of people's awareness, lack of conservation strategy specifically for red pandas, inadequate protection, etc. Both Pradhan et al. [8] and Choudhury [9] cited the above-mentioned two direct threats as the main threats to populations of red panda in Singalila and other parts of India. Habitat destruction in the Eastern Himalaya, which also holds a substantial part of potential red panda habitats, is often linked with unplanned development in the region. Developing large dams (Figure 20.12), road networks and industries have adverse impacts on red panda habitat. Red pandas are mostly distributed in temperate forests in the

FIGURE 20.13 Red panda habitat cleared for cattle grazing.

Eastern Himalaya and the pressures of a growing population in these areas are tremendous. People depend on forests for firewood and fodder, which are good enough reasons for habitat degradation. Forests are also used for cattle grazing which has been a major reason for degradation of red panda habitat (Figure 20.13). Though cattle grazing was not a major problem in Singhalila during recent times [8], it is still a problem in parts of Arunachal Pradesh and was a problem in Sikkim until recently.

DISCUSSION

Present knowledge of red panda distribution from its entire range in India is sparse. Choudhury [9] suggested that total habitat probably used by red panda in India is 12 500 km^2, with a state-wise break-up of 11 300 km^2 in Arunachal Pradesh, 300 km^2 in Meghalaya, 800 km^2 in Sikkim and 100 km^2 in Darjeeling district of West Bengal. Given that the red panda prefers a habitat between 2600 and 3600 m [8], we found that a total of 18 650 km^2 of forest still exists in Sikkim (1200 km^2), Arunachal Pradesh (17 200 km^2) and Darjeeling district (250 km^2) of West Bengal within this altitudinal range. However, it might be noted that not all of these forests are in a condition to hold red panda populations. For example, in Sikkim, confirmed red panda habitat has been estimated at around 700 km^2, which is much less than the potential habitat within the said altitudinal range. Choudhury [9] also reported red panda to occur in nine PAs of Arunachal Pradesh, all of which needs to be re-confirmed now. The case for Meghalaya is similar, where he reported red panda to occur in three PAs and in absence of any authentic recent records, this could not be validated. Choudhury [9] reported red panda from six PAs in Sikkim, out of which red panda is not presently reported from Fambong Lho; whereas a red panda stronghold has been reported from an additional PA, the Pangolakha Wildlife Sanctuary, Sikkim (Partha Ghose and Basant Sharma, WWF-India, personal communication). The two PAs which were reported by Choudhury [9] and Pradhan [8] to have red pandas in West Bengal still maintain red panda populations. The present paper suggests that an intensive study of red pandas should be carried out in other parts of Sikkim, in the remaining PAs and non-PA areas where red panda has been reported during the extensive surveys. Extensive field surveys are to be carried out in Arunachal Pradesh and Meghalaya to establish the baseline for red panda.

Chettri et al. [14] mentioned that red panda along with other globally threatened animals found in the Kanchenjunga Landscape, the transborder area shared between Kanchenjunga Conservation Area in Nepal and the PAs in Darjeeling and Sikkim, are extremely susceptible to the effects of habitat fragmentation because of low population densities of these species. They also suggested that the present PA network in this landscape is not enough for these species, which is reflected by the fact that most of these species also occur in areas outside the PAs. Research on red panda, as mentioned above, is important, however, the research findings are to be incorporated in the future management plans of the PAs so that the forest departments in the respective states could implement those recommendations in order to ensure the long-term survival of the red panda. Choudhury [9] mentioned that major parts of the larger PAs known to have red panda, i.e., Khangchendzonga and Namdapha NP and Dibang Wildlife Sanctuary were understaffed. The situation has not changed much. Though the Wildlife Wing of the Forest Department in different states has geared up protection in most of the PAs, in many instances, regular patrolling rendering adequate protection and monitoring of wildlife populations are wanting. Strategic support to the PAs is required for this – this may include capacity building of frontline staff of the Forest Department, scientific equipment and infrastructure for the Forest Department, etc. For areas outside the PA network in India, where red panda still occur, converting them into a National Park or Sanctuary, as suggested by Choudhury [9] might not be feasible anymore. The two new categories for PAs in India, i.e., Conservation Reserve (in land owned by government) and Community Reserve (in land owned by indigenous communities) might be explored. However, in recent times, communities are becoming more averse to the idea of any government-imposed legal notification on the land owned by them as this almost stops their customary rights. Therefore, creation of more Community Conservation Areas (CCAs) could be an option on the lines of WWF–India's present work in Western Arunachal Landscape. CCAs are entirely managed by the community representatives; therefore, they are the ones to decide on access and benefits from that area. Thus, this concept is more acceptable to the local communities.

Hilton-Taylor [15] mentioned that habitat loss is the predominant threat for the great majority of mammals. Red pandas are no exception. Yonzon and Hunter [5] found the red panda to be a habitat specialist. In the Langtang National Park, Nepal, these two biologists found that the effect of *chauri*, a crossbreed between yak (*Bos grunniens*) and hill cow (*B. taurus*) on red panda habitat was severe. Presence of *chauri*, their herders and pet dogs was clearly detrimental to the pandas. Though competition between *chauri* and red panda for bamboo leaves was not apparent as red panda fed higher on the bamboo than the cattle, the latter may have had a role in reducing overall bamboo abundance by trampling. In India, the first author also recorded herders clearing bamboo culms in temperate forests to create open grazing ground for their cattle during winter months. This was a major problem in Sikkim till the past decade. During 2006, the policy makers in the state banned cattle grazing in the PAs of Sikkim which was enforced by the FEWMD in collaboration with the local NGOs. This reduced pressures on red panda habitat, at least in four PAs of Sikkim – Khangchendzonga National Park and Pangolakha, Maenam and Barsey sanctuaries. In Singalila National Park, Darjeeling, India, forest fires in the late 20th and early 21st centuries damaged red panda habitat which was compounded by cattle grazing that caused a reduction in undergrowth, even trees were lopped to collect firewood [16] (Figures 20.14 and 20.15 show more recent fire damage).

Rai and Sundriyal [17] suggested that growing tourism may also lead to habitat destruction and resource depletion. Red panda, being a shy animal, will move away from areas where tourism impacts are adversely affecting their habitat and this might be the case in some areas within its range in India. Among the four states where red panda reportedly occur in India, Sikkim receives the maximum number of tourists. During the past 5 years, on average, half a million tourists visited Sikkim per annum. A great deal of these tourists also visit the PAs of the state, mostly for trekking in Khangchendzonga Biosphere Reserve and Barsey Rhododendron Sanctuary, and for sight-seeing in North Sikkim, passing through Singba Rhododendron Sanctuary. Though FEWMD has been working in these areas for increasing protection, still their reach is inadequate. Given the high turnover of tourists, it is not always possible to enforce the rules and regulations and there are instances when tourists and trekkers have been found to use rhododendron branches for firewood to cook meals. WWF–India has partnered with local NGOs like KCC, SKES, Kabi Endeavours, The Mountain Institute and some others to sensitize and train the trekking and tourism service providers in Sikkim for reducing pressures on red panda

FIGURE 20.14 Red panda habitat damaged by fire.

FIGURE 20.15 Red panda habitat damaged by fire.

habitats. Locals in high-altitude areas of Sikkim have also been found to use firewood for space-heating as well as cooking. WWF–India is presently collaborating with different departments of the Government of Sikkim for piloting alternative energy techniques in two areas within the east district of Sikkim where locals have been reported to depend on natural resources collected from red panda habitats. The objective is to wean the locals from forest dependence and reduce pressures on red panda habitats.

Poaching is the second direct threat to red panda survival in India. Within India, hundreds of red pandas were captured from areas which now form Singalila National Park during the 1960s, at least 300 individuals were captured from Singalila itself [16], however, this is not a problem in West Bengal and Sikkim anymore. This has been possible by the effective enforcement of the Wildlife (Protection) Act of India, 1972 which prohibits capture and hunting of red panda and also bans habitat destruction and alteration within the PAs. This problem is still persistent in Arunachal Pradesh where most of the forests are under the control of local communities and the Forest Department's reach is not far and wide in remote corners of the state. Motivation for hunting red panda is purely for sport and skin which is traded locally in a clandestine manner because of some awareness about the law of the land in recent times. Mishra et al. [12] found hunting to be the most serious threat to wildlife throughout the western part of Arunachal Pradesh. They mentioned that wildlife conservation, until recently, was not a priority of most government departments and realistic measures to reduce hunting will require participation of the local communities. In recent years, some red panda trapping has also been reported from Arunachal Pradesh where people tend to keep red panda cubs as pets, but the problem arises once the cub reaches adulthood; adult red pandas are mostly killed but sometimes released back into the forests with the intervention of the Forest Department. Keeping these in mind, WWF–India, since 2004, started working with the local communities in the two districts of Arunachal Pradesh for safeguarding large tracts of forests which harbour red panda in addition to other threatened species of plants and animals. This resulted in the creation of two CCAs – the Thembang Bapu CCA in West Kameng and the Pangchen–Lumpo–Muchat CCA in Tawang districts. Both these CCAs have potential red panda habitats, and efforts are in place to provide conservation benefits to the local communities, who are the owners of these areas, so that they become motivated to conserve wildlife and their habitats. Maharana et al. [18] mentioned that local communities from surrounding areas exploit the natural resources of Khangchendzonga National Park by grazing livestock and extraction of fuel, fodder and timber. However, most of these problems have been addressed in recent times by the FEWMD, Govt. of Sikkim and participatory conservation efforts are in place here too. *Himal Rakshak*, an elite patrolling group, created from among the community members was notified by the Sikkim Forest Department in 2006. *Himal Rakshaks* are responsible for monitoring wildlife populations and curb wildlife crime in Khangchendzonga Biosphere Reserve that includes the National Park. During 2008, one *Himal Rakshak* had even photographed a red panda along the Yambong trekking trail within the Reserve. Such participatory conservation efforts are to be increased for safeguarding existing populations of red panda in remaining areas.

Yonzon and Hunter [5] during their study on red panda at Langtang National Park, Nepal, found that most red panda deaths were human related and probably connected to the presence of cattle herders and dogs in the area during the monsoon birth season. Feral dogs killing red panda have also been reported from Pangolakha Wildlife Sanctuary, Sikkim in recent years. WWF–India along with the FEWMD, Govt. of Sikkim is working with the stakeholders, mostly the Indian Army units which are present in these areas, for mitigating this problem. However, the location of these areas is close to an international border, hence their strategic importance and local religious sentiments have been a hindrance for taking any active management steps.

According to Cardillo et al. [19], small geographic ranges and low population densities determine the maximum population size a species can attain. They also mentioned that gestation length is an important indicator of life-history speed that determines how quickly populations can recover from low levels. Given that red pandas occur in low population densities in most areas and that their gestation period is a little over 3 months and that they give birth to offspring only once a year, the threatened populations of this species are likely to recover rather slowly.

Moreover, there is speculation that climate change may have some effects on red panda habitat. A change in plant phenology has been noticed in parts of Eastern Himalaya which might change red panda behaviour accordingly. A study by Malcolm et al. [20] highlighted the potential seriousness of the impacts of global warming on the erstwhile Indo-Burma Biodiversity Hotspot within which most of the red panda habitats of India exist. This will most likely have adverse effects on red panda survival in India.

Extinction risk in the mammalian order Carnivora, of which red panda is a member, is predicted more strongly by biology than exposure to high-density human populations [19], however, that does not leave red pandas less susceptible to threats. Loss of corridors and small island populations of red panda mean it is going to face problems of inbreeding in the near future. Chettri et al. [21] found that the corridors between the different PAs in Kanchenjunga Landscape spread in Nepal, and areas of Darjeeling and Sikkim in India, which are necessary for maintaining connectivity within wider existing habitats for flagship species such as tiger, elephant and red panda, are practically non-existent. Therefore, conservation of large continuous tracts of temperate forests is the key to tackle this threat. WWF–India has been working for conservation of biodiversity with special reference to threatened wildlife in the Kanchendzonga Landscape, covering the northern part of West Bengal and Sikkim since 2005 and will continue its efforts to join hands with the neighbouring countries for conservation of large chunks of habitats that may serve as wider space for red panda and other threatened species, as it is said that existing PAs in this area cannot exist in isolation as islands [21].

ACKNOWLEDGEMENTS

We are indebted to the FEWMD, Govt. of Sikkim and Forest Dept., Govt. of Arunachal Pradesh for allowing us to conduct this work. In particular, the authors would like to thank the following: Mr T.R. Poudyal, Mr D.G. Shrestha, Mr S.T. Lachungpa, Mr M.L. Arrawatia, Mr N.T. Bhutia, Mr T. Chandy, Mr H. Pradhan, Mr C. Lachungpa, Mr C.S. Rao, Dr S. Tambe, Mr T.D. Rai, Mr J.B. Subba, Mr G. Lepcha, Mr Karma Legshey, Mr U. Gurung, Mr S. Thatal, Ms U.G. Lachungpa, Mr T.B. Subba and Dr Madan Shankar of FEWMD, Govt. of Sikkim; Mr N.C. Bahuguna, Dr P.T. Bhutia, Mr A.K. Jha, Mr Tapas Das, and Ms S. Ghatak of Forest Dept., Govt. of West Bengal; PCCF, CWLW, CCF (WL), CF and DFOs of Forest Dept., Govt. of Arunachal Pradesh, in particular Mr P. Ringu and Mr C. Loma; colleagues at WWF–India, namely Rajeev, Ambika, Lak Tsheden, Dwaipayan, Rudra, Rajarshi, Partha, Basant, Priya, B.K., Rakesh, Pema, Tanushree, Sanjay, Tariq, Jagdish, Izhar and Renu; members of the management committees of two CCAs in Western Arunachal Landscape; local communities in the study areas in Sikkim and northern West Bengal; members of KCC, SKES and TMI, Sikkim; Dr Axel Gebauer of Görlitz Zoo; Dr Angela Glatston of Rotterdam Zoo; Mr Roland Melisch and Mr Stefan Zieglar of WWF–Germany; The Tata Dorabji Trust, German Zoo Directors' Association and numerous other stakeholders who have been helpful. The authors would also like to sincerely thank WWF–India, Mr Ravi Singh and Dr Sejal Worah in particular for permitting them to write this article.

References

[1] D.M. Olson, E. Dinerstein, The Global 200: A representation approach to conserving the Earth's most biologically valuable ecoregions, Conservat. Biol. 12 (1998) 502–515.
[2] The IUCN Red List of Threatened Species. IUCN, Gland, Switzerland. www.iucnredlist.org. 2010.
[3] K.G. Johnson, G.B. Schaller, H. Jinchu, Comparative behaviour of the red panda and giant pandas in the Wolong Reserve, China J. Mammals 69 (1988) 5523–5564.
[4] D.G. Reid, H. Jinchu, Y. Huang, Ecology of the red panda *Ailurus fulgens* in the Wolong Reserve, China J. Zool. 225 (1991) 347–364.
[5] P.B. Yonzon, M.L. Hunter Jr., Conservation of the red panda *Ailurus fulgens*, Biol. Conservat. 57 (1991) 1–11.
[6] P.B. Yonzon, M.L. Hunter Jr., Cheese, tourists and red pandas in the Nepal Himalayas, Conservat. Biol. 5 (2) (1991) 196–202.
[7] F. Wei, Z. Feng, Z. Wang, J. Hu, Current distribution, status and conservation of wild red pandas *Ailurus fulgens* in China, Biol. Conservat. 89 (1999) 285–291.

[8] S. Pradhan, G.K. Saha, J.A. Khan, Ecology of the red panda *Ailurus fulgens* in the Singhalila National Park, Darjeeling, India, Biol. Conservat. 98 (2001) 11–18.
[9] A.U. Choudhury, An overview of the status and conservation of the red panda *Ailurus fulgens* in India, with reference to its global status, Oryx 35 (2001) 250–259.
[10] S.T. Buckland, D.R. Anderson, K.P Burnham, J.L. Laake, Distance Sampling: Estimating Abundance of Biological Populations, Chapman and Hall, London, 1993.
[11] F. Wei, Z. Feng, Z. Wang, J. Hu, Habitat use and separation between the giant panda and the red panda, J. Mammal. 81 (2) (2000) 448–455.
[12] C. Mishra, M.D. Madhusudan, A. Datta, Mammals of the high altitudes of Western Arunachal Pradesh, eastern Himalaya: an assessment of threats and conservation needs, Oryx 40 (1) (2006) 29–35.
[13] A.R. Glatston, Status Survey and Conservation Action Plan for Procyonids and Ailurids: The Red Panda, Olingos, Coatis, Raccoons, and their Relatives, IUCN, Gland, Switzerland, 1994.
[14] N. Chettri, B. Bajracharya, R. Thapa, Feasibility assessment for developing conservation corridors in the Kanchenjunga landscape, in: N. Chettri, D. Shakya, E. Sharma (Eds.), Biodiversity Conservation in the Kanchenjunga Landscape, International Centre for Integrated Mountain Development, Kathmandu, Nepal, 2008.
[15] C. Hilton-Taylor, 2000 IUCN Red List of Threatened Species, IUCN SSC, Gland, Switzerland, 2000, p. 61.
[16] N.C. Bahuguna, S. Dhaundyal, P. Vyas, N. Singhal, Red panda in Darjeeling at Singalila National Park and adjoining forest: A status report, Small Carniv. Conservat. 19 (1998) 11–12.
[17] S.C. Rai, R.C. Sundriyal, Tourism and biodiversity conservation: the Sikkim Himalaya, Ambio 26 (1997) 235–242.
[18] I. Maharana, S.C. Rai, E. Sharma, Environmental economics of the Khangchendzonga National Park in the Sikkim Himalaya, Geo J. 50 (2000) 329–337.
[19] M. Cardillo, A. Purvis, W. Sechrest, J.L. Gittleman, J. Bielby, G.M. Mace, Human population density and extinction risk in the world's Carnivora, Plos Biol. 2 (7) (2004) 909–914.
[20] J.R. Malcolm, C. Lin, R.P. Neilson, L. Hansen, L. Hannah, Global warming and extinctions of endemic species from biodiversity hotspots, Conservat. Biol. 20 (2) (2006) 538–548.
[21] N. Chettri, R. Thapa, B. Shakya, Participatory conservation planning in Kanchenjunga transboundary biodiversity conservation landscape, Tropic. Ecol. 48 (2) (2007) 163–176.

CHAPTER

21

Red Pandas in the Wild in China

Fuwen Wei[1] *and Zejun Zhang*[2]

[1]Key Laboratory of Animal Ecology and Conservation Biology, Institute of Zoology, The Chinese Academy of Sciences, Chaoyang, Beijing, People's Republic of China

[2]Institute of Rare Animals and Plants, China West Normal University, Nanchong, Sichuan, People's Republic of China

OUTLINE

INTRODUCTION

To date, fossils of the red panda and its relatives have been widely unearthed in Asia, Europe and North America. Among these fossil materials, *Sivanasua*, occurring in the upper Miocene of Europe and the lower Pliocene of Asia, seemed to be the earliest indisputable ailurine, which exhibited cranial and dental structures ancestral to and possibly contemporaneous with *Ailurus* [1]. *Parailus* from the lower Pliocene of England, Europe, and North America seemed closest to *Ailurus* in general cranial and dental

Red Panda. DOI: 10.1016/B978-1-4377-7813-7.00021-5

morphology [2,3], whose discovery indicated a European–Asian origin for the Ailurinae with a subsequent trans-Beringian radiation [2,4]. The history of current red pandas can be traced back to the middle Pleistocene at least, whose fossil materials were unearthed at several sites in South China, including Wufeng (Hubei province), Xichou and Fumin (Yunnan province) [5,6], and Zijin (Guizhou province) [7].

Based upon information available, in this chapter we introduce the distribution and status of wild red pandas in China, including their historical and current distribution, habitat, population, and threats potentially influencing their sustainable survival in future. The information primarily came from field surveys and published literatures. First, we introduce the methods adopted in the field surveys.

SURVEY METHODS IN THE FIELD

The animal is now confined to faraway mountain ranges in southwestern China, including Sichuan, Yunnan and Tibet provinces. During the past three decades, several field surveys were conducted to learn about its distribution and status, which were primarily organized and conducted by Sichuan, Yunnan and Tibet provincial forestry bureaus [7–11]. In addition, some localities of former distribution of the animal, such as those in Shaanxi, Gansu, Qinghai and Guizhou provinces were visited, too [12].

During these surveys, several methods were used to gather information on the animal in the wild [12], including:

1. Formal interviews with officials and local people who were familiar with the animal, to collect information on wild red panda distribution, resource dynamics, and conservation status
2. On-site visits to some special areas, and daytime surveys of habitats on foot. Concerning funds, manpower and extensive survey areas (more than 70 counties), two to three counties were sampled in each mountain range to estimate population distribution and abundance. Researchers entered red panda habitats and carefully searched for their traces, including spoors, feeding sites, and droppings (Figure 21.1) left on the ground, logs, trees or shrubs and other conspicuous locations. All feeding sites and droppings were counted to provide a crude measurement of population abundance
3. Data on forest distribution within the range of red pandas were obtained from Sichuan, Yunnan and Tibet forestry bureaus, which conducted a thorough forest resource survey during 1985–1991. In addition, some related books [13–15] were referred to
4. Red pandas imported from the wild in China for exhibition in zoos, nature reserves and parks were counted and compiled through sending questionnaires or on-site visiting to some important localities. Some information on poaching and trade of the animal was also collected from the Internet.

DISTRIBUTION

As for its global distribution, the red panda inhabits the Himalayan Mountains and its surrounding areas, including Nepal, India, Bhutan, Myanmar and China. It is primarily

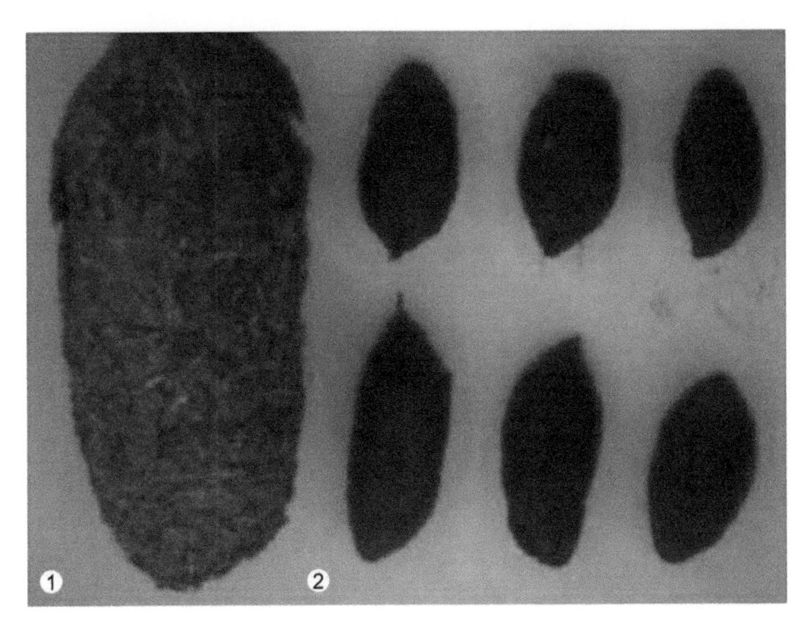

FIGURE 21.1 Droppings of red pandas, compared with giant pandas. 1: giant pandas; 2: red pandas. Although droppings of both pandas are primarily composed of bamboos, it is very easy to distinguish them in the wild. Droppings of red pandas are much smaller than giant pandas (4.4 cm × 2.2 cm vs 14.5 cm × 5.0 cm), even those of an infant giant panda (7.5 cm × 2.5 cm) are larger than those of an adult red panda). *Photo by Fuwen Wei.*

found in temperate and subtropical forests, which are closely associated with bamboo-thicket understoreys [12,16–18]. The only exception is in Meghalaya, where the animal is found to live in the tropical forest [16]. The westernmost occurrence is in Mugu district of western Nepal (82°E), and the easternmost in the upstream valley of Min River in Sichuan province, China (about 104°E). Meanwhile, its north–south distribution is narrower, from the southern parts of the Gaoligongshan Mountains on the Myanmar–China border (25°N) to the upper valleys in the Minshan Mountains (33°N) [19–22]. In China, the red panda occupies a mountainous range with characteristic high ridges and deep valleys (Figure 21.2).

The current distribution of *Ailurus* was considered a radiation outward from a central core in the Myanmar–Yunnan–Sichuan highlands along regions of recent orogenic activity, most notably the Himalayas [23–25]. The extensive mountainous ranges uplifted during the Pleistocene created substantial new habitats for the animal, enabling its successful survival from the Pleistocene inclement climate. Roberts and Gittleman [25] argued that the distribution of *Ailurus* should be a series of disjunctive, physically isolated populations rather than a continuous and interbreeding one, since erosive activities by rivers caused the partitioning of the Himalayan Mountains into blocks separated by deep gorges, which posed great physical and ecological barriers to the transmigration of individuals.

Two subspecies or species (see Chapter 7) of *Ailurus* are recognized. In China, the nominate form, *A. f. fulgens*, is primarily found in Tibet and northwestern Yunnan while *A. f. styani* occurs in Sichuan and northeastern Yunnan [12,26] (Figure 21.3). Traditionally, the Nujiang River is assumed to separate the two subspecies as an ecological barrier [12,27]. However, Colin Groves argues that such a barrier is doubtful, and that the red pandas occurring in east Tibet seem more like *A. f. styani* than *A. f. fulgens* (see Chapter 7).

FIGURE 21.2
Distribution range of red pandas, characteristic of high ridges and low valleys (photographed in the Xiangling Mountains, China). *Photo by Fuwen Wei.*

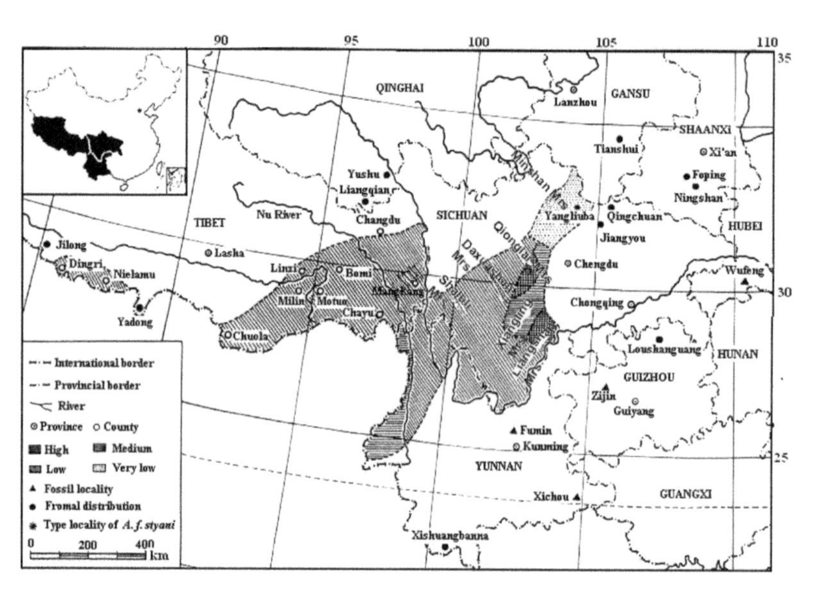

FIGURE 21.3
Historical distribution, current distribution and relative density of red pandas in China *(modified from [10]). Minshan Mrs.: Minshan Mountains; Qionglai Mrs.: Qionglai Mountains; Daxueshan Mrs.: Daxueshan Mountains; Shaluli Mrs.: Shaluli Mountains; Xiangling Mrs.: Xiangling Mountains; Liangshan Mrs.: Liangshan Mountains.*

Sichuan

Only one subspecies, namely *A. f. styani*, is found in the Sichuan province. Its current distribution range extends from the Shaluli Mountains eastward into the Xiangling Mountains (including the Daxiangling and Xiaoxiangling Mountains), southward into the Liangshan Mountains, westward into the Shaluli Mountains, and northward into the Minshan Mountains (see Figure 21.3), of which, the Qionglai, Liangshan and Xiangling mountains are their primary distribution habitats, inhabited by a large quantity of red pandas.

Liangshan Mountains

These are the southernmost mountains inhabited by the animal (see Figure 21.3). Altogether there are 17 counties with red panda distribution records, of which Mabian, Leibo, Ganluo, and Xide counties have the highest density, followed by Zhaojue, Dechang and Puge [26].

Xiangling Mountains

In the Daxiangling Mountains, the animal is primarily found in Hongya and Yinjing counties, and in the Xiaoxiangling Mountains, mainly in Shimian, Mianning, Xide and Dechang counties. Its density in the mountains keeps relatively high.

Qionglai Mountains

The animal is found in more than 10 counties in the mountains. However, its abundance is not equal in every part. In central parts of the mountains, for example, in Wolong and Fengtongzhai Nature Reserves, its density is much higher compared with that in the surrounding areas, for example, in Xiaojin, Lixian, Dayi, Lushan, Mingshan, Qionglai, Chongqing counties, Kangding, Luding, and Danba counties [26] (see Figure 21.3).

Minshan Mountains

These are the northernmost mountains inhabited by the animal, and its density is very low. Red pandas have been noted to disappear from some parts of its former distribution areas, for example, in Qingchuan and Jiangyou counties [12,28]. In addition, at Yangliuba, Pingwu counties, where the type specimen of *A. f. styani* was collected, no signs were found during our field survey in 1998.

Shaluli Mountains

In these mountains, red pandas are found in the southern part, including Xiangcheng, Daocheng, Derong, Muli, Yanyuan and Yanbian counties [12,26].

Daxueshan Mountains

The animal is only found in Yajian and Jiulong counties [26].

Yunnan

Both subspecies occur in the province. *A. f. fulgens* is found to the southern Nujiang River and in the Himalayan hills of the Gaoligong Mountains, which are connected with Chayu of Tibet in the northwest and with Myanmar in the southwest (see Figure 21.3). *A. f. styani* is distributed to the eastern Nujiang River where its range extends to 11 counties in the south and connects with the Shaluli Mountains of Sichuan province in the north. Red pandas were formerly reported from Xishuangbanna (Mengla county), in the southern part of the province [29]. However, this was not confirmed by our field investigations [12] (see Figure 21.3).

According to Hu and Du [30], red pandas are currently distributed in 14 counties of Yunnan Province (Table 21.1).

Tibet

Traditionally, only *A. f. fulgens* is found in the province [12,16]. However, Colin Groves argues that red pandas in east Tibet are more likely to belong to *A. f. styani* (see Chapter 7). Currently, the animal can only be found in the southern valleys of the Himalayan Mountains, including Mangkang, Dingri, Changdu, Chuona, Nielamu, Linzhi, Milin, Bomi, Chayu, and Motuo counties [11,26,29,31,32] (see Figure 21.3). Investigations indicated that no red pandas were found in Jilong and Yadong counties where they were reported in the past [29,31,32] (see Figure 21.3). In addition, there were no signs of red pandas in some areas of Chayu and Mangkang [12], which were the areas where red pandas could easily be seen formerly [11].

Qinghai

It has been reported that red pandas occurred in the southern part, such as in Langqian and Yushu counties [33]. However, no signs can be found there now [12] (see Figure 21.3).

Shaanxi

Red pandas were reported to be distributed in Ningshan county [34], which was solely based on a skin purchased there. This record seems disputable now as it received no further support from field surveys [12,26]. In addition, the animal was reported to occur in Foping Nature Reserve [35], but no traces were found there during our long-term field investigation.

Gansu

The animal has been reported to occur in the south of the province. However, after a massive investigation conducted in the 1970s, no traces were found [8,12,26]. Wang [36]

TABLE 21.1 Forest area (km^2), habitat area (km^2) and population estimates of red pandas in Yunnan province, China.

Density rank	Distribution county	Density (individuals/km^2)	Forest area (km^2)	Habitat area (km^2)	Quantity	Total
Crowded	Gongshan, Lushui, Fugong, Baoshan, Tengchong, Deqin, Zhongdian, Weixi	0.20–0.25	11693.16	5741.34	1150–1400	
General	Lanping, Lijiang, Yunlong, Ninglang	0.14–0.17	9661.02	4743.56	930–1180	2000–2600
Scarce	Yiliang–Weixin	0.11–0.13	304.07	149.3	20	

From [30]

reported that the animal was distributed in Tianshui county based on a skin purchased there (see Figure 21.3). This record seemed questionable, too, for no further information is available from the field to confirm it [12].

Guizhou

Fossils were unearthed in Zijin county of the province [7]. In the 1970s, there perhaps was a small population extant in Loushanguan [26]. However, no signs can be found there today [12] (see Figure 21.3).

HABITAT

According to Wei et al. [12], there are 76245.5 km^2 of forests remaining in their range, including 35088.3 km^2 in Sichuan (7596.8 km^2 in the Minshan Mountains, 7681.3 km^2 in the Qionglai Mountains, 7734.9 km^2 in the Liangshan Mountains, 4736.9 km^2 in the Xiangling Mountains, 2845.8 km^2 in the Daxueshan Mountains, and 4492.6 km^2 in the Shaluli Mountains), 21658.1 km^2 in Yunnan and 9499.1 km^2 in Tibet (Table 21.2).

Based on our recent studies [37–39], red pandas show a clear preference for particular habitats. Due to their small body size, they usually walk on elevated fallen logs and shrubs such as rhododendrons, *Rhododendron* spp., which gives them easy access to bamboo leaves [37–44]. Therefore, they prefer the microhabitat close to fallen logs or shrubs, and with a high density of shrubs and fallen logs [37,38,44] (Figure 21.4). In Sichuan province, red and giant pandas are sympatric in most mountainous ranges, and their macrohabitats are roughly similar [12]. Based on the estimate of habitats occupied by giant pandas in their forests (49.1%) [45], the area of red panda habitat in China was estimated to be about

TABLE 21.2 Forest area (km^2), habitat (km^2) and population size of red pandas in China

Province	Mountains	Area of forests (km^2)	Area of habitats (km^2)	Population
Sichuan	Minshan	7596.8	3730	3000–3400
	Qionglai	7681.3	3771.5	
	Xiangling	4736.9	2325.8	
	Liangshan	7734.9	3797.8	
	Daxueshan	2845.8	1397.3	
	Shaluli	4492.6	2205.9	
Yunnan		21658.1	10634.1	1600–2000
Tibet		19499.1	9574.1	1400–1600
Total		76245.5	37436.5	6000–7000

Combined from [12,30]

FIGURE 21.4 A wild red panda resting on shrubs in Yele Nature Reserve, the Xiaoxiangling Mountains, China. *Photo by Fuwen Wei.*

37436.5 km^2 (17228.3 km^2 in Sichuan, 10634.1 km^2 in Yunnan and 9574.1 km^2 in Tibet) [12] (see Table 21.2).

POPULATION

Although Pleistocene fossils of red pandas have been found in Wufeng of Hubei province, Xichou and Fumin of Yunnan province [5] and Zijin of Guizhou province [7], no pandas are found in these areas today. In addition, red pandas have become extinct in Shaanxi, Gansu, Qinghai and Guizhou provinces, and have disappeared from parts of Sichuan, Yunnan and Tibet [11,12,40]. The extant population was estimated at about 6000–7000 in China, including 3000–3400 in Sichuan, 1600–2000 in Yunnan, and 1400–1600 in Tibet (see Table 21.2) [12]. It is estimated that the population has decreased by as much as 40% over the past 50 years.

The above estimates on population size in China are perhaps underestimated. For example, in Yunnan province, the area of habitats with a high density (0.20–0.25/km^2) of red panda population was estimated to be 1505.80 km^2 by Wei et al. [12], including Gongshan, Lushui and Fugong, the area with medium density (0.14–0.17/km^2) estimated at 8979.10 km^2, incluing Deqin, Zhongdian, Weixi, Lanping, Lijiang, Yunlong, Baoshan and Tengchong, and the area with low density (<0.13/km^2) estimated to be 149.30 km^2, including Yiliang and Weixin. However, Hu and Du [30] argued that besides Lushan, Gongshan and Fugong, Baoshan and Tengchong in Gaoligongshan Nature Reserve, Deqing, Zhongdian and Weixi should also be ranked as high-density areas. Therefore, they estimated the population size of red pandas in Yunnan province to be 2000–2600, significantly higher than that by Wei et al. [30]. Correspondingly, the total number of red pandas in China may range from 6000 to 7600 (see Table 21.2).

THREATS

Deforestation

The red panda is a forest-dwelling animal. Deforestation, which causes fragmentation of forest, constitutes the fundamental threat to its long-term survival in China [12]. Forest fragmentation not only restricts the available habitat, but also weakens the viability of the fragmented population [46]. Historically, annual forest consumption was much higher than the growth for a long period in China. Up to 1985, there were 121 forestry enterprises with over 70 000 staff active in Sichuan province [28]. During 1958–1960, the lumber stock felled by Chuanxi Forestry Bureau of Sichuan province was 929 000 m^3, 910 000 m^3, and 822 000 m^3 respectively [12]. However, the timber production in the corresponding year only reached 359 800 m^3, 362 400 m^3, and 549 000 m^3 respectively [47]. In addition, 200 million m^3 of forest resources were consumed during a 25-year period in Aba Autonomous Region of Sichuan Province, constituting 58.8% of the total forest growth and gives annual forest consumption four times as high as the annual forest yield [12,47]. Liang [47] even predicted that, if forest felling continued at the above rate, the forest would be totally clear-felled in Aba by the end of last century. In Ganzi Autonomous Region of the same province, 120 million m^3 of forest resources were consumed over 30 years, which made up 28% of current forest growth [15]. In Liangshan Autonomous Region, 23.8 million m^3 of forest resources were swallowed up over 35 years, constituting 10% of current forest growth (227.07 million m^3) (Figure 21.5). In addition, it was estimated that 22 big forestry enterprises have swallowed up 3597.9 km^2 of red panda habitats over 25 years in Sichuan province, making up 20.9% of their current habitats [12].

Deforestation can directly change the context of a habitat, resulting in a decrease in habitat quality, and in turn resulting in a reduced population abundance, and even its local extinction. In Baofengping area, Mabian Reserve, Sichuan province, where we set up a field research station in early 1990, the population density remained high during our study period. However, after we left there, massive deforestation began. By 2000, when we went back to the same area for the Third National Survey for the Giant Panda, most of the original forests were felled, and a lot less faeces of the animal could be found, except at the end of the valleys, where deforestation could not easily be conducted due to the steep topography.

However, today, the situation seems to be gradually improving. To ensure the sustainability of forest resources, the Chinese government has paid much attention to afforestation. By 1990, the total afforested areas had covered about 200 000 hectares in western Sichuan [47]. In addition, the nationwide "Natural Forest Protection Project (NFPP)" was initiated by the Chinese government in 1998, which has been assumed extremely important for protecting wild animals, including the red panda, and their habitats. Even so, it

(A)

(B)

FIGURE 21.5 Forest felling in red panda's habitat (photographed in the Xiangling Mountains, China). (A) Forest logged by local people for cabin construction and heating; (B) hillside was almost clear-felled, leaving sparse forests in high valleys. *Photo by Fuwen Wei.*

must be remembered that forest renewal is a long process, and negative effects from deforestation will last a relatively long period. In addition, man-made forests may not easily lend themselves to red panda habitats [12].

Habitat Fragmentation and Population Isolation

Habitat fragmentation is another fundamental threat for the animal, which can be caused by many factors, for example, natural factors, such as rivers and ridges, and artificial ones, such as forest felling, road construction, massive bamboo flowering. Compared with natural ones, artificial factors contribute much more to habitat degradation and fragmentation. In Sichuan province, the mountainous ranges inhabited are all isolated from each other by rivers and alongside human activity, including resident settlement, road construction, and farm cultivation. For example, the Min River and human activity resulted in the isolation of red panda populations between the Minshan Mountains and the Qionglai Mountains, and for the isolation between the Qionglai Mountains and the Daxiangling Mountains, the Dadu River and human activity contributed. Even in some mountainous ranges, red panda populations are isolated, too. For example in the Qionglai Mountains, red panda populations are at least isolated into four parts by National Road No. 318, and the road from Baoxing county to Xiaojin county.

Habitat fragmentation can result in serious ecological consequences for population survival, especially for those with low density. Isolated small populations may be faced with many threats, including from the population itself and their inhabited habitats, for example, inbreeding depression, genetic stochasticity, environmental variation, and catastrophe as well [48,49]. More seriously, these threats can result in a continuous decrease in population size, and finally initiate a process called "extinction vortex" [49]. In China, giant pandas in the wild have been separated into 20–30 isolated populations, and many of these populations consist of less than 30 individuals [50]. Concerning its sympatry with giant pandas in Sichuan province and its special habitat requirements (see Chapter 11), red pandas should be fragmented and isolated more seriously than giant pandas. In fact, many small isolated populations perhaps have gone to extinction, for example, those in Qingchuan county, Jiangyou county, Wanglang Nature Reserve, etc. Even at the type locality of *A. f. styani*, we could not find their signs in the field when conducting an on-site survey in 1998.

Poaching, Trade and Exhibition

In the past, red panda fur was sometimes used to make hats and clothing by local people. The fur hat with its long luxurious tail at the back looks beautiful and warm. Wei et al. [12] reported that this type of hat was needed by some newly married couples as it was regarded as a talisman for a happy marriage. Such a fur hat is worth over 500 Yuan RMB (roughly equivalent to one year's salary for a common worker then) and a fur coat over 1000 Yuan RMB [12]. During our investigations in the field, we still found some local people wearing this type of fur hat (Figure 21.6).

The exact number of people with such hats is difficult to estimate. However, trade in the animal was quite prevalent in some ranges in the past. For example, 29 pelts were sold in 1979–1981 in Mianning county of Sichuan province [51]. In Tibet, some 200 skins were sold on an annual basis in the 1970s [11]. In Changdu in Tibet alone, 148 skins were sold in 1968–1971. In addition, more than 400 red pandas were caught in the eastern forests of Tibet in the 1970s [52]. Low [53] even found four specimens for sale in Guangdong and Fujian provinces. In 1999, we found two skins for sale at a small grocery store in Derong county of Sichuan Province, where the seller told us that they were imported from Yunnan province. Even recently, there have been several reports of red panda trade in China (Table 21.3).

F. Cuvier praised the animal as the most beautiful one in the world. Indeed, it has great value as an exhibit to the public. In China, exhibition of the animal started at Chengdu Zoo of Sichuan province in 1953. Since then, there has been a total of 78 zoos, parks and reserves to exhibit the animal and about 1500 red pandas have been imported from their native habitats [12], of which, more than 1000 individuals came from Sichuan province, and the rest from Yunnan and Tibet provinces. During 1985–1992, China exported 141 live red pandas to foreign countries [12,54].

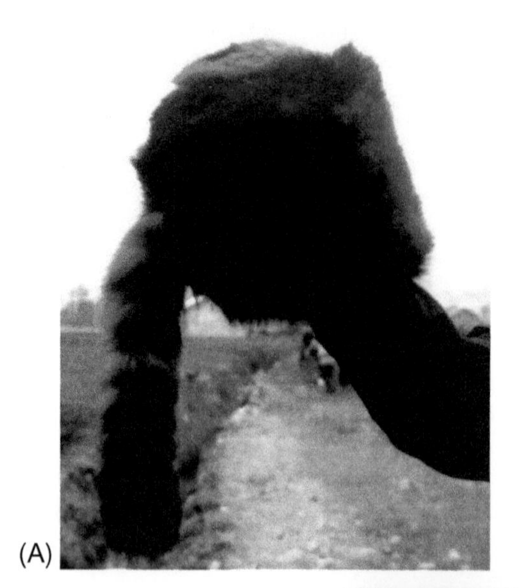

FIGURE 21.6 Red pandas sometimes are hunted by local people for their fur to make hats. (A) A hat made of red panda's fur in Mianning county, Sichuan; (B) People are in red-panda hats in Yunnan. *Photo by Fuwen Wei.*

The above data seem remarkable, indicating that pressures from poaching, trade and exhibition are still severe, especially to a declining population. This not only leads to an even greater decline of red panda numbers, but also to extinction in some areas. For example, today, it is difficult to discover any trace in Jilong and Yadong county, Tibet province [11].

Tourism and Road Construction

In some nature reserves in China, there is an increasing trend to carry out animal viewing and ecological tourism. Sometimes, tourism can bring not only disturbances to red

TABLE 21.3 Cases of illegal trade of red pandas in China during the last decade

Province	Year	Number of individuals involved
Yunnan	1999	34
Beijing	2001	3
Hainan	2001	1
Yunnan	2002	1
Sichuan	2002	9
Beijing	2002	2
Sichuan	2002	1
Sichuan	2003	2
Shaanxi	2005	5
Yunnan	2005	12
Henan	2007	3
Shaanxi	2007	1

pandas' normal life, but also negative influences on their habitats through road construction, timber collection for heating and modern trash as well. For example, road construction can cause habitat fragmentation and population isolation, potentially affecting individual movement, and resulting in even further inbreeding and loss of genetic diversity.

Bamboo Flowering

In the mid-1970s, the staple-food bamboo species (*Fargesia denudate* and *F. scabrida*) for the giant panda flowered extensively and died in the Minshan Mountains (Figure 21.7). Subsequently, 138 dead giant pandas were found in the wild [55]. Although no reports are available at present on the influence of bamboo flowering upon red pandas' survival, it can be inferred that they should be more severe than for giant pandas because:

(1) lower accessibility arising from their smaller body, red pandas usually feed on only one bamboo species (with shorter stem) in their habitats. Giant pandas, on the contrary, can utilize most bamboos in their habitats (see Chapter 11). Once their staple-food bamboos flower massively, red pandas may be faced with more serious famine

(2) compared with giant pandas, red pandas have relatively lower ability in movement. Once bamboos flowered, they are more prone to being confined to small patchy habitats, making them easier to become extinct. In addition, even if they do make long-distance movement to search for a new food source after bamboo flowering, a greater cost in survival can be expected than that for giant pandas due to more predators.

FIGURE 21.7 (A) Bamboos in flower and (B) dead bamboo. *Photo by Fuwen Wei.*

SUMMARY

Although fossils related to red pandas were widely unearthed in Asia, Europe and North America, extant red pandas are confined to the Himalayan Mountains and areas nearby, including Nepal, India, Bhutan, Myanmar and southwestern China. Their current distribution suggests a radiation outward from a central core in the Myanmar–Yunnan–Sichuan highlands along regions of recent orogenic activity. The extensive mountain ranges uplifted during the Pleistocene created substantial new habitats for the species, enabling its successful survival from the Pleistocene inclement climate. The extant distribution of *Ailurus* is a series of disjunctive, physically isolated populations caused by erosive activity of rivers partitioning the Himalayan Mountains into blocks.

The history of current red pandas can be traced back to the middle Pleistocene, whose fossil materials were unearthed at several sites in South China. Both subspecies are found in China, of which, the endemic *A. f. styani* extends from west Sichuan province to east Yunnan province, and the nominate *A. f. fulgens* is extant in south Tibet and North Yunnan. Traditionally, the Nujiang River is considered a natural barrier to separate the two subspecies.

Red pandas in China have disappeared from some former distribution localities, such as in Guizhou, Shaanxi, Gansu, and Qinghai. During the last 50 years, it was estimated that the red panda population has decreased as much as 40% in China. Their extant habitats are estimated about 37 436.5 km^2, and the total of individuals about 6000–76000, which has been separated into many isolated populations.

Many efforts have been made to promote the conservation of red pandas in the wild in China, including legislation, reserve establishment, local resident education and so on. However, the animal is still faced with many severe threats affecting its sustainable survival, such as deforestation, habitat fragmentation and population isolation, poaching, and bamboo flowering as well.

China is the primary distribution range of wild red pandas. However, compared with efforts made to conserve giant pandas, those for red pandas are much less. The animal seems to be "forgotten". In fact, much information on the status of wild red pandas is still vacant in China, especially in Tibet, Yunan and the Minshan Mountains in Sichuan. As such, a thorough nationwide survey for the animal will be expected in future, just like those that have been conducted for the giant panda, to collect data on its distribution, abundance, disturbances and threats, and the conflict between nature preservation and economic development of local communities. In addition, more reserves need to be established for the red panda, especially in areas where its populations are isolated and many human disturbances exist.

References

[1] G.E. Pilgrim, The fossil Carnivora of India, Paleontol. Indica 18 (1932) 1.
[2] R.H. Tedford, E.P. Gustravson, The first American record of the extinct panda, *Parailurus*, Nature 265 (1977) 621.
[3] S.C. Wallace, X. Wang, Two new carnivores from an unusual late Tertiary forest biota in eastern North America, Nature 432 (2004) 556.
[4] B. Kurtén, E. Anderson, Pleistocene Mammals of North America, Columbia University Press, New York, USA, 1980.
[5] Yunnan Bureau of Geology, Fossil Atlas of Yunnan Province, Yunnan People's Publishing House, Kunming, China, 1974.
[6] Institute of Vertebrate Paleontology and Paleoanthropology of Academia Sinica, Handbook of Vertebrate Fossils of China, Science Press, Beijing, China, 1979.
[7] Y. Xu, Y. Li, Mammal fossils in the Pleistocene at Zijin of Guizhou province, China, Vertebrata Palasiatica 5 (1957) 342.
[8] Expedition of Rare Animals in Gansu, The Rare Animal Species in Gansu, Gansu Forestry and Agriculture Bureau, Lanzhou, China, 1976.

[9] Expedition of Rare Animals in Sichuan, A Report on the Rare Animal Species in Sichuan Province, China, Sichuan Forestry Bureau, Chengdu, China, 1977.
[10] Kunming Institute of Zoology, Annals of Yunnan Province: Animal Fauna, Yunnan People's Press, Kunming, China, 1989.
[11] B. Yin, W. Liu, Precious and Rare Wildlife and its Protection in Tibet, China Forestry Press, Beijing, China, 1993.
[12] F. Wei, Z. Feng, Z. Wang, et al., Current distribution, status and conservation of wild red pandas *Ailurus fulgens* in China, Biol. Conservat. 89 (1999) 56.
[13] Yunnan Forestry Institute, Natural Reserves of Yunnan Province, Publishing House of Forestry, Beijing, China, 1987.
[14] H. Liu, Dictionary of Provinces, Cities and Counties of China, Tourism Press of China, Beijing, China, 1990.
[15] C. Li, Y. Yang (Eds.), Status and Assessments of Forests in Southern Sichuan Province, Sichuan Science and Technology Press, Chengdu, China, 1990.
[16] A. Choudhury, An overview of the status and conservation of the red panda *Ailurus fulgens* in India, with reference to its global status, Oryx 35 (2001) 250.
[17] M.S. Roberts, Demographic trends in a captive population of red pandas (*Ailurus fulgens*), Zoo Biol. 1 (1982) 119.
[18] J. Stainton, Forests of Nepal, John Murray, London, UK, 1972.
[19] A. Choudhury, Red panda *Ailurus fulgens* F. Cuvier in the north-east with an important record from Garo hills, J. Bombay Nat. Hist. Soc. 94 (1997) 145.
[20] G.B. Corbet, J.E. Hill, The Mammals of the Indomalayan Region: A Systematic Review, Oxford University Press, Oxford, UK, 1992.
[21] J.R. Ellerman, T.C.S. Morrison-Scott, Checklist of Palaearctic and Indian Mammals 1758–1946, British Museum (Natural History), 1996.
[22] D. Macdonald, The Encyclopaedia of Mammals, George Allen & Unwin, London, UK, 1984.
[23] A. Gansser, Geology of the Himalayas, John Wiley, London, UK, 1964.
[24] M.S. Manandhar, Nature Ann. 1 (1978) 1.
[25] M..S. Roberts, J.L. Gittleman, *Ailurus fulgens*, Mammalian Species 222 (1984) 1.
[26] F. Wei, J. Hu (Eds.), Status and Conservation of Red Pandas in Sichuan, China Science & Technology Press, Beijing, China, 1993.
[27] F. Wei, Z. Wang, Z. Feng, et al., Seasonal energy utilization in bamboo by the red panda (*Ailurus fulgens*), Zoo Biol. 19 (2000) 27.
[28] G. Li, Z. Yang, X. Huang, et al., Genetic diversity and influencing factors in captive red pandas, Acta Ecol. Sin. 20 (2000) 184.
[29] Y. Gao, Fauna Sinica, Mammalia, Science Press, Beijing, China, 1987.
[30] G. Hu, Y. Du, Current distribution, population and conservation status of *Ailurus fulgens* in Yunnan, China, J. Northwest Forest. Univ. 17 (2002) 67.
[31] Mammals of China and Mongolia, American Museum of Natural History, New York, USA, 1938.
[32] Z. Feng, G. Cai, C. Zheng, The Mammals of Tibet, Science Press, Beijing, China, 1986.
[33] Northwest Plateau Institute of Biology, Qinghai Fauna, Qinghai People's Press, Xining, China, 1989.
[34] Shaanxi Institute of Zoology, Rare Economic Mammals in Shaanxi Province, Shaanxi Science & Technology Press, Xi'an, China, 1974.
[35] W. Pan, Z. Gao, Z. Lu, The Refugia of Giant Pandas in the Qinling Mountains, Peking University Press, Beijing, China, 1988.
[36] X. Wang, Vertebrate Fauna of Gansu Province, Gansu Science and Technology Press, Lanzhou, China, 1993.
[37] F. Wei, Z. Feng, Z. Wang, et al., Habitat use and separation between the giant panda and the red panda, J. Mammal. 81 (2000) 448.
[38] Z. Zhang, F. Wei, M. Li, et al., Winter microhabitat separation between giant and red pandas in *Bashania faberi* bamboo forest in Fengtongzhai Nature Reserve, J. Wildlife Manage. 70 (2006) 231.
[39] Z. Zhang, F. Wei, M. Li, et al., Microhabitat separation during winter among sympatric giant pandas, red pandas, and tufted deer: the effects of diet, body size, and energy metabolism, Can. J. Zool. 82 (2004) 1451.
[40] J. Hu, F. Wei, Foraging behavior of the red panda *Ailurus fulgens*, Sichuan Teachers' College 13 (1992) 83.

[41] K.G. Johnson, G.B. Schaller, J. Hu, Comparative behavior of red and giant pandas in the Wolong Reserve, China, J. Mammal. 69 (1988) 552.
[42] D.G. Reid, J. Hu, Y. Huang, Ecology of the red panda *Ailurus fulgens* in the Wolong Reserve, China, J. Zool. 225 (1991) 347.
[43] F. Wei, Z. Feng, Z. Wang, et al., Feeding strategy and resource partitioning between giant and red pandas, Mammalia 63 (1999) 417.
[44] F. Wei, W. Wang, A. Zhou, et al., Food selection and foraging strategy of red pandas, Acta Theriol. Sin. 15 (1995) 259.
[45] M. Wang, Survey Reports on Habitat of Giant Pandas in China, Sichuan Forestry Bureau, Chengdu, China, 1989.
[46] R. Kirkpatrick, The natural history and conservation of the Snub-nosed monkeys (Genus *Rhinopithecus*), Biol. Conservat. 72 (1996) 363.
[47] H. Liang (Ed.), Overview of Deforestation and Afforestation in Southern Sichuan Province, Sichuan Science and Technology Press, Chengdu, China, 1990.
[48] R.W. Howe, G.J. Davis, The demographic significance of "sink" populations, Biol. Conservat. 57 (1991) 239.
[49] R.C. Lacy, A computer simulation model for population viability analysis, Wildlife Res. 20 (1993) 45.
[50] J. Hu, Research on the Giant Panda, Shanghai Science, Technology and Education Press, Shanghai, China, 2001.
[51] Z. Yu, Z. Yu, S. Sha (Eds.), Natural Reserves in Sichuan, Sichuan Social Science Press, Chengdu, China, 1983.
[52] Z. Feng, G. Cai, C. Zheng, The Mammals of Tibet, Science Press, Beijing, China, 1986.
[53] J. Low, The Smuggling of Endangered Wildlife Across the Taiwan Strait, Traffic International, Cambridge, USA, 1990.
[54] The Netherlands. Proposal of transfer of *Ailurus fulgens* from Appendix II to Appendix I. 1994, (unpublished).
[55] J. Hu, G.B. Schaller, W. Pan, et al., The Giant Panda of Wolong, Sichuan Science and Technology Press, Chengdu, China, 1985.

CHAPTER

22

Project Punde Kundo: Community-based Monitoring of a Red Panda Population in Eastern Nepal

Brian Williams[1], *Naveen K. Mahato*[2] *and Kamal Kandel*[2]

[1]Red Panda Network, Mountain View, California, USA

[2]Red Panda Network–Nepal, Kathmandu, Nepal

OUTLINE

HISTORY AND IMPORTANCE

In 2006, Red Panda Network began working in eastern Nepal to conserve the red panda (*Ailurus fulgens fulgens*) population located on the eastern Nepali-Indo border (Figure 22.1). Studies of this transboundary region and its red panda population by Pradhan [1,2] and Williams [3,4] demonstrated that the forests in this region are threatened by habitat

Red Panda. DOI: 10.1016/B978-1-4377-7813-7.00022-7

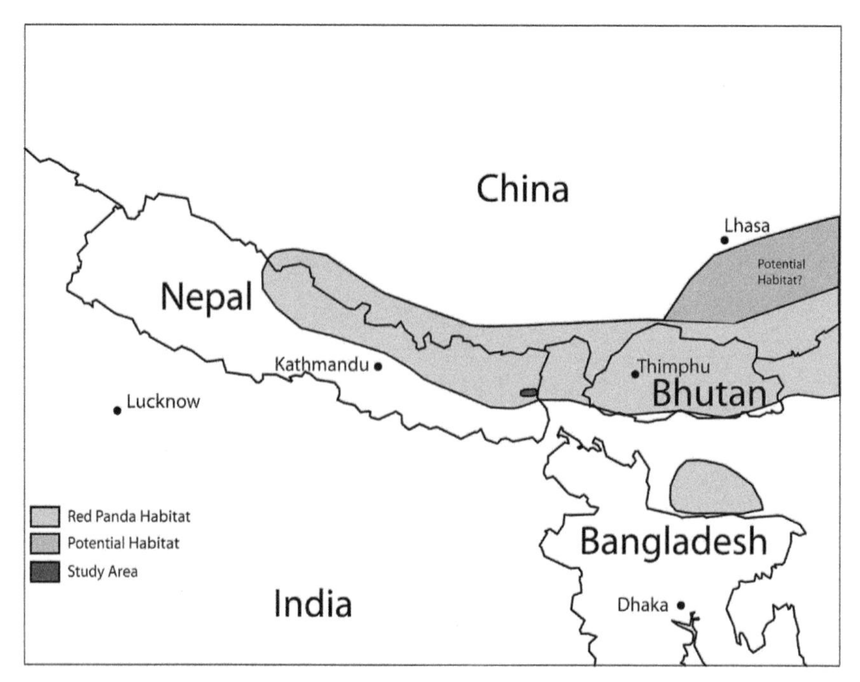

FIGURE 22.1 Distribution of red panda in Nepal and the location of Project Punde Kundo.

fragmentation and deforestation from yak herding, firewood and fodder collection, and road building. Pradhan and Williams recommended that conservation of the population was dependent on work with the Nepalese border settlements. Based on these recommendations, Red Panda Network (RPN) and its affiliate organization, Red Panda Network–Nepal (RPNN), initiated their signature programme, Project Punde Kundo (PPK), ("punde kundo" means "red panda" in the local Nepalese dialect). This project focuses on community-based monitoring within forests along the Nepali-Indo border (Figure 22.2).

Project Punde Kundo specifically focuses on the Singhalila range in the Panchthar, Ilam and Taplejung districts of eastern Nepal. This area, also referred to as the Panchthar-Ilam-Taplejung (PIT) Corridor, is recognized as a region of international importance for biodiversity conservation due to its species richness and diversity and transboundary connectivity between Kanchenjunga Conservation Area in Nepal and Singhalila National Park and Barsey Rhododendron Sanctuary in India. This is an important region, it is a part of the World Wildlife Fund's Eastern Himalayan Ecoregion and has been named on its Global 2000 list [5], Conservation International has labelled it the "Himalayan Hotspot" [6], Birdlife International and Bird Conservation Nepal have designated it as one of Nepal's Important Bird Areas [7], and World Wildlife Fund–Nepal and His Majesty's Government of Nepal have included it in the Sacred Himalayan Landscape [8].

The PIT corridor is important to red panda conservation because it contains 178 km^2 of red panda habitat, which supports approximately 25% of Nepal's red panda population, with an estimated 100 individuals [3,4,9]. Conservation of the PIT Corridor would connect the tri-national Kanchenjunga Conservation Area with India's Barsey Rhododendron

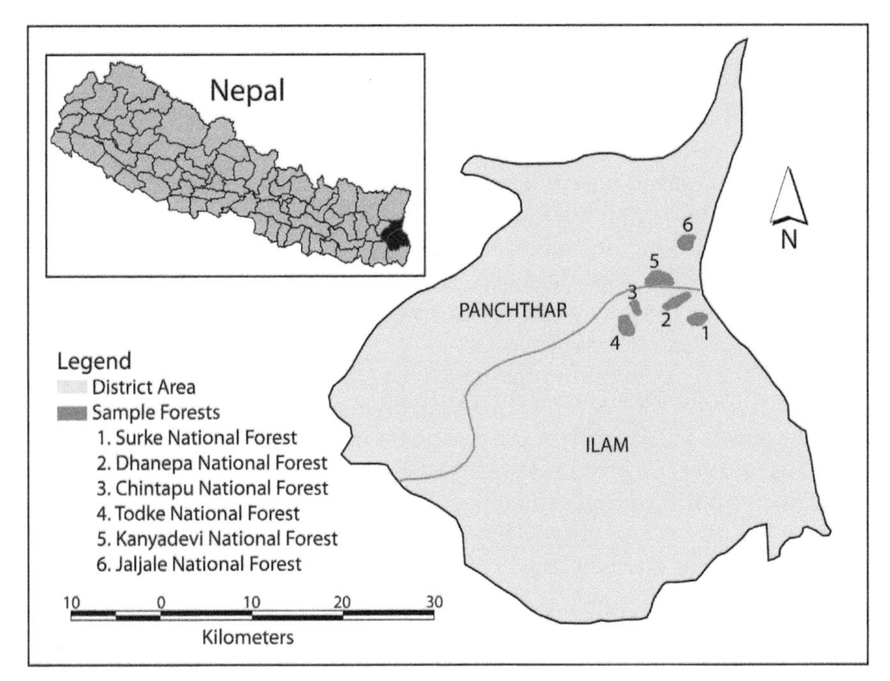

FIGURE 22.2 The location of Project Punde Kundo monitoring sites.

Sanctuary and Singhalila National Park, creating an uninterrupted stretch of protected land extending for 11 500 km^2. This area is critical not only to red pandas but also to a suite of other endangered species such as the clouded leopard (*Neofelis nebulosa*) and Himalayan thar (*Hemitragus jemlahicus*), as well as an exceptionally rich avian population [5,10]. The Project Punde Kundo work area is 3350 ha (33.5 km^2) of the southern portion of the PIT corridor (see Figure 22.2). This area has an estimated red panda population of 20 individuals, based on a crude relative density of one individual per 1.67 km^2 [1,2].

Project Punde Kundo provides a model for community-based conservation of landscape species. In order to maximize our ability to collect red panda population information and directly involve local communities, RPNN hires and trains forest users from communities located within the PIT Corridor to conduct long-term red panda population and habitat monitoring. Through this process, the community monitors have begun to shift their perception of the forest from a source of income or survival to one of a natural resource having inherent aesthetic value. They have also come to see their roles shift from forest users to forest guardians. This model can be used with other landscape-level species to develop a conservation ethic within local communities.

OVERVIEW

Project Punde Kundo focuses on utilizing local ecological knowledge and conservation to establish a community-based system of data collection, conservation education and stewardship. The intention of RPN and RPNN is to create a system that will be self-sustaining within 10 years. The broad objectives of Project Punde Kundo are to:

1. Identify and preserve 15 unprotected red panda "hotspots" by 2020: areas with red panda subpopulations of 50 or more individuals
2. Conduct community-based baseline research on red panda and its habitat
3. Assist communities in devising socially and environmentally responsible forest management plans and income alternatives
4. Increase awareness of the important ecosystem services provided to the region by the fragile eastern Himalayan broadleaf and conifer ecoregion among local villagers, decision makers, and government officials
5. By 2020, reduce the deforestation rate of eastern Himalayan broadleaf and conifer forests by 10% from its 2008 level. (The Department of Forest Resources and Survey stated that there was a 2.3% rate of deforestation in the mid-hill forests from 1979 to 1994 [11]. Thus, at this projected rate over a 10-year period the rate of deforestation would be 23% in mid-hill forests in Nepal. Our goal is that our work will cause a 10% decrease over 10 years (1% a year). This would be the equivalent to a 43% decrease in overall deforestation in our working area.)

The heart of PPK is its "Conservation in Action" strategy. This strategy is centred on a three-step iterative cycle. The first step is the creation of a community-based monitoring project. The second step is to create a forest conservation area managed by the local community or a community-conserved area. The third step is the creation of "Conservation in Action" contracts, in which communities define how they will sustain their protected areas in perpetuity. The model focuses on the active participation of local communities in every step of the process.

Project Punde Kundo "Conservation in Action" Cycle

1. Community-based Monitoring

Forest users called "forest guardians" are trained to monitor red panda and other wildlife populations within their community forests.

2. Community-based Conservation Areas

Local communities, through community forest networks, delineate community conservation areas that contain a red panda population of at least 50 individuals, for permanent protection.

3. Conservation in Action Contracts

Local communities create Conservation in Action contracts outlining roles and responsibilities for long-term monitoring and management of the community-based conservation areas.

To date, RPN and RPNN have implemented the first stage of this cycle, a community-based monitoring project in the Ilam and Panchthar districts of eastern Nepal, with the organizational support of the Critical Ecosystem Partnership Fund, the Government of

Nepal's Department of Forests, and the Government of Nepal's Department of National Parks and Wildlife Conservation.

LOCATION

Since 2006, RPN and RPNN have monitored the red panda population and its associated habitat in six forests having a combined area of 3350 hectares. These forests consist of one national forest (Surke) and five community forests (Dhanepa, Chhintapu, Todke, Kanyadevi, and Jaljale) in the Ilam and Panchthar districts of eastern Nepal (see Figure 22.2).

1. *Surke national forest* (400 ha; elevation: 2400–3000 m): The Surke forest is situated on the west-facing slopes of the Mai river valley in the Mabu Village Development Committee (VDC) of Ilam district.
2. *Dhanepa community forest* (550 ha; elevation: 2400–3300 m): The Dhanepa community forest is located in Maimajhuwa VDC of Ilam district. This south-east-facing forest is also located in the northern Mai river valley, and is part of the Mai Valley Important Bird Area.
3. *Chhintapu community forest* (350 ha; elevation: 2400–3100 m): The Chhintapu community forest is also in Maimajhuwa VDC of Ilam district. This is also a south-east-facing forest that lies at the western end of the Mai river valley.
4. *Todke community forest* (450 ha; elevation: 2400–3000 m): The fourth forest within Maimajhuwa VDC of Ilam district is the Todke community forest. This east-facing forest lies at the western end of the Mai river valley.
5. *Kanyadevi community forest* (850 ha; elevation: 2800–3200 m): The Kanyadevi community forest is in Siddin VDC of Panchthar district. The forest is situated on north-facing slopes.
6. *Jaljale community forest* (750 ha; elevation: 2700–3300 m): The Jaljale community forest is in Prangbung VDC of Panchthar district. This forest is on the Nepali–Indo border and connects with Singhalila National Park of the Indian state of West Bengal.

METHODOLOGY

RPN and RPNN utilize a six-step methodology for community-based monitoring:

Step 1 Community consultation: Before RPN enters a potential monitoring location, it consults with the local community to learn where people have and have not seen red pandas. During this process, it identifies potential local partners, and scouts the local political landscape.

Step 2 Preliminary assessment: Once RPN is confident about the anecdotal information relayed, it conducts a preliminary assessment of potential working sites and has a meeting with the associated forest managers to choose two forest guardians per forest. In community forests, RPN meets with the community forest user group (CFUG) while, in national forests, it meets with the District Forest Officer.

Step 3 Transect establishment: Next, RPN establishes elevational transects within the whole range of red panda habitat within each forest (>2200 m).

Step 4 Data collection training: Once transects are established, RPN conducts training with the selected forest guardians to train them on how to walk a transect, what to observe on the transect and how to record data on data sheets.

Step 5 Monthly monitoring: Next, RPN establishes a monthly monitoring schedule and a community ranger conducts monthly data collection with each forest guardian team.

Step 6 Community forest user group (CFUG) outreach and awareness building workshops: At the end of every field season, the forest guardian team facilitates a meeting with each CFUG to share their findings and their understanding of the importance of red pandas from their monitoring efforts.

In the Ilam and Panchthar districts, all six steps have been implemented. The activities implemented to date are outlined in the sections below.

Steps 1 and 2: Community Consultation and Preliminary Assessment

In November 2006, RPN initiated long-term monitoring of red pandas in Eastern Nepal through a preliminary field assessment and consultation with local communities and community-based organizations. RPN chose the six forests mentioned above so that they would cover all four aspects (North, East, South, West) and with community input, it also identified two forest guardians per forest (Figure 22.3).

Step 3: Transect Establishment

Within the six forests, RPN and RPNN established 43 transects totalling 63 040 linear metres between elevations of 2400 and 3300 m (Table 22.1). Transects were perpendicular to the elevational gradient and parallel to each other. The start and end point of each

FIGURE 22.3 Community consultation with Jogmai VDC. *Copyright Mr Sonam Lama.*

transect was determined by distinct natural features (e.g., ridges and gullies). Each survey transect was marked with paint, and the start and end points of each transect were georeferenced and recorded (Figure 22.4).

Data Collected

For each transect, the monitors collect data on habitat, vegetation structure, canopy cover, any direct or indirect (scat, pawprints or bamboo eaten) sign of red pandas, substrate of scat sighting, age of the scat, number of scat, ten nearest trees to the sign, distance of sign from the start point, bamboo density and disturbance by livestock, for firewood, for fodder and of bamboo cut within the ten trees around the sign sighting.

For each direct red panda sighting, monitors record the location, all observed characteristics of the animal sighted, their activities while they were observed, time and duration of the sighting, weather, visibility, number of individuals, and, if possible, age group, sex and group composition.

Step 4: Forest Guardian Training

Once each transect was established, RPN and RPNN initiated a forest guardian training programme. The main objective of the programme was to train forest guardians in red

TABLE 22.1 Summary of transect lengths in the six surveyed forests.

Forest	Elevation (masl) of transects Transect length (m)										Total length (m)
	2400 (masl)	2500 (masl)	2600 (masl)	2700 (masl)	2800 (masl)	2900 (masl)	3000 (masl)	3100 (masl)	3200 (masl)	3300 (masl)	
Dhanepa Community Forest	1530	1500	1400	1700	1600	1800	1500	1050	800	720	13600
Surke National Forest	1330	1780	1720	800	1160	900	620	–	–	–	8310
Chhintapu Community Forest	1400	2300	1560	1600	1500	2300	–	800	–	–	11460
Kanyadevi Community Forest	–	–	–	–	600	1250	2250	1500	150	–	5750
Todke Community Forest	2450	2400	2900	3100	2070	800	400	–	–	–	14120
Jaljale Community Forest	–	–	–	1300	1400	1800	1900	2300	600	500	9800
Grand total											63040

FIGURE 22.4 Establishing a transect. *Copyright Mr Sonam Lama.*

panda monitoring. The initial training lasted 3 days and was attended by 12 forest guardians and one forest guard from the District Forest Office in Ilam.

The main topics covered in the 3-day programme were:

1. Why and how to conserve wildlife in community forests
2. Nepalese national policy, acts and regulations that pertain to wildlife
3. Wildlife monitoring techniques.

At the beginning of the 2007–2008 and 2008–2009 field seasons, RPNN conducted 2-day follow-up workshops to gain input from the forest guardians regarding data collection techniques and challenges. RPNN also discussed data collection changes with the team and trained them on the use of these new techniques (Figure 22.5).

Step 5: Monthly Monitoring

In eastern Nepal, the monitoring season runs from mid-October through the beginning of June. Since the completion of the initial training, RPNN has conducted monthly monitoring of all six forests. In conjunction with monthly monitoring, RPN and RPNN also conducted a pilot study with camera traps to assess the potential use and effectiveness of camera trapping with red pandas (Figure 22.6).

Camera Trap Pilot Study

The study area was divided into three plots. Three camera traps were deployed in each plot for 14 days. Red pandas, like other procyonids, defaecate communally (i.e., different individuals use a central latrine to defecate). With this knowledge, the RPN and RPNN field team set camera traps in locations with the greatest amount of scat within a plot (Figure 22.7). At each camera trap location, three cameras were placed on the tree: one angled to the ground at the base of the trunk (Figure 22.8), where the animal could enter

FIGURE 22.5 Forest guardian training: all 12 forest guardians are learning how to take data in the field. *Copyright Red Panda Network.*

FIGURE 22.6 Direct red panda observation (Surke National Forest). *Copyright Mr Sonam Lama.*

the tree, and two angled at the latrine site. Each camera was programmed to take a photo every two minutes and to record the date and time of each exposure.

Figures 22.9–22.13 are a series of photographs from the camera trap pilot study. The final five images are, to our knowledge, the first images of a wild red panda captured in a

FIGURE 22.7 RPN-Nepal biologist Naveen Mahato setting up the Bushnell camera trap unit at the base of a tree. *Copyright Red Panda Network.*

FIGURE 22.8 The Bushnell camera trap unit at the base of the tree. *Copyright Red Panda Network.*

camera trap for scientific study. These images show a sequence of shots of a red panda defaecating in a latrine.

Step 6: CFUG and Local Community Outreach and Awareness Building Workshops

At the end of every field season, RPN and RPNN have conducted outreach and awareness-building workshops with each forest guardian team and their respective community-forest

FIGURE 22.9 Red panda climbing up to the latrine. *Copyright Red Panda Network and Red Panda Network–Nepal.*

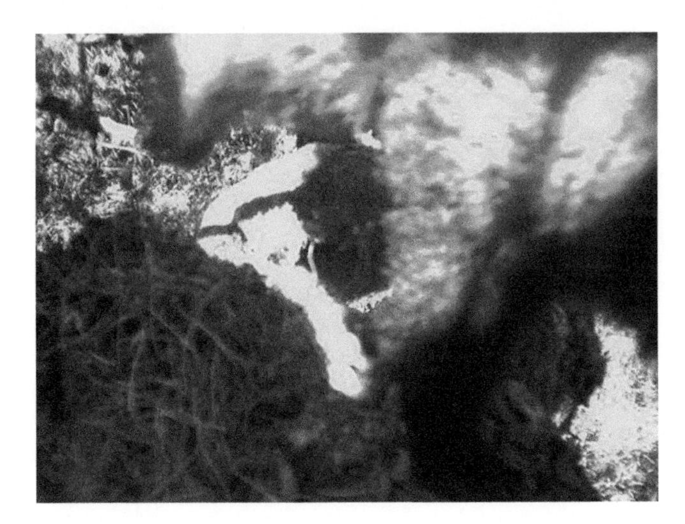

FIGURE 22.10 Red panda defecating over the latrine. *Copyright Red Panda Network and Red Panda Network-Nepal.*

user group. These workshops were created and led by the forest guardians themselves. Before these workshops, RPN and RPNN wildlife ecologists and social mobilizers conducted a brainstorming session with the forest guardian teams to elicit lessons learned and generate a unique educational activity for each CFUG. This activity has been crucial because it has produced examples from the entrusted CFUG member, the forest guardian. For example, last year, a forest guardian showed his CFUG members a sequence of pictures demonstrating the progression of deforestation in their community forest. After the demonstration, the group discussed the impact their daily activities have had on their forests and ways they could deter those activities but still sustain their lifestyles. This experience demonstrated to the CFUG members the importance of forest conservation for their lives.

FIGURE 22.11 Newly defecated red panda scat in the latrine. *Copyright Red Panda Network and Red Panda Network–Nepal.*

FIGURE 22.12 Red panda is sniffing the scat. *Copyright Red Panda Network and Red Panda Network–Nepal.*

FIGURE 22.13 Red panda is leaving the latrine. *Copyright Red Panda Network and Red Panda Network–Nepal.*

OUTCOMES

Red Panda Ecology

Red panda sightings have been recorded throughout the monitoring season, with more direct sightings occurring during the winter months of the breeding season (November–early February). In 2006–2007, we recorded a total of 11 events of red panda sightings in all six forests. Based on the location of the sightings, we concluded that at least five of the red pandas sighted were unique individuals. In 2007–2008, eight events of red panda sightings in different locations were recorded. Based on the location of these sightings, it was concluded that at least six of the red pandas sighted were unique individuals. During this time period, outside of our working area, local villagers observed a mother and two cubs in Surke National Forest, demonstrating that a breeding population is present.

Relative Abundance

The relative abundance from 2006 and 2007 data (number of red panda signs per km of transect length) in the six forests is presented in Table 22.2.

The total size of the study area is 3350 hectares or 33.5 km^2. Based on the relative density in Singhalila National Park of one red panda/1.67 km^2 [1], the relative population size of the study is 20 red pandas.

Camera Trapping

During the pilot camera trapping study, we captured what are believed to be the first images of wild red pandas being camera trapped for scientific study (see Figures 22.9–22.13). Having confirmed that red pandas can be captured with camera traps, RPN and RPNN suggest that a methodology for utilizing camera traps for scientific study with red pandas should be standardized.

Community Impact

Through the CFUG workshops, awareness-building campaigns and the continued presence of RPNN in the field, there is an observed change in local attitudes toward conservation. For example, the chairman of a local CFUG approached RPN and wanted to

TABLE 22.2 The relative abundance (number of red panda signs per km of transect length) based on 2006 and 2007 data in the six forests

Number	Forest	Red panda signs per km of transect length
1	Dhanepa Community Forest, Maimajhuwa	3.015/km
2	Chhintapu Community Forest, Maimajhuwa	1.134/km
3	Kanyadevi Community Forest, Siddin	1.74/km
4	Singhadevi Community Forest, Prangbung	0.51/km
5	Surke National Forest, Mabu	0.722/km
6	Todke Community Forest, Maimajhuwa	0/km

understand what he and his user group could do to preserve the biodiversity in his forest. He said, "We used to have many different bird species that lived in our forests. They have dwindled and we would like to know what we can do to bring them back." This demonstrates a large amount of trust and confidence that he would confide his desires to preserve his community's forests. Another incident occurred in the spring of 2008 demonstrating the impact of the Conservation in Action cycle on local awareness. A red panda was captured and caged by a local villager that lived near a Buddhist monastery. Because of the CFUG workshops and awareness-building campaigns, the local villagers knew that this was an endangered animal and it was illegal to capture and keep it. They took the animal from its captor and released it back into the wild.

Also, there have been observations that the impact of community-based monitoring extends beyond a change in attitudes to the understanding that conservation represents a potential alternative livelihood. Forest guardians receive a monthly stipend for conducting their transects, providing a stable source of income for a population that has otherwise been marginalized. The average PIT corridor household size is five people, thus by providing income for our 12 forest guardians, our work supports a total of 60 people.

These examples demonstrate the potential that exists from community-based efforts, where continued presence and integration into the local community enables villagers to take individual actions that improve their future and the future of its red panda population.

Future Activities

With the support of the Critical Ecosystem Partnership Fund, the Government of Nepal's Department of Forests, and Department of National Parks and Wildlife Conservation, in 2009 and 2010, RPN and RPNN will expand and assist six additional forest user groups to initiate low-cost, sustainable monitoring of red panda habitat. By the end of 2010, RPN and RPNN will have implemented steps two and three of our Conservation in Action cycle. We will delineate the world's first community-conserved area for red pandas and conduct "dreamshops", workshops in which communities outline their dreams from which will be produced locally driven Conservation in Action contracts that will ensure the conservation of these areas in perpetuity.

CONCLUSIONS

The Chinese have a bamboo tree that takes 4 years for its roots to grow before a single shoot grows above ground. In the fifth year the tree grows 80 feet (24.4 m). Community-based monitoring is like the bamboo tree, it takes years of building the roots of a conservation ethic before the shoots and thus fruits of its labour can be realized.

For the past two years, Project Punde Kundo has been laying the roots of conservation among six communities in eastern Nepal. The outcomes of this programme demonstrate that community-based monitoring of a landscape species like red pandas can be an effective tool for the creation of a conservation ethic. By involving communities directly in the conservation programme from the outset, and having them physically taking part in the collection of information on a landscape species, they begin to take ownership of the species. Also, they have an opportunity to experience their landscape from a different perspective. In eastern Nepal, cattle herders who were the main threat to red pandas and their habitat became ardent supporters of saving red pandas and their habitat. These individuals were able to view the forest for what it is, a forest, rather than a resource to be utilized. This new perspective generated a sense of ownership of the forest and pride in the red panda. Thus, building the deep roots from which RPN can grow its bamboo conservation tree.

In our second year, we began to see some of the signs that the conservation roots were mature and that shoots of conservation action were beginning to form. A red panda was captured by a local villager then released by his fellow villagers who understood that this was an illegal act. With this strong foundation, we plan to create community conservation visions from which each individual community can ensure the continued survival of the PIT red panda population and a more secure future for generations to come. With the strong conservation ethic instilled from 2 years of community-based monitoring, communities are more likely to trust our intentions. This trust was extended when local villagers asked for the creation of monitoring programmes in their community forests. The next steps are to ask villagers to create ways to reduce deforestation from overgrazing, excessive firewood and fodder cutting. This could be accomplished through the initiation of Conservation in Action contracts and lead to steps that will ensure the long-term conservation of the PIT Corridor. It could be a model for future conservation efforts. Through this process, our goal of having the PIT Corridor declared the world's first community-conserved area dedicated to red pandas will be achieved by 2010.

ACKNOWLEDGEMENTS

We would like to thank the Government of Nepal's Department of Forest and Department of National Parks and Wildlife Conservation for providing us the required permit to carry out Project Punde Kundo. The support and cooperation of District Forest Offices of Ilam and Panchthar are also highly appreciated. We would also like to thank The Mountain Institute, Shree High Altitude Herbal Production and Conservation Institute, Shree Deep Jyoti Youth Club, Panchthar, Namsaling Community Development Centre, Ilam Cooperation Council, ECCA, Bird Conservation Nepal, Biodiversity Conservation Coordination Committee, eco-clubs, and Community Forest Users Groups of Ilam and Panchthar for their support, participation and continuous cooperation in Project Punde Kundo.

We would also like to thank the following organizations for their generous support of our work: Critical Ecosystem Partnership Fund, Rotterdam Zoo, San Diego Zoological Society, Global Greengrants Fund, Ocean Park Conservation Foundation – Hong Kong, The Himalayan Fair, Melbourne Zoo, Wellington Zoo, San Diego Zoo AZA, Kansas City Zoo AZA, the Lew Family, Linda Yates and Paul Holland as well as the hundreds of individuals donors that have contributed to this effort. Your support is what has made all of this possible.

References

[1] S. Pradhan, G.K. Saha, J.A. Khan, Ecology of the red panda *Ailurus fulgens* in the Singhalila National Park, Darjeeling, India, Biol. Conservat. 98 (2001) 11–18.

[2] S. Pradhan, Studies on Some Aspects of the Ecology of the Red Panda, *Ailurus fulgens* (Cuvier 1825) in the Singhalila National Park, Darjeeling, India, Thesis, North Bengal University, India, 1998.

[3] B. Williams, Red panda in eastern Nepal: how does it fit into ecoregional conservation of the eastern Himalaya, in: J. McNeely, T. McCarthy (Eds.), Conservation Biology in Asia, Society for Conservation Biology and Resources Himalaya, Kathmandu, Nepal, 2006, pp. 236–251.

[4] B. Williams, Status of the Red Panda in Jamuna and Mabu Villages of Eastern Nepal, Thesis, San Jose State University, 2004.

[5] E. Wikramanayake, C. Carpenter, H. Strand, M. McKnight, Ecoregion-Based Conservation in the Eastern Himalaya, WWF and ICIMOD, Kathmandu, Nepal, 2001.

[6] Conservation International. Himalaya hotspot, http://www.biodiversityhotspots.org/xp/hotspots/himalaya/Pages/default.aspx, 2009.

[7] H. Baral, C. Inskipp, Important bird areas of Nepal: a report submitted to the Royal Society for the Protection of Birds (RSPB), 2001.

[8] World Wildlife Fund–Nepal, Biodiversity Governance of the Sacred Himalayan Landscape, Thematic Research Working brief. No. 2.

[9] P. Yonzon, Conservation of the red panda (*Ailurus-fulgens*), Biol. Conservat. 57 (1991) 1–11.

[10] P. Yonzon, Kanchenjunga Mountain Complex: Biodiversity Assessment and Conservation Planning, WWF–Nepal, Kathmandu, 2000Section 1, pp. 1–29.

[11] Department of Forest Resources and Survey (DFRS), Forest Resources of Nepal (1987–1998), Forest Resource Information System Project, Publication No. 74. Babar Mahal, Kathmandu, Nepal, 1999.

CHAPTER

23

Conservation Initiatives in China

Fuwen Wei[1], *Zejun Zhang*[2] *and Zhijin Liu*[1]

[1]Key Laboratory of Animal Ecology and Conservation Biology, Institute of Zoology, The Chinese Academy of Sciences, Chaoyang, Beijing, People's Republic of China

[2]Institute of Rare Animals and Plants, China West Normal University, Nanchong, Sichuan, People's Republic of China

OUTLINE

WHY DO WE PROTECT RED PANDAS?

The red panda is one of the earth's living fossils whose ancestor can be at least traced back to several millions years ago, with a wide distribution across Eurasia and North America [1]. However, it is now confined to the southern slopes of the Himalayas and the area nearby [2–6], extending through Nepal, Bhutan, India and Myanmar into southwestern China [6–8]. Generally, the animal exhibits a disjunctive distribution pattern, separated by high ridges, deep valleys and increasing human activities, confronted with many threats that would severely influence its sustainable survival (see Chapter 21). In China, Wei et al. [6] estimated that wild red pandas have decreased as much as 40% during the past 50 years. In the 2000 IUCN Red List, it was categorized as endangered on the basis of criteria C2a; i.e., that the population is estimated to be <2500 mature individuals (C), and that there is a continuing decline in numbers and population structure (2), with severe fragmentation, and with no subpopulation estimated to contain >250 mature individuals (a) [9].

Red Panda. DOI: 10.1016/B978-1-4377-7813-7.00023-9

The red panda is widely considered to own great values to be protected throughout its distribution range, including:

- saving red pandas is important because they are an ambassador for clean air and water for approximately one billion people on our planet. The forests where red pandas inhabit are the so-called "lungs of South Asia", and if they are not intact and functioning properly, the people, animals and plants there would be severely negatively influenced. In addition, the mountainous chains of Eastern Himalaya and parts of southwestern China, where red pandas occur, are the origin of the three largest rivers in Southeastern Asia, namely the Brahmaputra, Ganges and Yangtze Rivers, which provide continuous water for Nepal, Bhutan, Myanmar, northern and northwestern India, and central and southern China. Protecting red pandas and their inhabited forests can lend necessary insurance to the water supply for people living in these ranges
- red pandas have evolved to be highly specialized as a bamboo feeder, which makes it important from a scientific standpoint [5,8]. To meet its nutrient and energy demand from bamboo, which is low in protein and high in cellulose, the animal has developed a series of ecological traits. In addition, it has no close living relatives in the world, and is considered a living relict of times past
- in their habitats, there are many rare and endangered mammals in sympatry. For example, in western Sichuan, mammals sympatric with red pandas include giant panda (*Ailuropoda melanoleuca*), golden monkey (*Rhinopithecus roxellana*), *Macaca multatta*, *Macaca thibetana*, Asian black bear (*Selenarctos thibetanus*), *Panthera pardus*, *Moschus berezovskii*, *Cervus unicolor*, *Capricornis sumatraensis*, *Naemorhedus goral*, and takin (*Budorcas taxicolor*). These mammals can benefit much from the effective protection of red pandas and their inhabited forests.

Although the animal is still confronted with many severe threats, they have fortunately received more and more conservation efforts during the past two decades. For example, the Chinese government has launched a series of law and regulations, such as National Constitution, Criminal Laws, Wild Animal Protection Law, Forestry Law and Environmental Protection Law and so on, to preserve rare animals and plants, of course including the red panda. It is classed as a category II species under the Wild Animal Protection Law, meaning that it cannot be caught, hunted, sold, transported, imported or exported by anyone without a permit from the State Forestry Bureau or its duly delegated authorities [6].

In this chapter, we first review the conservation initiatives for the animal, and then discuss further its conservation perspectives in China.

CONSERVATION INITIATIVES

Laws and Regulations

Few will doubt that conserving nature requires support from those who make laws, shape policies, and dole out resources [10]. To date, the Chinese government has paid much attention to wildlife protection, and has launched a series of laws and regulations to preserve rare animals and plants in her homeland (Table 23.1).

TABLE 23.1 Laws and regulations of the People's Republic of China about wildlife protection.

Laws or regulations	Date
Constitution of the People's Republic of China	1949
Criminal Law of the People's Republic of China	1979
Forest Law of the People's Republic of China	1984
Management Methods for Natural Reserves of Forest and Wild Animals	1985
Environment Protection Law of the People's Republic of China	1989
Wild Animal Protection Law of the People's Republic of China	1989
Regulations of the People's Republic of China on the Implementation of Terrestrial Wildlife Protection	1992
Regulations of Natural Reserves of the People's Republic of China	1994
Regulations of the People's Republic of China on Administration of Import and Export of Endangered Wild Animals and Plants	2006

To a great extent, the laws and regulations are effectively protecting rare animals in China from being randomly captured or killed. However, some law-breakers still illegally hunt and smuggle the red panda after it has been under these protection laws and regulations. For example, it was reported that a number of red pandas had been illegally transported to Taiwan from the mainland of China [11]. In Mianning county of Sichuan province, 10 smuggled pandas were sequestrated between 1988 and 1996 [6]. The Yunnan Forestry Bureau confiscated 15 smuggled red pandas and 18 pelts between 1988 and 1989 at 37 trade markets on the frontier between China and Myanmar [6]. In 1990, the Sichuan Forestry Bureau confiscated 11 pelts from fur stores in Chengdu city, the capital of Sichuan Province [6]. In 1994–1995, a live red panda was confiscated by the forestry bureau on the Sino-Burma border of Yunnan province [12,13].

Public Education

Educational efforts constitute a special form of indoctrination that offers the prospect of shaping decisions by present and future leaders, whereas non-governmental organizations and other entities attempt through government contacts, programme development, financial inducements, and international conferences to reach the existing power structure. Usually, the primary goal of education is to indoctrinate young people with positive attitudes toward wildlife and environment when they are in their formative stages. In China, the State Environmental Protection Administration and the Department of Education jointly launched an "Action Plan for Environmental Publicity and Education" in 1997, emphasizing environmental knowledge in regular curricula. According to the action plan, most related administrators would be receptive to discussions on programming and curricula. In addition, some conservation organizations from the West, for example, the Wildlife Conservation Society (WCS), have developed environmental education programmes in China, too.

Natural Reserves

During the past several decades, the influences of human activities on biodiversity and landscapes have been widely recognized [14–17]. As the human population continues to increase, demands on natural resources will grow larger and larger. Establishment of nature reserves is a traditional and useful approach to biodiversity conservation which, to some extent, can be considered as the last homeland that humans leave for wildlife on the earth. From the 1950s onwards, the Chinese government has established about 2000 national and provincial reserves as refuges for wildlife, of which, about 46 reserves were established within red panda ranges, including 32 in Sichuan, eight in Yunnan and six in Tibet provinces [6]. These reserves cover about 25 668.36 km^2, including 16 121.52 km^2 in Sichuan, 7189.3 km^2 in Yunnan and 2357.54 km^2 in Tibet (Table 23.2). According to our survey, densities of red pandas in most reserves are much higher than those outside, perhaps indicating that these reserves have played an important role in on-the-spot protection of the animal.

However, enforcement, especially outside these protected areas, is virtually non-existent. A large proportion of red panda habitats have not been effectively protected outside reserves at present, where human activities are going on and red pandas are somewhat easier to be illegally hunted.

Conservation Projects

In China, there are many national projects launched by the Chinese government or its delegated agencies to protect or improve the ecological environment. Some are related to preservation of wild red pandas or their habitats. Here, we briefly introduce several projects.

Natural Forest Protection Project (NFPP)

Many nature reserves or "protected areas" are not well protected from human activities, and the effectiveness of the conservation is usually limited by increasing human pressure [18]. In China, it was said that population growth, economic development, and policy failures as well have resulted in severe environmental problems, such as loss of biodiversity, desertification and soil erosion. In particular, the 1998 floods along the Yangtze River and in northeastern China devastated large parts of China, leading to the loss of more than 3000 human lives and US $12 billion in property damage and output reduction.

In response, the Chinese government initiated a national ban to prohibit logging natural forests in 1998, namely "Natural Forest Protection Project" (NFPP), providing new opportunities to natural forests and biodiversity protection. Its primary goal is to comprehensively improve the ecological environment in China through entirely stopping commercial logging of natural forests along the middle or upstream of the Yangtze River and Yellow River, reducing timber production in some key national forest ranges, and extensively planting trees in logged areas.

Seventeen provinces, 734 counties and 167 forest enterprises are involved in the project (Figure 23.1), fortunately, covering all ranges inhabited by the red panda. As a forest-dwelling species, the animal depends on forests as their habitats. As such, it can be

TABLE 23.2 Nature reserves established within red panda ranges in China

Province	Reserve	Location(county)	Area (km^2)	Total (km^2)
	Jiuzhaigou	Jiuzhaigou	600.00	
	Baihe	Jiuzhaigou	200.00	
	Huanglongsi	Songpan	400.00	
	Xiaozhaizi	Beichuan	67.00	
	Wolong	Wenchuan	2000.00	
	Fengtongzhai	Baoxing	400.00	
	Labahe	Tianquan	125.00	
	Mabian Dafengding	Mabian	300.00	
	Meigu Dafengding	Meigu	130.00	
	Baiyang	Songpan	582.90	
	Sier	Pingwu	189.70	
	Piankou	Beichuan	197.30	
	Wujiao	Jiuzhaigou	371.00	
	Qianfushan	Anxian	172.30	
	Moshui	Lushan	317.90	
	Anzihe	Chongzhou	101.00	
Sichuan	Yele	Mianning	247.00	16121.52
	Wawushan	Hongya	160.00	
	Liziping	Shimian	409.30	
	Mamize	Leibo	388.00	
	Maanshan	Ganluo	408.26	
	Shenguozhuang	Yuexi	337.00	
	Heizhugou	Ebian	180.00	
	Hongba	Jiulong	360.00	
	Caopo	Wenchuan	1224.00	
	Heishuihe	Dayi	308.00	
	Tangjiahe	Qingchuan	400.00	
	Gonggashan	Luding, Kangding, Jiulong, Shimian	4069.00	
	Baishuihe	Pengzhou	303.86	
	Longxi-Hongkou	Dujiangyan	340.00	

(Continued)

TABLE 23.2 (cont'd)

Province	Reserve	Location(county)	Area (km²)	Total (km²)
	Jiudingshan	Shifang, Mianzhu	637.00	
	Baodinggou	Maoxian	196.00	
Yunnan	Gaoligongshan	Baoshan, Lushui	1239.00	
	Tianchi	Yunlong	66.30	
	Baimaxueshan	Deqin	1879.77	
	Nujiang	Gongshan, Fugong	3254.33	7189.3
	Habaxueshan	Zhongdian	219.08	
	Yulongxueshan	Lijiang	260.00	
	Haizhiping	Yiiang, Weixin	27.82	
	Samagong	Diqing	243.00	
Tibet	Gangxiang	Bomi	46.00	2357.54
	Motuo	Motuo	62.62	
	Chayu	Chayu	101.40	
	Zhangmukouan	Nielamu	68.52	
	Mangkang	Mangkang	1853.00	
	Linzhidongou	Linzhi	226.00	
Total	46			25868.36

imagined that the long-term survival of the animal will undoubtedly benefit from the project. Loucks et al. [19] argued that the NFPP could strengthen the wildlife's future in China's forests, especially giant pandas and red pandas, by enhancing protection and restoring corridors among the remaining forest fragments and increasing habitat preservation.

Grain to Green Project

The project was initiated first in 1999 in Sichuan, Gansu and Shaanxi provinces. In 2000, more provinces along the upstream of the Yangtze River and the middle or upstream of the Yellow River were involved, including Hubei, Shanxi, Henan, Qinghai, Ningxia, Xinjiang, and Hunan provinces. By 2002, the project covered about 25 provinces and more than 1800 counties, including Sichuan, Yunnan and Tibet where the red panda is currently distributed.

The project aims to adjust agricultural structure, increase local residents' income, and finally improve ecological conditions in China. To date, great efforts have been made by the Chinese government to carry out the project. It was reported that during 1999–2002, 23.157 billion Yuan (RMB) had been invested to encourage local people to return their ploughed lands to forests. The total budget for the project will exceed 430.00 billion Yuan (RMB).

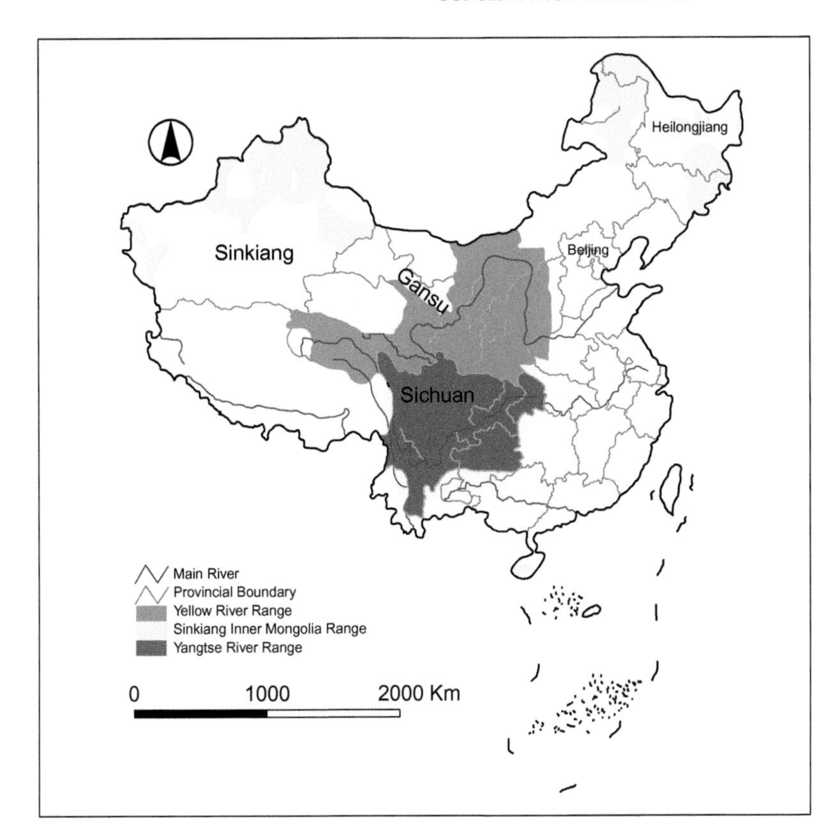

FIGURE 23.1 Ranges involved in the Natural Forest Protection Project (NFPP) in China. From Management Center of NFPP, National Forestry Bureau of China; http://www.tianbao.net/gcgh/index.aspx

The red panda can certainly benefit much from the project. On the one hand, through "grain to green", its habitat can be expected to expand gradually. On the other hand, human-caused disturbances in its habitat can be diminished now that dependence on natural resources from local residents will be reduced as the project is on-going.

National Wildlife Conservation and Natural Reserve Construction Project

The project was launched in December 2001, with the ultimate goal to save some national key-protected animals and plants in China. The primary content in the project includes wildlife protection, natural reserve construction, wetland conservation and species' gene preservation. Of which, species saving, ecological systems, and wetland are especially emphasized. To date, 15 species or categories have been chosen in the project as key-protected species, including the giant panda (*Ailuropoda melanoleuca*), *Nipponia nippon*, *Panthera tigris*, *Alligator sinensis*, *Rhinopithecus roxellanae*, *Elephas maximus*, *Cycas revolute*, etc.

The red panda is not currently listed as a key-protected animal in the project. However, concerning its wide sympatric with the giant panda and the golden monkey, the animal can be effectively protected from the conservation efforts made to these key-protected animals. In addition, as natural reserve construction goes on, more and more habitats and individuals (especially for those outside reserves at present) will be involved in reserves (it was reported that by 2050, the total of natural reserves will reach about 2500, covering about 18.0% of the land area of China).

Yunnan Alternative Energy Project

Beside the above-mentioned projects, some international non-governmental organizations conducted some projects to help develop local economies and protect wildlife and their environment, for example, the Yunnan Alternative Energy Project.

Previously, unsustainable collection and the use of fuel wood have had serious consequences for species and environment conservation in Yunnan province. Half a million households in its northwest area depend on fuel wood collection for cooking, heating and house construction. Meanwhile, indoor air pollution from constant wood burning could lead to serious respiratory health problems and diseases. To protect the 25 000 square miles of land in The Nature Conservancy's Yunnan Great Rivers Project area, it is necessary to address the needs of the local population in the area for reliable and safe energy. As such, The Nature Conservancy and the local government agencies are jointly providing subsidies and technical assistance to install energy alternatives in rural households and schools, such as biogas furnaces, fuel-efficient stoves, etc.

In addition, The Nature Conservancy also helped to carry out a project titled "China Rural Energy Enterprise Development (CREED)", providing funds to small entrepreneurs for their offering energy products and services to rural customers, and help to fund rural energy services, so that poor families will have a means to use modern and clean energy. To date, approximately 2500 household biogas, solar heating, micro-hydropower, and biogas-greenhouse units have been installed, and dependence on fuel wood collection from the surrounding forests is greatly decreased.

Part of the habitats inhabited by the red panda in Yunnan province falls in the range of the above project. As the project goes on, it can be expected that dependence upon fuel wood from forests will be reduced, resulting in the decreased human activities in the animal's habitats, favouring its survival in the future.

CONSERVATION PERSPECTIVES

Since the 1980s, the Chinese government has made more and more efforts to protect wildlife in her homeland, such as law legislation, reserve establishment and public education as well. Undoubtedly, the future of the red panda will greatly benefit from these measures. On the one hand, under the Wild Animal Protection Law, any illegal catching, taming, hunting, selling, purchasing and transportation of the animal and its products will be severely punished. On the other hand, as the above-mentioned ecological projects, such as Natural Forest Protection Project, Grain to Green Project, are going on, degradation of its habitats can be expected to cease, and human disturbances will be diminished. The red panda should have a more hopeful future than in the past.

The red panda, however, is still faced with many severe threats affecting its sustainable survival in future (see Chapter 21). During the last 50 years, its population has decreased as much as 40%, and it has disappeared from some former localities [6]. In China, the Wild Animal Protection Law was promulgated in November 1988 and enforced on March

1, 1989. Under the law, the red panda was classified as Category II. Concerning the change of the animal in distribution, population size, and conservation status, there is a need to promote it to be a Category I species under the law.

Regular monitoring of its habitat and population is extremely necessary, although such monitoring only began about 10 years ago in China (not for the red panda). Through regular monitoring, information, such as habitat, population and human disturbances as well, can be collected systematically, which is essential for effective conservation and management.

Public propagandization and education, especially in some faraway mountainous ranges inhabited by the red panda, should be reinforced. People living in these areas still heavily depend on the natural resources around them, including capturing animals for their fur or meat, felling timbers for heating or house construction, and collecting herb medicine as well. Of course, to reach the goal of such kind of public propagandization and education, it is necessary to help local people to improve their income from other sources.

In the end, reintroduction of captive individuals into wild populations will be needed, especially for small populations. Concerning the fragmentation of its habitats, many populations perhaps need "immigrated" individuals to assure their long-term survival. For example, it was reported that the population in Haiziping Nature Reserve, Yunnan Province only consisted of about 20 individuals [20]. However, great caution should be taken when reintroducing captive individuals into wild populations, concerning the low success rate (11–75%) in reintroduction activities and with little to no improvement over the past 20 years [21,22].

References

[1] R.H. Tedford, E.P. Gustravson, The first American record of the extinct panda, *Parailurus*, Nature 265 (1977) 621.

[2] S.C. Wallace, X. Wang, Two new carnivores from an unusual late Tertiary forest biota in eastern North America, Nature 432 (2004) 556.

[3] A. Choudhury, An overview of the status and conservation of the red panda *Ailurus fulgens* in India, with reference to its global status, Oryx 35 (2001) 250.

[4] M.S. Roberts, J.L. Gittleman, *Ailurus fulgens*, Mammalian Species 222 (1984) 1.

[5] A.R. Glatston (Ed.), Status Survey and Conservation Action Plan for Procyonids and Ailurids: the Red Panda, Olingos, Coatis, Raccoons and Their Relatives, IUCN, Gland, Switzerland, 1994.

[6] F. Wei, Z. Feng, Z. Wang, et al., Current distribution, status and conservation of wild red pandas *Ailurus fulgens* in China, Biol. Conservat. 89 (1999) 56.

[7] K.G. Johnson, G.B. Schaller, J. Hu, Comparative behavior of red and giant pandas in the Wolong Reserve, China, J. Mammal. 69 (1988) 552.

[8] F. Wei, Z. Feng, Z. Wang, et al., Habitat use and separation between the giant panda and red panda, J. Mammal. 81 (2000) 448.

[9] C. Hilton-Taylor (Ed.), 2000 IUCN Red List of Threatened Species, IUCN, Gland, Switzerland and Cambridge, UK, 2000.

[10] D.G. Lindburg, K. Baragona, Giant Pandas: Biology and Conservation, University of California Press, Berkeley, USA, 2004.

[11] J. Low, The Smuggling of Endangered Wildlife Across the Taiwan Strait, Traffic International, Cambridge, USA, 1990.

[12] Y. Li, Z. Gao, X. Li, et al., Biodivers. Conservat. 9 (2000) 901.

[13] S. Wang and Y. Li (Eds.), Illegal trade in the Himalayas, New York, USA, 1998.

[14] P.R. Ehrlich (Ed.), The Scale of Human Enterprise and Biodiversity Loss, Oxford University Press, New York, USA, 1995.
[15] J.A. McNeely, M. Gadgil, C. Leveque (Eds.), Human Influences on Biodiversity, Cambridge University Press, New York, USA, 1995.
[16] P.M. Vitousek, H.A. Mooney, J. Lubchenco, et al., Human domination of earth's ecosystem, Science 277 (1997) 494.
[17] E.O. Wilson, Biodiversity, National Academy of Science Press, Washington, USA, 1988.
[18] V. Dompka, Human Population, Biodiversity and Protected Areas: Science and Policy Issues, American Association for the Advancement of Science, Washington, USA, 1996.
[19] C.J. Loucks, Z. Lv, E. Dinerstein, et al., Giant pandas in a changing landscape, Science 294 (2001) 1465.
[20] G. Hu, Y. Du, Current distribution, population and conservation status of *Ailurus fulgens* in Yunnan, China, J. Northwest Forest. Univ. 17 (2002) 67.
[21] J. Fischer, D.B. Lindenmayer, An assessment of the published results of animal relocations, Biol. Conservat. 96 (2000) 1.
[22] J.A. Stamps, R.R. Swaisgood, Someplace like home: experience, habitat selection, and conservation biology, Appl. Anim. Behav. Sci. 102 (2007) 392.

CHAPTER

24

Records and Reports of Red Pandas *Ailurus fulgens* from Areas with Warm Climates

J.W. Duckworth

IUCN/SSC Small Carnivore Red List Authority Focal Point, Vientiane, Lao PDR

OUTLINE

INTRODUCTION

The Red panda, *Ailurus fulgens*, is well known to inhabit cold mountainous areas, chiefly at high altitude, in areas with heavy winter snow (particularly above 2200 m) (see Chapter 11), but the last hundred years have also provided a number of reports of it in areas atypical in several ways: south of the conventionally accepted range; lying at lower altitude, sometimes even in the lowlands; with hotter, even much hotter, climates; and supporting tropical broad-leafed evergreen forests. This short chapter reviews all such reports that have been traced, because their number and geographic spread suggest that this could be a genuine phenomenon of natural red panda distribution. However, no such population has been confirmed as involving wild animals.

The anomalous records, hereafter referred to as "warm-climate pandas", come, with varying levels of detail, from Vietnam, the Lao People's Democratic Republic (Lao PDR; Laos),

Red Panda. DOI: 10.1016/B978-1-4377-7813-7.00024-0

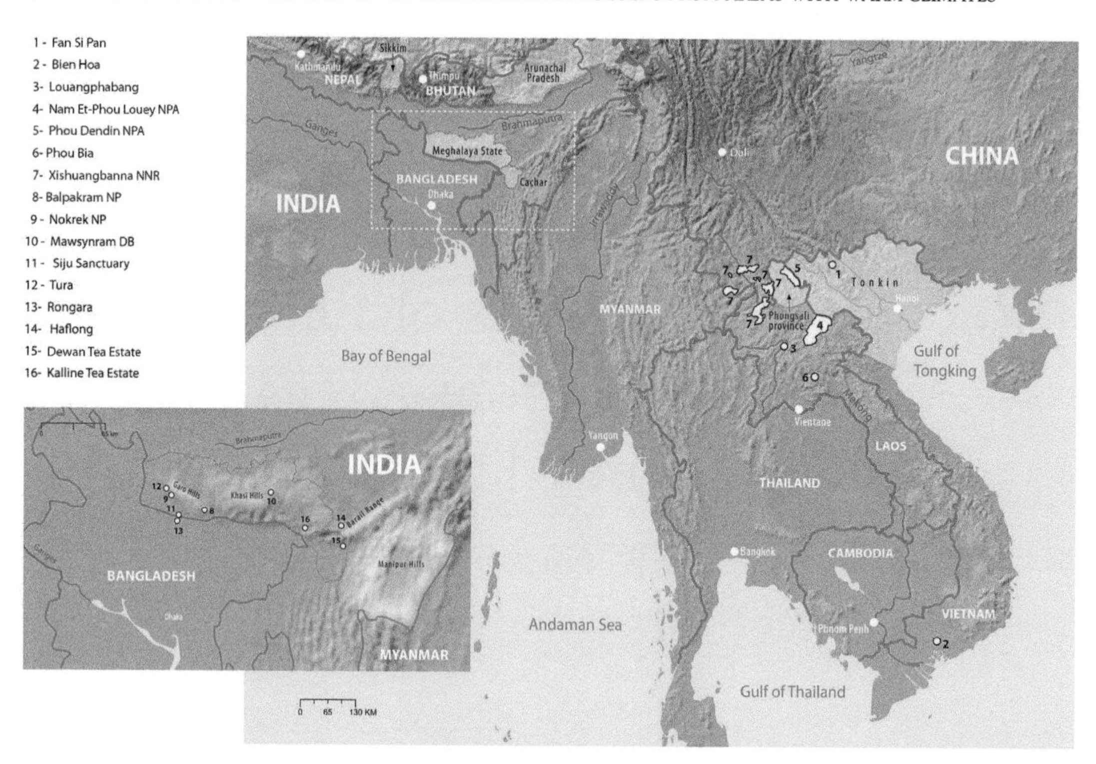

FIGURE 24.1 Locations of warm-climate panda reports, and other places mentioned in the text. *(Map produced by Leen Zuydgeest, Rotterdam Zoo).*

China and India (Figure 24.1). The Vietnamese reports faded a long time ago and are included here for their apparent role in "seeding" the Lao statements. These latter have largely been ignored in authoritative zoological works, but the country is generally listed as supporting the species in the conservation grey literature. The Chinese record of a warm-climate panda has recently been withdrawn. The Indian evidence has much more field detail than that from any other country, and is the only set with any compelling basis to be taken seriously.

VIETNAM

de Monestrol [1] apparently saw red panda(s), live or dead, whole or in part, in Indochina in a way sufficient for him to believe that the species occurred there naturally. He later [2] listed a Vietnamese name for the species (*Con Cáo; con* is simply the general Vietnamese substantive for "animal") and even later [3] implied that he had record(s) specifically from Tonkin, describing the animal as only rarely found, and nocturnal; but nowhere did he detail what he actually saw. Other authors have taken *Cáo* to refer to red fox, *Vulpes vulpes*, either alone [4] or modified (e.g., *Cáo Lúa* [5]), a species not mentioned by de Monestrol in his final work [3] at all. No explicit evidentiary basis for this linkage was traced, and because the red fox itself is an ace rarity in Vietnam [6,7], it is not clear how usage of an autochthonous Vietnamese name could have been validated. At this era, red panda pelts and tails were traded over long distances from Yunnan province, China, and at Talifu (= Dali; 25°42′N, 100°11′E), red panda was one of the commonest trade furs: the tails were used as in Canton as brushes and in Dali itself as dusters, while the skin was believed to be of little use [8]. The Legendre Indochina Expedition [9,10] procured two red panda skins in Hanoi in October 1931, which are now held, one each, at the American Museum of Natural History (AMNH) and at the Academy of Natural Sciences, Philadelphia (AMNH 87413 and 87414). The one at AMNH was recently re-examined and is an obvious commercial skin (D.P. Lunde *in litt.* 2008), and confirms that there was contemporary trade in Hanoi. Trade is the most plausible source for de Monestrol's record(s), if the identification as red panda was correct. Also referring to Vietnam, Bourret [11] compiled a highly inclusive list of Indochina's mammals, explicitly including anything which might occur, whether or not there was any actual indication of it doing so, soon after his arrival at l'École Supérieure des Sciences de l'Université indochinoise d'Hanoi. He devoted much of the following 20 years to aligning it more closely with reality. This list included red panda as already known from Indochina, without any locality being specified, or any Vietnamese name being given. No source was explicitly indicated, but the bibliography includes de Monestrol [1] and so this latter may have provided sufficient reason for him to list the species. Bourret [11] had already assigned *Cáo* as the Vietnamese name for Red Fox, despite saying (p. 29) that "aucun renard n'a été encore signalé en Indochine" [= no fox has yet been found in Indochina].

If red panda lived in Vietnam, it would presumably be in the northern highlands, which have a significant area of land above 2200 m, and rise to 3148 m (Fan Si Pan); but they are never snow-bound. Various Himalayan mountain forest birds syntopic with red panda are known from these Vietnamese highlands, e.g., hill partridge, *Arborophila torqueola*, Temminck's tragopan, *Tragopan temminckii*, Ward's trogon, *Harpactes wardi* and sundry babblers (Timaliidae) [12], but there has been no plausible claim there of red panda. This is despite a recent resurgence of wildlife surveys in the country. Some enigmas concerning carnivoran status remain in Vietnam, most notably a weasel taxon *Mustela* (*nivalis*) *tonkinensis* known only from the holotype collected in the northern highlands in 1939 [13–15]. However, a small weasel is a good deal more cryptic than a red panda, and it is implausible that the latter could have remained undetected if it persisted into the last few decades. Hunting levels are very high and montane forest is greatly fragmented in the higher altitudes [16]; it is just about plausible, given the perilous status there of remaining quarry species of mammal and bird sensitive to hunting-driven declines [17–20], that if red panda had occurred there, it could have been hunted out too early during the period of biological exploration to have been documented. However, there is no even weakly suggestive indication that it ever was present in Vietnam in historical or recent times, and the species is not discussed in Roberton's [21] comprehensive review of the country's small carnivorans.

One other source concerning Vietnam warrants mention. Roussel [22] noted "le panda blanc, aperçu dans la forêt de Sông Mai" [the white panda, recorded in the forest of Song Mai], in his cursory listing of mammals of Indochina. On this laconic account, it is impossible to determine what animal was seen, but a palm civet (Paradoxurinae) is an obvious possibility; most of the various other small carnivorans that he mentioned are too difficult to assign to biological species to spot any obvious gap to which "the white panda" might relate. Moreover, one of the few which can be identified ("le putois dit de Java à robe fauve, tête et extrémité de la queue blanches" [the Javan polecat, having a fulvous pelt with both head and tail-tip white]) is the Malay weasel, *Mustela nudipes* (as well as the fitting morphological description, the name Fûret de Java was often used for this species in French), never recorded from Indochina (or for that matter, Java) despite a generally confused history of distribution records [23]. Elsewhere in his book, Roussel [22] mentioned the Sông Mai forests as lying near Bien Hoa (10°57′N, 106°49′E), in far southern Vietnam: among the least plausible parts of Indochina to support red panda. Roussel's [22] statement is chiefly important because his book was one of the first accounts of hunting in French Indochina and, as such, his statement may have originated the idea that pandas occurred there; his writing probably had substantial influence upon the hunting community's "common knowledge", and is referred to by de Monestrol, among others.

LAO PEOPLE'S DEMOCRATIC REPUBLIC

All historical claims of red panda in Laos that allow determination of source lead back to one primary document, Cheminaud [24], who devoted Chapter 11 (pp. 129–131) of his second volume of *Mes chasses au Laos*, sub-titled *Les bêtes sauvages de l'Indochine*, to "le petit panda" [the small panda]. Of the 2½ page chapter, a good half-page explains how he broke his pipe and so needed to seek out "golden bamboo", which can, in emergencies, be used for cigarette papers. This anecdotal style typifies the author's work, and the following English précis includes all the panda-relevant content of the chapter.

Cheminaud [24] attached the Lao name "*Tô Mi Kham* ...[=] ours doré" (i.e., golden bear, in both languages; *Tô* is simply the general Lao substantive for animal) to "le petit panda", a very pretty small animal of northern Laos which could casually be taken for a plantigrade. It was golden red with a white head and black tail; had the size, and somewhat the habits, of a cat; was gentle, soft-natured and pleasant, being more prone to lick its captor's hand than to bite or scratch it. Colonial hunters called it "ours des bambous"' (i.e., bamboo bear); it enjoyed honeyed secretions of bamboo during the rainy season, and was particularly associated with golden bamboo (*may sang kham*; still a recognized name today). One afternoon (in daylight), Cheminaud saw an animal emerging, without difficulty, from the foot of a golden bamboo clump, he was going to shoot it; but his Lao companion averred that it could be captured alive, then ran and grabbed it by the scruff of the neck. The animal did not much protest, and its chief concern seemed to be to continue crunching on a stick of bamboo. It was taken back to camp in a knapsack. This panda, Cheminaud's first encounter, was a female. He attached a collar and chain and it adapted so quickly to captivity that within 8 days one would have assumed it had never known any other life. He fed it on bamboo and sweet-potatoes. It died when, the forestry work over, the party left the region.

Another time his crew caught a male "petit panda"', while felling a Teak *Tectona grandis*. The animal fell to the ground, broke a foot, scratched its captors a little, and ejected onto them a viscous secretion with a disagreeable odour. Its foot healed, this individual also adapted to captivity but died, despite the care lavished on it. Cheminaud concluded from these two experiences that the animal is rather delicate, and that were it not, it would displace the Siamese cat as one of the most agreeable companions of humankind. No date or location is given for either animal, although the first was captured within a few days' travel of Louangphabang (19°54′N, 102°08′E). Information elsewhere in Cheminaud [24,25] suggests that the encounters were within 1900–1910, potentially as late as 1927.

The casual morphological description is broadly consistent with red panda, although it is almost as applicable to masked palm civet, *Paguma larvata*, which is common in Laos [26,27]. The described behavioural traits are consistent with red panda, and the statement that the captive animals lived on a bamboo diet, evidently for some time, effectively proves red panda to be the animal under discussion. One wild capture could conceivably be attributed to an escaped pet or trade animal, but to try to do so with two, at this era of patchy connectivity and so far from the otherwise known range, is stretching the bounds of credibility. The only internal oddity concerns the occurrence within teak forest. Teak grows in Laos naturally in only a few places, all on the Mekong plain. These are areas which support lowland mixed deciduous forest, with under 2000 mm of rain per year, a marked 3–4-month dry season (in which teak is deciduous), temperature ranging mostly within 22–27°C and widespread annual fires [28]. This is a highly implausible habitat to support a red panda population or even, given its distance from mountainous areas, a dispersing individual. Various passages in Cheminaud [24,25] indicate that a lot of his work consisted of felling and transporting teak logs, so a mistake in identifying the tree is unlikely. Of the various wildlife surveyors recently active in Laos, none consulted has ever heard the given Lao name, *Tô Mi Kham*, in use, but nothing can be interpreted from this in the absence of information on who used the name to Cheminaud and in what context. Some of Cheminaud's other associations of Lao names with animals were misplaced, e.g., the use of *Taloung* for red-shanked douc *Pygathrix nemaeus* [24] when today it is used for grey leaf monkeys *Trachypithecus* within Laos, and douc has one of the most universally applied Lao names of any animal throughout its Lao range [29]. Read in isolation, Cheminaud's story seems to be highly suggestive of the presence of red panda in Laos.

However, the same can be said of the same author's accounts of such biogeographic improbabilities as chital, *Axis axis*, and sloth bear, *Melursus ursinus*, both of which he claimed to have hunted and examined in the flesh in Laos. In total, his writings listed 10 species of mammal not then known from Laos. Two have subsequently been found (Sunda colugo, *Cynocephalus variegatus*, and flying-fox, *Pteropus* sp. [30,31]), but the implausibility that multiple species such as chital and sloth bear, and even cheetah, *Acinonyx jubatus*, were all living in Laos a century ago forced Duckworth et al. [32] to review Cheminaud's overall work [24,25], building upon an earlier doubt expressed over Cheminaud's information about bears [33]. This revealed a mix of obvious misidentifications (sometimes shared by the general contemporary expatriate hunting community); opportunities for a credulous person, as Cheminaud evidently was, to draw erroneous conclusions; and internal inconsistencies. In addition, Cheminaud's own work gives multiple indications that actually he was not very familiar with the field status of mammals in Laos, e.g., his statements that arboreal civets and flying squirrels were rare and localized: they are actually widespread, common, and readily seen [26,34–36], and cannot plausibly have been anything else a century ago. The accounts were published, and so presumably written, decades after the experiences, and it is beyond doubt that the author supplemented his memories (which by then may have wandered somewhat) and notes with information from other books. Finally, Cheminaud's book was published in a commercial series of generally rather racy, even swashbuckling, travelogues. The extent of editorial tuning-up of the author's content cannot now be assessed, but cannot be assumed to have been negligible.

A century ago in Laos, the would-be mammalogist had no identification keys, no authoritative texts, and no comprehensive specimen reference collections; nor, until R. Bourret had established himself in Hanoi, did a single competent land vertebrate taxonomist reside in the region. Thus, Cheminaud's large number of mistakes [24,25] was to be expected. Indeed, Delacour [6] prefaced his overview of the mammals of Indochina thus: "À consulter les ouvrages sur la chasse et les quelques listes des mammifères de l'Indochine parus avant 1932, il apparaît que les indications sur ces animaux, mises jusqu'alors à la portée du public, étaient empreintes d'une extrême fantaisie. Cela se trouvait d'ailleurs fort excusable, car aucun travail scientifique d'ensemble n'avait publié sur le sujet jusqu'à cette époque. . .Osgood (1932) demeure inconnu de beaucoup de nos compatriotes" [The hunting memoirs and the several lists of the mammals of Laos, Cambodia and Vietnam produced before 1932 and available to the public are embedded in extreme fantasy. This is understandable given that nothing scientific on the subject had been published up until then. . .and the English-language Osgood [37] remains unknown by many French people]. Cheminaud's two-volume work [24,25] is a perfect example of the specious ramblings which goaded Delacour [6] to produce his list, in French, although he did not, in accordance with the gentlemanly spirit of the times, specify the works he considered dubious.

Despite the several factors that helped the information in Cheminaud's books [24,25], and in various other contemporary texts, to part company with reality without the authors necessarily intending them to do so, there remains a rump of Cheminaud species accounts which if not factually true must have been fabricated, including the red panda. Cheminaud's [25] stated aim was to educate overseas hunters about animals that they might encounter in Laos: it seems that he had no ethical scruples in attempting to do so by writing a readable book reflecting "common knowledge" of the contemporary hunting community. This would be fine were the generally held beliefs accurate, but because they were not all so, it allowed Cheminaud [24,25] to perpetuate major errors even with the best of intent. The individual chapters, which are mostly hunting anecdotes, should be seen not as factual records but as dramatizations: the modern parallel concerns natural history films on television today which manipulate animals, habitat and circumstances to generate desirable footage. Cheminaud [25] betrayed his ignorance of mammalian biogeography when writing that "et quand je dis Laos, je pourrais aussi bien lui substituer l'Asie tropicale entière, puisque la faune est sensiblement la même partout" [and when I say Laos, I might as well have included all of tropical Asia, because the fauna is effectively the same throughout it]. He "knew" that red pandas lived in Indochina through de Monostrol [1,2] and Bourret [11], and perhaps even took Roussel's [22] vague statement at face value. And because they lived in Indochina, red pandas would presumably inhabit Laos, and, therefore, should be in his book.

In conclusion, it is not logically defensible to extend any more credence to Cheminaud's [24] statements that red pandas occur in Laos than to his information about sloth bear or chital in the country [32]. Unfortunately, the works of J. and M. Deuve, culminating in Deuve [38], did just this. The Deuves accepted red panda as a Lao animal without caveat, although where they wrote most extensively [39] they sloppily (and ironically) stated that Cheminaud had never encountered red panda in Laos, but had simply been told by people in north Laos of an animal fitting its description. The Deuves' work is replete with trenchant errors concerning the status of Lao mammals, and they even managed utterly to mis-portray the Lao status of one of the most unmistakable of them all, the red-shanked douc [29]. Therefore, their acceptance of red panda's presence in Laos adds no authority. Unfortunately, their work has instilled in the subsequent conservation grey literature many erroneous beliefs, not only from Cheminaud. The *Lao PDR Biodiversity Country Report* [40], to inform the *Lao PDR National Biodiversity Strategy and Action Plan*, even gave a date for red panda extinction in Laos (1942: it is not difficult to infer the stimulus to select this date), and the international Red List of threatened species [41] lists Laos within the red panda's range, without caveat, as it has done for years [42,43].

Historically, mammals in Laos were not sufficiently well collected for the lack of red panda specimens to imply absence of the animal. Wildlife surveys across the country as late as the 1990s found various distinctive mammals, comparable in size to red panda or even larger, for the first time, including some new to science ([44] and references therein). The only mention of red panda in Laos independent of Cheminaud seems to be in Sayer [45]: "a reliable observer reported seeing a live lesser panda offered for sale in a market in Vientiane in early 1983. The animal may well have originated in S. China from whence it could have been transported by boat down the Mekong. The record also raises the interesting possibility that this species may occur in mountain forests in North Laos". It is difficult at this late remove to assess the relative likelihoods of the two origins. Nowadays, wildlife trade is much more from Laos to China (often via Vietnam) than the reverse [46,47]. Material from further north in Asia does sometimes occur [44] but in non-perishable form. Despite extensive observations from 1992 onwards, the author has not himself seen, nor traced a record from any other observer of, a live wild mammal for sale in a Lao market of a species that does not occur in Laos. Things may have been different in 1983: a decade previously, Laos was a receiving and redistributing node in the international trade in pet and zoo animals, particularly high-profile mammals under international protection [48], and this may have continued into the 1980s, in that CITES documents then often listed Lao PDR as the origin for species which do not occur there ([49] Annex 2). The extent to which this paper assignment was by that time still accompanied by actual movement of the animals themselves into and out of the country is unclear, but at least some animals were still genuinely passing through Laos, e.g., several orang-utans *Pongo pygmaeus* which were confiscated in 1987 in a major government operation which involved imprisoning the then director of the Government of Lao Wildlife Division (K. P. Berkmüller verbally 1992) [49,50]. Finally, a small doubt must remain over the identification of this 1983 animal, with no ability now to assess the observer's familiarity with possible confusion species such as masked palm civet. The only naturalist known to be in Laos at that time was W.W. Thomas, who died in 2004 [51]. When reviewing his bird records [52], no reference to red panda was found in his extensive Lao notes (although various other mammals seen in markets and as captives are, as is a meeting with Sayer). J. Sayer (*in litt.* 2008) has no notes additional to the report, so the possibility remains it was in fact another observer.

The northern highlands of Laos constitute the least unlikely area of the country to support red pandas. Recent field-work up there has included four intensive surveys, each of several weeks, of the birds, large mammals, habitat and human use of defined areas [53–58] and an ongoing longer-term carnivoran research project (A. Johnson verbally 2008) [27]. The latter achieved high camera-trap coverage of one of the northern highlands' largest forest blocks, Nam Et–Phou Louey National Protected Area, and found 20 species of carnivoran. The other four surveys lacked significant camera-trapping, so their chances of detecting living, wild, red pandas were low. But, captive and hunted animals were examined widely in villages and markets, and mammals were discussed extensively with villagers, even in the remotest settlements, on all surveys. In Phongsali province, northernmost Laos, 12 men in five villages in 1996, and residents of six villages in and around Phou Dendin National Protected Area in 2004–2005, were interviewed with red panda specifically in mind by R. J. Tizard and W. G. Robichaud, respectively. These interviews used free-ranging discussions, asking people with apparent forest knowledge to describe, for each sort of mammal living locally, its name, appearance and habits, and used pictures only selectively. In 2004–2005, residents of two villages reported local presence of the animal in the red panda colour photographs, but their further descriptions included anomalous features, e.g., a black body and a non-banded tail. An animal as distinctive and confiding as the red panda, if locally extant, could not be unknown to villagers. A large proportion of the northern highlands, notably Phou Bia (Laos's highest peak, although it rises only to 2819 m), has not been visited by a competent mammal surveyor speaking good Lao and/or local languages, so the current lack of positive village reports of red pandas is not yet strong evidence of their absence from Laos. That, however, is a different situation from there being any reliable suggestion that red panda does occur in the country.

PEOPLE'S REPUBLIC OF CHINA

Red panda was stated to live in Xishuangbanna National Nature Reserve in the far south of Yunnan province (21°47′N, 101°03′E) by various sources, e.g., Wang Yingxiang [59]. This was based on a single specimen held in the Kunming Institute of Zoology (Kunming, Yunnan, China), which is undoubtedly correctly identified and is labelled with this locality (JWD own data 1998). On-going work in Xishuangbanna NNR failed to find the species (e.g., [60]). Moreover, Wang Yingxiang (verbally, 2006) now considers that the specimen must have originated elsewhere, and recently [61] restricted the species' Yunnanese distribution to the north-western, northern and mid parts of the province. The specimen is not recent and even in the best-regulated collections, data are sometimes transposed between specimens. Possibly this happened here. There seem to have been no other reports of populations in climatically anomalous areas of China (Wei Fuwen *in litt.* 2008).

INDIA

Red panda is well known from Himalayan India (see Chapter 20), but Choudhury [62] described the "startling find" of the species in the Garo hills, Meghalaya (west of the

FIGURE 24.2 Dr John R. Lao (left) with the skin of the red panda that he shot in the Garo hills in the 1960s. Dr A. Choudhury sits to the right *(Photograph by Mr Hashim Choudhury, 10 October 1995, Tura town).*

Khasi hills, centred on 25°30′N, 90°20′E). This was *c.* 200 km south of the generally accepted range in India, and across the other side of the Brahmaputra valley. This is a dispersal barrier for various primates [63], and thus plausibly for small carnivorans. The first indication was the inclusion of the species' name in a local Forest Department brochure to Balpakram National Park (25°16′N, 90°51′E). Choudhury himself observed one skin in 1995, held by the original collector, Dr John R. Lao, who provided details about the incident. He shot an animal, which he initially took to be a black giant squirrel, *Ratufa bicolor,* in the Nokrek area of the Garo hills (25°27′N, 90°18′E) in the early 1960s (Figure 24.2). Dr Lao had subsequently observed single red pandas twice in Nokrek and once in Balprakam, without shooting them (A.U. Choudhury *in litt.* 2008). Two more skins, taken shortly before his visit, from this hill-range were reported to Choudhury [62], although he did not see them. Local officials indicated that the species was confined to the higher areas of Balpakram and Nokrek national parks, which rise only to 1023 m and 1412 m, respectively (Figures 24.3 and 24.4). The skin had a head-and-body length far exceeding published measurements from elsewhere; however, the tail was within the usual range (Table 24.1), and even larger skins have subsequently been found within nearby Arunachal Pradesh [66]. This was a well-processed skin, with no visible over-stretching (A.U. Choudhury *in litt.* 2008) [66]. A local name for the species, *Matchibel,* was provided by the Garo ethnic group. A few years later, through an ongoing mammal survey in northeast India [67], Choudhury [68] referred to having personally seen two skins in Meghalaya; identified a third area of the state holding red pandas, the Mawsynram development block of East Khasi Hills district (centred on 25°20′N, 91°35′E); gave an habitual altitudinal range in Meghalaya of 700–1400 m; and gave a second-hand sight-record of

FIGURE 24.3 Habitat in the gorge of Balpakram National Park, 770 m asl at point of photograph *(Photograph by Anwaruddin Choudhury, 10 November 2008).*

FIGURE 24.4 Subtropical forest in Nokrek National Park, 1400 m asl *(Photograph by Anwaruddin Choudhury, 13 November 2008).*

TABLE 24.1 Pelt measurements of red panda.

Head and body	Tail	Area	Source
505–630 (7)	370–480 (7)	Sichuan, Yunnan and Zangmu (Tibet)	[64]
574–610 (4)	366–406 (4)	Sikkim	[65]
589–622 (3)	355–472 (2)	Yunnan	[65]
720–790 (2)	430–500 (2)	Arunachal Pradesh, India	[66]
730	430	Meghalaya	[62]

Measurements in [67] and [66] were made in the flesh; those from Arunachal Pradesh were of recently prepared dried pelts; and that from Meghalaya was a tanned skin reportedly several decades old. The Arunachal Pradesh specimens were specifically selected for measurement through their large size [66].

one red panda on 28 January 1997 at the astonishingly low altitude of 200 m in Siju Wildlife Sanctuary (25°13′N, 90°25′E). Siju is contiguous with, and just west of the highest point in Balpakram National Park. The record from the East Khasi Hills relates to a party of British anglers familiar with the Himalayas who, while driving at night, saw a red panda and mentioned it in a travelogue account in a British newspaper. The local guide confirmed the sighting and identification to A.U. Choudhury (*in litt.* 2008). The second skin, which was partly damaged and was not photographed, was from an animal shot in Balpakram sometime in the 1980s (A.U. Choudhury *in litt.* 2009).

There was little historical mammal survey in the area. Tura (the station below Nokrek) was one of four Garo hills sites collected by H.W. Wells in 1919–1921 [69], but coverage was very patchy and nothing can be inferred from the absence of red panda in the list of species collected. Recently, information was gathered, during wider-ranging conservation activities, on mammals in 31 akhings (an akhing is a unit of land owned by a clan) in the Rongara block of South Garo Hills district (25°07′N, 90°28′E), focused on mammal population trends over time and human–wildlife conflict [70,71]. The two surveyors, one of them Garo, associated, through discussions with Garo elders and hunters, the Garo name *Matchibil* with Binturong *Arctictis binturong*, and gave red panda the Garo name of *Matcha Pantao* [71]. This survey included red panda in the animals covered during interviews with a field guide, but did not specifically focus on it. Only 25 (7%) out of the 334 respondents said that they had seen the species (N. Ved *in litt.* 2008), a low enough percentage to reflect background 'noise' stemming from mistaken matching of name to species, second-hand reports in error through relating to other areas, times or species, and other factors; but it is perfectly consistent with there being a small population of red pandas in these hills.

Choudhury's records turn out not to be the first case of "warm-climate pandas" in India. When asked about them, Alan Lane (*in litt.* 2008) provided two records from the 1950s–1960s when he and his father, John Lane, worked in the tea estates of Cachar. John Lane, who was an honorary forest officer in the north-west Cachar area (appointed by the Assam Forestry Service), saw a couple of red pandas at the Dewan Tea Estate (north-east Cachar; 24°55′N, 93°03′E), at the outgarden of Burtoll where the Barail range (=Mikir hills) comes down to the edge of the garden, one December in the early 1950s. Later, both Lanes saw a red panda at Sundura outgarden, Kalline Tea Estate (24°59′N, 92°33′E) in December 1964, high up in a forest tree in an area quite thickly covered with tea. Sundura, through which the Kalline River flows, lies right at the foot of the Barail range and was then surrounded by bamboo jungle, sal, *Shorea robusta*, forest (deciduous) and evergreen forest. It is noteworthy that both these sightings were in the coldest months of the year, when red pandas are likely to visit the lowest altitudes. The exact altitudes of the sightings were not recorded; they must have been between about 300 and 1800 masl, but both areas are reasonably close to much higher-lying areas, with which they were then linked by natural habitat. In addition, Cyril Booth, District Commissioner for the Barail range district (then called the United North Cachar & Mikir Hills district) in 1962, reported (per A. Lane *in litt.* 2008) red pandas in the hills above Haflong (25°12′N, 92°59′E; 700 m; the hills around rising to 1800 m).

The Garo and Khasi hills are extensions of the North Cachar Hills and Barail range. The reports of red pandas spanning several decades at several sites within these hill ranges, backed up by two skins (albeit the only reportedly dated one being killed 30 years before documentation) and one first-hand and several second-hand sight-records of wild-living animals, and perhaps one or more specific names in the local language, are strong indications that there is, or at least was, a red panda population in these ranges. What is less clear is its origin. A. Lane (*in litt.* 2008) considers it unlikely that the animals that he and his father observed were directly escaped or released animals. He stated, however, that "one could certainly understand the view that maybe these were the descendants of previous captives that tea planters may have acquired and released back to the jungle when the planters departed"; at this time, many planters kept wildlife as pets, ranging from big cats and primates to muntjacs and birds. A.U. Choudhury (*in litt.* 2008) observed red pandas kept occasionally as pets in Sikkim and Arunachal Pradesh [68], although he has never seen one in or around Meghalaya. Unwanted pets were commonly, in the 1950s–1960s, released into the Meghalaya forests; no species (other than red panda) which might be considered non-local were seen in this area during this period (A. Lane *in litt.* 2008). Given the great difference in ecological conditions between the red panda's main range and the Garo hills, acceptance of a truly wild population there on the basis of the information so far published would be premature. A field survey specifically to assess the current distribution and status of red panda in the area, using people familiar with their signs, is needed.

CONCLUDING REMARKS

The areas with these reports of warm-climate pandas include some of the least biologically explored parts of Asia and, on this basis, it cannot be excluded that there could be such red panda populations. However, if the species could occupy northern Laos, Xishuangbanna and the Garo hills at an altitudinal range of 700–1400 m and upwards, there is a huge area of suitable habitat in northern South-east Asia within which there is no obvious reason why red panda would have a highly localized distribution. Even small carnivorans which should be much less readily found, e.g., stripe-backed weasel, *Mustela strigidorsa*, with its somewhat similar range in the eastern Himalayas and northern South-east Asia, actually turn out on investigation to be known by many more records than is generally realized [72]. There is no other small carnivore known by so few specimen records or so few reports across the large geographic area of the hills south of the Himalayas and south-east to northern Vietnam. Two explanations, not mutually exclusive, suggest themselves for this pattern: that the panda records result from mistaken identification and/or confusion of trade or pet animals for wild ones; and that there are a handful of relict populations reflecting a formerly more extensive range: perhaps now only one, that in the Garo hills of Meghalaya. While the Vietnamese, Lao and Xishuangbanna reports are groundless, a recent, and on-going, contraction in range, which could result in abandonment of climatically atypical areas, is certainly plausible: within recent decades red pandas have retreated from parts of their Chinese range, e.g., some areas of Chayu and Mangkang counties (Tibet), where they could formerly be seen easily, and the fossil record extends the Chinese range over an even bigger area [60,73,74]. It is likely that the distribution of red panda has shown several major expansions and contractions related to glaciation-driven climate change [75] and thus would have had the potential to leave disjunctive populations in areas now atypical in habitat for the species. Why and how such populations would have survived, if any have, is not clear.

It is urgent to clarify whether there truly are wild red pandas in Meghalaya: if there are, these astonishing populations are likely to be small and very threatened. Partly this could be assumed through the continuation of whatever trends (presumably climate-related) that have caused the wider decline in red panda range. Moreover, the loss of forest cover in Meghalaya has been reported to be very rapid: reportedly, almost half that present in 1980–1982 was lost by 1993 (33.1% to 18%) [68]. This is almost impossibly high (if such rates had continued, by now there would be no significant forest left in the state) and the extent to which methodological differences in stand assessment (it seems that the figures for the different years came from two different agencies) could account for this change is not discussed.

ACKNOWLEDGEMENTS

As in all evaluations of fragmentary information where incompatible viewpoints exist, my conclusions are unlikely to please everybody, but I have tried to indicate the full range of tenable views in this tantalizing situation. I particularly thank those who provided the information necessary to discuss the Indian records, Anwaruddin Choudhury; Dipankar Ghose; Kashmira Kakati; Alan Lane and John Lane; Divya Mudappa; and Nimesh Ved, Yash Shethia and Benson Sangma, all of the Samrakshan Trust (Garo Hills). Klaus Berkmüller, Gary Galbreath, Angela Glatston, Colin Groves, Arlyne Johnson, Barney Long, Darrin Lunde, Bill Robichaud, Scott Roberton, Jeff Sayer, Jan Schipper, Gerry Schroering, Rob Timmins, Rob Tizard, Dirk Van Gansberghe, the late Nico van Strien, Steven Wallace, Wang Ying-xiang and Wei Fuwen all generously provided unpublished information or ideas to help in the evaluation of these records. Kim Yooree checked my translations from French.

References

[1] H. de Monestrol, Les Chasses et la Faune d'Indochine, Imprimerie d'Extrême Orient, Hanoi, Vietnam, 1925.
[2] H. de Monestrol, Les Chasses et la Faune d'Indochine, 2me edn., Imprimerie d'Extrême Orient, Hanoi, Vietnam, 1931.

[3] H. de Monestrol, Chasses et Faune d'Indochine, A. Portail, Saigon, Vietnam, 1952.
[4] [MOSTE] Ministry of Science, Technology and Environment, [Red Data Book of Vietnam. Volume 1. Animals.], Science and Technics Publishing House, Hanoi, Vietnam, 1992 (in Vietnamese).
[5] H.H. Dang, [Checklist of Mammals in Vietnam.], Publishing House 'Science and Technics', Hanoi, Vietnam, 1994 (in Vietnamese).
[6] J. Delacour, Liste provisoire des mammifères de l'Indochine française, Mammalia 4 (1940) 20–39, 46–58
[7] V.T. Dao, Sur quelques rares mammifères au nord du Vietnam, Mitteil. Zoologisch. Museum Berlin 53 (1977) 325–330.
[8] R. Mell, Beiträge zur Fauna sinica. I. Die Vertebraten südchinas; Feldlisten und Feldnoten der Säuger, Vögel, Reptilien, Batrachier, Arch. Naturgesch. 88 (A, 10) (1922) 1–134.
[9] S.J. Legendre, Land of the White Parasol and the Million Elephants: a Journey through the Jungles of Indochina, Dodd, Mead & Co., New York, USA, 1936.
[10] E.C. Dickinson, Birds of the Legendre Indochina expedition 1931–1932, Am. Museum Novit. 2423 (1970) 1–17.
[11] R. Bourret, La Faune de l'Indochine. Vertébrés, Imprimerie d'Extrême Orient, Hanoi, Vietnam, 1927.
[12] C. Robson, A Field Guide to the Birds of South-East Asia, New Holland, London, UK, 2000.
[13] B. Björkegren, On a new weasel from northern Tonkin, Ark. Zool. 33B (15) (1941) 1–4.
[14] A.V. Abramov, Taxonomic remarks on two poorly known Southeast Asian weasels (Mustelidae, *Mustela*), Small Carniv. Conservat. 34&35 (2006) 22–24.
[15] C.P. Groves, On some weasels *Mustela* from eastern Asia, Small Carniv. Conservat. 37 (2007) 21–24.
[16] D.C. Wege, A.J. Long, K.V. Mai, V.D. Vu, J.C. Eames, Expanding the Protected Areas Network in Vietnam for the 21st Century: an Analysis of the Current System with Recommendations for Equitable Expansion, Birdlife International Vietnam Programme, Hanoi, Vietnam, 1999.
[17] Q.T. Nguyen, V.S. Nguyen, X.T. Ngo, T.S. Nguyen, Evaluation of the Wildlife Trade in Na Hang District. PARC Project VIE/95/G31&031, Govt of Viet Nam (FPD)/UNOPS/UNDP/Scott Wilson Asia Pacific Ltd., Hanoi, Vietnam, 2003.
[18] X.D. Nguyen, V.S. Nguyen, V.L. Truong, X.T. Ngo, Q.T. Nguyen, Evaluation of the Wildlife Trade in Ba Be and Cho Don Districts. PARC Project VIE/95/G31&031, Govt of Viet Nam, (FPD)/UNOPS/UNDP/Scott Wilson Asia Pacific Ltd, Hanoi, Vietnam, 2003.
[19] S.R. Swan, S.M.G. O'Reilly, Van Ban: a Priority Site for Conservation in the Hoang Lien Mountains. Community-Based Conservation in the Hoang Lien Mountains: Technical Report No. 1, Fauna and Flora International Vietnam Programme, Hanoi, Vietnam, 2004.
[20] S.R. Swan, S.M.G. O'Reilly, Mu Cang Chai Species/Habitat Conservation Area. Community-Based Conservation in the Hoang Lien Mountains: Technical Report No. 2, Fauna and Flora International Vietnam Programme, Hanoi, Vietnam, 2004.
[21] S.I. Roberton, The Status and Conservation of Small Carnivores in Vietnam. PhD thesis, School of Biological Sciences, University of East Anglia, UK, 2007.
[22] L. Roussel, La Chasse en Indochine, Plon-Nourrit et Cie, Paris, France, 1913.
[23] J.W. Duckworth, B.P.Y.-H. Lee, E. Meijaard, S. Meiri, The Malay Weasel *Mustela nudipes*: distribution, natural history and a global conservation status review, Small Carniv. Conservat. 34&35 (2006) 2–21.
[24] G. Cheminaud, Mes Chasses au Laos, [Tome II] Payot, Paris, France, 1942.
[25] G. Cheminaud, Mes Chasses au Laos, [Tome I] Payot, Paris, France, 1939.
[26] J.W. Duckworth, Small carnivores in Laos: a status review with notes on ecology, behaviour and conservation, Small Carniv. Conservat. 16 (1997) 1–21.
[27] A. Johnson, C. Vongkhamheng, M. Saythongdum, The diversity, status and conservation of small carnivores in a montane tropical forest in Northern Laos, Oryx 43 (2006) 626–633.
[28] J. Vidal, La végétation du Laos: 2ème partie: groupements végétaux et flore. Travaux du Laboratoire Forestier de Toulouse, Tome v, 1ère sect., vol I, art III, 1960.
[29] R.J. Timmins, J.W. Duckworth, Status and conservation of Douc Langurs (*Pygathrix nemaeus*) in Laos, Internatl. J. Primatol. 20 (1999) 469–489.
[30] N. Ruggeri, M. Etterson, The first records of Colugo (*Cynocephalus variegatus*) from the Lao PDR, Mammalia 62 (1998) 450–451.

[31] C.M. Francis, A. Guillén, M.F. Robinson, Order Chiroptera: bats, in: J.W. Duckworth, R.E. Salter, K. Khounboline (Eds.), Wildlife in Lao PDR: 1999 Status Report, IUCN–The World Conservation Union/ Wildlife Conservation Society/Centre for Protected Areas and Watershed Management, Vientiane, Lao PDR, 1999, pp. 225–235.
[32] J.W. Duckworth et al., Cheminaud's mammals of Laos: limitations to the use of a hunter's memoirs, 2010 (in prep).
[33] G.J. Galbreath, Science and moon bears: a scientific addendum, in: S.M Montgomery (Ed.), Search for the Golden Moon Bear, Simon and Schuster, New York, USA, 2002, pp. 309–310.
[34] J.W. Duckworth, R.J. Timmins, R.C.M. Thewlis, T.D. Evans, G.Q.A. Anderson, Field observations of mammals in Laos, 1992–1993, Nat. Hist. Bull. Siam Soc. 42 (1994) 177–205.
[35] J.W. Duckworth, A survey of large mammals in the central Annamite mountains of Laos, Zeits. Säugetierkunde 63 (1998) 239–250.
[36] T.D. Evans, J.W. Duckworth, R.J. Timmins, Field observations of larger mammals in Laos, 1994–1995, Mammalia 64 (2000) 55–100.
[37] W.H. Osgood, Mammals of the Kelley–Roosevelts and Delacour Asiatic Expeditions, Publ. Field Museum Nat. Hist. Zool. Ser. 18 (10) (1932) 193–339.
[38] J. Deuve, Les Mammifères du Laos, Ministère de l'Education Nationale, Vientiane, Laos, 1972.
[39] J. Deuve, M. Deuve, Contribution à la connaissance des mammifères du Laos: deuxième partie, Bull. Soc. Roy. Hist. Natur. Laos 9 (1963) 35–46.
[40] V. Holmgren, S. Savathvong, T. Kvitvik, K. Vainio-Mattila, G. Meyer, T. Redl, Lao People's Democratic Republic Biodiversity Country Report, Ministry for Agriculture and Forestry, and Science, Technology and Environment Agency, Vientiane, Lao PDR, 2004.
[41] IUCN, 2006 IUCN Red List of Threatened Species, <www.iucnredlist.org>, 2007.
[42] B. Groombridge, 1994 IUCN Red List of Threatened Animals, IUCN, Gland, Switzerland, and Cambridge, UK, 1993.
[43] IUCN, 1996 IUCN Red List of Threatened Animals, IUCN, Gland, Switzerland, 1996.
[44] J.W. Duckworth, R.E. Salter, K. Khounboline, Wildlife in Lao PDR: 1999 Status Report, IUCN–The World Conservation Union/Wildlife Conservation Society/Centre for Protected Areas and Watershed Management, Vientiane, Lao PDR, 1999.
[45] Sayer, J.A. (1983). Nature Conservation and National Parks. Final report. Forest Management Project, Office National de Protection de l'Environnement, Office National de Chasse et Pêche, and Food and Agriculture Organisation of the United Nations, Vientiane, Lao PDR.
[46] H. Nooren, G. Claridge, Wildlife Trade in Laos: the End of the Game, Netherlands Committee for IUCN, Amsterdam, Netherlands, 2001.
[47] Z. Li, H. Ning, S. Shan, Wildlife trade, consumption and conservation awareness in southwest China, Biodivers. Conservat. 17 (2008) 1493–1516.
[48] J. Domalain, The Animals Connection, William Morrow, New York, USA, 1977.
[49] R.E. Salter, Wildlife in Lao PDR: a Status Report, IUCN, Vientiane, Lao PDR, 1993.
[50] S. Singh, Contesting moralities: the politic of wildlife trade in Laos, J. Polit. Ecol. 15 (2008) 1–20.
[51] C. Poole, Obituary: William Wayt Thomas, Birding Asia 3 (2005) 95.
[52] J.W. Duckworth, R.J. Tizard, W.W. Thomas's bird records from Laos, principally Vientiane, 1966–1968 and 1981–1983, Forktail 19 (2003) 63–84.
[53] R.J. Tizard, P. Davidson, K. Khounboline, K. Salivong, A Wildlife and Habitat Survey of Nam Ha and Nam Kong Protected Areas, Luang Namtha Province, Lao PDR, Centre for Protected Areas and Watershed Management/Wildlife Conservation Society, Vientiane, Lao PDR, 1997.
[54] D.A. Showler, P. Davidson, K. Khounboline, K. Salivong, A Wildlife and Habitat Survey of Nam Xam NBCA, Houaphanh Province, Lao PDR, Centre for Protected Areas and Watershed Management/Wildlife Conservation Society, Vientiane, Lao PDR, 1998.
[55] P. Davidson (Ed.), A Wildlife and Habitat Survey of Nam Et and Phou Loeuy NBCAs, Houaphanh Province, Lao PDR, Centre for Protected Areas and Watershed Management/Wildlife Conservation Society, Vientiane, Lao PDR, 1998.

[56] P. Davidson, A Wildlife and Habitat Survey of Nam Et and Phou Loeuy NBCAs, Houaphanh Province, Lao PDR: addendum, Centre for Protected Areas and Watershed Management/Wildlife Conservation Society, Vientiane, Lao PDR, 1999.

[57] J.W. Duckworth, W.G. Robichaud, Yellow-bellied weasel *Mustela kathiath* sightings in Phongsaly province, Laos, with notes on the species's range in South-East Asia, and recent records of other small carnivores in the province, Small Carniv. Conservat. 33 (2005) 17–20.

[58] J. Fuchs, A. Cibois, J.W. Duckworth, et al., A review of bird records from Phongsaly province and the Nam Ou (Laos), Forktail 23 (2007) 22–86.

[59] Y. Wang, Mammals in Xishuang Bann area and a brief survey of its fauna, in: Yongchun Xue (Ed.), Report of Expedition to Xichuangbanna Nature Reserve, Yunnan Science and Technology Press, Kunming, China, 1987, pp. 289–310 (in Chinese with English abstract and title).

[60] F. Wei, Z. Feng, Z. Wang, Ju. Hu, Current distribution, status and conservation of wild red pandas *Ailurus fulgens* in China, Biol. Conservat. 89 (1999) 285–291.

[61] Y. Wang, A Complete Checklist of Mammal Species and Subspecies in China. A Taxonomic and Geographic Reference, China Forestry Publishing House, Beijing, China, 2002.

[62] A. Choudhury, Red panda *Ailurus fulgens* F. Cuvier in the north-east with an important record from the Garo hills, J. Bombay Nat. Hist. Soc. 94 (1997) 145–147.

[63] A. Choudhury, Primates of Assam: their Distribution, Habitats and Status. PhD thesis, Gauhati University, Guwahati, India, 1989.

[64] M. Zhang, Procyonidae, in: Gao Yaoting, et al. (Eds.), Fauna Sinica: Mammalia, vol. 8. Carnivora, Science Press, Beijing, China, 1987, pp. 103–110 (in Chinese).

[65] R.I. Pocock, The Fauna of British India, including Ceylon and Burma, second edn, Mammalia, vol. II, Taylor and Francis, London, UK, 1941.

[66] A. Choudhury, On some large-sized red pandas *Ailurus fulgens* F. Cuvier, J. Bombay Nat. Hist. Soc. 99 (2002) 285–286.

[67] A. Choudhury, A Systematic Review of the Mammals of North-east India, with Special Reference to the Non-human Primates. DSc thesis, Gauhati University, Guwahati, India, 2001.

[68] A. Choudhury, An overview of the status and conservation of the red panda *Ailurus fulgens* in India, with reference to its global status, Oryx 35 (2001) 250–259.

[69] M.A. Hinton, H.M. Lindsay, Mammal survey of India. Burma and Ceylon. Report No. 41, Assam and Mishmi Hills, J. Bombay Nat. Hist. Soc. 31 (1926) 383–403.

[70] N. Ved, B. Sangma, Wildlife Distribution, Hunting and Conflict: a Preliminary Survey, Samrakshan Trust, Meghalaya Field Office, Baghmara, Meghalaya, India, 2007.

[71] Samrakshan Trust, Wildlife Distribution and Hunting. South Garo Hills. Unpublished Report to the Rufford Small Grants Foundation, 2008.

[72] A.V. Abramov, J.W. Duckworth, Y-x. Wang, S.I. Roberton, The Stripe-backed Weasel *Mustela strigidorsa*: taxonomy, ecology, distribution and status, Mammal Rev. 38 (2008) 247–266.

[73] Y. Xu, Y. Li, X. Xue, Mammal fossils in the Pleistocene at Zijin of Guizhou province, Verteb. Palasiatic. 5 (1957) 342–350 (in Chinese).

[74] B. Yin, W. Liu (Eds.), [Precious and Rare Wildlife and its Protection in Tibet.], China Forestry Publishing House, Beijing, China, 1993.

[75] M. Li, F. Wei, B. Goossens, et al., Mitochondrial phylogeography and subspecific variation in the red panda (*Ailurus fulgens*): implications for conservation, Molec. Phylogenet. Evol. 36 (2005) 78–89.

CHAPTER

25

Release and Reintroduction of Captive-bred Red Pandas into Singalila National Park, Darjeeling, India

Alankar K. Jha

IFS, Padmaja Naidu Himalayan Zoological Park, Darjeeling, India

OUTLINE

INTRODUCTION

The red panda, *Ailūrus fulgens,* is an endangered species, its range extends from Nepal in the west to a few provinces of China in the east. In India, it is found in the Darjeeling and the Sikkim Himalayas and in the north-eastern state of Arunachal Pradesh. Its image is represented on one of a new series of postage stamps issued by the Indian government

Red Panda. DOI: 10.1016/B978-1-4377-7813-7.00025-2

FIGURE 25.1 Indian postage stamp depicting a red panda.

to represent the rare fauna of that region (Figure 25.1). The decline of the red panda population in the past has been directly proportional to the human population explosion, which has resulted in increases in hunting, trapping, trading, habitat disturbance and fragmentation. To preserve the red panda for the future requires us to make conservation efforts which, fortunately, the local population supports.

THE STATUS OF RED PANDAS IN THE DARJEELING AREA

There are few data available on the status of the red panda in the wild in India. It is believed to be very rare, and a recent study in the Singalila National Park near Darjeeling in West Bengal showed that a small population of about 78 animals was surviving in the area. Red pandas have also been sighted on a number of occasions in the Neora Valley National Park, also in the vicinity of Darjeeling, and one animal was sighted in the Kainjaley area in 2008. This area is not a national park and also lies outside of the usual range of the red panda.

Darjeeling Zoo

The Padmaja Naidu Himalayan Zoological Park (PNHZP) in Darjeeling is dedicated to the conservation of endangered Himalayan fauna and it is seriously concerned about the possible extinction of Himalayan fauna, in particular of the red panda. This zoo is situated at Birch hill in a patch of virgin forest. The town of Darjeeling itself is situated in the lower Himalayas at 27°3′N and 88°18′E, at a distance of 640 km from Kolkata. The town is situated on a long spur, which projects to the north from the Senchal-Singalila range of mountains. This spur rises somewhat abruptly from Ghoom to an elevation of 7886 ft (2404 m) at Katapahar, and then gradually descends to 7520 ft (2292 m) at Jalapahar and to 7002 ft (2134 m) at Chowrasta, the centre of the town. The left fork of the spur leads to Lebong and the right to the Birch Hill and the Zoological Park, which is situated at an altitude, at the highest point, of 6874 ft (2095 m) above sea level. The annual rainfall in this area varies between 100 and 115 inches (2500–2925 mm) and the daily temperatures range from nearly freezing in the winter to about 20°C in summer. Winter snowfall can be quite heavy at times (about once every 3–4 years) and frosts are common. This type of temperate climate is well suited to the sub-alpine fauna and flora and is therefore appropriate for the breeding and cultivation of Himalayan fauna and flora. The natural ranges of a number of indigenous species such as the snow leopard, *Panthera unica,* the marbled cat, *Pardofelis marmorata*, the Indian fox, *Vulpes bengalensis*, the clouded leopard, *Felis nebulosa*, the leopard *Panthera pardus* and the bharal, *Pseudois nayaur,* all overlap with the altitudinal zone at which the Zoological Park is situated. The forest type at this altitude corresponds to Champion's type 11B/C1 or Northern Montane/East Himalayan wet temperate forest [1]. This type of forest is found at altitudes between 5000 and 8000 ft (1524–2440 m), and is dominated by oaks, *Quercus* spp., laurels, *Lauraceae,* magnolias *Magnolioideae,* alders, *Alnus* spp., maples, *Acer* spp., birches, *Betulaceae,* and bucklandia.

The 80 acres of the Birch Hill Forest where the zoo is located are the remnants of the original woodlands of the region. History shows that, as recently as 50 years ago, this area was still home to a variety of Eastern Himalayan fauna including Himalayan black bear, *Selenarctos thibetanus laniger,* barking deer, *Muntijacus,* plus various species of Himalayan cats and squirrels including Hogsons's flying squirrel, *Petaurista magnificus,* and the orange-bellied Himalayan squirrel. The Indian fox, the red fox, *Vulpes vulpes,* and the Asiatic jackal, *Canis aureus,* were also a common sight [2].

The Darjeeling Zoo first opened on 14 August 1958. On 21 November 1975, it was renamed the Padmaja Naidu Himalayan Zoological Park, by the then prime minister of India, Mrs Indira Gandhi, in honour of the memory of the late Padmaja Naidu, the former governor of West Bengal, the state where the zoo is situated, although the zoo is generally better known simply as Darjeeling Zoo. PNHZ Park was established with the primary objective of studying and preserving Himalayan fauna and efforts are still on-going to achieve these objectives by means of the following goals:

1. Providing appropriate facilities for husbandry and veterinary care of captive wild animals
2. Captive breeding of endangered Himalayan fauna, preferably those originating from the Eastern Himalaya, with the aim of providing suitable individuals for reintroduction/restocking in their natural habitat
3. Educating local people, students and visitors about their wildlife and raising awareness about the importance of preserving the Himalayan ecosystem
4. Undertaking both pure and applied research into wildlife biology, behaviour, veterinary care and conservation.

Conservation Breeding Programme in Darjeeling Zoo

There are only two zoos in India which currently hold red pandas; the Padmaja Naidu Himalayan Zoological Park, Darjeeling and the Himalayan Zoological Park in Gangtok, Sikkim. Of these, only the Darjeeling Zoo holds a good-sized, breeding population of red pandas. However, the Gangtok Zoo has also started breeding with the species and is also a part of the Red Panda Conservation Breeding Programme that was instigated by the Indian Government.

The IUCN–SSC Action Plan for Ailurids and Procyonids recommended the establishment of captive populations of red panda in its range states. The breeding programme in India conforms to this recommendation. Therefore a captive population of red pandas in Indian zoos formed an integral part of a global red captive breeding masterplan [3]. As a result of the masterplan recommendations, PNHZP received five red pandas from several European zoos to augment the group of four wild-caught red pandas that were already living in the zoo at that time.

At that time, the PNHZP was the only zoo located in the vicinity of red panda habitat that was holding red pandas, the Himalayan Zoo opened at a later date. It was an ideal partner for the international red panda breeding effort as it lay within the natural range of the red panda and had suitable facilities for managing red pandas in captivity. Furthermore, there were at least two protected areas with wild red panda populations close to the zoo; the Singalila National Park and the Neora Valley National Park.

The Indian captive breeding programme for red pandas was initiated in the early 1990s in the PNHZP in response to the IUCN recommendations and the global captive-breeding masterplan. The main objectives of the Indian breeding programme were as follows:

1. To establish a viable captive population of red pandas in the Indian sub-continent held in zoological collections based within the natural range of the species
2. To distribute surplus captive-bred red pandas, from the prime breeding centre at PNHZP, to other subsidiary conservation breeding centres in suitable locations in Eastern Himalaya
3. To provide surplus captive-bred red pandas for restocking the dwindling populations of the species in the Singallila and Neora Valley National Parks and to reintroduce red pandas into the Senchal Wildlife Sanctuary
4. To provide scientists and naturalists with the opportunity to study various aspects of the biology and behaviour of this rare species
5. To stimulate public awareness of the plight of this endangered species and to provide the people with information on this species.

The History of the Current Captive Population of Red Pandas in Indian Zoos

The Indian zoo population of red pandas was originally founded on nine animals; four wild-caught animals already living in the zoo in the early 1990s and five zoo-bred animals imported from Europe. The details of these are presented in Table 25.1.

In addition to these founding individuals, one further red panda was brought from Rotterdam Zoo in April 1993. Unfortunately, this individual did not survive. He was less than one year old when he was sent to India and, in retrospect, it was felt he may have been too young for the stresses of transport.

The first successful (planned) breeding of red panda occurred on 20 June 1994 when two cubs 'Ekta' and 'Friend' were born to 'Basant' and 'Amita' before the arrival of the new founders from Europe. The next births occurred the next year when new arrivals

TABLE 25.1 Red pandas founding the Indian breeding programme

House Name	Stud Book number	Sex	Date of arrival and source	Date of birth
Anita	8221	F	91–92 Wild	±1986
Basant	8649	M	91–92 Wild	±1986
Chanda	8222	F	91–92 Wild	±1986
Divya	8648	F	91–92 Wild	±1986
Gora	9305	M	10.11.94 Cologne Zoo	25.6.93
Hari	9302	M	10.11.94 Rotterdam Zoo	30.6.93
Indira	9330	F	10.11.94 Madrid Zoo	26.6.93
Omin	9404	M	25.12.96 Antwerp Zoo	17.7.94
Prity*	9430	F	25.12.96 Rotterdam Zoo	26.6.94

**Prity together with 'Jugal', a Darjeeling born male, was later transferred to the Himalayan Zoo in Gangtok to start a new breeding group there.*

'Hari' and 'Indira' produced their first cubs followed by more cubs from 'Basant' and 'Amita' and 'Basant' also produced his first cubs with 'Divya', one of the other wild-caught females.

Between 1994 and 31 December 2008 around 50 red pandas have been born at the PNHZP. Table 25.2 presents an overview of the births in Darjeeling and Figure 25.2 shows a young panda in a nest box in the zoo enclosure. Furthermore, it is good to note that the pair of red pandas established in Gangtok also started breeding.

THE REINTRODUCTION

By 2003, the Indian zoo population of red pandas had increased substantially and there were 22 red pandas living in the PNZHP in Darjeeling. The population was therefore considered to be well enough established to be able to take the next step in the programme, namely to release two zoo-born red pandas back into the wild. Two young females, Sweety (born 25 June 1997) and Mini (born 17 June 1998), were selected for release, neither of which has been used for breeding while in the zoo. Females were chosen for release as it was felt that females would be more likely to contribute to the wild population through giving birth. After their selection, the acclimatization process began; their diet was slowly changed from the zoo diet which included milk, sugar, fruit and eggs provided at regular intervals to a more natural diet based largely on bamboo which was provided at more irregular intervals. This change in their diet took about 6 months and meant that, when the animals arrived at the release facility, they were ready for total dependence on a natural diet of wild fruits, bamboos and berries. In the meanwhile, during this acclimatization process, the required health checks of the two females were undertaken so that the necessary clearances from the Government of India and Central Zoo Authority could be

TABLE 25.2 Red panda births in PNHZP, Darjeeling

Year	No. born	No. litters	No. deaths (<30 days)
1994	2	1	0
1995	5	3	1
1996	6	3	3
1997	5	3	2
1998	6	2	0
1999	7	3	0
2000	2	1	0
2001	5	2	0
2002	1	1	1
2003	3	2	0
2004	3	2	0
2005	0	0	0
2006	1	1	0
2007	1	1	0
2008	2	1	0

FIGURE 25.2 A young panda in a nest box in PNHZP.

obtained. In addition, genetic studies were undertaken by the Center for Cellular and Molecular Biology, Hyderabad, India to confirm the taxonomic status of the pandas and to record their genetic fingerprints.

The Soft Release Protocol

The females were first transferred to a special soft release facility that had been constructed in the Gairibas area of the Singalila National Park. The flora of this area comprised *Castanopsis hystrix, Quercus lamellata, Machilus odoritissima, Evodea* spp., *Michelai excelsa,* Rhododendrons, *Eurya japonica, Arundinaria maling, Rubus* spp., *Daphne cannabina.* The soft release facility had an area of 5 hectares and was situated at 27°03′N and 88°01′E, at an altitude of 2626 m. Sufficient care was taken to watch and protect this facility against predators. Straight iron sheets surrounded the area, and the shrubs and trees near the perimeter fence were removed to prevent any accidental escape by one of the animals (Figure 25.3). The two females were brought to the facility in mid-April. During the first month they were housed in a small enclosure (10 m^2) situated within the soft release facility (Figure 25.4) and then gradually released into the whole area. The animals were kept in the facility for a period of 7 months where they were observed and acclimatized prior to their final release on 14 November 2003. By the time of release the animals were completely dependent on the natural food available in the enclosure.

The Singalila National Park was chosen as the place for the reintroduction as it is the only national park in the vicinity of the zoo that supported a wild population of red pandas. The choice of Gairibas as the release site was based on an earlier pre-release survey conducted by the Wildlife Wing of the Forest Department, Government of West Bengal, in

FIGURE 25.3 The enclosure used for the soft release.

FIGURE 25.4 The small early release enclosure.

FIGURE 25.5 Forest cover at release site.

collaboration with staff of the PNHZP. It was the area with the highest density of red pandas in the region. This was deemed to be an important factor as it increased the likelihood that the two females would find mates. Moreover, the forest there had a dense vegetation of maling bamboo (*Arundinaria maling*), the red pandas' preferred diet. Figure 25.5 shows the typical habitat found in the area. In addition, the location of an office of the Forestry Service in the vicinity and the presence of very small villages were also important factors in the decision as they were considered to provide both accessibility and thus constant monitoring.

Post-Release Monitoring of the Pandas

Prior to release each female was fitted with a radio collar (Telenoics, USA). After release the pandas were monitored using the non-triangulation location technique known as "Homing in on the Animal Method" [4]. The method is simple and the positional data are obtained by following the transmitted signal's increasing strength until the radio-collared animal is seen. The Wildlife Wing of the Forest Department of Government of West Bengal monitored the females on alternate days. This method provided a good overview of the movements of the released pandas as well as information on their behaviour, breeding and eventual death. Figure 25.6 is a typical data sheet showing the tracking locations of a single female over one month.

The two females behaved quite differently, one female, Mini, remained close to the release site and settled in an area referred to here as the Middle Area (average altitude 2800 m) which lay between PWD road and the Nepal border. Although she did explore the adjacent areas (Pulkhola and Plantation area) in the weeks following her release, however, she spent 80% of her time in the Middle Area. Sweety, on the other hand, was considerably more mobile than Mini.

Mini also started interacting with the wild pandas much earlier than did Sweety. The first wild panda sighted in her area was on 18 November, just 4 days after the release. She was sighted again with wild pandas in the Middle Area on 4 December and then on December 13, 21 and 31. The sightings were repeated on three dates in January; 1, 16 and 23 and a further six dates in February; 9, 11, 13, 20, 23 and 25.

Despite all these positive indications of Mini's adjustment and survival in the wild, the project lost her in March when she was predated, probably by a clouded leopard. Her remains, the skull, part of her tail and a paw, were found together with the radio collar by a member of the monitoring team on 15 March 2004.

Sweety, on the other hand, as mentioned earlier, was very mobile. She remained close to the release site for 6 days and then travelled a distance of about 2 km. Her relocation over this distance from the release site, to a comparatively unknown area, made tracking and monitoring her difficult initially. In December, she settled in an area about 1–1.5 km from Gairibas known locally as MR Road. She remained there throughout January but, from February, she started moving further exploring the areas towards Kaiyakatta. She spent a lot of her time in this area between February and June. The first wild red panda was seen in the MR road area on 4 December. However, Sweety was not seen with a wild panda until 17 February. She was seen with wild pandas again on 26 February, 11 March, 1 April and 3 April. Mating was recorded on 12 March 2004. The courtship and mating with the wild male were successful, Sweety became pregnant and, on 7 July 2004, she gave birth to a single cub in a tree hollow nest. This cub unfortunately went missing from the nest a month later.

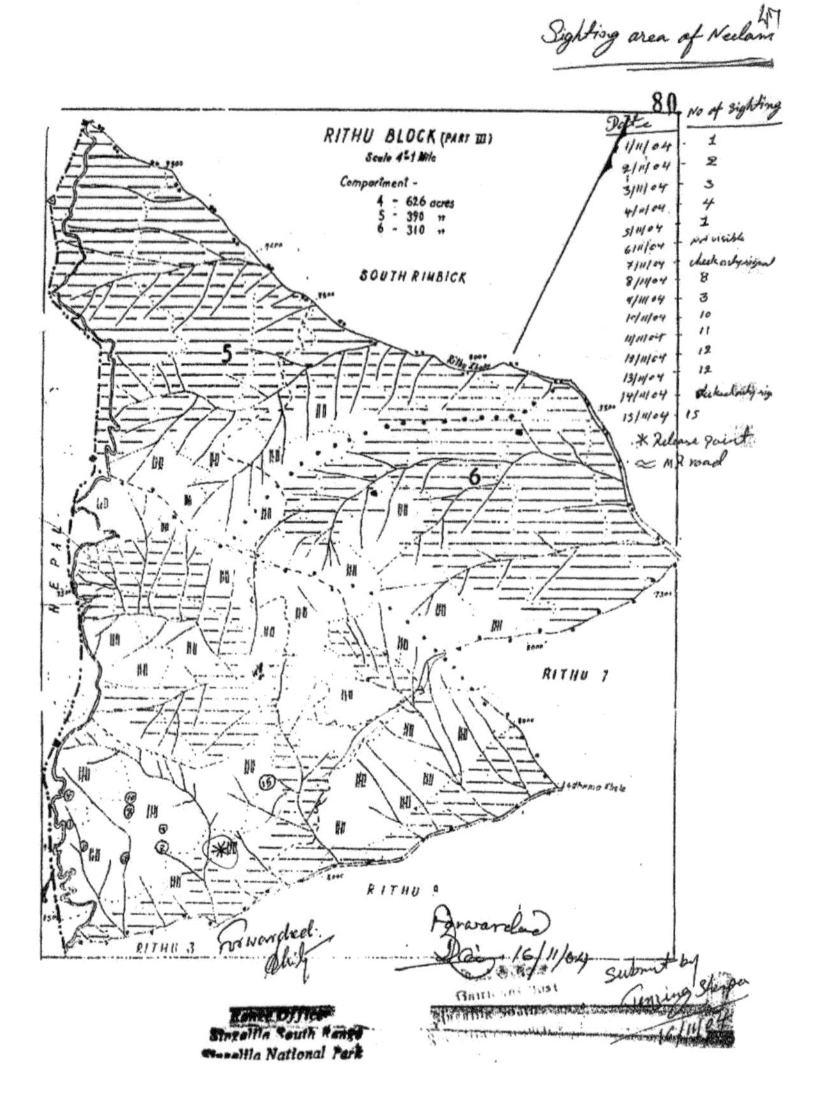

FIGURE 25.6 An example of the tracking results for a released female.

Behavioural Observations

Behavioural observations of the collared animals showed that Sweety was not only more mobile than Mini but she also urinated more frequently. This could indicate that she was putting substantially more effort in establishing herself in the wild and in an area of habitat where she would not encounter other pandas as easily as Mini. The high frequency of urination and movement seem to indicate that Sweety needed to scent mark to communicate her presence to conspecifics in and around the area more vigorously than Mini did.

CONCLUSION

The release of the red pandas into the Singalila National Park conformed to the Reintroduction Guidelines of the IUCN SSC Reintroduction Specialist group [5]. A pre-release study of the area was undertaken, the conclusion of which was that the habitat was suitable for a release and that there was a high density of wild red pandas. These conclusions were confirmed by the observations; Mini did not venture far from the release site, which would indicate that she was satisfied with the habitat. Also the attention that Mini received from potential mates as soon as she arrived in the Middle Area would indicate that the release site was apparently a good location for a potential breeding female. The density of wild red pandas was clearly an appropriate consideration when releasing a potentially breeding female. However, two points of concern emerged, these were the proximity of both a trekking route and the Nepal border. This meant that the area was subject to disturbance from trekkers and that led to concerns for the safety of the animals. The international border meant that there were problems for the trackers on the four occasions when Mini entered Nepal. These factors should be taken into account for future re-stocking projects in the area.

These results and observations confirm previous findings that the MR road area, oak forest habitat, only supports a very low density of red pandas [6]. The few pandas that were sighted and reported from this region were transient animals; there is no clear evidence of permanent use of this habitat by pandas. Sweety's temporary residence would seem to indicate that the MR road area could support red pandas, although her later preference for the Kaiyakatta Area could be taken to indicate that it is not ideal habitat. Information collected during the project on the use of this area was based on direct sightings and, as such, may be an artifact resulting from infrequent chance sightings rather than a reflection of actual habitat use. An earlier study in the Singalila National Park questioned if this area was good red panda habitat or whether it had just not yet been colonized by the species [6]. Some animals do not necessarily occupy their entire potential habitat even though they are able to disperse into unoccupied areas; behaviour can limit the distribution of a species [7]. Alternatively, the density of red pandas in this area was so low that sub-adults may not have needed to disperse far from the area where they were born.

The method for locating the source of the radio signal when tracking the pandas that was used in this project was fairly simple. However, the location of the animal by this technique was followed by a period of direct observation that meant it was possible to see how the animals were behaving and to ensure their welfare. On the other hand, actually getting close to the animals could be disruptive. This disadvantage had to be weighed against the benefits of direct observation. Also, the work required meant that Mini and Sweety could only be monitored on alternate days.

A common concern when placing a radio tracking device on a relatively small animal, is the weight of the transmitter [4]. The transmitter used in this project weighed approximately 95 g which was considered as an acceptable weight for an animal of the size of a red panda (correspondence with Telenoics). Indeed, during the project, the animals gave no indication that they were hindered in any way by the radio collars. This was further confirmed by Sweety's survival and the successful reproduction.

A second point of concern was that life in captivity might have led a zoo-born animal to lose some of the basic instincts and behaviours that are essential for survival in the wild. However, the behaviour of these pandas and the eventual birth of a cub clearly demonstrate that zoo-born female pandas were capable of surviving in the wild despite their captive origin.

Behavioural observations conducted in the pre-release facility showed that Sweety was more active than Mini even prior to actual release. If this activity is predictive of exploratory behaviour after release, observations of behaviour in the pre-release facility could be used to select which individuals are better suited behaviourally for release back into the wild. In addition, tests and simulations to evaluate and train anti-predator behaviour could be used to improve post-release success.

This reintroduction project was a pioneering one in this part of world. This was the first time in India, or even south-east Asia, that a captive-born animal has been purposely reintroduced into the wild as part of a re-stocking project. The full impact of its success and failure cannot be measured at this stage. The reintroduction of Mini and Sweety was just the first phase of a longer programme. A second introduction of two more females (Neelam and Dolma) took place shortly afterwards. These animals were taken to the soft release facility in November 2003 and were released into the wild in August 2004. They were monitored after their release but no breeding occurred. The information collected from these releases has been analysed and the project has been reviewed. The conclusions and recommendations of this will be taken into account in the planning of the second phase of the project. The PNHZP and the Wildlife Wing of Forest Department are planning to release two male red pandas, in the same location. The Wildlife Institute of India is in the process of conducting a habitat study of the location and, in the meantime, two animals have already been identified by the zoo for this reintroduction, their diet is currently being adapted and the plan is to implement this introduction in the very near future.

References

[1] H.G. Champion, S.K. Seth, A Revised Survey of the Forest Types of India, Manager of Publications, Delhi, 1968.

[2] Anon, Census of fauna in North Bengal hills, West Bengal India. Unpublished document, Government of West Bengal, 2000.

[3] A.R. Glatston, F.P.G. Princee, A Global Master Plan for the Captive Breeding of the Red Panda. Part 1: *Ailurus fulgens fulgens*, Stichting Koninklijke Rotterdamse Diergaarde, Rotterdam, The Netherlands, 1993.

[4] G.C. White, R.A. Garrot, Analysis of Wildlife Radio-tracking Data, Academic Press, 1990.

[5] IUCN–SSC Reintroduction Specialist Group, IUCN Guidelines for Reintroduction, IUCN, Gland, Switzerland, 1998.

[6] S. Pradhan, Studies on some aspects of the ecology of the red panda (*Ailurus fulgens*, Cuvier 1825) in the Singhalila National Park Darjeeling, India. Unpublished PhD thesis, North Bengal University, India, 1998.

[7] C.J. Krebs, Ecology; the Experimental Analysis of Distribution and Abundance, Harper & Row, New York, USA, 1985.

CHAPTER

26

Synthesis

Angela R. Glatston
Rotterdam Zoo, Rotterdam, The Netherlands

OUTLINE

This book on the red panda is intended to present the reader with an overview of our current knowledge about this unfamiliar and relatively unknown species. When preparing a volume such as this, there are obviously choices to be made; some aspects of the red panda's biology, where little (recent) research has been published, are conspicuous by their absence while others are summarized in review chapters. For example, I have not included chapters dealing specifically with physiology or genetics, although these topics are discussed in other chapters, on the other hand, I have included a review chapter focused on red panda reproduction. It is hoped, therefore, that the reader will find that even those topics that are not listed in the table of contents will be covered to some extent in the chapters that are presented here.

In the introduction, it was clearly stated that one of the main aims of this book was to bring the red panda to the attention of the scientific and conservation communities and the wider public. It has often been my experience that when I mention the name "red panda" to the average person, they will look at me blankly, trying to decide what kind of animal I am talking about. I have the impression that many are imagining a type of giant

Red Panda. DOI: 10.1016/B978-1-4377-7813-7.00026-4

panda where the black part of the body patterning is replaced by red, only very few will be thinking about the subject of this book.

The culture chapter (Chapter 2), by Axel Gebauer and myself, looks at how familiar the red panda is to people in its range states as well as in the wider global community and how its name and image are perceived by various cultures. When the IUCN–SSC Action Plan for Procyonids and Ailurids [1] was published more than a decade ago, it was stated that, although people in the range state countries knew the red panda, its role in traditional cultures was almost non-existent. However, since a few more research data on the topic have become available it would seem that, although limited, there are some local traditions associated with red pandas. As far as we can tell these seem to indicate that the red panda is associated with good luck and that it has the power to ward off evil spirits. The only report cited in which the red panda is deemed to be unlucky seems to be one where there is confusion between red pandas with other red-coloured species such as the red fox. Even so, the impact of the red panda seems to be limited for what we would consider a conspicuously beautiful species. The answer to this conundrum probably lies, as is suggested, in the red panda's natural biology: its habitat is relatively remote and inaccessible; its colours, which seem so bright in the zoo environment, are well camouflaged against the tree lichens in the wild and, finally, it is shy and usually more active at dawn, dusk or at night than during the day, although some reports seem to contradict each other here. All in all, you have to be fairly lucky to see a red panda in the wild, which may be why they feature so little in local custom and tradition.

On the other hand, it is interesting to note that the red panda seems to be attracting far more attention and interest in the modern world than it ever did in traditional cultures. It is certainly becoming better appreciated in its range states as demonstrated by the use of its image in marketing various products and services. What is perhaps more remarkable is the extent to which the red panda's name and image has permeated our western culture while, as a group, we seem to be largely unaware of its existence. The name, red panda, appears to have a certain resonance, and it is used to provide cachet to jewellery makers, acrobats, restaurants, bands, radio plays, etc., which have nothing to do with the animal concerned either in fact or in the minds of the public. How many of the estimated 270 million people [2] who use Firefox as their web browser are aware that the firefox, another evocative name, is in fact a red panda. It is also fascinating to see how the image of the red panda is becoming more and more widespread on the Internet and even in movies. The question needs to be asked how can the red panda image and name become so familiar in the media while the broader public remains unaware of the real animal's existence.

IS IT A RACCOON? IS IT A BEAR? – NO, IT'S A RED PANDA

The main problem which has beset taxonomists since the red panda was discovered has been where to place it within the Carnivores; this is an issue that has been very much confused by its possible relationship to the giant panda. It has variously been considered that the red panda is the only Old World member of the procyonid family, a very small bear (with a long tail) or that it belongs to a separate family, either on its own or together with the giant panda. One of the primary objectives of this volume has been to provide a conclusive answer to this question. To this end, we have presented three chapters (Chapters 3–5) dealing with the evolution of the red panda and a fourth providing a review of current thinking on these issues; the evidence presented seems to be compelling that we are dealing with a unique family (Ailuridae) containing one, but more probably two, species. The first of these chapters looks at the early fossils and places the Ailuridae into evolutionary context with the other Carnivore families while the second and third chapters discuss the fossil Ailurinae, the subfamily that includes the current red pandas. We are presented with descriptions of the New World *Pristinailurus,* and the first description of the new find of a late fossil *Parailurus* from Slovakia. From these chapters, we see that there have been a number of remarkable fossil discoveries in recent years that are finally clarifying the position of the Ailurids within the fossil record and with respect to the giant panda. It is clear that the red panda firmly belongs in the canid lineage; the group containing the dogs, bears, procyonids, pinnipeds, mustelids, etc., rather than the felid lineage. The origins of the Ailurid line are to be found in Europe where the earliest members of this family arose; only later did their descendants disperse to the New World to give rise to an isolated group of now extinct Ailurids. These early Ailurids were very generalized carnivores, showing none of the dental adaptations to a herbivorous diet which arose in later forms. It is interesting to note that the false thumb was developed in the early, more generalized Ailurids and seems to be an adaptation for climbing trees rather than one for manipulating food.

The distribution of the Ailurine group seems to have been correlated with the distribution of temperate forests, and the decline of these forests, in turn, is associated with a decline in the range of the Ailurines. The Slovakian fossil *Parailurus,* discussed in the third of these chapters, may represent the terminal phase of Ailurines in Europe. It is smaller than other *Parailurus* found to date and the community where this fossil occurred was different also; the habitat had changed to one of humid forest interspersed with steppe-like plains and it had been invaded by new competitors and predators. These factors, rather than climate change, were the cause of the demise of the Ailurines in Europe. The fate of the last *Parailurus* species in Europe may even teach us a lesson relevant to today's red pandas regarding the importance of retaining closed forest and problems caused by the encroachment of new predators such as domestic and feral dogs.

From the fossil record, it would appear that the Ailurids are more closely related to the Mephitidae or skunks than to the other members of the carnivora. In evolutionary history, it would appear that the dogs and bears were the first to split from the caniformia followed by the pinnipeds. The Ailurids and the Mephitids diverged later while the Mustelids and Procyonids separated at an even later stage. This interpretation of the fossil record is supported by recent molecular studies and therefore Groves (Chapter 7) concludes that the red panda belongs in its own separate family, the Ailuridae.

It is not only the fossil record that can confirm the exact phylogeny of a species; anatomical features can also play an important role. Fisher (Chapter 6) provides us with a very complete and clear overview of red panda anatomy based on her own work as well as that obtained from the literature. In general, red panda anatomy can be said to feature a mixture of primitive and adaptive characteristics favouring different carnivore families from which no clear phylogenetic connection can be deduced. Equally well, the reproductive physiology of the red panda which was discussed by Northrop and Czekala

(Chapter 8) shows no clear affinities with a particular family of carnivores and neither does the structure of the placenta as discussed by Benirschke (Chapter 9), although, interestingly, there are some features of the placenta that show clear affinities with those of members of the Mustelid and Metphitid families.

A second question, which Groves discussed in his chapter, is the taxonomic relationship between the Himalayan and Styan's red pandas. *Ailurus fulgens fulgens* and *Ailurus fulgens styani* are generally considered as subspecies, although, Thomas [3], the first author to describe Styan's panda, eventually considered they were, in fact, two separate species. Roberts [4], on the other hand, deemed them to be so similar that he thought zoo workers should consider the option of interbreeding these two "races". However, he did advocate that an aggressive programme, looking for differences between the two forms, should be undertaken before zoos embarked on mixing them. To date, there has been no concerted effort to mix the two red pandas, which is fortunate as Groves considers them to be very clearly distinct. Furthermore, he deems that the division between these forms has been underrated in the past and that we are indeed dealing with two distinct species; *Ailurus fulgens* and *Ailurus styani*. He also goes on to remark that most of the specimens studied to date have been very localized and that there are large areas of the red panda's range from which there are no data, so there may indeed be other species or subspecies of red panda which have not yet been described, not to mention the enigmatic reports of red pandas in Meghalaya which will be discussed later.

RED PANDA REPRODUCTIVE BIOLOGY

The next chapters of the book (Chapters 8–10) deal with various aspects of red panda biology: reproduction, placentation, and parental behaviour. Reproduction in the red panda has always been a topic of interest. We know the species breeds once per year, that reproduction seems to be under control of photoperiod, as indicated by the fact that pandas in the southern hemisphere are 6 months out of phase with their relatives in the northern hemisphere, but we have never been certain if pandas are mono- or polyoestrous and whether or not they exhibit delayed implantation; gestation seems to be around 130 days but the developmental state of the young is more similar to that of a cat or a dog after 8 or 9 weeks' gestation. Northrop and Czekala provide us with an overview of red panda reproduction and answer a number of these questions. They conclude that red panda reproductive strategy shows similarities to raccoons, skunks and other carnivores as well as a number of adaptations to their environment and solitary life style. Induced ovulation is certainly indicated by the hormone data and supported by the long copulation rituals portrayed in Gebauer's photographic essay of mating behaviour. The birth season is such as to allow the red panda to give birth in the period of milder climatic conditions and optimum food quality. The actual mechanism of control is not certain although Princée (personal communication) has demonstrated that birth occurs significantly later at higher latitudes.

The authors suggest that hormone monitoring might be one method to address the low survival rates of neonates found in the zoo population. It would allow zoos to determine the probable date of parturition and so be better prepared for the event that in turn might improve neonate survival. In addition, hormonal evaluation of faeces could provide a valuable tool for non-invasive monitoring of reproduction in the wild. We still have a lot to learn about red panda reproduction and zoo populations provide an important resource to help us learn more about oestrus, mate choice and delayed implantation. This would allow for both better captive management and a better understanding of the problems of disturbance in the wild.

Gebauer's description of parental behaviour is unique; these are the first detailed observations made of the behaviour in the nest as well as the first sonograms for certain vocalizations. The image presented is one of a highly altricial species which spends most of its early life in the nest. The young panda learns much of its behaviour before it even leaves the nest for the first time. The mother plays an important role in this, she plays with the young and brings objects to the nest for it to manipulate, aiding its development. This means that, by the time that the young pandas leave the nest for the first time in the wild, they are much better prepared to survive in a harsh and complex environment. The observations also show how easily parental care is disturbed by external events, a factor which we need to remember when keeping red pandas in zoos, but also perhaps in community conservation programmes where there is more contact between wild red pandas and (local) people.

RED PANDAS IN ZOOS – WILL THEY SURVIVE?

Red pandas have been maintained in captivity in zoos since the mid-1800s, however, few were exported to western zoos before the 1940s and births remained a sporadic occurrence until the late 1970s. Nevertheless, in his chapter (Chapter 12), Marvin Jones describes a period in the 1940s when San Diego Zoo enjoyed a prolonged period of successful red panda breeding which persisted until the deaths of the original wild-caught animals in the early 1950s. With their demise, the period of captive breeding success ended and was not repeated again until the 1970s when a few zoos in Europe and North America started to breed red pandas with some degree of regularity. All the same, this period of success in San Diego clearly demonstrates the importance of good keeper care to the successful reproduction of red pandas in zoos.

The red panda international studbook was initiated in 1977 and since that time records have been maintained of all the red pandas held in zoos around the world. In the early years of the studbook, the captive population was not performing very well, indeed regular, reliable breeding only really began in the mid-1980s after the publication of husbandry and management guidelines and the start of the regional breeding programmes. Since that time the zoo populations of red pandas have grown, however, we need to know for certain whether that growth is continuing and if it is sufficient to provide the basis for a viable captive population for the foreseeable future. Superficial examination of the statistics for current captive *fulgens* and *styani* populations would seem to indicate that they are fairly safe, however, in my chapter on the status of the current population (Chapter 17) I question that assumption. Closer examination of the data reveals that the populations have recently stopped growing and that some of the founder bloodlines have been lost. Although these are global problems, some of the regional populations are suffering more than others. The chapter pinpoints problems of fertility and infant mortality which have not been adequately addressed as two of the core issues for future research programmes. However, there is a suspicion that breeding programme coordinators may not be strict enough in vetting aspiring red panda zoos and that some zoos are not adequately adhering to the husbandry and management guidelines, and that these may be the major reasons behind the decrease in performance of the populations. Still, the data presented in this chapter only represent a snapshot of the situation in the captive population. In order to investigate the future potential of the captive populations in greater detail, Kristin Leus discusses the results of a number of simulations which have been undertaken to investigate how the populations might perform in the future (Chapter 19).

Leus recommends the best course of action will be to manage the zoo populations as semi-autonomous regional groups with limited exchange of animals between them. She concludes that there is sufficient potential available to form the basis of viable, global populations of red pandas in our zoos. In the case of the *fulgens* population, this would be facilitated by the occasional exchange of animals with the Indian zoo population as a means of obtaining new founders and we will need to make decisions regarding how to proceed with the smaller regional populations. The situation of the *styani* population is worse despite the greater genetic variation available in this population. Breeding success is too low, particularly in the North American zoo population. If this is not improved, the Japanese zoo population will need to expand by at least 60 more animals if the global zoo population is to survive and become self-sustaining.

The situation for both the *fulgens* and *styani* populations would be improved if the Chinese captive populations were fully managed and integrated into the global programme. However, Kati Loeffler's report on red pandas in Chinese zoos (Chapter 18) indicates that this situation is still a long way off. The changes in husbandry and management that have improved the health and well-being of red pandas in western zoos have not reached China. So far as Loeffler can tell, the zoo population is continuously being supplemented with new animals coming from the wild but, as record keeping in China is both incomplete and de-centralized, it is difficult to determine how prevalent this practice has become. Also the incomplete nature of their studbook records means that the determination of genetic variation is problematic. The situation is further complicated by the fact that most of the zoos and animal keepers are unaware of the existence of two (sub)species of red pandas which means that inadvertent cross-breeding is a distinct possibility. However, without the inclusion of the Chinese population of zoo red pandas, the future of the *styani* population looks distinctly insecure.

The Indian Authorities are committed to the conservation of the red panda and they have proposed the establishment of an *ex situ* population of at least 200 animals of which 100 should be maintained in Indian institutions [5]. This proposal to manage a sizeable Indian captive population as part of a larger global population is both imaginative and a very positive way forward. However, to achieve this visionary ideal, the holding capacity in India will need to be substantially increased and distributed over more institutions to spread the risk. Furthermore, changes will be needed to speed up the bureaucratic processes required for animal exchange; it has currently taken more than 2 years to organize an exchange of zoo-born red pandas between India and Australia/New Zealand. This length of delay is not helpful when organizing transfers of animals with an average reproductive lifespan of around 6–7 years. To survive and function as part of the global population, the Indian zoo red panda population will have to grow and it would benefit from a regular inflow of new founders, either from the wild or from one of the other regional populations. In particular, the problem of the skewed sex ratio needs to be urgently addressed. If these problems are not addressed, the negative changes could persist and take the population beyond the point of no return and it will be lost.

India, like China, is a range state and, as such, the Indian zoo population of red pandas, together with the planned reintroduction programme, could provide a unique opportunity for the flow of genetic material between wild and captive populations. This would mean that some new founders will be taken from the wild; but where will these new founders be caught? Singalila National Park would be an obvious choice given its close proximity to the Padmaja Naidu Himalayan Zoo in Darjeeling which coordinates the red panda breeding programme in India. However, as Jha reports in his chapter (Chapter 25), there is only a small population of 78 animals (presumably including non-breeding individuals) in Singalila and the question arises whether such a small population could sustain any harvesting to supplement the zoo population. On the other hand, such a small population would undoubtedly benefit from the input of fresh blood resulting from the successful reintroduction of unrelated, captive-born animals.

As mentioned above, the viability of the zoo populations of red pandas suffers from inadequate fertility and high infant mortality. The source of these problems arises from our husbandry of this species in our zoos. We know from the field that the red panda is a shy animal originating from remote, inaccessible regions. We know that it is a Carnivore adapted to a diet of bamboo which requires that it spends much of its daily activity in feeding in order to meet its nutrient requirements. We know that the red pandas in zoos originate from temperate forests where the climate is cool and that they are better adapted to cold temperatures than they are to heat, for example, thermographic images of the red panda illustrate just how well the thick coat insulates; the only warm parts to be seen are around its eyes and muzzle (Figure 26.1).

The current husbandry and management guidelines for the red panda, which were compiled in 1989 [6], take these factors into account. In her chapter (Chapter 13), Kati Loeffler examines how these guidelines are currently being implemented and how current husbandry and management practices impact population parameters. First, it would seem that not all zoos are equally serious about applying the guidelines. In particular, insufficient attention is paid to the aspects intended to reduce stress. For example, the guidelines recommend that zoo visitors should only be given access to one, or at most two, sides of a red panda exhibit with the aim of providing the animals with more opportunity to withdraw. Yet, half the zoos surveyed allowed the visitors access to more than half the perimeter of the enclosure. It is the responsibility of both the zoos holding red pandas and the coordinators of the regional breeding programmes to ensure that the husbandry and management guidelines are more strictly adhered to. This is one step that could easily be taken to improve red panda breeding results.

Other choices are less clear. Some zoos consider that red panda mothers perform better when they are removed to an off-display area to give birth and raise their cubs. Other zoos consider it more disruptive to the mother to move her than to leave her in her usual exhibit. Some zoos believe that red panda cubs should be checked and weighed regularly

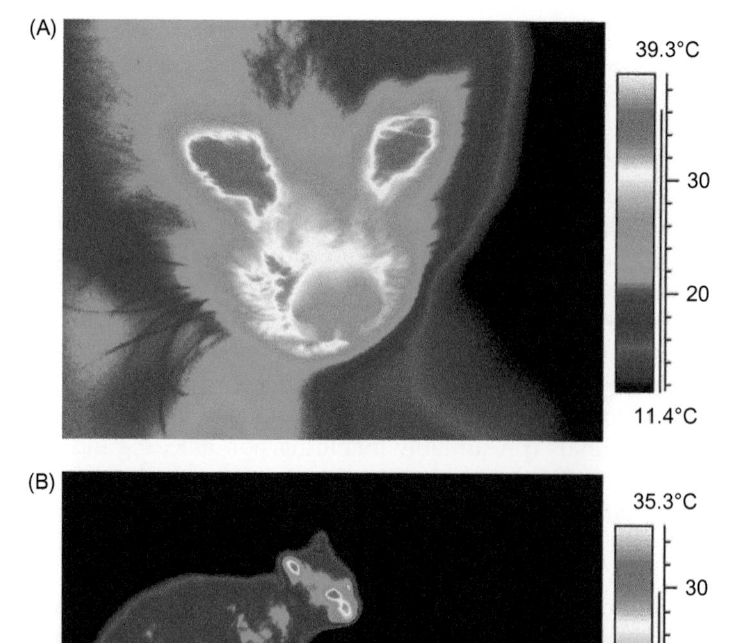

FIGURE 26.1 Thermographic images of a red panda in Rotterdam Zoo. (A) Face, (B) whole body. *(Image: Arno Vlooswijk, Nutscode Thermographics).*

so that keepers can intervene to care for the young if things start to go wrong. Other zoos believe that mother and cubs should be left totally undisturbed. All these tactics have their advantages and disadvantages. It is not necessarily clear from the infant mortality results which are the more successful, especially as there are several other factors which undoubtedly play a role, such as the relationship between the keeper and the pandas and the character of the animal concerned, etc. This is clearly an important area for future research.

Providing a nutritionally adequate, palatable diet can pose a complicated problem for a zoo holding a vegetarian Carnivore species such as the red panda. However, this is one area of red panda management that has undeniably improved over recent years. This can clearly be seen from post-mortem reports. Causes of death in the captive population of the red pandas have been monitored regularly since the studbook was established in 1978. In the early years, there was a prevalence of nutrition-related pathologies. As we can see from Brian Preece's chapter (Chapter 16), this is no longer the case. Today, most zoos provide their red pandas with a diet based on a high-fibre concentrate that is supplemented with bamboos, grass, leaves, vegetables and some fruit, as has been described in the nutrition chapter. This has largely replaced the original feeding regimens that were based around a sweetened, milk-based gruel [7], which was largely responsible for the fatty livers and dental problems of the past. Current diets are nutritionally much more similar to the natural bamboo diet described by Wei Fuwen, although no attempts have been made to make captive diets mirror the seasonal changes in the nutritional content of bamboo but perhaps this is not necessary. However, there are still some problems that do need to be resolved such as the palatability of the diet to young, weanling pandas and the fact that eating captive diets does not occupy enough of the red panda's daily activity. Pandas may compensate for this activity "vacuum" by over-grooming and eating hair. Pathology is important as it can confirm if there have been real changes in the patterns of mortality related to diet or to improved husbandry, etc. Unfortunately, despite efforts to the contrary, analysis of deaths is still compromised by non-standardized post-mortem reports combined with the failure of pathologists to note critical information such as age, weight and sometimes even sex of the dead animal on their reports. This is again something that can be easily remedied and will assist in improving husbandry and veterinary care in the future.

A separate chapter (Chapter 15) in this book is dedicated to the veterinary care. It is clear from this chapter that if a red panda survives its first weeks, it will likely continue to survive to a good age; mature pandas do not seem to be susceptible to many health problems. In fact, Philippa and Ramsay speculate that future veterinarians will need to pay more attention to geriatric problems in this species now that the nutrition-related problems seem to have been largely overcome. Again, in this chapter, the authors emphasize the need to pay more attention to infant survival and the problems resulting from mis-mothering. This seems to be a theme running through all the chapters discussing captive management. Partial solutions to this problem can be found in hand-rearing or supplemental feeding but these tools only alleviate the symptoms rather than address the causes of mis-mothering. It is sad to reflect that after so many years of captive management we still have not managed to substantially reduce the level of these early deaths.

The extraordinary susceptibility of red pandas to canine distemper is one of the most important issues raised in the veterinary chapter. In zoos, this is not so much a disease caught from passing dogs but rather one induced through vaccination: red pandas are so sensitive to canine distemper that they can be killed by the standard, commercial live or attenuated vaccines. If zoo veterinarians need to vaccinate their red pandas against this disease they have to find alternatives which, depending on the country where they are located, can be extremely difficult. However, we should be more concerned about the threat that canine distemper could pose in the wild. Human populations are expanding and people are moving closer and closer to the red panda's habitat. These people are accompanied by domestic animals including dogs that are often used to protect their herds. The chance that wild red pandas catch distemper from these domestic animals is increasing year on year.

So, in answer to the question, can the red panda zoo population survive, the answer is yes but we should not take this survival for granted; we need to be more stringent in the management of the population, in the husbandry and we need to conduct the necessary research to combat the problem of mis-mothering.

RED PANDAS IN THE WILD – CAN THEY SURVIVE?

In his comprehensive review chapter on red panda ecology (Chapter 11), Wei Fuwen provides us with a clear picture of the habitat requirements of red pandas, based largely on his data for Styan's panda. Typical red panda habitat is temperate forest, with the possible exception of Meghalaya where a tropical forest red panda has been reported (see Chapter 24), with good-quality bamboo and running water. In addition, red pandas have a preference for steeper slopes with structures such as rocks and fallen logs which enable a small-bodied animal, such as a red panda, to reach the young bamboo leaves high above the ground. Throughout their range, red pandas have one or two favourite species of bamboo, however, these preferences vary across the range.

Two authors have given us an impression of how red pandas and their habitat are surviving in the two main range states, China and India (Chapters 20 and 21). The picture which Wei provides of red pandas in China is mixed. It would seem that red pandas are well distributed in both Sichuan and Yunnan, however, across the rest of China we are given a picture of declining numbers and/or regional extinctions. Wei also tells us that there are still more than 75 000 km^2 of forest remaining in China and he estimates that about half of that is suitable for red pandas. These figures would agree with Choudhury's estimates [8]. Based on this potential area of forest habitat, Wei estimates there could be between 6000 and 7600 red pandas still living in China; this figure encompasses both *fulgens* and *styani*. In view of these numbers and the regional declines and extinctions, Wei estimates that the population has declined by 40% in the last 50 years. If these trends continue, the animal will be virtually extinct in China by 2050. Many of the threats which have caused this decline are still present; much of the forest area inhabited by red pandas has been subject to deforestation which reduces and fragments the red panda's habitat. While the rate of tree loss may have been reduced, the damage may have already been done because, as Wei reminds us, deforestation changes the context of the remaining habitat. In addition, poaching, capture and illegal trade remain constant threats to the remaining red panda population.

Ghose provides a comprehensive overview of the distribution of red pandas in India and indicates that the density of red pandas is very variable. He suggests that, based on the information available, about 85% of the red pandas in India occur in Arunachal Pradesh. He agrees with Wei Fuwen that forest destruction and fragmentation are the primary threat to the red panda. The random felling of live trees is prohibited in the forest reserves. However, enforcement of the national legislation is difficult to implement and prosecutions are rare. In addition, livestock grazing, particularly yaks, within forests reserves has always been established practice. Herding has an impact on the quality of red panda habitat, overgrazing negatively affects certain types of bamboo, which is an essential food for the red panda. This is a threat to the red panda and could be the major reason why it has disappeared from large parts of its expected range in Sikkim. Luckily, as Ghose reports, this practice has effectively stopped in the last decade. Unfortunately, a new threat in the form of tourism has arisen to take its place. Trekkers disturb red panda habitat and take wood for cooking. However, steps are currently underway to mitigate this by training the organizations dealing with tourists.

Illegal hunting also remains a challenge, particularly in Arunachal Pradesh, the major red panda stronghold in India. However, the creation of community conservation areas promotes awareness and so helps to reduce the problem. The involvement of the local community in environmental protection seems the key to preserving the red panda and its forest home in India as it does in Nepal.

Sikkim remains an important area for red pandas because, although it only covers a small percentage of the potential habitat in India, it is an important ecological corridor that links the main distribution areas in Arunachal Pradesh and Bhutan with that in Nepal. However, economic development, in combination with the rapid population growth, threatens the habitat of the red panda in Sikkim. In the long term, the red panda habitat, however, may only have a future if energy alternatives to fuel wood are introduced and applied widely in Sikkim.

Unfortunately, there are no reports currently available on the situation in Bhutan although fieldwork is now underway but the results are not currently available. Bhutan is considered to have the most untouched panda habitat that remains today but the country is small and we are discussing an area of potential red panda habitat of about 5400 km^2. Myanmar, on the other hand, has around 6400 km^2 of potential red panda habitat. This is situated in the north of the country, in Kachin state close to the Chinese border. Recent surveys have been undertaken in the Hkakaborazi National Park (see Appendix I), these indicate that the red panda still lives in the area but the authors conclude numbers may not be that high. However, we do know that red pandas in Myanmar are subject to hunting (Frank Momberg, FFI, personal communication).

In 2001, Choudhury published a paper in *Oryx* [8] in which he estimated the amount of red panda forest habitat remaining and then used this figure to compute the potential number of red pandas surviving. He calculated that there were some 70 000 km^2 of remaining habitat distributed across the red panda's range, much of which is not in protected areas. From these figures, he deduced that there were about 16 000–20 000 red pandas surviving in the wild of which about 5000–6000 were in India and a further 6000–7000 in China. These figures refer to *fulgens* and *styani* combined, thus two separate species. However, even these combined estimates may be optimistic; Choudhury judges about 49.1% of the potential forest habitat is actually used by red pandas. He then approximates the numbers of animals based on a density of one red panda per 4.4 km^2. However, according to Yonzon's work in the Langtang National Park in Nepal [9], only 68 of the parks 470 km^2 of forest were core habitat areas for red pandas. These animals had clear preferences for altitude, slope aspect, etc., which substantially reduced the area of forest they used. Therefore, the 49.1% figure for used habitat, which is based on giant panda research, may be an overestimate.

Choudhury's estimate of the number of red pandas in China seems to be in agreement with that of Wei in this volume, however, his figure for India could be too high. Although Ghose was unable to provide any hard figures for the number of red pandas in his paper, we do have some indications of numbers. Jha (Chapter 25) states that there are 78 animals in the Singalila National Park, an area of comparatively high red panda density [9]. The Neora Valley National Park, which is a similar size, can be assumed to have a similar-sized population. The estimate for Sikkim is of the order of 250–300 individuals

[10]. Currently, we do not have any published data for Arunachal Pradesh but Choudhury estimates that there are about 11 300 km^2 of forest habitat used by red pandas. If we look at the number of red pandas which Wei Fuwen estimates to be in Yunnan, a province with a similar amount of red panda habitat, we could fairly estimate that there are around 3000 red pandas in Arunachal (this is approximately the 85% suggested by Ghose). With these figures the total number of red pandas in India is likely to be closer to 3500 animals than the 5000–6000 estimated by Choudhury.

If we extrapolate this to the rest of the red pandas' range, to include the separate estimations that we have for Nepal (under 100 animals according to Brian Williams) and China (up to 7600 animals according to Fuwen) and then add an approximation for Myanmar and Bhutan together of 3000 animals (again based on the Yunnan figures due to the similar amount of red panda habitat), we get a maximum population of 14 000–14 500 individuals, and this is an optimistic estimate. The IUCN red data book indicates that Choudhury and Yonzon consider the number may be as low as 10 000 [11]. These are worrying figures, especially in light of the decline and fragmentation of forests across the region.

Despite his larger estimated population of 16 000–20 000 red pandas, Choudhury recommended that the red panda be classified as endangered, not on the basis of its population size, but on the basis of its rapid decline in numbers; 50% over the last 50 years. Wei supports this view on the basis of his observations in China. It is therefore surprising to realize that the status of the red panda was downgraded by the IUCN from endangered to vulnerable, which was confirmed as recently as 2008 [11]. The data presented in this volume suggest that with this classification, the prognosis for its survival is not good. Yonzon and Hunter predicted that the red panda would likely be extinct in the Langtang National Park in 30 years, this prediction was made nearly 20 years ago, they may nearly be gone by now. On the basis of the data presented in this volume, we might well be wise to reconsider the conservation status of the red panda, particularly in light of the fact that we could be dealing with two species rather than the single one formerly assumed.

We have not included the possible numbers of red pandas in Meghalaya, in the preceding discussion. Although wildlife experts have often dismissed the observations of red pandas in this unlikely habitat, Duckworth's evaluation of the situation demonstrates that there is sufficient evidence for us to take these reports seriously. A warm climate Ailurine is not necessarily a contradiction; the fossil *Parailurus*, described by Kundrát in Chapter 5, did not live in the typical Ailurine temperate forest environment. This animal in Meghalaya also lives in a very different type of habitat and may also be larger in size than the usual red panda. It is therefore possible that we might be dealing with a new Ailurine species, a possibility that needs investigating further. This Meghalaya panda population, if it still survives, is undoubtedly very small and, therefore, rapid action is needed to confirm its existence and to preserve it if necessary.

Despite these low numbers, the situation of the red panda has not stabilized, they are still subject to a number of threats. The most important of these are loss of habitat, habitat fragmentation and disturbance. These are problems that are not likely to disappear in the near future; human populations are growing and people need land for herds and crops and wood for cooking and heating. As the human population spreads closer to the red panda's habitat, the threat of dogs increases. Dogs not only kill pandas and disturb nesting females but they also bring with them the very real threat of canine distemper, a disease that could decimate small, isolated populations of red pandas. The growth of human populations also means more infrastructure; roads are needed which in turn open up more remote regions to human exploitation, and the need for (renewable) energy leads to the construction of hydroelectric dams all of which cause further habitat destruction and disturbance. As we have seen in various chapters in this volume, illegal hunting and trapping of red pandas occur across the whole of its range, however, this cannot be said to be a serious threat. The red panda is protected under Appendix 1 of CITES, this means legal international trade is effectively prohibited. In appendix II, a short report on the illegal trade in red pandas is presented. This report reflects only that trade which is discovered and so may only represent a small percentage of the trade that actually occurs, but numbers are nevertheless very low. In the past, the zoo trade used to take its toll; Bahaguna and his colleagues estimated that at least 300 red pandas were trapped for trade in the 1960s in Singalila [12]. This practice still persists in China where, according to Loeffler, wild red pandas are regularly captured and brought into zoos. Live pandas may not be captured for zoos but some are still captured and sold as pets and, in a recent development, red pandas are turning up in restaurants in China.

Maybe we need to learn a lesson from the fossil history, the last *Parailurus* species in Europe succumbed to a combination of loss and fragmentation of their forest habitat, competition from newcomers and predation. If we are not careful history could repeat itself with the red pandas that today are also threatened with a similar combination of circumstances.

The red panda is offered a degree of protection across the whole of its range, for example, it is illegal to hunt them in Nepal, India, Myanmar and China. The red panda also occurs in a number of protected areas across its range states. Wei tells us that the red panda is very much forgotten in Chinese conservation plans, nevertheless, there are a number of conservation initiatives in China that indirectly benefit the species; the creation of nature reserves and other policies which encourage reforestation or the re-zoning of agricultural land benefit red pandas, as do the green energy initiatives that reduce the need for firewood. So despite not being a key protected species, the red panda does profit from initiatives helping the giant panda and golden monkey. However, the most important requirement for conservation, as Wei tells us, is to educate people not to exploit wildlife.

In his chapter (Chapter 22), Brian Williams, reports on a very different approach to conservation, this time directed specifically towards the red panda, which is being developed in Nepal. This is a community conservation initiative that involves the local people and makes conservation an economically attractive alternative for them. It is a holistic approach to conservation benefiting local people, the environment and local species and seems to be a good model for the future conservation initiatives across the region as it allows the collection of scientific data, encourages education and so can change attitudes to wildlife. The only matter of concern is that by such initiatives that bring local people into more contact with red pandas and their habitat, the possible danger from canine distemper carried by domestic dogs could become a greater issue.

The volume concludes with a chapter dealing with the reintroduction of zoo-born red pandas into the wild (Chapter 25). Leus also discussed this issue in her chapter (Chapter 19) on the future of zoo red pandas. She envisaged a situation where range state zoos act as a conduit for genetic material between the wild and captive populations. In this situation, range state zoos would occasionally obtain wild-caught animals and integrate them into the captive population, they would also return genetic material to the wild through reintroductions of zoo-born animals. If this interchange of wild-caught and captive-born individuals were to be feasible it could benefit both wild and captive populations in the future. However, one prerequisite is that it must be possible to successfully habituate zoo-born animals to the wild. Jha's report on the reintroduction of two, zoo-born, red panda females into the Singalila National Park shows that this is a viable proposition. Although a predator killed one of these two females, the other lived and bred with a wild male. Even though her young did not survive, its very birth gives some hope that reintroduction is a reasonable option. Nevertheless, the fact that the fate of this female, as well as those of two other females, not discussed in the chapter but which were later reintroduced into the same region, is unknown indicates that better long-term monitoring will be needed in the future.

However, if reintroduction is to become an integral part of conservation policy, there needs to be a sustainable, genetically diverse captive population regionally available to provide candidates for this. In the case in question, it is questionable whether the Indian red panda population was in a state to provide four females for reintroduction during 2004. The fact that the regional population has been critically short of females since that time would argue to the contrary. However, it is possible to increase the Indian zoo population either through captures from the wild or through import of captive red pandas from zoos in other regions. This latter option is preferable not only because the wild population is fragile but also because Leus recommends that reintroduced animals should be representative of different gene lines than those occurring in the wild population.

The Chinese authorities are also interested in using the reintroduction of captive-born individuals as a tool to boost wild populations. Here the approach is different; animals are captured from the wild specifically to participate in these breeding efforts. When I was in Yunnan in 2002, I visited a breeding centre where a group of wild-caught red pandas were held with the intention of breeding them in the centre and then reintroducing the offspring back into the wild (Figure 26.2). However, the facility did not seem to be achieving its aims. Its approach to red panda management was very similar to that described by Kati Loeffler in her chapter (Chapter 18) on Chinese zoos and breeding results were apparently not very good. However, many experts are not in support of using captive-bred animals for reintroduction in any case, they prefer the option of translocation and restocking with wild-caught individuals, the survival rate of translocated wild individuals is better than that of zoo-born ones [13]. However, this probably depends on the species involved. At present, there are not enough data available on reintroduced red pandas and none at all on translocated red pandas to make such a comparison.

So, in answer to the question, can red pandas survive in the wild, the situation is very much in the balance, it depends on conservation initiatives such as those outlined above, on winning the support of the local population and creating undisturbed protected areas where red pandas can thrive away from people. Maybe the reintroduction of zoo-bred red pandas will be a tool to boost the genetic variability of some of these small fragmented populations. However, above all, it is necessary to re-evaluate the red panda's situation in the light of Groves' recommended taxonomy.

(A)

(B)

(C)

FIGURE 26.2 Red panda breeding station in Yunnan, China. (A) Layout of breeding runs. On the right hand side is a large enclosure for use by the whole group of red pandas; (B) An individual breeding run; (C) Nest box in breeding run, with two red pandas.

CONCLUSION

So, to summarize, the red panda belongs to its own family, it is a terminal relic of a once flourishing group, a living fossil not closely related to any other extant species and, as such, it is extremely significant biologically and of high conservation value. We are only just beginning to understand its biology and appreciate its adaptations to the very specialized niche of bamboo-eating carnivore and we still have much to learn. Unfortunately, this unique species is vulnerable to extinction both in captivity and in the wild and it may even disappear before we have a chance to fully understand it.

The red panda is a very beautiful, appealing species and one which is becoming something of a cultural icon in the modern world. This would indicate that there is potentially a lot of public interest in the species and we need to harness this and exploit it to conserve the red panda and its environment. Habitat loss, destruction and fragmentation not only threaten the red panda, these are issues facing the whole of the Himalayan region.

For centuries, the towering peaks and secluded valleys of the Himalayas have inspired naturalists, adventure seekers and spiritualists. The Himalayan mountain range forms a 1500-mile-long barrier that separates the lowlands of the Indian subcontinent from the high, dry Tibetan Plateau. More than 10 000 plant species, 300 mammal species and 900 bird species coexist with millions of people from diverse cultures and religions, forming a rich tapestry of life across Pakistan, Northern India, Nepal, Bhutan and Southwest China. The temperate forests sustain Himalayan black bears, red pandas and golden langurs, while the rhododendron forests harbour the satyr tragopan. Above the tree line, alpine meadows and cliffs offer refuges for snow leopards and takins. As a result of continued explosive economic growth, the pressure on the Himalayas and its natural resources have increased dramatically throughout the last decades: over-harvesting of forests for food and timber, intensive grazing, agricultural expansion, deforestation, wildlife poaching and tourism already threaten a number of species. If these threats continue unabated, much of the unique wildlife of the Himalayas will soon be at the brink of extinction, not just the red pandas.

The red panda is in an exceptional position to function as one of the flagships for Himalayan conservation. We can use its uniqueness, its charm and its current popularity to generate interest and concern for the whole region. Zoo red pandas could be instrumental in this whole process; as a research resource to help us learn more about this elusive animal, as a focus for awareness to make people conscious of environmental issues in the Himalayas, as a fund-raising tool and as a genetic asset; a reserve population which can also be used as needed for reintroduction and restocking in the wild.

We still have the chance to save the red panda for future generations to enjoy and, by using it as a flagship for Himalayan conservation, we can hopefully motivate people to protect and preserve its spectacular habitat. We can do this and we should, and I hope that in some small way this book will encourage people to do so.

ACKNOWLEDGEMENTS

I would like to thank Dr Geoff Hosey for his helpful comments on this chapter and Arno Vlooswijk for providing the thermographic images.

References

[1] A.R. Glatston (compiler), The Red Panda, Olingos, Coatis, Raccoons and Their Relatives. Status Survey and Conservation Action Plan, IUCN, Gland, Switzerland, 1994.

[2] Firefox at 270 million users. <http://weblogs.mozillazine.org/asa/archives/2009/05/firefox_at_270.html> (Viewed 12.03.10).

[3] O. Thomas, On Mammals from the Yunnan Highlands collected by Mr George Forrest and presented to the British Museum by Col. Stephenson R. Clarke, DSO, 1922.

[4] M. Roberts, On the subspecies of *Ailurus fulgens*, in: A.R. Glatston (Ed.), The Red of Lesser Panda Studbook No. 2, Sitchting Koninklijke Rotterdamse Diergaarde, Rotterdam, The Netherlands, 1982, pp. 13–24.

[5] B.R. Sharma, N. Akhtar, B.K. Gupta (Eds.), Proceedings of International Conference on India's Conservation Breeding Initiative, Central Zoo Authority, Delhi, India, 2008.

[6] A.R. Glatston, Husbandry and management guidelines, in: A.R. Glatston (Ed.), The Red or Lesser Panda Studbook No. 5, Stichting Koninklijke Rotterdamse Diergaarde, Rotterdam, The Netherlands, 1989, pp. 33–52.

[7] N. Nijboer, M.C.K. Bleijenberg, A. Glatston, The European red panda diet survey, in: A.R. Glatston (Ed.), The Red or Lesser Panda Studbook No. 6, Stichting Koninklijke Rotterdamse Diergaarde, Rotterdam, The Netherlands, 1991, pp. 41–45.

[8] A. Choudhury, An overview of the status and conservation of the red panda *Ailurus fulgens* in India, with reference to its global status, Oryx 35 (2001) 250–259.

[9] P. Yonzon, R. Jones, J. Fox, Geographic information systems for assessing habitat and estimating population of red panda in Langtang National Park, Nepal. Ambio 20 (1991) 285–288.

[10] S. Ziegler, A. Gebauer, R. Melisch, et al., Sikkim – Im Zeichen des Roten Panda, Zeitschrift Kölner Zoos, 2010, in press.

[11] IUCN red list of threatened species. <http://www.iucnredlist.org/>

[12] C. Bahaguna, S. Dhaundyal, P. Vyas, N. Singhal, The red panda in Singalila National Park and adjoining forest: a status report, Small Carniv. Conservat. 19 (1998) 11–12.

[13] K.R. Jule, Effects of captivity and implications for ex situ conservation: with special reference to the red panda (*Ailurus fulgens*). PhD thesis, University of Exeter, UK, 2008.

APPENDIX I RED PANDA *AILURUS FULGENS*; STATUS AND DISTRIBUTION IN MYANMAR

Extracted from:
Status and distribution of small carnivores in Myanmar
Than Zaw, Saw Htun, Saw Htoo Tha Po, Myint Maung, A.J. Lynam, Kyaw Thinn Latt and J.W. Duckworth
Published in *Small Carnivore Conservation*, 38, 2–28, April 2008

Geographical Distribution

Red panda was found in the northernmost sites, Hkakaborazi National Park and Hponkanrazi Wildlife Sanctuary (see Table). No animals were camera-trapped, despite high effort at Hkakaborazi. Faeces provisionally identified as red panda were found twice in Hkakaborazi within dwarf bamboo–pine forest (28°08′17″N, 97°38′14″E, 3890 m; 28°05′26″N, 97°37′53″E; 3080 m) in late October 2003. These were confidently assigned to the species by experienced local hunters, and comprised, largely, bamboo leaves. Their appearance seemed identical to faeces produced by a captive panda caught at Zalahtu

(3390 m) by staff of the Nature and Wildlife Conservation Division (NWCD), Hkakaborazi National Park, which was sent to Yangon Zoological Gardens. Three other live red pandas caught in 2002–2003 by staff of the same unit were sent to the head office of Hkakaborazi National Park, in Putao, but died along the way (see Table). There are previous records from this area. Dollman (1932) and Pocock (1941) listed 150 miles north of Myitkyina, near the Yunnan, China, border; two sites on the Nam Tamai (one at 27°50′N, 97°55′E); and the Taron Valley. Subsequently, two were collected on Janraung Bum at 8000′ in February 1962 (Tun Yin 1967). Red pandas also inhabit north-east Kachin state: two skins reputedly came from Sakkauk, on the north flank of Emaw Bum (Anthony 1941), four different individuals were seen during a bird-surveying visit to Mount Majed in early 2005 (Eames 2005), and a freshly-hunted animal was seen on Emaw Bum in early 2007 (Eames 2007).

Habitat and Altitude

Signs were recorded only at high altitudes (over 3000 m), above the timber line, among dwarf bamboo (5–8′ tall). Pocock (1941) reported Myanmar specimens from the range 3500–7000′ (1070–2130 m; Nam Tamai Valley) and at 9000′ (2740 m; Taron Valley). Local people reported that pandas in Hkakaborazi move down-slope in winter (Rabinowitz & Saw Tun Khaing 1998).

Behaviour

Assuming the identification of faeces is correct, the animals make latrines, where there are multiple piles of faeces of different ages.

Threats and Conservation Status

The lack of camera-trap records suggests that red pandas might be scarce in Hkakaborazi, because there was signifcant survey effort over 3000 m (nine camera positions totalling about 340 trapnights) and effort was high within 2000–3000 m, but it could simply be that by chance none was photographed. However, red pandas, which forage primarily on the ground (Roberts & Gittleman 1984), are surely vulnerable to snaring (for, e.g., musk deer *Moschus*), which is widespread in the area. Rabinowitz & Saw Tun Khaing (1998) found that in Hkakaborazi, local people did not actively target pandas, but did kill or collect them opportunistically, and sold the skins. Red panda skins were seen for sale in markets on the Thai–Myanmar border at Tachilek in 1998 (AJL own data), an area far from likely wild pandas and a known trading point (e.g. Davidson 1999). The threat of harvest for the international captive animal trade is difficult to assess: there is ample opportunity through the markets along the Hkakaborazi–China border. Choudhury (2001) identifed habitat degradation as the chief threat in India to red panda. Hence, it is noteworthy that habitat in Hkakaborazi and Hponkanrazi, especially at mid and high altitudes, is relatively stable (Renner et al. 2007). In some other areas, e.g. Emaw Bum, forests are much degraded (Eames 2007), and some populations within the species' small Myanmar range are no doubt in decline.

TABLE Records of red panda in Myanmar.

Survey area	Site	Latitude (N)	Longitude (E)	Date	State
Hkakaborazi	Zalahtu, near Madein	28°08′	97°24′	March 2003	Live
Hkakaborazi	near Tahundam	28°11′	97°38′	Jan 2002	Live
Hkakaborazi	near Tahundam	28°14′	97°37′	Feb 2003	Live
Hkakaborazi	Tahundam	28°10′	97°40′	13 Feb 2004	Dead
Hkakaborazi	Tazundam	28°02′	97°34′	2 Nov 2003	Dead
Hkakaborazi	Gushin-1	27°38′	98°13′	Apr 2004	Dead
Hponkanrazi	Ziadam	27°34′	7°06′	9 Feb 2002	Skin

References

H.E. Anthony, Mammals collected by the Vernay–Cutting Burma expedition, Publications of the Field Museum of Natural History, Zoology Series 27 (1941) 37–123.

A. Choudhury, An overview of the status and conservation of the Red panda *Ailurus fulgens* in India, with reference to its global status, Oryx 35 (2001) 250–259.

P. Davidson, Spotcheck of wildlife on sale in Myanmar market, TRAFFIC Bulletin 17 (1999) 98.

G. Dollman, Mammals collected by Lord Cranbrook and Captain F. Kingdon Ward in upper Burma, Proceedings of the Linnean Society of London 145 (1932) 9–11.

J.C. Eames, Ornithological expedition to Kachin state, Myanmar, yields new finds. The Babbler, BirdLife International in Indochina 13 (2005) 7–8.

J.C. Eames, Return to Imawbum. The Babbler, BirdLife International in Indochina 21 (2007) 31–34.

R.I. Pocock, The fauna of British India, including Ceylon and Burma. Mammalia. (2nd ed., vol. 2). (1941). London: Taylor & Francis Ltd.

A. Rabinowitz, Khaing, Saw Tun, Status of selected mammal species in north Myanmar, Oryx 32 (1998) 201–208.

S.C. Renner, J.H. Rappole, P. Leimgrüber, D.S. Kelly, Nay Myo Shwe, Thein Aung, Myint Aung, Land cover in the Northern Forest Complex of Myanmar: new insights for conservation, Oryx 41 (2007) 27–37.

M.S. Roberts, J.L. Gittleman, *Ailurus fulgens*, Mammalian Species 222 (1984) 1–8.

Tun Yin, Wild animals of Burma. Rangoon Gazette, Rangoon, 1967.

APPENDIX II RED PANDAS – AN ANALYSIS OF THE THREATS POSED BY TRADE

Roland Melisch, TRAFFIC

Red pandas *Ailurus fulgens* have long appealed to people owing to their very attractive fur coloration. This, together with low or undetermined population figures of the species in the wild was certainly the reason of the species receiving full protection in all ranges states. Traditional uses of their furs have been the reasoning why the species already became listed during the inception of the Convention on International Trade in Endangered Species of Wild Fauna and Flora (CITES) in 1975. First listed under its Appendix II, which allows commercial trade under a strict control regime, the species became uplisted to Appendix I in 1994, after a proposal submitted by the Netherlands (CITES 1994) was agreed by the Parties to CITES. The uplisting of 1994 came into force on

16th February 1995, and since then all commercial trade in the species and its parts has been strictly prohibited. Exemptions are only granted for non-commercial purposes, e.g., such as exchanges of living specimens for conservation breeding. However, the import of Appendix-I listed species is not allowed if the purpose of the import is for commercial purposes. The CITES (Res Conf. 5.10) provides a general definition of 'commercial purposes' as "economic benefit, including profit (whether in cash or in kind) . . .".

Indications of circumventing these strict conditions of the international ban on commercial trade were published by O'Connell-Rodwell & Parry-Jones (2002) who raised concerns that not all reported trade transactions involving "zoos" appear to have been conducted in accordance with the provisions of CITES. The proliferation of safari parks in Asia since the 1980s and in China since the mid-1990s does make it very difficult to assess the conservation purposes and benefits of the trade in several species, including red pandas. Regardless of the difficulties in applying 'non-commercial purposes' to facilities where the public must pay a fee for viewing the animals, at least one safari park in China, the Guangzhou Panyu Safari Park, has an animal exchange programme under which it had exported four red pandas to Malacca Zoo in Malaysia. The authors concluded that it would appear that commercial trade in CITES Appendix I-listed species is being conducted where the profit is 'in kind'.

Since the listing of the red panda under CITES Appendix I, however, only a few incidental and anecdotal records of trade in live red pandas or their parts have been noted. A review of the UNEP-WCMC CITES trade database revealed only one seizure, a shipment of bodies from Japan to the US in 1995, that has been reported by the Parties to CITES. Almost all other entries for the species refer to trade in captive live red pandas between zoos.

As one rare published example of illegal trade in red pandas, a study of wildlife trade along the Yunnan (China)–Vietnam border comparing markets in 1997 to studies conducted three years earlier, Li & Wang (1999) found only one case of a red panda fur on sale. Another one of these typical incidents is depicted by a confiscation in 2005, when a single specimen of red panda fur was seized in the Kanchenzonga Landscape at Gola La Pass on the Nepalese side. A local villager attempting to smuggle and sell the fur in China's Tibetan Autonomous Region (TAR) was pushed back by the TAR police to Nepal. Enforcement in the Chinese–Nepalese Kanchenzonga Landscape border areas very much depend on the enforcement efforts of TAR, since no enforcement personnel is active on the Nepalese side of this border (D. Chapagain, WWF Nepal, 6th Sep 2006, *pers. comm.* to the author).

To understand threats to a species posed by trade it is important to understand any potential underlying causes. As opposed to e.g., many other larger carnivores, ungulates, primates and elephants, red pandas are not amongst those animals which cause any economic damage or act as any sort of competitors to the local human population. According to Glatston (1994), Ghose (2005–2009), Ziegler *et al.* (in print), and experiences of WWF in India, Nepal, Bhutan and China (*pers. comm.* to the author), human–red panda conflicts are totally unknown.

Red pandas have rather been hunted traditionally, e.g. for use of furs for hats and coat applications worn by men and women of the Yi minority in Yunnan Province (China) north of Lijiang. (Fuwen Wei & Zhang Zejun, Chinese Academy of Sciences, 6th Sep 2006, *pers. comm.* to the author). The habit of wearing red panda furs hats still continues and has been documented between 2001 to 2006 (Oxford), however it is unclear whether the hats are traditional old wear, or whether they have been more recently made.

It was only in July 2009 that a new, formally unknown aspect was revealed, when a business traveller visiting Zhongshan (Guangdong, China) reported the offer of red pandas for human consumption in restaurants (Del Castillo, *in litt.* 20th July 2009). The sale of wild meat is common practice in China's southern provinces, however government controls to combat the sale of protected species had been increased significantly in recent years, and the commercial breeding of red pandas for domestic sale is so far unknown in China (Xu Hongfa, TRAFFIC East Asia China, *in litt.* 22nd July 2009).

In conclusion, when comparing with habitat alteration, destruction and other biome-related threats, trade seems currently not to pose a significant threat to red pandas. However, recently revealed trends relating to traditional clothing, human consumption and exchange for "zoos" and safari parks needs to be closely monitored. Furthermore, the opening up of new trade routes between India and China via Sikkim's Nathu La border on 6th July 2006 will require further attention (Ziegler *et al.*, in print). As a consequence, TRAFFIC, will continue to conduct enforcement work targeting illegal wildlife trade in Asia along with the mandated government staff, and will maintain its efforts to train judicial personnel and government enforcement staff involved in implementing, prosecuting and enforcing wildlife trade laws in red panda range states.

Acknowledgements

TRAFFIC's wildlife trade enforcement work in Asia's red panda range states is build on the strong commitments of its governments, and enabled through financial support by the US and UK governments, and WWF. Many thanks to John Caldwell of UNEP–WCMC for supporting this note with data, and to Steven Broad of TRAFFIC for editorial remarks.

References

CITES, 1994. CoP 9 Proposal 11: Transfer from Appendix II to Appendix I of Procyonidae: *Ailurus fulgens*, submitted by The Netherlands. In: CITES (1994): Proposals for Amendment of Appendices I and II. Ninth Meeting of the Conference of the Parties to CITES. http://www.cites.org/eng/cop/09/prop/index.shtml (viewed on 22.02.10).

D. Ghose (compiler), Report of the Red Panda Pre-PHVA Workshop, 17–19 February 2007, WWF India–Sikkim Programme/Blijdorp Zoo, Gangtok, Rotterdam, 2007.

D. Ghose, (2005–2009). Red Panda Conservation Project Technical Progress Reports. Unpublished. WWF India–Sikkim Programme, Gangtok.

A.R. Glatston (compiler), Status Survey and Conservation Action Plan for Procyonids and Ailurids. The Red Panda, Olingos, Coatis, Raccoons, and their Relatives, IUCN/SCC Mustelid, Viverrid, and Procyonid Specialist Group, Gland, 1994.

W. Li, H. Wang, Wildlife Trade in Yunnan Province, China, at the Border with Vietnam, TRAFFIC Bulletin 18 (1) (1999) 21–30.

C. O'Connell-Rodwell, R. Parry-Jones, An Assessment of China's Management of Trade in Elephants and Elephant Products. TRAFFIC Online Report Series No. 3., TRAFFIC East Asia, Hong Kong, 2002.

P. Oxford, Various photographs depicting men and women of the Yi minority wearing red panda fur coat hats in Yunnan Province (China). (2001–2006). http://www.naturepl.com/ (viewed on 22.02.10).

S. Ziegler, A. Gebauer, R. Melisch, B.K. Sharma, P.S. Ghose, R. Chakraborty, P. Shresta, D. Ghose, K. Legshey, H. Pradhan, N.T. Bhutia, S. Tambe, & S. Sinha, Sikkim – im zeichen des roten panda, Zeitschrift des Kölner Zoos (in print).

Index

D

E

CPI Antony Rowe
Chippenham, UK
2016-01-25 18:35